Boulevard of Broken Symmetries

Effective Field Theories of Condensed Matter

Theo van Doesburg, Composition in Gray (Rag-time), 1919,
Peggy Guggenheim Collection, Venice
(Solomon R. Guggenheim Foundation, NY)

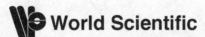

Boulevard of Broken Symmetries

Effective Field Theories of Condensed Matter

Adriaan M. J. Schakel

Freie Universität Berlin, Germany

World Scientific

NEW JERSEY · LONDON · SINGAPORE · BEIJING · SHANGHAI · HONG KONG · TAIPEI · CHENNAI

Published by

World Scientific Publishing Co. Pte. Ltd.
5 Toh Tuck Link, Singapore 596224
USA office: 27 Warren Street, Suite 401-402, Hackensack, NJ 07601
UK office: 57 Shelton Street, Covent Garden, London WC2H 9HE

Library of Congress Cataloging-in-Publication Data
Schakel, Adriaan M. J.
Boulevard of broken symmetries : effective field theories of condensed matter / Adriaan M.J. Schakel.
p. cm.
Includes bibliographical references and index.
ISBN-13: 978-981-281-390-9 (hardcover : alk. paper)
ISBN-10: 981-281-390-X (hardcover : alk. paper)
1. Condensed matter. 2. Quantum field theory. I. Title.
QC173.454.S33 2008
530.4'1--dc22

2008033321

British Library Cataloguing-in-Publication Data
A catalogue record for this book is available from the British Library.

Printed in Singapore.

To the living memory of my mother Anna M. Konings,
to my wife Constanze Musterer,
and to our three-year old son Roman.

Preface

This book describes the main topics in contemporary condensed matter physics in a modern and unified way, using quantum field theory in the functional integral approach. Rather than developing general formalisms and giving formal derivations, an informal presentation is chosen based on analogies and generalizations of standard results in classical and quantum mechanics and in statistical physics.

The book highlights symmetry aspects in acknowledging that much of the collective behaviors of condensed matter systems at low temperatures *emerge* above a nontrivial ground state which spontaneously breaks the symmetry of the system. Such a breakdown of a continuous symmetry usually leads to gapless Nambu-Goldstone modes, which endow the system with long-range correlations, and to topologically stable defects. These emerging properties are responsible for most startling phenomena in ordered states.

Field theories that provide an *effective* description of the condensed matter system under study take the central stage in this book, where the word effective is used in the double sense of being efficient, i.e., productive with minimal effort, and powerful. Such a description is viable because the details at microscopic length and time scales are often irrelevant for phenomena at long wavelength and low frequency. Such phenomena are often dictated by universal laws instead. The topics are studied typically by starting from a microscopic model. Irrelevant degrees of freedom are subsequently integrated out perturbatively to arrive at the effective theory. The physical properties of the system under consideration are finally extracted from the effective description, emphasizing their emerging nature special to the long wavelength, low frequency corner.

To implement this program, simple, yet powerful calculation tools rooted in the functional integral approach to quantum fields are presented. One such tool is the derivative expansion method which is repeatedly used to integrate out irrelevant degrees of freedom. Often these are fermions. For fermionic condensed matter systems at low temperatures, one-fermion-loop contributions to the effective action can have profound implications which the derivative expansion method efficiently summarizes in an effective action. The method is completely equivalent to evaluating Feynman diagrams, and has the advantage of being straightforward to implement.

In conjunction with the emphasis on effective theories, the modern approach toward renormalization is taken, in which a wave number cutoff is introduced to set the scale beyond which the microscopic model under consideration ceases to be valid. In this approach, the cutoff is taken seriously and kept finite (as opposed to taken to infinity at the end as is traditionally done in particle physics). The effect of the unknown physics above the cutoff is incorporated by redefining the parameters of the original theory.

The independent presentation, free of historical constraints and based on functional integrals, allows for a compact and self-contained treatment of the main topics in contemporary condensed matter physics, most of which have been recognized by Nobel Prizes in Physics. It was deemed not appropriate to include exercises. Often these are used to hide technical details. In this book, which aims to also be practical, many technical details are explicitly worked out instead to strengthen and widen the calculation skills of the students. Including these details was also deemed appropriate as some of the techniques used have their origin in particle physics, and despite their virtues, are not commonly applied to condensed matter physics. Their use reflects the author's joy in obtaining results more easily and faster than the more standard derivations. References are relegated to the Notes at the end of each chapter and are not included in the running text to avoid interrupting the flow of the presentation. Although original references are sometimes given, no attempt is made to systematically cite the original sources. The prime purpose for including references in this book is to point students to sources that may be helpful to them. Frequently, those sources are cited where a similar approach is taken as advocated here. Excellent and accessible summaries of most of the topics covered in this book can be found on the official web site of the Nobel Foundation, as well as in the special Centennial Issue of Reviews of Modern Physics, **71**, pp. S1–S488 (1999).

This book has been formatted with the use of LaTeX, an extended document preparation system that derives from TeX, on several computer platforms (HP, IBM, and SUN) running some version of UNIX (NEXTSTEP, LINUX, and SOLARIS). Without the powerful tools that come with the UNIX operating system, this project would never haven been realized. This is especially true for the GNU EMACS editor, or better, programming environment, which is probably one of the most formidable computer programs ever written. Most of the figures have been created with XFIG, a drawing program for the X Windows System, and with GNUPLOT, an interactive plotting program. Internet search engines, especially GOOGLE, and the online dictionary and thesaurus provided by the DICT Development Group were indispensable for this project. I am deeply indebted to all the people that contributed to the amazing pieces of software used in this project.

This book originates from lecture notes written for an undergraduate course on condensed matter physics that I taught at the *Freie Universität Berlin*, Germany during the winter semester 1995/1996. The lectures were repeated as a crash course at the *Universidade Federal de Pernambuco*, Recife, Brazil in March 1996. I am grateful to Professor G. Vasconcelos for inviting me and for a grant from the *Conselho Nacional de Desenvolvimento Científico e Tecnológico* (CNPq). I thank him and the other members of the Physics Department for their warmheartedness.

Parts of the material were presented at the Workshop on *Cooperative Phenomena in Condensed Matter* at Pamporovo, Bulgaria, March 1998 and at the XL Cracow School of Theoretical Physics on *Quantum Phase Transitions in High Energy and Condensed Matter Physics*, Zakopane, Poland, June 2000. It is a pleasure to thank the organizers, Professor D. I. Uzunov, Bulgarian Academy of Sciences, Sofia and Professor J. Spałek, *Uniwersytet Jagielloński, Kraków*, Poland for their invitation to lecture and their warm hospitality, see [Uzunov and Shopova (1999)] and [Spałek (2000)] for the Proceedings. These more advanced lectures were in part based on my *Habilitationsschrift*, which carries the same title as the main title of this book, and which can be found on the ARXIV, an e-print service at Cornell University. I would like to thank Dr. K. K. Phua, Executive Chairman of World Scientific Publishing Company, for suggesting to turn my *Habilitationsschrift* into a textbook, and my editor, Mr. Alvin Chong, for his help and support on this project.

Chapters 2 and 4 served as a basis for a two-hour undergraduate course on Bose-Einstein condensation that I taught at the *Universität Leipzig*, Germany during the summer semester 2004. I am indebted to Professor W. Janke for arranging a visiting professorship and for providing a stimulating and pleasant environment.

Finally, the support of the Institute of Theoretical Physics at the *Freie Universität Berlin* is greatly acknowledged.

Adriaan M. J. Schakel
Berlin
June 2008

Synopsis

1. Classical Field Theory

This first introductory chapter treats essential aspects of classical field theory, with the emphasis on symmetries, their spontaneous breaking by a nontrivial ground state, and consequences of that, such as the emergence of gapless Nambu-Goldstone modes and topological defects. Special attention is given to Galilei symmetry for most of the theories considered are nonrelativistic in nature. As an example, the hydrodynamics of an ideal classical fluid is presented from this perspective. Homotopy theory to classify topologically stable defects is introduced next. The chapter ends with a discussion of vortex solutions as an application.

2. Quantum Field Theory

In this second introductory chapter, some standard results from statistical physics and quantum mechanics are recalled. Generalizing ordinary integrals, the chapter introduces functional integrals to describe quantum fields. Feynman propagators and diagrams are derived from the functional integrals together with rules to compute the diagrams. A modern treatment of linear response theory is given, and sum rules as well as Ward identities are derived.

3. Calculation Tools

In this third and last introductory chapter, various calculation tools, such as the Matthews-Salam functional determinant and the derivative expansion method, which is ideally suited to compute effective actions, are introduced for use in later chapters. The geometrical, or Berry phase is shown to play an important role in the effective field program, giving a term linear in time derivatives known as the Wess-Zumino term. Finally, the Landau-levels technique is presented to intuitively calculate one-loop quantum corrections in charged systems.

4. Bose-Einstein Condensation

This chapter gives a concise description of Bose-Einstein condensation in weakly interacting Bose gases, starting from an ideal Bose gas both in the absence and in the presence of a harmonic trap. The microscopic Bogoliubov theory of weakly interacting Bose gases and its physical predictions, including renormalization aspects, are studied in detail. In line with the general approach taken in this book, the Bogoliubov theory is then used as a basis to derive the effective theory, both at zero and at finite temperature. With the effective theory at hand, superfluid hydrodynamics and collective modes are discussed. Other topics include the localization effect of impurities, the interaction-induced shift in the condensation temperature in a large-N expansion, and the Bose-Hubbard model to describe atomic gases in optical lattices.

5. Vortices in 2D

This chapter is concerned with superfluid films. It starts with a discussion of the unusual dynamics of vortices in such films at the absolute zero of temperature and the corresponding geometric phase. The chapter continues with a detailed discussion, which includes renormalization aspects, of the Berezinskii-Kosterlitz-Thouless, or vortex-unbinding phase transition in superfluid ^{4}He films. This transition provides an example of a defect-mediated phase transition. As starting point the two-dimensional (2D) classical Coulomb gas is taken. The dual formulation of a superfluid film, where the vortices are no longer represented by singular objects, is given and shown to be provided by the sine-Gordon model.

6. Fermi Gases

This chapter considers weakly interacting Fermi gases. The effect of impurities is studied and the conductivity is computed by means of the derivative expansion method. The chapter continues with a discussion of magnetic impurities (Kondo effect) which includes renormalization aspects. The second part of the chapter is concerned with degenerate electron gases.

7. Magnetic Order in Fermi Systems

This chapter investigates the emergence of magnetism in fermionic systems. The study is based on the microscopic Hubbard model. Besides displaying ferromagnetic order, the model also shows antiferromagnetic order for certain choices of the parameters. The derivative expansion method is used in both regimes to perturbatively integrate out the fermionic degrees of freedom to arrive at the effective theory. It describes the gapless spin waves which constitute the Nambu-Goldstone modes accompanying the spontaneous breakdown of the spin rotation symmetry in both regimes.

8. Superconductivity

This chapter provides a modern, self-contained account of the theory of superconductivity. With the microscopic model of Bardeen, Cooper, and Schrieffer (BSC) as starting point, the effective theories both near the transition temperature as well as near the absolute zero of temperature are derived by integrating out the fermionic degrees of freedom and applying the derivative expansion. The emphasis is on the spontaneously broken global phase symmetry in the neutral system and its physical consequences such as the appearance of a gapless Anderson-Bogoliubov collective mode and vortices. Most startling features of a superconductor (Meissner effect and persistent currents) are shown to arise from the Anderson-Higgs mechanism, according to which the coupling of the neutral system to an electromagnetic field produces a massive vector field. Subjects usually regarded as relatively difficult, such as the time-dependent Ginzburg-Landau theory and gapless superconductivity are easily accounted for in this scheme.

9. Duality

The description of a superfluid film is extended in this chapter to charged systems. First, a genuine 2D theory is considered, followed by the more realistic case where the magnetic field is not confined to the superconducting film, but can spread in the third direction perpendicular to the film. The same steps leading to the dual formulation of the neutral theory (sine-Gordon model) are shown to lead to the bosonization rules in the genuine 2D charged theory, provided one of the spatial directions is taken as representing (Euclidean) time. The chapter closes with a modern description of the celebrated Peierls instability in quasi-one-dimensional conductors at the absolute zero of temperature, and with a discussion of charge fractionalization that emerges in this condensed-matter system.

10. From BCS to BEC

This chapter considers the pairing theory in the so-called composite-boson limit, where Cooper pairs form tightly bound states. In this limit, which has recently been accessed experimentally in trapped Fermi gasses close to a Feshbach resonance, the pairing theory directly maps onto the Bogoliubov theory. A possible application in the context of superconductor-insulator transitions at very low temperatures, an example of a quantum phase transition, is discussed. The scaling theory for quantum critical phenomena in pure and disordered systems is set up, and conclusions for the superconductor-insulator transitions are drawn.

11. Superfluid ^{3}He

In this chapter, the BCS theory, describing Cooper pairing in the channel with relative orbital angular momentum quantum number $L = 0$, is extended to pairing in the $L = 1$ channel to describe superfluid ^{3}He. The resulting order parameter transforms as a vector under rotations both in spin and orbital space. The possible superfluid ^{3}He phases are characterized by the different ways these rotational symmetries combined with the global phase symmetry are spontaneously broken. The superfluid ^{3}He-A phase, one of the experimentally most relevant phases, is analyzed in detail because of its surprising feature of having a topological line defect in momentum space. The fermionic energy gap vanishes at the two nodes where this line defect intersects the Fermi surface of the liquid state. Around these so-called boojums, the spectrum of the fermionic excitations vanishes linearly, as for massless relativistic fermions. The presence of gapless fermionic excitations, or zero modes implies that even at zero temperature, superfluid ^{3}He-A has a normal component. They are shown to be at the heart of various special effects such as a chiral anomaly and the Callan-Harvey effect, usually associated with relativistic theories. The boojums are diabolic points, marking the points on the Fermi surface of the liquid state where two energy levels cross and produce a geometrical phase which has direct physical consequences.

12. Quantum Hall Effect

The topic of this chapter is the quantum Hall effect, with both integer and fractional filling factor. The chapter starts with noninteracting electrons at low temperatures confined to move in a plane and placed in a strong magnetic field perpendicular to the plane. Discussed next are the changes resulting from inserting an infinitely thin solenoid, or fluxon into the system. The essential role of impurities in the quantum Hall effect is highlighted. The Landau levels of the noninteracting 2D electron gas are treated in a way that allows for including the Coulomb repulsion which is essential to the fractional quantum Hall effect. The Laughlin wavefunctions for the ground state and the excited state of the fractional quantum Hall effect are treated in detail and shown to describe composite fermions, carrying electric charge and magnetic flux. Also the effective field theory, which involves a so-called Chern-Simons term of the fractional quantum Hall effect, is presented. The chapter closes with the half-integer quantum Hall effect in graphene—a single layer of carbon atoms.

Notation and Conventions

Although studying condensed matter systems, we will often use a notation commonly employed in particle physics. A spacetime point will be denoted by $x = x^\mu = (t, \mathbf{x})$, $\mu = 0, 1, \cdots, D$, with D the number of space dimensions. The frequency ω and wave vector $\mathbf{k}$ of a field will be denoted by the vector $k^\mu = (\omega, \mathbf{k})$. Greek indices label spacetime coordinates, while space coordinates x^i, $i = 1, \cdots, D$ are labeled by Latin indices. The time derivative $\partial_0 = \partial_t = \partial/\partial t$ and the gradient ∇ are sometimes combined in a single vector $\partial_\mu = (\partial_t, \nabla)$. We will also often write $k \cdot x = k_\mu x^\mu = k^\mu x_\mu$ for $\omega t - \mathbf{k} \cdot \mathbf{x}$, and use Einstein's summation convention, stating that an index which appears twice in a term is to be summed over. Indices are raised and lowered with the help of the diagonal metric

$$\eta^{\mu\nu} = \begin{pmatrix} 1 & & & \\ & -1 & & \\ & & \ddots & \\ & & & -1 \end{pmatrix} \tag{0.1}$$

so that, for example,

$$\partial^\mu = \eta^{\mu\nu} \partial_\nu = (\partial_t, -\nabla). \tag{0.2}$$

The Fourier transform is defined as

$$f(x) = \int \frac{\mathrm{d}^d k}{(2\pi)^d} f(k)\, \mathrm{e}^{-ik \cdot x}, \tag{0.3}$$

with the inverse transform

$$f(k) = \int \mathrm{d}^d x\, f(x)\, \mathrm{e}^{ik \cdot x}. \tag{0.4}$$

For a function $f(\mathbf{x})$ defined in the finite volume $-L/2 \leq x^1, x^2, \cdots, x^D \leq L/2$ and satisfying periodic boundary conditions $f(x^1, \cdots, x^i + L, \cdots x^D) =$

$f(x^1, \cdots, x^i, \cdots x^D)$ for $i = 1, 2, \cdots D$, the wave vector $\mathbf{k}$ only takes the discrete values

$$k^i = \frac{2\pi}{L} m^i, \tag{0.5}$$

with $m^i = 0, \pm 1, \pm 2, \cdots$. The Fourier transform then becomes a series

$$f(\mathbf{x}) = \frac{1}{V} \sum_{\mathbf{k}} f(\mathbf{k})\, \mathrm{e}^{\mathrm{i}\mathbf{k}\cdot\mathbf{x}}, \tag{0.6}$$

where $\sum_{\mathbf{k}} \equiv \sum_{m^1, m^2, \cdots, m^D}$ and $V \equiv L^D$ is the volume. The inverse transform reads

$$f(\mathbf{k}) = \int_{-L/2}^{L/2} \mathrm{d}^D x\, f(\mathbf{x})\, \mathrm{e}^{-\mathrm{i}\mathbf{k}\cdot\mathbf{x}}. \tag{0.7}$$

The functions $\mathrm{e}^{\mathrm{i}\mathbf{k}\cdot\mathbf{x}}$ satisfy the orthogonality condition

$$\int_{-L/2}^{L/2} \mathrm{d}^D x\, \mathrm{e}^{-\mathrm{i}(\mathbf{k}-\mathbf{k}')\cdot\mathbf{x}} = V \delta_{\mathbf{k},\mathbf{k}'}, \tag{0.8}$$

with $\delta_{i,j}$ the Kronecker delta which yields unity if $i = j$ and zero otherwise, as well as the completeness relation

$$\sum_{\mathbf{k}} \mathrm{e}^{\mathrm{i}\mathbf{k}\cdot(\mathbf{x}-\mathbf{x}')} = V\delta(\mathbf{x}-\mathbf{x}'), \tag{0.9}$$

with $\delta(\mathbf{x}) \equiv \delta(x^1)\delta(x^2)\cdots\delta(x^D)$ the Dirac delta function in D dimensions.

The infinite-volume limit $V \to \infty$, where the wave vector $\mathbf{k}$ becomes a continuous variable, is recovered by letting

$$\sum_{k^i} = \sum_{m^i} \Delta m^i = \frac{L}{2\pi} \sum_{m^i} \Delta k^i \to L \int \frac{\mathrm{d}k^i}{2\pi}, \tag{0.10}$$

where $\Delta k^i \equiv (2\pi/L)\Delta m^i$ with $\Delta m^i = 1$ the minimal (positive) change in m^i, so that

$$\sum_{\mathbf{k}} \to V \int \frac{\mathrm{d}^D k}{(2\pi)^D}. \tag{0.11}$$

In this limit,

$$V\delta_{\mathbf{k},\mathbf{k}'} \to (2\pi)^D \delta(\mathbf{k}-\mathbf{k}'). \tag{0.12}$$

Finally, for a function $f_{\mathbf{x}}$ defined on a hypercubic lattice with lattice spacing a and lattice points labeled by their position vector $\mathbf{x}$, the Fourier transform is formally given by Eq. (0.6) with the *proviso* that the wave vector is now restricted to the first Brillouin zone defined by $-\pi/a \leq k^i \leq \pi/a$. The total number of values the wave vector can take is given by $N = V/a^D$. The inverse transform now reads,

$$f_{\mathbf{k}} = a^D \sum_{\mathbf{x}} f_{\mathbf{x}}\, \mathrm{e}^{-\mathrm{i}\mathbf{k}\cdot\mathbf{x}}, \tag{0.13}$$

where the sum is over all lattice sites. The orthogonality condition and completeness relation assume the form

$$\sum_{\mathbf{x}} e^{-i(\mathbf{k}-\mathbf{k}')\cdot\mathbf{x}} = N\delta_{\mathbf{k},\mathbf{k}'},$$
$$\sum_{\mathbf{k}} e^{i\mathbf{k}\cdot(\mathbf{x}-\mathbf{x}')} = N\delta_{\mathbf{x},\mathbf{x}'}, \tag{0.14}$$

respectively, where the restriction to the first Brillouin zone in the sum $\sum_{\mathbf{k}}$ is understood.

Gaussian units are adopted for electromagnetic quantities unless otherwise stated. In these units, the Maxwell equations in $D = 3$ take the form

$$\nabla \cdot \mathbf{D} = 4\pi\rho, \quad \nabla \cdot \mathbf{B} = 0, \tag{0.15}$$

$$\nabla \times \mathbf{E} + \frac{1}{c}\frac{\partial \mathbf{B}}{\partial t} = 0, \quad \nabla \times \mathbf{H} = \frac{4\pi}{c}\mathbf{J} + \frac{1}{c}\frac{\partial \mathbf{D}}{\partial t}, \tag{0.16}$$

with c the speed of light. Moreover, ρ and $\mathbf{J}$ denote the electric charge and current densities, $\mathbf{H}$ and $\mathbf{B}$ stand for the magnetic and induction fields, while $\mathbf{D}$ and $\mathbf{E}$ are the displacement and electric fields. The electric fields are related through $\mathbf{D} = \mathbf{E} + 4\pi\mathbf{P}$, with $\mathbf{P}$ the polarization (dipole moment pro unit volume), and the magnetic fields are related through $\mathbf{H} = \mathbf{B} - 4\pi\mathbf{M}$, with $\mathbf{M}$ the magnetization (magnetic moment per unit volume). Finally, the Lorentz force $\mathbf{F}$ acting on a point particle of charge q moving with the velocity $\mathbf{v}$ reads

$$\mathbf{F} = q\mathbf{E} + \frac{q}{c}\mathbf{v} \times \mathbf{B} \tag{0.17}$$

in Gaussian units.

Contents

Chapter 1

Classical Field Theory

The concept of a (classical) field as used in contemporary physics originates from Faraday's intuitive picture of what he called "lines of force" to describe magnetism and electricity. Maxwell turned this profound picture, which was based on analogies with fluid flow and elastic media, into the modern field theory of electromagnetism. The concept proved useful and found applications in much of physics.

1.1 One-Dimensional Crystal

One of the simplest examples of a classical field theory describes a one-dimensional crystal in the harmonic approximation. To derive it, consider a chain of equal mass points arranged in a straight line, each connected to the next one through identical springs, see Fig. 1.1. Assume that the masses, whose equilibrium positions are separated by a distance a, can vibrate only in the direction of the chain, and let $\phi_r(t)$ denote the displacement of the rth mass point from its equilibrium position at time t. The kinetic energy T of the chain is

$$T = \frac{m}{2} \sum_r \dot{\phi}_r^2(t), \tag{1.1}$$

where the sum extends over all particles of mass m, and $\dot{\phi} \equiv \mathrm{d}\phi/\mathrm{d}t$. The potential energy V of the chain is the sum of the contributions of the individual springs,

$$V = \frac{k}{2} \sum_r [\phi_{r+1}(t) - \phi_r(t)]^2, \tag{1.2}$$

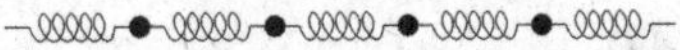

Fig. 1.1 A chain of equal mass points connected by springs in equilibrium.

with k the force constant. A spring contributes to V when it is stretched or compressed from its equilibrium length. The Lagrangian L describing the chain is defined by the difference of the two energies $L = T - V$.

Rather than working with this discrete system, we wish to describe the crystal as a continuum by letting the lattice spacing a tend to zero. In this limit, the Lagrangian

$$L = \frac{a}{2}\sum_r \left[\frac{m}{a}\dot{\phi}_r^2(t) - ka\left(\frac{\phi_{r+1}(t) - \phi_r(t)}{a}\right)^2\right] \tag{1.3}$$

can be written as an integral over the line

$$L = \frac{1}{2}\int \mathrm{d}x^1 \left\{\mu\,[\partial_t\phi(x)]^2 - Y\,[\partial_1\phi(x)]^2\right\}, \tag{1.4}$$

where $\mu \equiv m/a$ is the equilibrium mass per unit length of the continuous system, $Y \equiv ka$ is the Young modulus of elasticity, and $\phi(x)$, with $x = (t, x^1)$, is the displacement field. The corresponding action $S \equiv \int \mathrm{d}t\, L$ is seen to be given by an integral over space, which is one-dimensional in this example, and time

$$S = \int \mathrm{d}^d x\, \mathcal{L}, \tag{1.5}$$

with $d = 2$ and $\mathcal{L}$ the Lagrangian density. In the harmonic approximation, $\mathcal{L}$ is quadratic in the displacement field, and the continuum model is a free field theory.

As a side remark, note that for a real field, a term in the action of the form $\phi\partial_t\phi$ linear in time derivatives is a total derivative and can be ignored.

1.2 Action Principle

As for a free theory, fields with local interactions are also governed by an action that can be written as a spacetime integral of some Lagrangian density. Let $\mathcal{L}$ be a function of a set of fields ϕ_a ($a = 1, 2, \cdots$) and their derivatives $\partial_\mu\phi_a$. In principle, $\mathcal{L}$ can also contain higher-order derivatives of the fields, but we shall not discuss these cases here. Consider an arbitrary infinitesimal variation $\delta^0\phi_a$ in the fields,

$$\delta^0\phi_a(x) \equiv \phi'_a(x) - \phi_a(x). \tag{1.6}$$

The superscript 0 on the variation δ is to indicate that the spacetime coordinates of the original and varied fields, $\phi(x)$ and $\phi'(x)$, respectively, are the same. Under this variation, the action restricted to some region Ω of spacetime varies as:

$$\begin{aligned}\delta^0 S &= \int_\Omega \mathrm{d}^d x\, \delta^0\mathcal{L} = \int_\Omega \mathrm{d}^d x \left[\frac{\partial\mathcal{L}}{\partial\phi_a}\delta^0\phi_a + \frac{\partial\mathcal{L}}{\partial(\partial_\mu\phi_a)}\partial_\mu(\delta^0\phi_a)\right] \\ &= \int_\Omega \mathrm{d}^d x \left(\frac{\partial\mathcal{L}}{\partial\phi_a} - \partial_\mu\frac{\partial\mathcal{L}}{\partial(\partial_\mu\phi_a)}\right)\delta^0\phi_a + \int_{\partial\Omega}\mathrm{d}S_\mu \frac{\partial\mathcal{L}}{\partial(\partial_\mu\phi_a)}\delta^0\phi_a,\end{aligned} \tag{1.7}$$

where the last line follows by integrating by parts. The last term, with $\partial\Omega$ bounding the spacetime region Ω and dS_μ a surface element on the boundary, is the resulting boundary term. The equations governing the fields are obtained by requiring that the action be stationary under any variation (1.6) that vanishes on the boundary. This variational principle of least action yields the field equations

$$\frac{\partial \mathcal{L}}{\partial \phi_a} - \partial_\mu \frac{\partial \mathcal{L}}{\partial(\partial_\mu \phi_a)} = 0, \tag{1.8}$$

known as the *Euler-Lagrange equations*.

For the free theory of the preceding section, this gives the wave equation

$$\partial_t^2 \phi - c_s^2 \partial_1^2 \phi = 0 \tag{1.9}$$

with speed of propagation $c_s = \sqrt{Y/\mu}$. It is common to absorb the constant μ appearing in the Lagrangian (1.4) by defining $\phi' = \sqrt{\mu}\phi$ so that the coefficient of the kinetic term becomes $\frac{1}{2}$, and

$$\mathcal{L} = \frac{1}{2}(\partial_t \phi)^2 - \frac{1}{2} c_s^2 (\partial_1 \phi)^2, \tag{1.10}$$

where the primes on the field have been dropped again. This is the standard Lagrangian density describing longitudinal elastic waves in one space dimension.

1.3 Noether's Theorem

Symmetries play an important role in physics. A theorem due to Noether connects symmetries of the action to conserved charges, i.e., physical quantities that do not change in time. Examples of such conserved charges are energy, momentum, and particle number.

To derive the Noether theorem, consider the change in the action to yet unspecified infinitesimal transformations of the fields and coordinates,

$$\delta\phi_a(x) \equiv \phi'_a(x') - \phi_a(x) \tag{1.11}$$

and

$$\delta x^\mu \equiv x'^\mu - x^\mu, \tag{1.12}$$

respectively. In contrast to the variations $\delta^0\phi_a$ in Eq. (1.6), the present variations also include a change of the coordinates. To indicate this difference, the present variations do not carry a superscript 0. The coordinate transformation (1.12) leads to a change in the integration measure given by the Jacobian of the transformation:

$$d^d(x + \delta x) = \det\left[\partial_\mu(x^\nu + \delta x^\nu)\right] d^d x \approx (1 + \partial_\mu \delta x^\mu) d^d x. \tag{1.13}$$

Specifically,

$$\delta(\mathrm{d}^d x) = \mathrm{d}^d x\, \partial_\mu \delta x^\mu. \tag{1.14}$$

As a result, the change in the action derives from two different sources

$$\delta S = \int \mathrm{d}^d x [\delta \mathcal{L} + \mathcal{L} \partial_\mu \delta x^\mu], \tag{1.15}$$

which is to be distinguished from the previous one (1.7) arising from the variation (1.6) in the fields alone.

The infinitesimal transformation $\delta\phi_a(x)$ in the fields (1.11) can be written in terms of the variation (1.6) as

$$\begin{aligned} \delta\phi_a(x) &= \phi'_a(x+\delta x) - \phi_a(x) \\ &= \phi'_a(x) - \phi_a(x) + \partial_\mu \phi_a(x) \delta x^\mu \\ &= \delta^0 \phi_a(x) + \partial_\mu \phi_a(x) \delta x^\mu, \end{aligned} \tag{1.16}$$

and similarly

$$\delta \mathcal{L} = \delta^0 \mathcal{L} + \partial_\mu \mathcal{L} \delta x^\mu. \tag{1.17}$$

More explicitly,

$$\begin{aligned} \delta \mathcal{L} &= \frac{\partial \mathcal{L}}{\partial \phi_a} \delta^0 \phi_a + \frac{\partial \mathcal{L}}{\partial(\partial_\mu \phi_a)} \partial_\mu (\delta^0 \phi_a) + \partial_\mu \mathcal{L} \delta x^\mu \\ &= \left(\frac{\partial \mathcal{L}}{\partial \phi_a} - \frac{\partial \mathcal{L}}{\partial(\partial_\mu \phi_a)} \right) \delta^0 \phi_a + \partial_\mu \left(\frac{\partial \mathcal{L}}{\partial(\partial_\mu \phi_a)} \delta^0 \phi_a \right) + \partial_\mu \mathcal{L} \delta x^\mu, \end{aligned} \tag{1.18}$$

where in the first line it is used that the coordinates do no change under the variation δ^0, i.e., $\partial_\mu(\delta^0\phi_a) = \delta^0 \partial_\mu \phi_a$. For fields satisfying the Euler-Lagrange equations (1.8), the change in the action (1.15) assumes the form

$$\delta S = \int \mathrm{d}^d x\, \partial_\mu \left(-T^\mu{}_\nu \delta x^\nu + \frac{\partial \mathcal{L}}{\partial(\partial_\mu \phi_a)} \delta\phi_a \right), \tag{1.19}$$

where $T^\mu{}_\nu$ defines the *energy-momentum tensor*,

$$T^\mu{}_\nu \equiv \frac{\partial \mathcal{L}}{\partial(\partial_\mu \phi_a)} \partial_\nu \phi_a - \eta^\mu{}_\nu \mathcal{L}, \tag{1.20}$$

and Eq. (1.16) is used to replace $\delta^0\phi_a$ with $\delta\phi_a$. Note that in contrast to the variations (1.6), the infinitesimal transformations (1.11) are not required to vanish on the boundary. Now, if the action is invariant under the transformations (1.11) and (1.12), the current density j^μ, defined through

$$j^\mu \equiv -T^\mu{}_\nu \delta x^\nu + \frac{\partial \mathcal{L}}{\partial(\partial_\mu \phi_a)} \delta\phi_a, \tag{1.21}$$

is conserved for field configurations satisfying the Euler-Lagrange equations, i.e., $\partial_\mu j^\mu = 0$. This conservation in turn implies that the associated charge

$$Q \equiv \int \mathrm{d}^D x\, j^0, \tag{1.22}$$

remains unchanged in the course of time:

$$\frac{\mathrm{d}Q}{\mathrm{d}t} = \int \mathrm{d}^D x\, \partial_t j^0 = -\int \mathrm{d}^D x\, \partial_i j^i = 0. \tag{1.23}$$

The last equation follows from Gauss' theorem and the assumption that the currents decrease sufficiently fast at spatial infinity. The observation that a symmetry of the action implies a conservation law constitutes *Noether's theorem*. Symmetries that do not involve changes of the coordinates ($\delta x^\nu = 0$) are called *internal symmetries*.

In terms of the canonical conjugate $\pi_a(x)$ to the field $\phi_a(x)$, which is defined as

$$\pi_a \equiv \frac{\partial \mathcal{L}}{\partial(\partial_t \phi_a)}, \tag{1.24}$$

the charge density $j^0(x)$ assumes the compact form

$$\begin{aligned} j^0 &= \mathcal{L}\delta x^0 + \pi_a(\delta\phi_a - \partial_\nu \phi_a \delta x^\nu) \\ &= \mathcal{L}\delta x^0 + \pi_a \delta^0 \phi_a, \end{aligned} \tag{1.25}$$

by Eq. (1.16).

The connection between continuous symmetries of the action and conserved charges becomes even more pronounced when the symmetry transformations (1.11) are expressed in terms of the corresponding charges as follows. Consider the *Poisson bracket* $\{F, G\}$ of two functionals $F[\phi_a, \pi_a]$ and $G[\phi_a, \pi_a]$ of the fields and their canonical conjugates at a given time t defined as (no sum over a)

$$\{F, G\} \equiv \int \mathrm{d}^D x \left(\frac{\delta F}{\delta \phi_a(t, \mathbf{x})} \frac{\delta G}{\delta \pi_a(t, \mathbf{x})} - \frac{\delta F}{\delta \pi_a(t, \mathbf{x})} \frac{\delta G}{\delta \phi_a(t, \mathbf{x})} \right), \tag{1.26}$$

where the integral is over space coordinates only. Here, $\delta F / \delta \phi_a(x)$ denotes a *functional derivative*. In the same way that a function $f(x)$ assigns a number to its argument x:

$$f : x \to f(x), \tag{1.27}$$

a functional $F[g]$ also assigns a number to its argument, which happens to be a function $g(x)$:

$$F : g \to F[g]. \tag{1.28}$$

The ordinary derivative of a function $f(x)$

$$\frac{df(x)}{dx} = \lim_{\varepsilon \to 0} \frac{1}{\varepsilon} [f(x+\varepsilon) - f(x)] \tag{1.29}$$

is readily generalized to a functional derivative as follows:

$$\frac{\delta F[g]}{\delta g(y)} = \lim_{\varepsilon \to 0} \frac{1}{\varepsilon} \{F[g(x) + \varepsilon\delta(x-y)] - F[g(x)]\}, \tag{1.30}$$

with $\delta(x)$ the Dirac delta function. The rules for functional derivatives are very much like the ones for ordinary derivatives. As an elementary example, consider the functional

$$F[g] = \int dx\, g^2(x), \tag{1.31}$$

which indeed assigns a number to the function $g(x)$, namely the value of the integral. With the definition (1.30), the functional derivative gives

$$\begin{aligned}\frac{\delta F[g]}{\delta g(y)} &= \lim_{\varepsilon \to 0} \frac{1}{\varepsilon} \int dx \left\{[g(x) + \varepsilon\delta(x-y)]^2 - g^2(x)\right\} \\ &= 2 \int dx\, g(x)\delta(x-y) = 2g(y).\end{aligned} \tag{1.32}$$

More complicated functionals can be treated similarly. Other rules are

$$\frac{\delta F^n[g]}{\delta g(y)} = nF^{n-1}[g] \frac{\delta F[g]}{\delta g(y)} \tag{1.33}$$

and

$$\frac{\delta}{\delta g(y)} \exp(F[g]) = \exp(F[g]) \frac{\delta F[g]}{\delta g(y)}. \tag{1.34}$$

An important bracket is of a field and its canonical conjugate:

$$\{\phi_a(t, \mathbf{x}), \pi_b(t, \mathbf{x}')\} = \delta_{ab}\, \delta(\mathbf{x} - \mathbf{x}'), \tag{1.35}$$

which yields a delta function by the definition (1.26). For symmetry transformations with $\delta x^0 = 0$, the charge density (1.25) reduces to

$$j^0 = \pi_a \delta^0 \phi_a, \tag{1.36}$$

and

$$\{\phi_a(x), Q\} = \delta^0 \phi_a(x), \tag{1.37}$$

by the basic bracket (1.35). The charge Q is said to generate the symmetry transformation $\delta^0 \phi_a(x)$ specified in Eq. (1.11).

Consider, as an example, a translation $\delta x^i = \alpha^i$ in space. Such a transformation is generated by the charge $-\alpha^i P_i$ with P_i the total momentum

$$P_i \equiv \int d^D x\, \pi_a \partial_i \phi_a, \tag{1.38}$$

through

$$\left\{\phi_a(x), -\alpha^i P_i\right\} = -\alpha^i \partial_i \phi_a(x), \tag{1.39}$$

where the right side denotes $\delta^0 \phi_a$ as follows from Eq. (1.16) with $\delta\phi_a(x) = 0$.

The case of translations $\delta x^0 = \alpha^0$ in time, to which Eq. (1.37) does not apply, is somewhat special. The charge density (1.25) reduces to ($-\alpha^0$ times) the Hamiltonian density $\mathcal{H}$,

$$j^0 = \alpha^0(\mathcal{L} - \pi_a \partial_t \phi_a) = -\alpha^0 \mathcal{H}, \tag{1.40}$$

and

$$\left\{\phi_a(x), -\alpha^0 H\right\} = -\alpha^0 \partial_t \phi_a(x), \tag{1.41}$$

where $H \equiv \int \mathrm{d}^D x \mathcal{H}$ is the Hamiltonian and the right side denotes $\delta^0 \phi_a$ as follows again from Eq. (1.16) with $\delta\phi_a(x) = 0$. In deriving Eq. (1.41), use is made of the Hamilton equation

$$\frac{\delta H}{\delta \pi_a(x)} = \partial_t \phi_a(x). \tag{1.42}$$

Despite their different origins, Eqs. (1.38) and (1.41) can be summarized by the single equation

$$\left\{\phi_a(x), P_\mu\right\} = \partial_\mu \phi_a(x), \tag{1.43}$$

where we introduced the notation $P^\mu \equiv (H, \mathbf{P})$.

As an application, we investigate in detail the symmetry content of a nonrelativistic theory in the next section.

1.4 Nonrelativistic Field Theory

Consider the nonrelativistic classical theory specified by the Lagrangian density

$$\mathcal{L} = \mathrm{i}\hbar\psi^* \frac{\partial}{\partial t}\psi - \frac{\hbar^2}{2m}\nabla\psi^* \cdot \nabla\psi - \mathcal{V}(\psi^*\psi), \tag{1.44}$$

featuring the complex field ψ. The potential energy density $\mathcal{V}$ is assumed to be a function of $\psi^*\psi$. To be specific, we choose the simple form

$$\mathcal{V} = -\mu\psi^*\psi + \frac{g}{2}(\psi^*\psi)^2, \tag{1.45}$$

with positive coefficient g, while the coefficient μ can be either negative or positive. This classical theory plays an important role in condensed matter physics, as we will see in the following. In the context of Bose-Einstein condensation in weakly interacting Bose gases, it is known as the *Gross-Pitaevskii* theory.

The field ψ is a classical, complex field satisfying the Euler-Lagrange equation

$$i\hbar\frac{\partial}{\partial t}\psi = -\frac{\hbar^2}{2m}\nabla^2\psi + \frac{\partial V(\psi^*\psi)}{\partial\psi^*}. \tag{1.46}$$

Its canonical conjugate is the field ψ^*,

$$\pi(x) = \frac{\partial\mathcal{L}}{\partial(\partial_t\psi)} = i\hbar\psi^*(x) \tag{1.47}$$

so that

$$\{\psi(t,\mathbf{x}), i\hbar\psi^*(t,\mathbf{x}')\} = \delta(\mathbf{x}-\mathbf{x}'). \tag{1.48}$$

The two fields ψ and ψ^* are therefore to be considered as independent. In the same way that the electromagnetic vector potential A_μ satisfies the classical Maxwell equations, ψ satisfies the classical field equation (1.46), which because of our choice of parameters, formally assumes the form of a Schrödinger equation. After quantizing, the electromagnetic vector potential describes photons—the quanta of light, while the displacement field ϕ featuring in the Lagrangian (1.10) of a one-dimensional crystal describes acoustic phonons—the quanta of sound. In a similar fashion, the field ψ describes after quantizing nonrelativistic bosons of mass m. With this in mind, the specific forms of the coefficients in the classical theory (1.44) were chosen. The classical field ψ is the average field produced by many bosons in the same way that the classical vector potential of electrodynamics describes the average behavior of many photons.

For the time being, we take $\mu < 0$ so that the minimum of the potential is at $\psi = 0$. In a classical setting, where it is more appropriate to have Planck's constant not appear explicitly, one introduces the *dispersion constant* $\gamma \equiv \hbar/m$ and $\omega_0 \equiv -\mu/\hbar > 0$. The dispersion relation, which can be obtained from the Euler-Lagrange equation (1.46) by Fourier transforming the field and expanding the expression to linear order around $\psi = 0$, then reads

$$\omega(\mathbf{k}) = \omega_0 + \frac{1}{2}\gamma\mathbf{k}^2, \tag{1.49}$$

with $\mathbf{k}$ the wave vector and ω_0 denoting the lowest attainable frequency. This shows that the Lagrangian (1.44) describes a quadratically dispersing mode with frequencies larger than the cutoff ω_0.

We next investigate the symmetry content of the theory. Consider first a translation of the spacetime coordinates by an infinitesimal constant vector α^ν, $\delta x^\nu \equiv x'^\nu - x^\nu = \alpha^\nu$, and $\delta\psi(x) = 0$. It is readily checked that the action is invariant under this coordinate transformation so that by Eq. (1.21), the current density

$$j^\mu \equiv -T^\mu{}_\nu\alpha^\nu \tag{1.50}$$

and, since α^ν is a constant vector, also the energy-momentum tensor is conserved,

$$\partial_\mu T^{\mu\nu} = 0. \tag{1.51}$$

The energy $E = H = P^0$ and momentum $\mathbf{P}$ of the system are given by

$$E = \int \mathrm{d}^D x\, T^{00}, \quad P^i = \int \mathrm{d}^D x\, T^{0i}, \tag{1.52}$$

respectively. For the nonrelativistic theory (1.44), the general expression (1.20) for the energy-momentum tensor reduces to

$$T^{00} = \frac{\hbar^2}{2m}\nabla\psi^* \cdot \nabla\psi + \mathcal{V}(\psi^*\psi), \quad T^{0i} = -\frac{1}{2}\mathrm{i}\hbar\psi^* \overleftrightarrow{\partial}_i \psi, \tag{1.53}$$

where $\overleftrightarrow{\partial}_i \equiv \partial_i - \overleftarrow{\partial}_i$ stands for the the right minus left derivative. Note that $\partial^i = -\partial_i$, with ∂_i denoting the components of ∇, see Notation and Conventions. To arrive at the symmetric form (1.53) for T^{0i}, the first term of the Lagrangian (1.44) has been recast in the equivalent form $\frac{1}{2}\mathrm{i}\hbar\psi^*\overleftrightarrow{\partial}_t\psi$, which differ only by an irrelevant total derivative.

The theory (1.44) is also invariant under phase transformations

$$\psi(x) \to \psi'(x) = \mathrm{e}^{\mathrm{i}\alpha}\psi(x), \quad \psi^*(x) \to \psi'^*(x) = \mathrm{e}^{-\mathrm{i}\alpha}\psi^*(x), \tag{1.54}$$

with α a constant transformation parameter. Because the same value for α is used throughout spacetime, these transformations are referred to as *global* transformations. With α small, the transformations (1.54) take the infinitesimal forms

$$\delta^0\psi(x) = \mathrm{i}\alpha\psi(x), \quad \delta^0\psi^*(x) = -\mathrm{i}\alpha\psi^*(x). \tag{1.55}$$

By Eq. (1.21), this symmetry leads to the conservation of the current density

$$j^\mu = \frac{\mathrm{i}}{\hbar}\left(\frac{\partial\mathcal{L}}{\partial(\partial_\mu\psi^*)}\psi^* - \frac{\partial\mathcal{L}}{\partial(\partial_\mu\psi)}\psi\right), \tag{1.56}$$

where the constant $-\hbar\alpha$ has been divided out for convenience. Specifically,

$$\partial_\mu j^\mu = \partial_t j^0 + \nabla\cdot\mathbf{j} = 0, \tag{1.57}$$

with

$$j^0 = n = \psi^*\psi, \quad \mathbf{j} = -\mathrm{i}\frac{\hbar}{2m}\psi^*\overleftrightarrow{\nabla}\psi. \tag{1.58}$$

After quantization, the conservation of the charge $N \equiv \int \mathrm{d}^D x\, n = \int \mathrm{d}^D x\, \psi^*\psi$ physically denotes the conservation of particle number. The symmetry (1.54) generated by this charge,

$$\delta^0\psi(x) = \{\psi(x), -\hbar\alpha N\} = \mathrm{i}\alpha\psi(x), \tag{1.59}$$

plays an important role in understanding superfluidity in interacting systems.

A final symmetry enjoyed by the nonrelativistic theory (1.44) is the invariance under Galilei transformations. Under a Galilei boost with a constant velocity $\mathbf{u}$, the coordinates transform as

$$t \to t' = t, \quad \mathbf{x} \to \mathbf{x}' = \mathbf{x} - \mathbf{u}t \tag{1.60}$$

so that

$$\frac{\partial}{\partial t} \to \frac{\partial}{\partial t'} = \frac{\partial t}{\partial t'}\frac{\partial}{\partial t} + \frac{\partial \mathbf{x}}{\partial t'} \cdot \nabla = \partial_t + \mathbf{u} \cdot \nabla, \quad \nabla \to \nabla' = \nabla, \tag{1.61}$$

while the fields pick up an extra phase factor

$$\psi(x) \to \psi'(x') = e^{i(m/\hbar)(-\mathbf{u}\cdot\mathbf{x} + \frac{1}{2}\mathbf{u}^2 t)}\,\psi(x). \tag{1.62}$$

Both the field equation (1.46) and the action are invariant under these Galilei transformations, as can be explicitly checked. In infinitesimal form, the transformations become

$$\delta x^\mu = -u^\mu t, \quad \delta\psi(x) = i\frac{m}{\hbar}u^\mu x_\mu \psi(x), \tag{1.63}$$

and

$$\delta^0\psi(x) = i\frac{m}{\hbar}u_\mu x^\mu \psi(x) + u_\mu t\partial^\mu\psi(x) \tag{1.64}$$

with $u^\mu \equiv (0, \mathbf{u})$. The conserved current density reads by the general formula (1.21)

$$g^{\mu i} = T^{\mu i}t + i\frac{m}{\hbar}\left(\frac{\partial\mathcal{L}}{\partial(\partial_\mu\psi)}\psi - \frac{\partial\mathcal{L}}{\partial(\partial_\mu\psi^*)}\psi^*\right)x^i, \tag{1.65}$$

where the index i arises because we dropped the constant boost vector u_i. The conservation $dG^{0i}/dt = 0$ of the charges

$$G^{0i} = \int d^D x\, g^{0i} = \int d^D x(T^{0i}t - mnx^i) = P^i t - mNX^i, \tag{1.66}$$

signifies that the center of mass

$$\mathbf{X}(t) \equiv \frac{\int d^D x\, \mathbf{x}\, n(x)}{\int d^D x\, n(x)} \tag{1.67}$$

of the system moves with constant momentum,

$$\mathbf{P} = mN\frac{d\mathbf{X}}{dt}. \tag{1.68}$$

It can be explicitly checked that

$$\left\{\psi(x), u_\mu G^{0\mu}\right\} = \delta^0\psi(x), \tag{1.69}$$

with the right side given by Eq. (1.64).

In closing this section, we point out the remarkable relation

$$j^i = \frac{1}{m}T^{0i}, \tag{1.70}$$

linking the particle number current density j^i to the momentum density T^{0i}. The relation is remarkable because of its complete asymmetry. Whereas j^i is a vector and a current density, T^{0i} is part of a tensor and a charge density. The relation is the hallmark of a Galilei-invariant theory and states that every particle has the same particle-number-to-mass ratio.

1.5 Spontaneously Broken Symmetries

Instead of taking the coefficient μ in Eq. (1.45) as negative, we now assume it to be positive. The shape of the potential energy density then becomes as depicted in Fig. 1.2, showing that $\mathcal{V}$ develops a minimum away from the origin $\psi = 0$ at

$$|\psi|^2 = v^2 \equiv \mu/g. \tag{1.71}$$

The uniform system will, in the absence of an outside agent, settle at some point chosen at random along the circle forming the minimum of the potential. Because the entire system spontaneously settles on the same phase, the system is said to *order* itself. In the context of a weakly interacting Bose gas, the condition $\psi \neq 0$ signifies the formation of a Bose-Einstein condensate.

The dispersion relation (1.49) was obtained by expanding around the value $\psi = 0$, which for $\mu > 0$ we recognize as the false ground state. To obtain the dispersion relation for $\mu > 0$, we must expand around the true ground state specified by the minimum value (1.71). To this end, we introduce two new real fields φ and η by writing

$$\psi(x) = [v + \eta(x)]\, e^{i\varphi(x)}. \tag{1.72}$$

In terms of these new variables, the Lagrangian density becomes

$$\mathcal{L} = -(v+\eta)^2 \left[\hbar\partial_t\varphi + \frac{\hbar^2}{2m}(\nabla\varphi)^2\right] - \frac{\hbar^2}{2m}(\nabla\eta)^2 + \mu(v+\eta)^2 - \frac{g}{2}(v+\eta)^4, \tag{1.73}$$

omitting a total derivative. The real field η is seen to have no time derivative. It therefore is not dispersing or, put differently, η does not represent a propagating degree of freedom.

In this respect, the nonrelativistic theory (1.44) differs fundamentally from its relativistic counterpart, where the field η represents a genuine propagating mode.

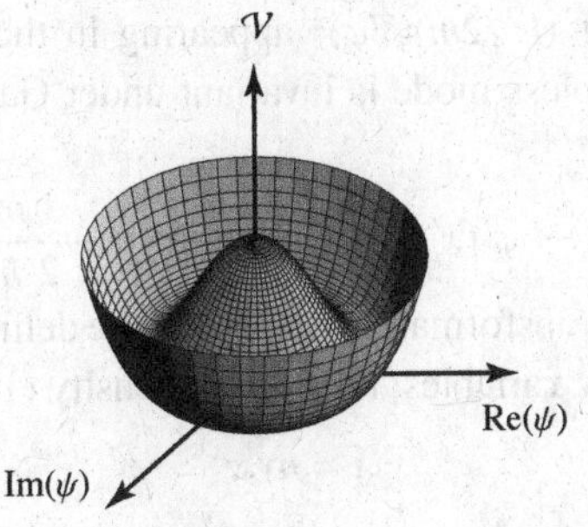

Fig. 1.2 Graphical representation of the potential energy density (1.45).

That mode, which has a frequency cutoff, corresponds to radial oscillations that climb the walls of the potential energy density in Fig. 1.2. When taking the nonrelativistic limit, this mode becomes nondispersing.

For spatial variations of wave number $|\mathbf{k}|$ small compared to $1/\xi$, where

$$\xi \equiv \frac{\hbar}{\sqrt{2m\mu}} \tag{1.74}$$

defines the typical length scale for such variations, the gradient term $(\nabla\eta)^2$ can be neglected, and η satisfies the field equation

$$[v + \eta(x)]^2 = -\frac{1}{g}\left\{\hbar\partial_t\varphi(x) + \frac{\hbar^2}{2m}\left[\nabla\varphi(x)\right]^2\right\} + \frac{\mu}{g}. \tag{1.75}$$

To remove the interaction between η and φ, we substitute this field equation back into the Lagrangian density, giving

$$\mathcal{L} = -\frac{\mu}{g}\left[\hbar\partial_t\varphi + \frac{\hbar^2}{2m}(\nabla\varphi)^2\right] + \frac{1}{2g}\left[\hbar\partial_t\varphi + \frac{\hbar^2}{2m}(\nabla\varphi)^2\right]^2, \tag{1.76}$$

where an irrelevant additive constant is dropped. The dispersion relation can now be read off from the quadratic terms,

$$\omega^2(\mathbf{k}) = c^2\mathbf{k}^2, \tag{1.77}$$

where the parameter $c \equiv \sqrt{\mu/m}$ defines the phase velocity. Instead of a *quadratically* dispersing mode found in Eq. (1.49) for $\mu < 0$, for $\mu > 0$ a *linearly* dispersing mode emerges. More importantly, in contrast to what was found for $\mu < 0$, this mode is gapless. It physically represents sound waves, i.e., propagating density perturbations. In terms of the parameters v and c, the coefficients in the Lagrangian density (1.76) assume the form

$$\frac{\mu}{g} = v^2, \qquad \frac{1}{2g} = \frac{1}{2}\frac{v^2}{mc^2}. \tag{1.78}$$

The combination $\hbar\partial_t\varphi + (\hbar^2/2m)(\nabla\varphi)^2$ appearing in the effective theory (1.76) of the nonrelativistic gapless mode is invariant under Galilei transformations for which

$$\varphi(x) \to \varphi'(x') = \varphi(x) - \frac{m}{\hbar}\mathbf{u}\cdot\mathbf{x} + \frac{1}{2}\frac{m}{\hbar}\mathbf{u}^2 t \tag{1.79}$$

according to the Galilei transformation (1.62) and the definition (1.72) of the phase field. In terms of the new variables, the current density (1.58) assumes the form

$$\mathbf{j} = n\mathbf{v}_\mathrm{s}, \tag{1.80}$$

with $n = \psi^*\psi = (v + \eta)^2$ the particle number density and

$$\mathbf{v}_\mathrm{s} = \frac{\hbar}{m}\nabla\varphi \tag{1.81}$$

a velocity field. This last expression shows that the phase field φ physically plays the role of a velocity potential.

We next come to an important observation concerning the symmetry content of the theory. In the previous section it was observed that the theory is invariant under the phase transformations (1.54) generated by the charge N, see Eq. (1.59). These transformations constitute the group U(1). The nontrivial ground state, characterized by the constant v, introduced in Eq. (1.71), is however not invariant under this symmetry group, since v transforms as

$$v \to v' = \mathrm{e}^{\mathrm{i}\alpha} v \neq v. \tag{1.82}$$

The ground state is invariant under the symmetry group only when v is zero, corresponding to the trivial ground state. A finite value v is said to *spontaneously break* the global U(1) symmetry.

A symmetry that is spontaneously broken is not completely lost. The invariance of the action implies that v' satisfies the same field equation as the constant field v, i.e., the transformed field v' also characterizes the ground state. The different ground states, all minimizing the potential energy density, are degenerate and related by a phase transformation. The complete set of ground states, known as the space of degeneracy or *ground-state manifold* is obtained by operating with the symmetry group on a given ground state. Such a set is called the orbit of the group. For the case at hand, the orbit is given by $v\mathrm{e}^{\mathrm{i}\theta}$ and is represented by the circle at the bottom of the potential energy density depicted in Fig. 1.2.

According to a theorem due to Goldstone, the spontaneous breakdown of a continuous symmetry in higher than two dimensions gives rise to a gapless mode—known as the *Nambu-Goldstone mode*. Because they are gapless, Nambu-Goldstone modes are the dominant degrees of freedom in an effective description of the system valid at low frequency and long wave length. Examples of Nambu-Goldstone modes are spin waves in ferro- and antiferromagnets, and sound waves in superfluids and crystals. In the case under study, the presence of a Nambu-Goldstone mode is signaled by the gaplessness of the dispersion relation (1.77). It corresponds to excitations that lie at the bottom of the potential energy density shown in Fig. 1.2. A little thought reveals that Nambu-Goldstone fields always parametrize the ground-state manifold. Under the symmetry transformation (1.54), φ is shifted,

$$\delta\varphi(x) = \{\varphi(x), -\hbar\alpha N\} = \alpha, \tag{1.83}$$

which is the typical transformation property of a Nambu-Goldstone field under the action of a spontaneously broken symmetry.

1.6 Effective Theory of Hydrodynamics

Probably the most familiar *nonrelativistic* field theory is provided by hydrodynamics. As a second application of the field concept we in this section give a modern field theoretic description of the hydrodynamics of an ideal, classical fluid, which was already well understood in the 19th century. The case of *homentropic* flow, for which the entropy per unit mass is constant, is particularly simple. The pressure P is then a function of the mass density ρ only, and the flow is automatically a potential flow. A feature of such a fluid is that it supports unattenuated sound waves, i.e., propagating density oscillations. The waves are unattenuated because viscosity and thermal conductivity, which usually serve to dissipate the energy of a propagating mode, are absent. More important to our present considerations is that sound waves are gapless. In this and the next section, we will identify them as the Nambu-Goldstone mode associated with the spontaneously broken Galilei symmetry. This identification then explains their gaplessness as an *emergent* property.

As starting point to describe the hydrodynamics of a homentropic fluid, we take the Lagrangian density

$$\mathcal{L} = \frac{1}{2}\rho\mathbf{v}^2 - \rho e + \phi[\partial_t\rho + \nabla\cdot(\rho\mathbf{v})], \tag{1.84}$$

where $\mathbf{v}$ is the velocity field, ρ the mass density, and e the internal energy per unit mass. For homentropic flow, e is a function of ρ alone. The first and second term in (1.84) represent the kinetic and potential energy density, respectively. The variable ϕ is a Lagrange multiplier introduced to impose the conservation of mass:

$$\partial_t\rho + \nabla\cdot(\rho\mathbf{v}) = 0. \tag{1.85}$$

The action principle yields for $\mathbf{v}$ the field equation

$$\mathbf{v} = \nabla\phi, \quad \text{or} \quad v^i = -\partial^i\phi, \tag{1.86}$$

showing that, indeed, a homentropic flow is automatically a potential flow. It also identifies the Lagrange multiplier ϕ, whose dimension is $[\phi] = \text{m}^2\text{s}^{-1}$, as the velocity potential. With Eq. (1.86), the Lagrangian density (1.84) becomes

$$\mathcal{L} = -\rho[\partial_t\phi + \tfrac{1}{2}(\nabla\phi)^2 + e] \tag{1.87}$$

after integrating by parts. A second field equation is obtained from the action principle by considering variations in ρ. This yields the Bernoulli equation

$$\partial_t\phi + \frac{1}{2}(\nabla\phi)^2 + h = 0, \tag{1.88}$$

with

$$h \equiv \frac{\partial(\rho e)}{\partial \rho} \tag{1.89}$$

the so-called enthalpy density. For a homentropic fluid, the enthalpy density coincides with the chemical potential μ per unit mass, $h = \mu$, and

$$\mu = -\left(\partial_t \phi + \tfrac{1}{2}\mathbf{v}^2\right) \tag{1.90}$$

by Eq. (1.88). From the definition (1.89), the thermodynamic relation for homentropic flow

$$\nabla h = \frac{1}{\rho}\nabla P, \tag{1.91}$$

with $P = \rho^2 \partial e/\partial \rho$ the pressure, readily follows. On taking the gradient of Eq. (1.88) and using Eq. (1.91), we obtain *Euler's equation*

$$\partial_t \mathbf{v} + \frac{1}{2}\nabla \mathbf{v}^2 + \frac{1}{\rho}\nabla P = 0 \tag{1.92}$$

governing the flow of the fluid. Since $\nabla \times \mathbf{v} = 0$ in the absence of vortices, this equation with $(1/\rho)\nabla P = \nabla \mu$ can be equivalently written as

$$\frac{\mathrm{d}\mathbf{v}}{\mathrm{d}t} = -\nabla \mu, \tag{1.93}$$

where the total derivative

$$\frac{\mathrm{d}}{\mathrm{d}t} \equiv \partial_t + \mathbf{v} \cdot \nabla \tag{1.94}$$

is the derivative following the flow.

From the Lagrangian density (1.87), the canonical conjugate π_ϕ to ϕ follows as

$$\pi_\phi = \frac{\partial \mathcal{L}}{\partial \partial_t \phi} = -\rho, \tag{1.95}$$

implying the Poisson bracket

$$\{\phi(t, \mathbf{x}), \rho(t, \mathbf{x}')\} = -\delta(\mathbf{x} - \mathbf{x}'). \tag{1.96}$$

The following symmetries can be identified in classical hydrodynamics:

(i) Invariance under spacetime translations, $x^\mu \to x^\mu + \alpha^\mu$, with α^μ a constant vector. By Noether's theorem, this invariance implies the conservation (1.51) of the energy-momentum tensor with components

$$T^{0j} = \frac{\partial \mathcal{L}}{\partial \partial_t \phi}\partial^j \phi = \rho v^j \tag{1.97}$$

$$T^{ij} = \frac{\partial \mathcal{L}}{\partial \partial_i \phi}\partial^j \phi - \mathcal{L}\eta^{ij} = \rho v^i v^j + P\delta^{ij} \tag{1.98}$$

$$T^{00} = \frac{\partial \mathcal{L}}{\partial \partial_t \phi}\partial_t \phi - \mathcal{L} = \frac{1}{2}\rho \mathbf{v}^2 + \rho e \tag{1.99}$$

$$T^{i0} = \frac{\partial \mathcal{L}}{\partial \partial_i \phi}\partial_t \phi = v^i(T^{00} + P), \tag{1.100}$$

as follows from Eq. (1.87). A few remarks are in order. First, time derivatives $\partial_t\phi$ have been eliminated through the field equation (1.88) so that, for example, $\mathcal{L}$ in the last equation is replaced with

$$\mathcal{L} \to \rho h - \rho e = \left(\rho\frac{\partial}{\partial\rho} - 1\right)(\rho e) = P. \tag{1.101}$$

Second, the energy or Hamiltonian density $\mathcal{H} \equiv T^{00}$ is the sum of the kinetic and potential energy density, as required. By Eq. (1.89) with $h = \mu$, it gives

$$\frac{\partial}{\partial\rho}\mathcal{H} = \frac{1}{2}\mathbf{v}^2 + \mu, \tag{1.102}$$

which is the standard definition of the chemical potential. Finally, yielding the complete set of equations of hydrodynamics, the Lagrangian density (1.87) encodes all the relevant information for the description of a homentropic ideal fluid.

(ii) Invariance under global shifts of the velocity potential,

$$\phi(x) \to \phi'(x) = \phi(x) + \alpha, \tag{1.103}$$

with α a constant. This symmetry of the action leads to the conservation law (1.85), or $\partial_\mu g^\mu = 0$, with g^0 the mass density and $\mathbf{g}$ the mass current density,

$$g^0 = -\frac{\partial\mathcal{L}}{\partial\partial_t\phi} = \rho \tag{1.104}$$

$$g^i = -\frac{\partial\mathcal{L}}{\partial\partial_i\phi} = \rho v^i. \tag{1.105}$$

(iii) Invariance under Galilei boosts (1.60), which leads to the conservation law, cf. Eq.(1.65),

$$\partial_\mu g^{\mu j} = 0, \tag{1.106}$$

with $g^{0j} = T^{0j}t - g^0 x^j$ and $g^{ij} = T^{ij}t - g^i x^j$ the corresponding charge and current densities. Note that the equivalence of the mass current g^i and the momentum density T^{0i}, which is the hallmark of Galilei invariance, is satisfied by the theory.

1.7 Sound Waves

We next turn to a description of sound waves. We restrict ourselves to waves of small amplitude. These generate only small deviations in the mass density $\bar{\rho}$ and pressure $\bar{P}$ of the uniform fluid at rest so that the Lagrangian density (1.87) can be expanded in powers of $\tilde{\rho} \equiv \rho - \bar{\rho}$, with $|\tilde{\rho}| << \bar{\rho}$ as

$$\mathcal{L} = -\left(\partial_t\phi + \tfrac{1}{2}\mathbf{v}^2\right)(\bar{\rho} + \tilde{\rho}) - \bar{e}\bar{\rho} - \bar{h}\tilde{\rho} - \frac{1}{2}\bar{h}'\tilde{\rho}^2 + O(\tilde{\rho}^3). \tag{1.107}$$

Here, the prime denotes the derivative with respect to ρ which is to be evaluated at $\rho = \bar{\rho}$. Since for a uniform system at rest, ϕ is constant, it follows from Eq. (1.88) that $\bar{h} = 0$. Denoting the thermodynamic derivative $\mathrm{d}P/\mathrm{d}\rho$ by c^2, which has the dimension of velocity squared, we can write the coefficient of the quadratic term in $\tilde{\rho}$ as

$$\bar{h}' = \frac{1}{\bar{\rho}}\bar{P}' = \frac{c^2}{\bar{\rho}}. \tag{1.108}$$

Apart from an irrelevant additive constant $(-\bar{e}\bar{\rho})$, the Lagrangian density thus becomes to this order

$$\mathcal{L} = -\left(\partial_t\phi + \tfrac{1}{2}\mathbf{v}^2\right)(\bar{\rho} + \tilde{\rho}) - \frac{1}{2}\frac{c^2}{\bar{\rho}}\tilde{\rho}^2. \tag{1.109}$$

We next eliminate $\tilde{\rho}$ from the Lagrangian density (1.109) by substituting

$$\tilde{\rho} = -\frac{\bar{\rho}}{c^2}\left(\partial_t\phi + \tfrac{1}{2}\mathbf{v}^2\right), \tag{1.110}$$

which follows from expanding the field equation (1.88). Physically, this equation with the minus sign on the right reflects Bernoulli's principle: in regions of rapid flow, the mass density $\rho = \bar{\rho} + \tilde{\rho}$ and therefore the pressure is low. It also shows that the expansion in $\tilde{\rho}$ involves derivatives $\partial_\mu\phi$. At low frequency and long wave length, the higher-order terms can therefore be safely ignored. After eliminating $\tilde{\rho}$, we obtain as effective theory governing the velocity potential ϕ:

$$\mathcal{L}_{\text{eff}} = -\bar{\rho}\left(\partial_t\phi + \tfrac{1}{2}\mathbf{v}^2\right) + \frac{\bar{\rho}}{2c^2}\left(\partial_t\phi + \tfrac{1}{2}\mathbf{v}^2\right)^2, \tag{1.111}$$

which is of exactly the same form as the effective theory (1.76) obtained for a nonrelativistic theory with spontaneously broken global U(1) symmetry, with the velocity potential ϕ replacing $(\hbar/m)\varphi$. The main difference is that while the Nambu-Goldstone field φ of the spontaneously broken U(1) symmetry is compact, the velocity potential ϕ featuring in classical hydrodynamics is not.

The field equation for the velocity potential ϕ that follows from the effective theory (1.111) is nonlinear:

$$\bar{\rho}\left(\partial_t^2\phi + \tfrac{1}{2}\partial_t\mathbf{v}^2\right) - \rho c^2 \nabla\cdot\mathbf{v} + \frac{1}{2}\bar{\rho}(\partial_t\mathbf{v}^2 + \mathbf{v}\cdot\nabla\mathbf{v}^2) = 0. \tag{1.112}$$

The information contained in this equation cannot be more than the conservation of mass because ϕ was initially introduced in Eq. (1.84) as a Lagrange multiplier precisely to enforce this conservation law. Indeed, Eq. (1.110), with $\bar{\rho}$ denoting the constant mass density of the uniform fluid at rest, implies that the field equation (1.112) reproduces Eq. (1.85) in this approximation. To simplify the field equation, we replace $\bar{\rho}$ in the first and last term with the full mass density ρ

(which is justified to this order) to arrive at the complete, but somewhat unfamiliar field equation

$$\partial_t^2\phi - c^2\nabla^2\phi = -\partial_t \mathbf{v}^2 - \frac{1}{2}\mathbf{v}\cdot\nabla\mathbf{v}^2 \tag{1.113}$$

of sound waves. It can be cast in the succinct form

$$\frac{\mathrm{d}}{\mathrm{d}t}\mu + c^2\nabla\cdot\mathbf{v} = 0, \tag{1.114}$$

with $\mathrm{d}/\mathrm{d}t$ denoting the derivative (1.94) following the flow, and μ the chemical potential (1.90). If the nonlinear terms are ignored, this equation reduces to the more familiar linear wave equation

$$\partial_t^2\phi - c^2\nabla^2\phi = 0, \tag{1.115}$$

implying a gapless linear dispersion relation, and identifying c, which was introduced through the thermodynamic derivative $\mathrm{d}P/\mathrm{d}\rho = c^2$, as the speed of sound.

As for the combination $\hbar\partial_t\varphi + (\hbar^2/2m)(\nabla\varphi)^2$ in Sec. 1.5, the combination $\partial_t\phi + \frac{1}{2}(\nabla\phi)^2$ appearing here is dictated by Galilei invariance, with the velocity potential transforming as

$$\phi(x) \to \phi'(x') = \phi(x) - \mathbf{u}\cdot\mathbf{x} + \frac{1}{2}\mathbf{u}^2 t. \tag{1.116}$$

Note that the chemical potential (1.90) is invariant under the Galilei transformations (1.61) and (1.116), while $\mathbf{v} \to \mathbf{v} - \mathbf{u}$, as required. It is readily checked that also the complete field equation (1.113) is invariant under Galilei boosts—that is, sound waves in a classical fluid enjoy Galilei invariance. The linearized wave equation (1.115) is, of course, not invariant because essential nonlinear terms have been dropped.

From the effective Lagrangian density (1.111), the various Noether charge and current densities can again be computed. They are, inevitably, of the same form as the exact expressions (1.97)–(1.100), but now with the approximations

$$\rho \approx \bar{\rho} - \frac{\bar{\rho}}{c^2}\left(\partial_t\phi + \tfrac{1}{2}\mathbf{v}^2\right) \tag{1.117}$$

by Eq. (1.110),

$$\mathcal{H} \approx \frac{\rho}{2}\mathbf{v}^2 + \frac{c^2}{2\bar{\rho}}(\rho - \bar{\rho})^2, \tag{1.118}$$

and

$$P \approx -\bar{\rho}\left(\partial_t\phi + \tfrac{1}{2}\mathbf{v}^2\right) \approx c^2(\rho - \bar{\rho}). \tag{1.119}$$

This last equation is consistent with the expression one obtains from directly expanding the pressure: $P(\rho) = \bar{P} + \tilde{\rho}\bar{P}'$ since $\bar{P} = 0$ and $\bar{P}' = c^2$.

To sum up, we arrived at the effective theory describing a gapless nonrelativistic mode, *viz.* sound waves, starting from the Lagrangian density (1.87) which entails the complete hydrodynamics of a homentropic fluid. The effective theory (1.111) displays a property typical for systems with spontaneously broken symmetry, namely $\mathcal{L}_{\rm eff}$ is a function not of the velocity potential itself, but only of derivatives of the field. The energy is minimal if ϕ is uniform in space, i.e., the system is rigid. This is a direct consequence of the shift symmetry (1.103) generated by the total mass $M \equiv \int \mathrm{d}^D x \rho(\mathbf{x})$,

$$\delta\phi(x) = \{\phi(x), -\alpha M\} = \alpha, \tag{1.120}$$

which in turn is a direct consequence of the Poisson bracket (1.96), stating that ϕ and ρ are canonically conjugate. A finite mass density automatically breaks Galilei invariance. Indeed, under an infinitesimal Galilei boost $\delta x^\mu = -u^\mu t$, ϕ transforms as

$$\delta^0\phi(x) = \left\{\phi(x), u_\mu G^{0\mu}\right\} = u_\mu x^\mu + u_\mu t \partial^\mu \phi, \tag{1.121}$$

where, as before, $u^\mu \equiv (0, \mathbf{u})$. The first term on the right shows that the velocity potential is shifted under a Galilei boost. As already remarked below Eq. (1.83), such a shift is typical for a Nambu-Goldstone field under the action of the spontaneously broken symmetry group.

1.8 Topological Defects

The emergence of gapless Nambu-Goldstone modes is a general consequence of spontaneously broken continuous symmetries. Here, we shall be concerned with a second, intimately related, manifestation of broken symmetries, namely the emergence of topological defects. The core of defects are regions where the symmetry is realized differently than in the bulk of the system. Often, the defect core is in the normal state so that here the symmetry is restored.

As an example, consider the Gross-Pitaevskii theory with spontaneously broken global U(1) symmetry. To be specific, we study static field configurations which are independent of the third coordinate, i.e., $\psi(\mathbf{x}) = \psi(\rho, \theta)$ in cylindrical coordinates. For such a field configuration to be of finite energy, it must take values in the ground-state manifold at spatial infinity

$$\lim_{\rho\to\infty} \psi(\rho, \theta) = \mathrm{e}^{\mathrm{i}\varphi(\theta)} v, \tag{1.122}$$

with φ the Nambu-Goldstone field, parametrizing the ground-state manifold, and v given in Eq. (1.71). From the mathematical point of view, $\psi(\rho \to \infty, \theta)$ defines a smooth mapping from the boundary of two-dimensional space, which is a circle

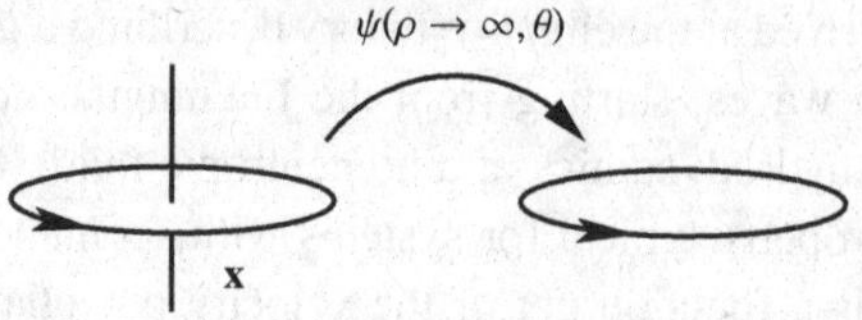

Fig. 1.3 Mapping of the boundary of two-dimensional space into the ground-state manifold.

at spatial infinity and will be denoted by S^1_x, into the ground-state manifold U(1), which also has the topology of a circle and will be denoted by S^1_ψ:

$$\psi(\rho \to \infty, \theta) : \mathrm{S}^1_x \to \mathrm{S}^1_\psi, \tag{1.123}$$

see Fig. 1.3. Such a mapping is characterized by a *winding number* w, giving the number of times the mapping wraps the boundary of space around the ground-state manifold. If the loop in coordinate space is traversed once so that the path returns to its starting point, then the phase $\varphi(\theta)$ must wind around the loop constituting the ground-state manifold exactly w times for ψ to be single valued. That is, the phase must change by $2\pi w$. A nonzero winding number is the signature of a topological defect—a vortex in this case. The vortex is represented by the straight line in Fig. 1.3.

The winding number is a topological invariant. All the mappings (1.123) fall into distinct *homotopy classes* labeled by the winding number w. It is impossible to deform a mapping of a given class with winding number w smoothly into another mapping belonging to a class with winding number $w' \neq w$. [Two smooth functions f and f' are said to be homotopic (Greek for "same place"), or to belong to the same homotopy class if f can be smoothly deformed into f'.] Now, consider shrinking the loop around the vortex core to become infinitesimally small. Because the winding number is a topological invariant, it must keep the same value w it had initially. This requires $\varphi(\theta)$ to turn through an angle $2\pi w$ no matter how small the loop becomes. The derivative of φ therefore diverges at the vortex core and $v \exp[\mathrm{i}\varphi(\theta)]$ becomes singular. In this region, the ordered state is destroyed and the system reverts to the normal state. As $\varphi(\theta)$ changes by $2\pi w$ on going around the vortex core, the circulation κ, defined as

$$\kappa \equiv \oint \mathrm{d}\mathbf{x} \cdot \mathbf{v}_\mathrm{s}, \tag{1.124}$$

is nonvanishing. Specifically,

$$\kappa = \frac{\hbar}{m} \oint \mathrm{d}\mathbf{x} \cdot \nabla\varphi = w\frac{2\pi\hbar}{m}, \tag{1.125}$$

showing that the circulation is quantized in units of $\kappa_0 \equiv 2\pi\hbar/m$. The *topological charge* defined as

$$Q \equiv \frac{1}{2\pi}\oint \mathrm{d}\mathbf{x}\cdot\nabla\varphi = w \tag{1.126}$$

is simply the winding number of the mapping. A field configuration of finite energy evolves in time such that the winding number remains unchanged, i.e., the topological charge is conserved.

The above reasoning can be extended to defects of dimension D' in $D > D'$ space dimensions. The way to characterize such defects is to surround them by a hypersphere S_x^r of dimension r, such that

$$r = D - D' - 1. \tag{1.127}$$

The minus one on the right represents the radial distance from the defect core to the surrounding hypersphere. Vortices are detected by surrounding them by a circle, i.e., $r = 1$. They are therefore pointlike ($D' = 0$) in two space dimensions and linelike ($D' = 1$) in three space dimensions, according to Eq. (1.127). Domain walls are detected by considering a point to the far left and to the far right of the wall, forming the hypersphere S_x^0. They are therefore pointlike ($D' = 0$) in one space dimension, linelike ($D' = 1$) in two space dimensions, and surfacelike ($D' = 2$) in three space dimensions. Point defects ($D' = 0$) in three space dimensions can be detected by surrounding them by an ordinary sphere ($r = 2$).

There also exist defects for which the entire space is to be treated as the surrounding hypersurface. To visualize such defects, coordinate space is compactified to a hypersphere S_x^D. This is justified when the ground state tends to a unique value at spatial infinity. The defect can then be pictured as a point defect in $(D+1)$-dimensional space, surrounded by the hypersphere S_x^D representing coordinate space.

In all these cases, a mapping can be defined from the hypersphere S_x^r into the ground-state manifold. Singularities arise when the image of this mapping cannot be smoothly shrunk to a point on the ground-state manifold. The mappings can be classified by winding numbers as in our example, telling how many times the ground-state manifold is covered when the hypersphere S_x^r is covered once.

1.9 Homotopy Groups

The homotopy classes characterizing a vortex form a (discrete) group, called the first homotopy, or fundamental group $\pi_1[U(1)]$ defined on the ground-state manifold $M = \mathrm{U}(1)$. To demonstrate this, consider for each homotopy class c a representative path $p_c(z)$ on the manifold parametrized by $0 \leq z \leq 1$, starting and ending at some base point p_0, i.e., $p_c(0) = p_c(1) = p_0$. The product $c_1 \circ c_2$ of two homotopy classes is defined as the class containing the composite path obtained by first traversing the path $p_{c_1}(z)$ and then, after returning to the base point, traversing the path $p_{c_2}(z)$. That is, the representative $p_{c_1 \circ c_2}(z)$ is defined as

$$p_{c_1 \circ c_2}(z) = \begin{cases} p_{c_1}(2z), & 0 \le z \le \frac{1}{2} \\ p_{c_2}(2z-1), & \frac{1}{2} \le z \le 1. \end{cases} \tag{1.128}$$

With this multiplication rule, associating to any two homotopy classes c_1 and c_2 a third homotopy class $c_3 = c_1 \circ c_2$, all the requirements for these classes to form a group are satisfied.

Indeed, the multiplication rule is associative $c_1 \circ (c_2 \circ c_3) = (c_1 \circ c_2) \circ c_3$ because the order in which the three representative paths, each starting and ending at the base point p_0, are transversed is unimportant. The unit element of the fundamental group $\pi_1(M)$ is defined as the homotopy class containing the path that remains at the base point, $p_e(z) = p_0$ for $0 \le z \le 1$. To verify that $e \circ c = c$, consider the smooth function $P(z, t)$,

$$P(z,t) = \begin{cases} p_0, & 0 \le z \le t/2 \\ p_c[(2z-t)/(2-t)], & t/2 \le z \le 1, \end{cases} \tag{1.129}$$

with $0 \le t \le 1$. Its definition is such that $P(z, 0) = p_c(z)$ and $P(z, 1) = p_{e \circ c}(z)$ by Eq. (1.128) with c_1 replaced with e and c_2 replaced with c. The existence of such a function, called a *homotopy*, shows that the function $p_c(z)$ can be smoothly deformed into $p_{e \circ c}(z)$, implying that they belong to the same homotopy class and $e \circ c = c$. The inverse c^{-1} of the homotopy class c is defined as the homotopy class containing the path $p_{c^{-1}}(z)$ obtained by traversing the representative path $p_c(z)$ of the homotopy class c in the opposite direction, i.e.,

$$p_{c^{-1}}(z) = p_c(1-z). \tag{1.130}$$

To show that $c^{-1} \circ c = e$, note that by the multiplication rule (1.128)

$$p_{c^{-1} \circ c}(z) = \begin{cases} p_c(1-2z), & 0 \le z \le \frac{1}{2} \\ p_c(2z-1), & \frac{1}{2} \le z \le 1. \end{cases} \tag{1.131}$$

The following homotopy

$$P(z,t) = \begin{cases} p_c(1-2tz), & 0 \le z \le \frac{1}{2} \\ p_c(2tz+1-2t), & \frac{1}{2} \le z \le 1 \end{cases} \tag{1.132}$$

reduces to $p_{c^{-1} \circ c}(z)$ in Eq. (1.130) for $t = 1$ and to $p_e(z) = p_0$ for $t = 0$. Both paths can thus be smoothly deformed into each other and are part of the same homotopy class. This shows that the homotopy classes on the manifold $M = \mathrm{U}(1)$ indeed form a group, $\pi_1(M)$. This is in fact true for an arbitrary manifold M, not just for group manifolds.

The nature of the fundamental group $\pi_1[\mathrm{U}(1)]$ characterizing vortices can be easily established. Consider two widely separated vortices with winding number w_1 and w_2. The finite energy solution ψ_1, describing the first vortex then belongs to the homotopy class c_{w_1}, while the solution ψ_2, describing the second vortex belongs to the homotopy class c_{w_2}. The composite $\psi_1 \circ \psi_2$, obtained by patching together the two widely separated solutions, is readily seen to belong to the homotopy class $c_{w_1+w_2}$ labeled by winding number $w_1 + w_2$ as follows. Deform the boundary of two-dimensional space, which encircles both vortices and belongs to the homotopy class $c_{w_1+w_2}$, into two very large circles, each encircling one of the vortices, such that they almost touch midway between the vortices. The first circle belongs to the homotopy class c_{w_1}, while the second belongs to c_{w_2}, and hence $c_{w_1+w_2} = c_{w_1} \circ c_{w_2}$. That is, the "multiplication" rule is just addition and the fundamental group forms the additive group of the integers,

$$\pi_1[\mathrm{U}(1)] = \pi_1(\mathrm{S}^1) = \mathrm{Z}. \tag{1.133}$$

As it is impossible to "lasso a basketball", the fundamental group of a two-sphere consists of only the unit element, $\pi_1(\mathrm{S}^2) = 0$. More generally,

$$\pi_1(\mathrm{S}^n) = 0 \tag{1.134}$$

for all $n \geq 2$.

The manifold M has, in the cases of interest to us, the particular form of a ground-state manifold. To see what this implies, let G be the symmetry group of the theory, and let ϕ_0 be a given ground state. Then the state $g\phi_0$ obtained by subjecting ϕ_0 to a symmetry transformation $g \in G$ also belongs to the ground state. Bearing accidental degeneracies, the ground-state manifold is therefore given by the set $\{g\phi_0\}$. Assume that certain transformations leave ϕ_0 invariant. The set

$$H = \{h; h \in G, h\phi_0 = \phi_0\} \tag{1.135}$$

is readily verified to form a subgroup H of G, called the *residual symmetry group*. With $g \in G$ parametrized as

$$g = kh, \tag{1.136}$$

it follows that $\{g\phi_0\} = \{k\phi_0\}$, showing that the ground-state manifold is identical in structure, i.e., *isomorphic* (Greek for "equal shape"), to the coset space G/H. The mappings from the hypersphere S_x^r in coordinate space into the ground-state manifold G/H form the rth homotopy group $\pi_r(G/H)$. Two mappings in the same equivalent class can be smoothly deformed into each other.

The proof that $\pi_r(M)$ falls into equivalent classes which for $r > 1$ form a group proceeds along the same line of arguments given for the fundamental group ($r = 1$). It is convenient to represent the hypersphere S^r as a r-dimensional unit hypercube, with all the points on the boundary identified as a single point. This implies that the mapping from this hypercube into M must take all the points of the boundary into the same base point p_0 of M. Let $0 \leq z_i \leq 1$, $i = 1, \ldots, r$ denote the parameters on the hypercube. A representative of the homotopy class c is denoted by $p_c(z_1, \ldots, z_r)$. Two mappings $p(z_1, \ldots, z_r)$ and $p'(z_1, \ldots, z_r)$ of S^r into M belong to the same homotopy class if $p(z_1, \ldots, z_r)$ can be smoothly deformed into $p'(z_1, \ldots, z_r)$, while keeping the value of the interpolating function on the boundary of the hypercube fixed at p_0. The product $c_1 \circ c_2$ of two homotopy classes c_1 and c_2 is defined as the homotopy class represented by

$$p_{c_1 \circ c_2}(z_1, \ldots, z_r) = \begin{cases} p_{c_1}(2z_1, \ldots, z_r), & 0 \leq z_1 \leq \frac{1}{2} \\ p_{c_2}(2z_1 - 1, \ldots, z_r), & \frac{1}{2} \leq z_1 \leq 1, \end{cases} \tag{1.137}$$

where the intervals are patched together side by side. As before, this multiplication rule can be readily shown to be associative. The unit element and the inverse c^{-1} of c can be defined as for $r = 1$. Whereas the fundamental group $\pi_1(M)$ need not be Abelian, all higher-order homotopy groups $\pi_r(M)$, $r > 1$ are. In general it is hard to visualize the higher homotopy groups. However, some special cases can be understood by analogy with the fundamental group, such as

$$\pi_r(\mathrm{S}^r) = \mathrm{Z}, \quad \pi_r(\mathrm{S}^m) = 0 \tag{1.138}$$

for $m > r$.

Finally, the zeroth homotopy group $\pi_0(M)$ may be defined as the set of disconnected components of the manifold M. For the cyclic group Z_n of order n, for example, $\pi_0(\mathrm{Z}_n) = \mathrm{Z}_n$. Despite its name, $\pi_0(M)$ need not be a group.

A useful tool for determining homotopy groups is the homotopy sequence

$$\cdots \to \pi_n(H) \to \pi_n(G) \to \pi_n(G/H) \to \pi_{n-1}(H) \to \pi_{n-1}(G) \to \cdots \tag{1.139}$$

Each of the mappings in this sequence from one homotopy group to the next preserves group multiplication, i.e., each mapping is a *homomorphism* (Greek for "same shape"). In formula, if f denotes such a mapping, then for two homotopy classes c_1 and c_2, $f(c_1 \circ c_2) = f(c_1) \circ f(c_2)$. The sequence (1.139) is also *exact*, which means that the kernel (Ker) of a mapping is the image (Im) of the previous mapping, where the kernel of a homomorphism is defined as the set of all elements that are mapped into the unit element 1.

As a first example of the use of the homotopy sequence (1.139), consider the Hopf map $f : \mathrm{S}^3 \to \mathrm{S}^2$ with $\mathrm{G} = \mathrm{S}^3$, $\mathrm{G/H} = \mathrm{S}^2$, and $\mathrm{H} = \mathrm{S}^1$. For these manifolds, the homotopy sequence takes the form

$$\pi_n(\mathrm{S}^1) \to \pi_n(\mathrm{S}^3) \to \pi_n(\mathrm{S}^2) \to \pi_{n-1}(\mathrm{S}^1). \tag{1.140}$$

Since $\pi_n(S^1) = 0$ for all $n > 1$, and $\mathrm{Im}[0 \to \pi_n(S^3)] = 0$, the exactness of the sequence gives for the first two mappings

$$0 = \mathrm{Ker}[\pi_n(S^3) \to \pi_n(S^2)], \tag{1.141}$$

implying that the second mapping is one to one,

$$\mathrm{Im}[\pi_n(S^3) \to \pi_n(S^2)] \simeq \pi_n(S^3). \tag{1.142}$$

We next shift one mapping to the right in the sequence to obtain for $n - 1 > 1$, i.e., $n > 2$

$$\pi_n(S^3) \simeq \mathrm{Ker}[\pi_n(S^2) \to 0]. \tag{1.143}$$

But $\mathrm{Ker}[\pi_n(S^2) \to 0] \simeq \pi_n(S^2)$ by definition, and hence

$$\pi_n(S^3) \simeq \pi_n(S^2). \tag{1.144}$$

For the Hopf map, this implies the somewhat surprising result

$$\pi_3(S^2) \simeq \pi_3(S^3) = Z. \tag{1.145}$$

As a second example of the use of the homotopy sequence (1.139), consider a symmetry group G with a trivial second homotopy group $\pi_2(G) = 0$, meaning that the image of the mapping from any closed surface in coordinate space into G can be smoothly shrunk to a point. By the exactness of the homotopy sequence (1.139),

$$\mathrm{Im}[0 \to \pi_2(G/H)] \simeq \mathrm{Ker}[\pi_2(G/H) \to \pi_1(H)]. \tag{1.146}$$

With $\mathrm{Im}[0 \to \pi_2(G/H)] = 0$, this implies that the kernel of the homomorphism $\pi_2(G/H) \to \pi_1(H)$ consists of only the unit element,

$$0 = \mathrm{Ker}[\pi_2(G/H) \to \pi_1(H)]. \tag{1.147}$$

In other words, the mapping is one to one,

$$\mathrm{Im}[\pi_2(G/H) \to \pi_1(H)] \simeq \pi_2(G/H). \tag{1.148}$$

Using the exactness of the homotopy sequence (1.139) a second time, we conclude that

$$\pi_2(G/H) \simeq \mathrm{Ker}[\pi_1(H) \to \pi_1(G)]. \tag{1.149}$$

If the symmetry group is in addition simply connected, i.e., $\pi_1(G) = 0$, then

$$\pi_2(G/H) \simeq \pi_1(H), \tag{1.150}$$

and the computation of the second homotopy group of a coset space is reduced to the computation of π_1 of a group.

As a physical application of this last result, consider a three-dimensional ferromagnet. Below the Curie temperature, the system spontaneously develops a finite magnetization. The resulting preferred direction breaks the SO(3) rotational symmetry down to spin rotations about that axis, which form the group SO(2) or U(1). It is convenient to consider the simply connected covering group SU(2) instead of SO(3) which is not simply connected. The result (1.150) then gives

$$\pi_2[\mathrm{SU}(2)/\mathrm{U}(1)] \simeq \pi_1[\mathrm{U}(1)] = \mathrm{Z}, \tag{1.151}$$

showing that a ferromagnet may exhibit topologically stable point defects. As an aside, since $\mathrm{SO}(3) = \mathrm{SU}(2)/\mathrm{Z}_2$, the arguments leading to the result (1.150) can be adapted to yield $\pi_1[\mathrm{SO}(3)] = \pi_0(\mathrm{Z}_2) = \mathrm{Z}_2$.

1.10 Quantized Vortices

Although topological considerations reveal what defects may arise in a given system, the actual existence of finite-energy solutions depends on the details of the field equations governing the system. To see what is involved, we in this section explicitly construct the topological solution describing a static straight vortex of winding number w in the Gross-Pitaevskii theory. The symmetry axis is taken to define the x^3-axis. As *Ansatz* for this classical solution, we assume that it factorizes as

$$\psi(\mathbf{x}) = v f(\rho/\xi)\, \mathrm{e}^{\mathrm{i}w\theta}, \tag{1.152}$$

with θ the azimuthal angle and f a function of only the distance $\rho \equiv \sqrt{(x^1)^2 + (x^2)^2}$ from the vortex axis. Here, v is the value minimizing the potential energy density $\mathcal{V}$, and ξ is the typical length scale for spatial variations introduced in Eq. (1.74). By Eq. (1.46), the amplitude function $f(\varrho)$, with ϱ the dimensionless variable $\varrho \equiv \rho/\xi$, satisfies the equation

$$f'' + \frac{1}{\varrho} f' - \frac{w^2}{\varrho^2} f + f - f^3 = 0. \tag{1.153}$$

For small ϱ, where the last two terms can be neglected in this differential equation, the function $f(\varrho)$ vanishes for $\varrho \to 0$ as

$$f(\varrho) \sim \varrho^{|w|}, \tag{1.154}$$

with a power of ϱ determined by the winding number. It shows that on approaching the vortex center, ψ tends to zero in a region of radius $\rho \sim \xi$, which defines the vortex core. For $\varrho >> 1$, the amplitude function tends to the asymptotic value $f = 1$ of a spatially uniform system as

$$f(\varrho) \sim 1 - \frac{w^2}{2}\frac{1}{\varrho^2}, \tag{1.155}$$

as can be verified by setting $f = 1 - \delta f$ and linearizing Eq. (1.153) with the derivatives set to zero. The density profile in the presence of a straight vortex follows as

$$n(\mathbf{x}) = |\psi(\mathbf{x})|^2 \sim v^2 \left(1 - w^2 \frac{\xi^2}{\rho^2}\right), \tag{1.156}$$

while the condensate velocity (1.81) assumes the form

$$\mathbf{v}_s(\mathbf{x}) = \frac{\hbar w}{m} \nabla\theta = \frac{\hbar w}{m} \frac{\mathbf{e}_\theta}{\rho} = \frac{\kappa}{2\pi} \frac{\mathbf{e}_3 \times \mathbf{x}}{\rho^2}, \tag{1.157}$$

with $\mathbf{e}_\theta$ denoting the unit vector in the azimuthal direction, and κ the (quantized) circulation (1.125).

1.11 Villain Vector Potential

An essential difference between spontaneously broken Galilei and U(1) symmetries is that the latter group is compact. More specifically, the transformation parameter α of the U(1) group is a compact variable, taking values in the finite interval $0 \leq \alpha < 2\pi$, while the transformation parameter $\mathbf{u}$ of the Galilei group is noncompact and can take any value in $\mathbb{R}^D$. It is consequently impossible to represent the velocity potential of classical hydrodynamics as the phase of a complex field. As a result, whereas a system with a spontaneously broken global U(1) symmetry supports topologically stable vortices, a system where only Galilei symmetry is spontaneously broken does not. This is not to say that vortices are absent in classical hydrodynamics. It merely states that their stability is not guaranteed by topological conservation laws. Closely connected to this is that while quantized in superfluids, the circulation around a vortex is not quantized in classical hydrodynamics. Yet, the circulation is conserved also in homentropic fluids. This is again not for topological, but for dynamical reasons, and can be proved by invoking Euler's equation (3.136).

Vortices in a classical fluid can be easily observed by punching a hole in the bottom of a vessel containing the fluid. As the fluid pours out, a vortex is formed in the remaining fluid—a phenomenon daily observed by people unplugging a sinkhole. The vortex core consists of air, i.e., the fluid mass density ρ is zero there and Galilei invariance restored.

In the eye of a tropical cyclone—another example of a vortex—nature does its best to restore Galilei symmetry, record low atmospheric pressures being measured there. A complete restoration would imply the absence of air and thus zero pressure.

To describe vortices in a potential flow, we introduce a vector potential $A^{V,\mu} = (\Phi^V, \mathbf{A}^V)$ in the Lagrangian density (1.87) through minimal coupling to the Nambu-Goldstone field:

$$\partial_\mu \phi \to \partial_\mu \phi + A^V_\mu \tag{1.158}$$

with the requirement

$$\nabla \times \mathbf{A}^V = -2\omega, \tag{1.159}$$

where ω denotes the vorticity of the fluid induced by the vortices,

$$\omega = \frac{1}{2} \nabla \times \mathbf{v}. \tag{1.160}$$

Equation (1.159) relates the vector potential $\mathbf{A}^V$ to the vorticity in the same way that Ampere's law relates the magnetic induction to the electric current which produces it. The vector potential is the continuum analog of the integer-valued field featuring in the so-called Villain formulation of certain lattice models, and will be referred to as *Villain potential*. The combination $\partial_\mu \phi + A^V_\mu$ is invariant under the *local* gauge transformation

$$\phi(x) \to \phi(x) + \alpha(x), \quad A^V_\mu(x) \to A^V_\mu(x) - \partial_\mu \alpha(x), \tag{1.161}$$

with the Villain potential playing the role of a gauge field. The transformation is local as the transformation parameter $\alpha(x)$ now depends on spacetime. The left side of Eq. (1.159) can be thought of as defining the "magnetic induction" associated with the Villain vector potential, $\mathbf{B}^V \equiv \nabla \times \mathbf{A}^V$. Similarly, the corresponding "electric field" $\mathbf{E}^V$ is defined as

$$\mathbf{E}^V \equiv -\frac{\partial}{\partial t} \mathbf{A}^V - \nabla \Phi^V. \tag{1.162}$$

The field equation for ϕ obtained after the minimal substitution (1.158) can be cast in the same succinct form (1.114) as in the absence of the vector potential, provided the expressions (1.90) and (1.86) for the chemical potential (per unit mass) μ and the velocity field $\mathbf{v}$ in the presence of a vortex are rendered gauge invariant by defining

$$\mu = -\left(\partial_t \phi + \Phi^V + \tfrac{1}{2}\mathbf{v}^2\right), \quad \mathbf{v} = \nabla \phi - \mathbf{A}^V. \tag{1.163}$$

By taking the gradient of the first equation, we arrive at the Euler equation in the presence of vortices

$$\partial_t \mathbf{v} + \frac{1}{2} \nabla \mathbf{v}^2 = \mathbf{E}^V - \nabla \mu. \tag{1.164}$$

By Eq. (1.159), this can be cast in the equivalent form

$$\frac{\mathrm{d}}{\mathrm{d}t} \mathbf{v} = \mathbf{E}^V + \mathbf{v} \times \mathbf{B}^V - \nabla \mu, \tag{1.165}$$

which by construction has the form of the Euler equation of a charged system.

To illustrate the use of Villain potentials, we consider a pinned, static vortex in three space dimensions with circulation κ located along an (infinitely thin) line L, which may be closed or infinitely long. Then the vorticity vector is given by

$$\omega(\mathbf{x}) = \frac{1}{2}\kappa\boldsymbol{\delta}_L(\mathbf{x}), \tag{1.166}$$

where $\boldsymbol{\delta}_L(\mathbf{x})$ is a delta function on the line L,

$$\delta_L^i(\mathbf{x}) = \int_L \mathrm{d}x'^i\,\delta(\mathbf{x}-\mathbf{x}') \tag{1.167}$$

so that

$$\kappa = \oint_\Gamma \mathrm{d}\mathbf{x}\cdot\mathbf{v} = \int_{S(\Gamma)} \mathrm{d}\mathbf{S}\cdot(\nabla\times\mathbf{v}) = 2\int_{S(\Gamma)} \mathrm{d}\mathbf{S}\cdot\omega, \tag{1.168}$$

where Γ is a closed path around the vortex and $S(\Gamma)$ a surface spanned by this loop with surface element $d\mathbf{S}$. The Villain potential then satisfies the equations $\Phi^{\mathrm{V}} = 0$ and

$$\nabla\times\mathbf{A}^{\mathrm{V}}(\mathbf{x}) = -\kappa\boldsymbol{\delta}_L(\mathbf{x}). \tag{1.169}$$

Ignoring the higher-order terms, we obtain as equation for the flow in the presence of a static vortex:

$$\nabla\cdot\mathbf{v} = 0, \quad \text{or} \quad \nabla\cdot\left(\nabla\phi - \mathbf{A}^{\mathrm{V}}\right) = 0, \tag{1.170}$$

which is solved by

$$\phi(\mathbf{x}) = -\int \mathrm{d}^3x'\,G(\mathbf{x}-\mathbf{x}')\nabla'\cdot\mathbf{A}^{\mathrm{V}}(\mathbf{x}'), \tag{1.171}$$

with $G(\mathbf{x})$ denoting the Green function of the Laplace operator

$$G(\mathbf{x}) = \int \frac{\mathrm{d}^3k}{(2\pi)^3}\frac{\mathrm{e}^{\mathrm{i}\mathbf{k}\cdot\mathbf{x}}}{\mathbf{k}^2} = \frac{1}{4\pi|\mathbf{x}|}. \tag{1.172}$$

Straightforward manipulations then yield the well-known Biot-Savart law for the velocity field in the presence of a static vortex

$$\mathbf{v}(\mathbf{x}) = \frac{\kappa}{4\pi}\int_L \mathrm{d}\mathbf{x}'\times\frac{\mathbf{x}-\mathbf{x}'}{|\mathbf{x}-\mathbf{x}'|^3}, \tag{1.173}$$

where the integration is along the vortex line L. For an infinite straight vortex line along the x^3-axis, this more general solution reduces to the form (1.157). This exemplifies the use of Villain potentials to describe topological defects in the continuum.

Notes

(i) Standard introductions to classical fields can be found in [Goldstein (1980)] and [Landau and Lifshitz (1975)].

(ii) The action principle and Noether's theorem are covered in any modern textbook on quantum field theory, see, for example, [Ramond (1981)].

(iii) The variational principle applied to hydrodynamics was developed by Eckart (1938). For a survey of variational principles, see [Yourgrau and Mandelstam (1979)].

(iv) The emergence of gapless modes in systems with spontaneously broken continuous symmetries was pointed out in [Nambu and Jona-Lasinio (1961)] and in [Goldstone (1961); Goldstone *et al.* (1962)].

(v) Anderson (1984) provides original and deep insights into the physical consequences of spontaneously broken symmetries in condensed matter, such as rigidity and the emergence of topological defects.

(vi) The nonlinear effective theory of a nonrelativistic gapless mode was proposed independently by Kemoklidze and Pitaevskii (1966), Takahashi (1988), and Greiter *et al.* (1989). The nonlinear wave equation describing sound in hydrodynamics can be found in, for example, the textbook [Shivamoggi (1998)]. Sections 1.6 and 1.7 are based on [Schakel (1996)].

(vii) There exist various excellent introductions to topological defects and homotopy theory for physicists. See, for example, [Coleman (1977); Volovik and Mineev (1977); Mermin (1979); Mineev (1980); Trebin (1982); Rajaraman (1982); Weinberg (1996)], and [Nakahara (2003)]. Coleman's Erice lecture is reprinted in [Coleman (1988)].

(viii) The classical field theory (1.44) was put forward by Gross (1958) and Pitaevskii (1958).

(ix) Vortices in the Gross-Pitaevskii theory were first studied in [Gross (1961)] and in [Pitaevskii (1961)].

(x) A more conventional way to describe vortices in hydrodynamics is through the use of Clebsch rather than Villain potentials, see [Yourgrau and Mandelstam (1979)]. The integer-valued fields to describe topological defects in lattice models were introduced by Villain (1975). For reviews, see [Savit (1980)] and [Polyakov (1987)]. The continuum formulation was developed by Kleinert (1989), see also Dirac's (second) seminal paper [Dirac (1948)] on magnetic monopoles.

Chapter 2

Quantum Field Theory

This chapter introduces quantum field theory using the functional integral approach. The method is a generalization of Feynman's intuitive sum-over-paths approach to quantum mechanics, where all possible field configurations are summed over. Each configuration is given a weight $\exp(iS/\hbar)$ determined by the (classical) action S. The functional integral approach to field quantization has the advantage over canonical quantization that most results of interest to us can be derived more easily and more quickly, and with considerably less overhead. The method is illustrated by first considering ideal quantum gases.

2.1 Quantum Statistical Mechanics

Consider an ideal nonrelativistic quantum gas consisting of identical particles. The system is assumed to be in contact with a heath bath at temperature T and also with a particle reservoir with which it can freely exchange particles. In this ensemble, the temperature and chemical potential μ, characterizing the particle reservoir, as well as the volume V are fixed. The energy E and particle number N of the system, being conjugate to T and μ, respectively, are indeterminate. Only their averages can be calculated. Such an ensemble is termed a *grand canonical ensemble*.

A central role in statistical mechanics is played by the partition function Z. For a grand canonical ensemble it is defined by the weighted sum

$$Z = \sum_{N=0}^{\infty} \sum_{\{E\}} e^{-\beta(E-\mu N)}, \tag{2.1}$$

where the sum is over N and over all possible energy values E the system can take. A state of N particles and energy E carries a statistical weight $\exp[-\beta(E - \mu N)]$, where $\beta = 1/k_B T$, with k_B the Boltzmann constant, denotes the inverse temperature. This weighing factor is a Boltzmann factor, modified to include a chemical potential. Since the energy and particle number are indeterminate, only the probability $P(E,N)$ for the system to be in a definite state of energy E and with particle number N can be computed. This probability is given by

$$P(E,N) = \frac{1}{Z}\,\mathrm{e}^{-\sum_r \beta(\epsilon_r-\mu)n_r}, \tag{2.2}$$

where the energy is written as a sum over discrete energy levels

$$E = \sum_r \epsilon_r n_r, \tag{2.3}$$

with n_r the occupation number of the energy level ϵ_r, and

$$N = \sum_r n_r. \tag{2.4}$$

The occupation number n_r can take the values $n_r = 0, 1, 2, \cdots$ for bosons, while $n_r = 0, 1$ for fermions as multiple occupancy is forbidden by the Pauli principle. A continuum of energy levels can be handled by taking the appropriate continuum limit (see below).

Let us work out the sums in Eq. (2.1) for a Bose system by writing the partition function as

$$Z = \sum_{N=0}^{\infty} \sum_{\{n_r\}} \mathrm{e}^{-\sum_r \beta(\epsilon_r-\mu)n_r}. \tag{2.5}$$

In order that Z makes sense, the chemical potential must satisfy the bound $\mu \leq \epsilon_0$, with ϵ_0 indicating the lowest energy level. When μ reaches the critical value

$$\mu_c \equiv \epsilon_0, \tag{2.6}$$

a macroscopic occupation of the ground state, which is allowed by Bose statistics, is possible. Although the set $\{n_r\}$ of occupancy numbers must satisfy the constraint (2.4), the sum over the total particle number N removes this constraint, and

$$Z = \prod_r \frac{1}{1-\mathrm{e}^{-\beta(\epsilon_r-\mu)}}. \tag{2.7}$$

The thermodynamic potential Ω, which is related to the partition function through

$$\Omega(\alpha,\beta) = -\beta^{-1}\ln(Z), \tag{2.8}$$

then takes the form

$$\Omega(\alpha,\beta) = \frac{1}{\beta}\sum_r \ln\left(1-\mathrm{e}^{-\beta\epsilon_r-\alpha}\right), \tag{2.9}$$

with $\alpha \equiv -\beta\mu$. The quantity

$$z \equiv \mathrm{e}^{-\alpha} = \mathrm{e}^{\beta\mu} \tag{2.10}$$

appearing in the thermodynamic potential is called the *fugacity*. In a grand canonical ensemble, α and β are to be considered independent. For given temperature and chemical potential, the average energy $\langle E \rangle$ and particle number $\langle N \rangle$ of the system can be obtained from the thermodynamic potential by taking the derivatives with respect to β and α, respectively:

$$\langle E \rangle = \frac{\partial}{\partial \beta}(\beta \Omega) = \sum_r \epsilon_r n(\epsilon_r) \tag{2.11}$$

$$\langle N \rangle = \frac{\partial}{\partial \alpha}(\beta \Omega) = \sum_r n(\epsilon_r) \tag{2.12}$$

with

$$n(\epsilon_r) = \frac{1}{e^{\beta(\epsilon_r - \mu) - 1}} \tag{2.13}$$

the Bose-Einstein distribution function.

For a Fermi system, where the occupation number n_r takes only the values 0 or 1, the partition function (2.5) assumes the form

$$Z = \prod_r \left(1 + e^{-\beta(\epsilon_r - \mu)}\right), \tag{2.14}$$

while the thermodynamic potential as a function of the independent variables α and β becomes

$$\Omega(\alpha, \beta) = -\frac{1}{\beta} \sum_r \ln\left(1 + e^{-\beta \epsilon_r - \alpha}\right). \tag{2.15}$$

The average energy and particle number at fixed temperature and chemical potential are now given by

$$\langle E \rangle = \sum_r \epsilon_r f(\epsilon_r) \tag{2.16}$$

$$\langle N \rangle = \sum_r f(\epsilon_r) \tag{2.17}$$

with

$$f(\epsilon_r) = \frac{1}{e^{\beta(\epsilon_r - \mu)} + 1} \tag{2.18}$$

the Fermi-Dirac distribution function.

If the thermodynamic potential is considered to be a function of the chemical potential μ and temperature T rather than of α and β, then

$$\langle N \rangle = -\frac{\partial \Omega}{\partial \mu} \tag{2.19}$$

and

$$\langle E \rangle = \Omega - T\frac{\partial \Omega}{\partial T} - \mu\frac{\partial \Omega}{\partial \mu}, \tag{2.20}$$

where it is used that

$$\beta\frac{\partial \Omega}{\partial \beta}\bigg|_{\alpha} = -T\frac{\partial \Omega}{\partial T}\bigg|_{\mu} - \mu\frac{\partial \Omega}{\partial \mu}\bigg|_{T}, \tag{2.21}$$

with the variables kept fixed explicitly indicated.

Specialized to ideal quantum gases, the sum over discrete states $\sum_r$ becomes an integral over wave vectors $\mathbf{k}$,

$$\sum_r \to V\int \frac{\mathrm{d}^D k}{(2\pi)^D}, \tag{2.22}$$

with V the volume of the system, while the energy ϵ_r reduces to the kinetic energy $\epsilon(\mathbf{k}) \equiv \hbar^2\mathbf{k}^2/2m$ of free particles of mass m and momentum $\hbar\mathbf{k}$. It follows that

$$\Omega = \frac{V}{\beta}\int \frac{\mathrm{d}^D k}{(2\pi)^D}\ln\left(1 - \mathrm{e}^{-\beta\epsilon(\mathbf{k})-\alpha}\right) \tag{2.23}$$

for bosons, ignoring for the moment a possible macroscopic occupancy of the $\mathbf{k} = 0$ ground state, and

$$\Omega = -\frac{V}{\beta}\int \frac{\mathrm{d}^D k}{(2\pi)^D}\ln\left(1 + \mathrm{e}^{-\beta\epsilon(\mathbf{k})-\alpha}\right) \tag{2.24}$$

for fermions. The thermodynamic pressure P is related to Ω through

$$P = -\partial\Omega/\partial V. \tag{2.25}$$

Later in this chapter, we introduce finite-temperature field theory and rederive these expressions for Ω with the help of functional integrals.

A final quantity that will frequently show up in the following is the *density of states* $\nu(\epsilon)$. To define it, the sum over discrete states $\sum_r$ is rewritten as

$$\sum_r = V\int_0^\infty \mathrm{d}\epsilon\, \nu(\epsilon), \tag{2.26}$$

where $\nu(\epsilon)$ gives the number of states in the energy interval ϵ and $\epsilon + \mathrm{d}\epsilon$ per unit volume. In spherical coordinates, this gives for an ideal Bose gas

$$\sum_r = V\int \frac{\mathrm{d}^D k}{(2\pi)^D} = V\frac{\Omega_D}{(2\pi)^D}\int_0^\infty \mathrm{d}k\, k^{D-1} = V\int_0^\infty \mathrm{d}\epsilon\, \nu(\epsilon) \tag{2.27}$$

where $\Omega_D = 2\pi^{D/2}/\Gamma(D/2)$ is the surface of a unit hypersphere embedded in D dimensions and

$$\nu(\epsilon) = \frac{\Omega_D}{(2\pi)^D}\frac{m}{\hbar^2}\left(\frac{2m\epsilon}{\hbar^2}\right)^{D/2-1}. \tag{2.28}$$

That is, for an ideal Bose gas the density of states behaves as $\nu(\epsilon) \propto \epsilon^{D/2-1}$. When the spectrum $\epsilon(\mathbf{k})$ of the elementary excitations differs from the ideal case, as for a Bose gas in a harmonic trap, the density of states will have a different dependence on ϵ.

For a dilute uniform ideal gas, the fugacity is small, $z = e^{-\alpha} < 1$, so that it can be used as an expansion parameter. With the logarithm expanded in a Taylor series as $\ln(1-x) = -x - \frac{1}{2}x^2 + \cdots$, this gives for the thermodynamic potential

$$\Omega = -\frac{V}{\beta} \int \frac{d^D k}{(2\pi)^D} \left(z e^{-\beta\hbar^2 \mathbf{k}^2/2m} \pm \frac{1}{2} z^2 e^{-\beta\hbar^2 \mathbf{k}^2/m} + \cdots \right). \tag{2.29}$$

The upper sign applies to a Bose gas, while the lower sign applies to a Fermi gas. The integrals are simple Gaussians of the form

$$\int \frac{dk}{2\pi} e^{-ak^2} = \sqrt{\frac{1}{4\pi a}} \tag{2.30}$$

and give $\Omega = \Omega_0 + \Omega_1$ with

$$\Omega_0 = -\frac{V}{\beta \lambda_T^D} z \tag{2.31}$$

and

$$\Omega_1 = \mp \frac{1}{2^{(D+2)/2}} \frac{V}{\beta \lambda_T^D} z^2, \tag{2.32}$$

where λ_T is the de Broglie *thermal wave length*,

$$\lambda_T \equiv \sqrt{\frac{2\pi\hbar^2\beta}{m}}. \tag{2.33}$$

The first term Ω_0 yields for the average particle number, energy, and pressure:

$$\begin{aligned} \langle N \rangle_0 &= \frac{V}{\lambda_T^D} z \\ \langle E \rangle_0 &= -V \frac{\partial}{\partial \beta} \frac{1}{\lambda_T^D} z = \frac{D}{2\beta} \langle N \rangle_0 \\ P_0 &= \frac{1}{\beta \lambda_T^D} z = \frac{1}{\beta V} \langle N \rangle_0, \end{aligned} \tag{2.34}$$

where the subscript 0 indicates that the averages are taken with respect to Ω_0. The relations of $\langle E \rangle_0$ and P_0 with $\langle N \rangle_0$ are the same as for an ideal *classical* gas.

With the second term Ω_1 included, the pressure becomes

$$\beta P = \frac{\langle N \rangle_0}{V} \left(1 \mp \frac{1}{2^{(D+2)/2}} \lambda_T^D \frac{\langle N \rangle_0}{V} \right), \tag{2.35}$$

where the minus sign applies to a Bose and the plus sign to a Fermi gas. The sign difference shows that because of its statistics, a Bose gas has a pressure lower than its classical counterpart, while because of the Pauli principle, a Fermi gas has a higher pressure. The last formula also shows that the condition $z = \mathrm{e}^{-\alpha} < 1$ for a dilute gas is equivalent to the condition

$$\lambda_T^D \frac{\langle N \rangle}{V} < 1. \tag{2.36}$$

In other words, quantum statistics becomes noticeable only when the average interparticle spacing becomes smaller than the thermal wave length λ_T.

2.2 Finite-Temperature Field Theory

Our goal is to describe the grand canonical ensembles of noninteracting bosons or fermions of the preceding section by fields. As a first step towards such a quantum field theory of ideal quantum gases, consider the nonrelativistic free theory

$$\mathcal{L} = \mathrm{i}\hbar\psi^* \frac{\partial}{\partial t}\psi - \frac{\hbar^2}{2m}\nabla\psi^* \cdot \nabla\psi + \mu\psi^*\psi, \tag{2.37}$$

which is quadratic in the fields. The chemical potential μ is introduced to account for a finite density of particles. It can be considered a Lagrange multiplier which implements the constraint

$$\int \mathrm{d}^D x\, \psi^*\psi = N. \tag{2.38}$$

The μ-dependent term in the Lagrangian density (2.37) yields, after integrating over space, the term μN appropriate for a grand canonical ensemble of particles.

After having included a chemical potential in the field theory, we proceed to include temperature. The easiest way to achieve this is by going over to imaginary times $\tau = \mathrm{i}t$, and restricting the values τ can take to the finite interval $0 \leq \tau \leq \hbar\beta$, with β the inverse temperature. With this rule, the Minkowski action S goes over into

$$S = \int \mathrm{d}^d x\, \mathcal{L}(t, \mathbf{x}) \to -\mathrm{i}\int_0^{\hbar\beta} \mathrm{d}\tau \int \mathrm{d}^D x\, \mathcal{L}(-\mathrm{i}\tau, \mathbf{x}) \equiv \mathrm{i}S_{\mathrm{E}}, \tag{2.39}$$

with $S_{\mathrm{E}} = \int \mathrm{d}\tau \mathrm{d}^D x\, \mathcal{L}_{\mathrm{E}}$ and $\mathcal{L}_{\mathrm{E}}(\tau, \mathbf{x}) = -\mathcal{L}(-\mathrm{i}\tau, \mathbf{x})$. Here, the subscript "E" on the action on the right is to indicate Euclidean rather than Minkowski spacetime.

Because the time axis is now of finite extent, the Fourier transform becomes a series. With periodic boundary conditions

$$\psi(0, \mathbf{x}) = \psi(\hbar\beta, \mathbf{x}), \tag{2.40}$$

appropriate for a bosonic field $\psi(\tau, \mathbf{x})$, the Fourier transform of the field ψ takes the form

$$\begin{aligned}\psi(\tau, \mathbf{x}) &= \frac{1}{\hbar\beta} \sum_n \int \frac{\mathrm{d}^D k}{(2\pi)^D} \psi_n(\mathbf{k})\, \mathrm{e}^{-\mathrm{i}\omega_n \tau + \mathrm{i}\mathbf{k}\cdot\mathbf{x}} \\ \psi^*(\tau, \mathbf{x}) &= \frac{1}{\hbar\beta} \sum_n \int \frac{\mathrm{d}^D k}{(2\pi)^D} \psi_n^*(\mathbf{k})\, \mathrm{e}^{+\mathrm{i}\omega_n \tau - \mathrm{i}\mathbf{k}\cdot\mathbf{x}},\end{aligned} \tag{2.41}$$

where $\psi_n(\mathbf{k}) \equiv \psi(\omega_n, \mathbf{k})$ and ω_n are the so-called *Matsubara frequencies*,

$$\omega_n = \frac{2\pi}{\hbar\beta} n \tag{2.42}$$

with n an integer. The transition $t \to -\mathrm{i}\tau$ to imaginary time is thus accompanied by the substitution

$$\int \frac{\mathrm{d}\omega}{2\pi} g(\omega) \to \frac{\mathrm{i}}{\hbar\beta} \sum_n g(\mathrm{i}\omega_n) \tag{2.43}$$

in Fourier space, where g is an arbitrary function. The $n = 0$ contribution, stemming from the $\psi_0(\mathbf{k})$ mode in the Fourier series (2.41), plays a special role as the corresponding Matsubara frequency vanishes, $\omega_0 = 0$. This zero-frequency mode, specific to bosonic systems, is the Fourier transform of the time-averaged field $\psi(\mathbf{x})$:

$$\psi(\mathbf{x}) \equiv \int_0^{\hbar\beta} \mathrm{d}\tau\, \psi(\tau, \mathbf{x}) = \int \frac{\mathrm{d}^D k}{(2\pi)^D} \psi_0(\mathbf{k})\, \mathrm{e}^{\mathrm{i}\mathbf{k}\cdot\mathbf{x}}. \tag{2.44}$$

To describe fermions instead of bosons, as was done in the first part of this section, the field ψ in the Lagrangian density (2.37) must be a Grassmann rather than an ordinary field. Grassmann fields have the property that interchanging two of them yields a minus sign:

$$\psi_1(x)\psi_2(x') = -\psi_2(x')\psi_1(x), \tag{2.45}$$

i.e., they anticommute. It immediately follows that $\psi^2(x) = 0$, meaning that already at the classical level a property akin to the Pauli exclusion principle is present. The derivation of the corresponding finite-temperature field theory proceeds in the same way as for bosons, save for the fact that antiperiodic boundary conditions for the Grassmann field $\psi(\tau, \mathbf{x})$ must be imposed:

$$\psi(0, \mathbf{x}) = -\psi(\hbar\beta, \mathbf{x}). \tag{2.46}$$

This results in replacing the bosonic Matsubara frequencies (2.42) with fermionic ones

$$\omega_n = \frac{\pi}{\hbar\beta}(2n + 1), \tag{2.47}$$

but otherwise the rules given above also apply to fermions.

2.3 Functional Integrals

In the functional integral approach, quantizing a field theory amounts to allowing the field ψ to take any configuration, not just the one determined by its field equations. As in statistical physics, all possible field configurations are to be accounted for with the appropriate weighing factor. Specifically, the partition function Z is given by

$$Z = \int \mathrm{D}\psi^* \mathrm{D}\psi \, \mathrm{e}^{-S_\mathrm{E}/\hbar}, \tag{2.48}$$

where the functional integral $\int \mathrm{D}\psi$ denotes the sum over all possible field configurations. A functional integral is a generalization of the ordinary integral $\int \mathrm{d}x$, where the integral is over function space rather than over coordinate space. For an ideal Bose or Fermi gas, S_E in Eq. (2.48) denotes the Euclidean action

$$S_\mathrm{E} = \int \mathrm{d}^d x \left(\hbar\psi^* \frac{\partial}{\partial\tau}\psi + \frac{\hbar^2}{2m}\nabla\psi^* \cdot \nabla\psi - \mu\psi^*\psi \right), \tag{2.49}$$

obtained from the Lagrangian density (2.37) formulated in Minkowski spacetime using the rule (2.39). The spacetime coordinate x now stands for $(\tau, \mathbf{x})$.

Equation (2.48) constitutes the basic postulate of the functional integral approach, stating that a quantum field theory is obtained from its classical counterpart by summing over all field configurations weighed with a Boltzmann-like factor determined by the Euclidean action.

Being quadratic in the fields, the functional integral (2.48) with the action (2.49) can be evaluated exactly by generalizing the Gaussian integral

$$\int \prod_{i=1}^{n} \mathrm{d}x_i \, \mathrm{e}^{-\frac{1}{2}x\cdot M\cdot x} = (2\pi)^{n/2} \det{}^{-1/2}(M), \tag{2.50}$$

with $x \cdot M \cdot x \equiv \sum_{i,j=1}^{n} x_i M_{ij} x_j$, M a symmetric positive-definite $n \times n$ matrix, and $\det(M)$ its determinant. For $n = 1$ this formula is easily checked. Since the integration variables in Eq. (2.48) are complex, we also record the complex generalization of Eq. (2.50). Consider $2n$ integration variables $x_1, y_1, x_2, y_2, \ldots,$ x_n, y_n, and introduce the complex coordinates

$$z_i = \frac{1}{\sqrt{2}}(x_i + \mathrm{i}y_i), \quad z_i^* = \frac{1}{\sqrt{2}}(x_i - \mathrm{i}y_i), \tag{2.51}$$

where z_i and z_i^* are to be considered independent integration variables. The complex counterpart of Eq. (2.50) then reads

$$\int \prod_{i=1}^{n} \mathrm{d}z_i^* \mathrm{d}z_i \, \mathrm{e}^{-z^*\cdot M\cdot z} = (2\pi)^n \det{}^{-1}(M), \tag{2.52}$$

with M now denoting a positive-definite Hermitian matrix.

By analogy, one obtains for the partition function (2.48) up to an irrelevant multiplicative factor

$$Z = \mathrm{Det}^{-1}\left(-p_\tau + \frac{1}{2m}\mathbf{p}^2 - \mu\right), \tag{2.53}$$

where $p^\mu = (p_\tau, \mathbf{p})$ with $p_\tau \equiv -\hbar\partial/\partial\tau$ and $\mathbf{p} \equiv -i\hbar\nabla$. The thermodynamic potential becomes

$$\Omega = -\beta^{-1}\ln(Z) = \beta^{-1}\,\mathrm{Tr}\ln\left(-p_\tau + \frac{1}{2m}\mathbf{p}^2 - \mu\right), \tag{2.54}$$

where use is made of the identity valid for any matrix M

$$\det(M) = \exp\left[\mathrm{tr}\ln(M)\right] \tag{2.55}$$

with tr denoting the trace over the discrete indices.

To give a precise meaning to the trace appearing in Eq. (2.54), we return to the ordinary Gaussian integral (2.52) for the special case which turns out to be relevant to us where the matrix M is diagonal, $M_{ij} = \lambda_i\delta_{ij}$ (no sum over i). We then have

$$\mathrm{tr}\ln(M) = \sum_{i=1}^{n}\ln\left(\lambda_i\delta_{ij}\big|_{j=i}\right) = \sum_{i=1}^{n}\ln(\lambda_i)\,\delta_{ij}\big|_{j=i}, \tag{2.56}$$

where the notation $\delta_{ij}\big|_{j=i}$ indicates that only the diagonal matrix elements are picked out. The reason to put the right side in this particular form becomes clear shortly when generalizing this expression to functional traces. The matrix elements in Eq. (2.56) can be obtained from the exponent appearing at the left side of Eq. (2.52) by taking derivatives,

$$\frac{\partial^2}{\partial z_k^*\partial z_l}\sum_{i,j=1}^{n} z_i^* M_{ij} z_j = \lambda_k\delta_{kl}, \tag{2.57}$$

with no sum over k. By analogy, the operator appearing in Eq. (2.54) is obtained by taking functional derivatives of the action,

$$\frac{\delta^2 S_{\mathrm{E}}}{\delta\psi^*(x)\delta\psi(x')} = \left(\hbar\frac{\partial}{\partial\tau} - \frac{\hbar^2}{2m}\nabla^2 - \mu\right)\delta(x - x'), \tag{2.58}$$

according to the definition (1.30). Because of the delta function, the right side is nonzero only for $x' = x$, i.e., the operator is diagonal. Again by analogy, the trace in Eq. (2.54) then reads explicitly

$$\begin{aligned}
\mathrm{Tr}\ln&\left(-p_\tau + \frac{1}{2m}\mathbf{p}^2 - \mu\right) \\
&= \int \mathrm{d}^d x\,\ln\left[\left(-p_\tau + \frac{1}{2m}\mathbf{p}^2 - \mu\right)\delta(x - x')\big|_{x'=x}\right] \\
&= \int \mathrm{d}^d x\,\ln\left(-p_\tau + \frac{1}{2m}\mathbf{p}^2 - \mu\right)\delta(x - x')\big|_{x'=x},
\end{aligned} \tag{2.59}$$

where the discrete sum over i is replaced with an integral over spacetime, while the Dirac delta function replaces the Kronecker delta in Eq. (2.56). As that equation, the last step in Eq. (2.59) is justified because the logarithm of a diagonal matrix is again a diagonal matrix with logarithms as elements.

For a noninteracting system, the trace of the logarithm is best evaluated by Fourier transforming the delta function

$$\delta(x - x') = \frac{1}{\hbar\beta} \sum_n \int \frac{\mathrm{d}^D k}{(2\pi)^D} \, \mathrm{e}^{-\mathrm{i}k\cdot(x-x')}, \tag{2.60}$$

with $k \cdot x \equiv \omega_n \tau - \mathbf{k} \cdot \mathbf{x}$ and $k^\mu \equiv (\mathrm{i}\omega_n, \mathbf{k})$ in addition to $x^\mu \equiv (-\mathrm{i}\tau, \mathbf{x})$ in Euclidean spacetime. That is, when going from Minkowski to Euclidean spacetime $x^0 = t \to -\mathrm{i}\tau$ and $k^0 = \omega \to \mathrm{i}\omega_n$ in accord with the substitution rule (2.43). The exponential function $\exp(\mathrm{i}k \cdot x')$ can be shifted to the left since the derivatives $p_\tau = -\hbar\partial_\tau$ and $\mathbf{p} = -\mathrm{i}\hbar\nabla$ do not operate on it. The variable x' can then be set equal to x and after the derivatives have been taken, the two exponential factors cancel,

$$\begin{aligned} \mathrm{Tr}\ln\left(-p_\tau + \frac{1}{2m}\mathbf{p}^2 - \mu\right) &= \int \mathrm{d}^d x \, \frac{1}{\hbar\beta} \sum_n \int \frac{\mathrm{d}^D k}{(2\pi)^D} \mathrm{e}^{ik\cdot x} \ln\left(-p_\tau + \frac{1}{2m}\mathbf{p}^2 - \mu\right) \mathrm{e}^{-ik\cdot x} \\ &= \int \mathrm{d}^d x \, \frac{1}{\hbar\beta} \sum_n \int \frac{\mathrm{d}^D k}{(2\pi)^D} \ln\left[-\mathrm{i}\hbar\omega_n + \epsilon(\mathbf{k}) - \mu\right], \end{aligned} \tag{2.61}$$

with $\epsilon(\mathbf{k}) = \hbar^2\mathbf{k}^2/2m$. This result can be interpreted as being obtained by using plane waves $\langle x|k\rangle \equiv \mathrm{e}^{-ik\cdot x}$ as a basis to evaluate the trace. Here, we introduced the Dirac bra-ket notation, with $p^\mu|k\rangle = \hbar k^\mu|k\rangle$ in the momentum representation. The plane waves are eigenfunctions of the operator $K(p) \equiv -p_\tau + \mathbf{p}^2/2m - \mu$ with eigenvalues $\lambda_k \equiv -\mathrm{i}\hbar\omega_n + \epsilon(\mathbf{k}) - \mu$, and

$$\begin{aligned} \mathrm{Tr}\ln K &= \frac{1}{\hbar\beta} \sum_n \int \frac{\mathrm{d}^D k}{(2\pi)^D} \langle k| \ln\left[K(p)\right] |k\rangle \\ &= \int \mathrm{d}^d x \frac{1}{\hbar\beta} \sum_n \int \frac{\mathrm{d}^D k}{(2\pi)^D} \langle k|x\rangle \ln\left[K(k)\right] \langle x|k\rangle \\ &= \int \mathrm{d}^d x \frac{1}{\hbar\beta} \sum_n \int \frac{\mathrm{d}^D k}{(2\pi)^D} \ln\left[K(k)\right], \end{aligned} \tag{2.62}$$

where in the second step we inserted the unit operator $\int \mathrm{d}^d x |x\rangle\langle x|$. For notational convenience we suppress $\hbar$ in the argument of $K(k)$.

The integral over spacetime in Eq. (2.61) simply yields a factor $\hbar\beta V$, with V the volume of the system. For the thermodynamic potential (2.54), we thus find

$$\frac{\Omega}{V} = \frac{1}{\beta} \sum_n \int \frac{\mathrm{d}^D k}{(2\pi)^D} \ln\left[-\mathrm{i}\hbar\omega_n + \epsilon(\mathbf{k}) - \mu\right]. \tag{2.63}$$

The summation over the Matsubara frequencies can be carried out by introducing the Schwinger proper-time representation of the logarithm,

$$\ln(a) = \lim_{z \to 0} \frac{1}{z} - \int_0^\infty \frac{\mathrm{d}s}{s} \mathrm{e}^{-sa}, \tag{2.64}$$

with the first term an irrelevant diverging contribution. This gives

$$\sum_n \ln(-\mathrm{i}\hbar\omega_n + \xi) = -\int_\delta^\infty \frac{\mathrm{d}s}{s} \sum_n \mathrm{e}^{(\mathrm{i}\hbar\omega_n - \xi)s}, \tag{2.65}$$

where we introduced the infinitesimal cutoff δ that is set to zero at the end of the calculation. The frequency sum as well as the integral over proper time can then be readily evaluated with the help of the Poisson summation formula

$$\sum_{n=-\infty}^{\infty} \mathrm{e}^{2\pi\mathrm{i}na} = \sum_{w=-\infty}^{\infty} \delta(a - w), \tag{2.66}$$

yielding

$$\sum_n \ln(-\mathrm{i}\hbar\omega_n + \xi) = -\sum_{w=1}^{\infty} \frac{1}{w} \mathrm{e}^{-\beta\xi w} = \ln\left(1 - \mathrm{e}^{-\beta\xi}\right). \tag{2.67}$$

The cutoff introduced in Eq. (2.65) suppresses the diverging $w = 0$ contribution here. This contribution, which is temperature independent, corresponds to the zero-point energy and is irrelevant for our present purposes.

To compare with more standard regularizations, we take the derivative of Eq. (2.67) with respect to ξ to obtain the *regularized* expression

$$\frac{1}{\beta} \sum_n \frac{1}{-\mathrm{i}\hbar\omega_n + \xi} = n(\xi), \tag{2.68}$$

with $n(x) = 1/[\exp(\beta\xi) - 1]$ the Bose-Einstein distribution function. When the sum over Matsubara frequencies on the left is evaluated directly with the help of contour integration, an extra constant term appears:

$$\frac{1}{\beta} \sum_n \frac{1}{-\mathrm{i}\hbar\omega_n + \xi} = n(\xi) + \frac{1}{2}. \tag{2.69}$$

Formally, the regularized expression (2.68) corresponds to including a convergence factor $\exp(\mathrm{i}\omega\tau)$ in the summand and taking the limit $\tau \to 0^+$ from above after the sum over the Matsubara frequencies has been carried out.

Using Eq. (2.67), with $\xi = \epsilon(\mathbf{k}) - \mu$ the kinetic energy measured relative to the chemical potential, in the expression (2.63) for the thermodynamic potential, we recover the old result (2.23) obtained from standard quantum statistical mechanics, where the integral over the wave vectors $\mathbf{k}$ is still to be carried out.

We continue to discuss a Fermi gas as described by the same Lagrangian density (2.37) used to study a Bose gas, with the *proviso* that ψ now denotes an anticommuting Grassmann field. To this end, consider a set of $2n$ independent Grassmann variables $\theta_1^*, \theta_1, \theta_2^*, \theta_2, \ldots, \theta_n^*, \theta_n$, satisfying

$$\theta_i \theta_j + \theta_j \theta_i = 0. \tag{2.70}$$

The Grassmann analog of the Gaussian integral (2.52) reads

$$\int \prod_{i=1}^{n} \mathrm{d}\theta_i^* \mathrm{d}\theta_i \, \mathrm{e}^{-\theta^* \cdot M \cdot \theta} = \det(M), \tag{2.71}$$

where it is important to note that the determinant, not its inverse, appears on the right. By analogy, we obtain for the fermionic partition function (2.48)

$$Z = \mathrm{Det}\left(-p_\tau + \frac{1}{2m}\mathbf{p}^2 - \mu\right), \tag{2.72}$$

instead of the bosonic expression (2.53). The thermodynamic potential now becomes

$$\Omega = -\beta^{-1} \ln(Z) = -\beta^{-1} \operatorname{Tr} \ln\left(-p_\tau + \frac{1}{2m}\mathbf{p}^2 - \mu\right). \tag{2.73}$$

Most of the steps taken for bosons can be copied here provided the bosonic Matsubara frequencies $\omega_n = 2n\pi/\hbar\beta$ are replaced with the fermionic frequencies (2.47). Instead of Eq. (2.67), we then obtain the regularized expression

$$\sum_n \ln\left(-\mathrm{i}\hbar\omega_n + \xi\right) = \ln\left(1 + \mathrm{e}^{-\beta\xi}\right), \tag{2.74}$$

where $\xi = \epsilon(\mathbf{k}) - \mu$ is the kinetic energy relative to the chemical potential as before. With this result, we recover the thermodynamic potential (2.24) of a gas of free fermions, illustrating that a field theory built from Grassmann fields gives a correct description of a Fermi gas.

2.4 Feynman Propagator

Consider the two-point correlation function

$$G_\mathrm{E}(x, x') \equiv \langle \psi(x) \psi^*(x') \rangle \tag{2.75}$$

in Euclidean spacetime, where the average is taken with respect to the partition function (2.48). Explicitly, the correlation or *Green function* reads

$$G_\mathrm{E}(x, x') = \frac{1}{Z} \int \mathrm{D}\psi^* \mathrm{D}\psi \, \psi(x) \psi^*(x') \, \mathrm{e}^{-S_\mathrm{E}/\hbar}. \tag{2.76}$$

For a noninteracting Bose gas specified by the Euclidean action (2.49), this functional integral can be evaluated in closed form. To this end, consider first the Gaussian integral (2.50) with an additional source term included

$$\int \prod_{i=1}^{n} \mathrm{d}x_i \, \mathrm{e}^{-\frac{1}{2}x\cdot M\cdot x + J\cdot x} = (2\pi)^{n/2} \det{}^{-1/2}(M)\, \mathrm{e}^{\frac{1}{2}J\cdot M^{-1}\cdot J}, \tag{2.77}$$

with $J \cdot x = \sum_{i=1}^{n} J_i x_i$, or its complex extension

$$\int \prod_{i=1}^{n} \mathrm{d}z_i^* \mathrm{d}z_i \, \mathrm{e}^{-z^*\cdot M\cdot z + J\cdot z^* + J^*\cdot z} = (2\pi)^n \det{}^{-1}(M)\, \mathrm{e}^{J^*\cdot M^{-1}\cdot J}, \tag{2.78}$$

where $J_i = (J_{1,i} + \mathrm{i}J_{2,i})/\sqrt{2}$. By taking derivatives with respect to the sources and setting these to zero at the end, Gaussian integrals with arbitrary powers of z and z^* can be generated. When needed, factors of i can be included in these formulas by appropriate changes in M and the sources, such as $M \to -\mathrm{i}M$ and $J \to \mathrm{i}J$, $J^* \to \mathrm{i}J^*$.

For a noninteracting Bose gas specified by the action (2.49), the generating functional (2.78) is generalized as follows:

$$\begin{aligned} Z[J, J^*] &= \int \mathrm{D}\psi^* \mathrm{D}\psi \exp\left\{-\frac{1}{\hbar}\left[S_{\mathrm{E}} + \int \mathrm{d}^d x\,(J\psi^* + J^*\psi)\right]\right\} \\ &= Z[0,0] \exp\left[\frac{1}{\hbar}\int \mathrm{d}^d x\, \mathrm{d}^d x'\, J^*(x)\Delta_{\mathrm{E}}(x,x')J(x')\right], \end{aligned} \tag{2.79}$$

with $Z[0,0] = Z$, and $\Delta_{\mathrm{E}}(x,x') = \Delta_{\mathrm{E}}(x-x')$ the so-called *Feynman propagator*

$$\Delta_{\mathrm{E}}(x,x') \equiv \frac{1}{\hbar\beta}\sum_{n=-\infty}^{\infty}\int \frac{\mathrm{d}^D k}{(2\pi)^D}\, \mathrm{e}^{ik\cdot x'} \frac{1}{-\mathrm{i}\hbar\omega_n + \epsilon(\mathbf{k}) - \mu}\, \mathrm{e}^{-ik\cdot x}. \tag{2.80}$$

We recall that $k \cdot x = \omega_n \tau - \mathbf{k}\cdot\mathbf{x}$ and $\epsilon(\mathbf{k}) = \hbar^2\mathbf{k}^2/2m$ is the kinetic energy. By taking the functional derivative of this expression with respect to the sources J and J^* and setting them to zero, we obtain for the Green function (2.76)

$$\begin{aligned} G_{\mathrm{E}}(x,x') &= \frac{1}{Z}\left(-\hbar\frac{\delta}{\delta J^*(x)}\right)\left(-\hbar\frac{\delta}{\delta J(x')}\right) Z[J,J^*]\Big|_{J=J^*=0} \\ &= \hbar\Delta_{\mathrm{E}}(x-x'), \end{aligned} \tag{2.81}$$

identifying the propagator as the two-point Green function (apart from a conventional prefactor).

The analog for the vacuum theory ($\mu = 0$) at zero temperature obtained via the transcription (2.43) applied backwards reads

$$G(x,x') = \mathrm{i}\hbar\Delta_{\mathrm{F}}(x-x'), \tag{2.82}$$

with the Feynman propagator

$$\Delta_{\rm F}(x-x') = \int \frac{d^d k}{(2\pi)^d}\, e^{ik\cdot x'} \frac{1}{\hbar\omega - \epsilon(\mathbf{k}) + i\eta}\, e^{-ik\cdot x}, \tag{2.83}$$

satisfying

$$\left(i\hbar\frac{\partial}{\partial t} + \frac{\hbar^2}{2m}\nabla^2\right)\Delta_{\rm F}(x) = \delta(x). \tag{2.84}$$

In Eq. (2.83, an infinitesimal term $i\eta$ is included in the denominator to avoid the singularity on the real axis at $\omega = \hbar\mathbf{k}^2/2m$. The small positive constant $\eta = 0^+$ is to be set to zero after the frequency integral has been performed. To show that this prescription yields the correct causal result, we evaluate the integrals explicitly using the integral representation of the step function

$$\theta(t-t') = \lim_{\eta\to 0^+} i \int \frac{dx}{2\pi} \frac{e^{-ix(t-t')}}{x + i\eta}, \tag{2.85}$$

which equals 1 for $t > t'$ and vanishes for $t < t'$. This representation can be checked by contour integration. The correlation function then becomes

$$G(x,x') = \theta(t-t') \int \frac{d^D k}{(2\pi)^D} e^{i\mathbf{k}.(\mathbf{x}-\mathbf{x}') - i\hbar\mathbf{k}^2(t-t')/2m} \tag{2.86}$$

$$= \theta(t-t')\left(\frac{m}{2\pi i\hbar(t-t')}\right)^{D/2} \exp\left(i\frac{m(\mathbf{x}-\mathbf{x}')^2}{2\hbar(t-t')}\right), \tag{2.87}$$

where the last line follows from a simple Gaussian integration of the form (2.77). Because of the step function, this correlation function has the correct causal behavior in that the positive frequency ($\omega = \hbar\mathbf{k}^2/2m$) modes are carried forward in time. In other words, the prescription of replacing the real frequency ω with the complex frequency $\omega + i\eta$ where $\eta = 0^+$ selects the *retarded* correlation function. For the action formulated in Minkowski spacetime, this $i\eta$ prescription implies an additional term $i\eta \int d^d x\, |\psi|^2$, which when inserted in the functional integral

$$Z = \int D\psi^* D\psi\, e^{(i/\hbar)S}, \tag{2.88}$$

leads to the convergence factor $\exp\left(-\eta \int d^d x\, |\psi|^2\right)$ for large ψ.

Below, we will also encounter cases where the frequency appears quadratic in the propagator. The simplest example where this happens is provided by a nonrelativistic particle in a one-dimensional harmonic potential. It is described by the classical action

$$S = \frac{1}{2}\int dt \left(\dot{x}^2 - \omega_c^2 x^2\right) = -\frac{1}{2}\int dt\, x\left(\frac{d^2}{dt^2} + \omega_c^2\right)x, \tag{2.89}$$

with oscillator frequency $\omega_c > 0$. For convenience, we set the mass of the particle to unity. This quantum mechanical problem can be interpreted as a (0+1)-dimensional field theory, i.e., one in zero space dimensions. The partition function written as a functional integral takes the form

$$Z[J] = \int \mathrm{D}x \exp\left[\frac{\mathrm{i}}{\hbar}\int \mathrm{d}t \left(\tfrac{1}{2}\dot{x}^2 - \tfrac{1}{2}\omega_c^2 x^2 + Jx\right)\right] \tag{2.90}$$

where we added a source term. Implementing the iη prescription by including the term $\mathrm{i}\eta \int \mathrm{d}t\, x^2$ in the action, we obtain

$$Z[J] = Z[0] \exp\left[-\frac{\mathrm{i}}{2\hbar}\int \mathrm{d}t\, \mathrm{d}t'\, J(t) D_F(t,t') J(t')\right], \tag{2.91}$$

with

$$\begin{aligned} D_F(t,t') &= \int \frac{\mathrm{d}\omega}{2\pi} \mathrm{e}^{\mathrm{i}\omega t'} \frac{1}{\omega^2 - \omega_c^2 + \mathrm{i}\eta} \mathrm{e}^{-\mathrm{i}\omega t} \\ &= -\frac{\mathrm{i}}{2\omega_c}\left[\theta(t-t')\mathrm{e}^{-\mathrm{i}\omega_c(t-t')} + \theta(t'-t)\mathrm{e}^{\mathrm{i}\omega_c(t-t')}\right] \\ &= -\frac{\mathrm{i}}{2\omega_c}\mathrm{e}^{-\mathrm{i}\omega_c|t-t'|} \end{aligned} \tag{2.92}$$

the Feynman propagator satisfying

$$-\left(\frac{\mathrm{d}^2}{\mathrm{d}t^2} + \omega_c^2\right) D_F(t) = \delta(t). \tag{2.93}$$

The second line in Eq. (2.92) is obtained by writing

$$\frac{1}{\omega^2 - \omega_c^2 + \mathrm{i}\eta} = \frac{1}{2\omega_c}\left(\frac{1}{\omega - \omega_c + \mathrm{i}\,\mathrm{sgn}(\omega)\eta} - \frac{1}{\omega + \omega_c + \mathrm{i}\,\mathrm{sgn}(\omega)\eta}\right), \tag{2.94}$$

where $\mathrm{sgn}(\omega)$ denotes the sign of ω at the location of the poles. For $\omega_c > 0$, the sign is positive in the first term of the propagator and negative in the second. Figure 2.1 shows these locations in the complex ω-plane. A more explicit evaluation of the functional integral (2.91), or using canonical quantization (see below) shows that Eq. (2.92) is indeed the correct form. As before, this Feynman propagator is seen to carry positive frequency modes forward in time (first term in the second line). In addition, however, it carries negative frequency modes backwards in time (second term)—that is, unlike Eq. (2.86), a propagator with a quadratic dependence on the frequency is an admixture of retarded and advanced correlation functions. Without further justification we will apply the iη prescription by letting $\omega \to \omega + \mathrm{i}\,\mathrm{sgn}(\omega)\eta$ whenever a causal correlation function is needed. If, on the other hand, a retarded correlation function is needed, as in response theory, we simply replace the real frequency ω by the complex frequency $\omega + \mathrm{i}\eta$.

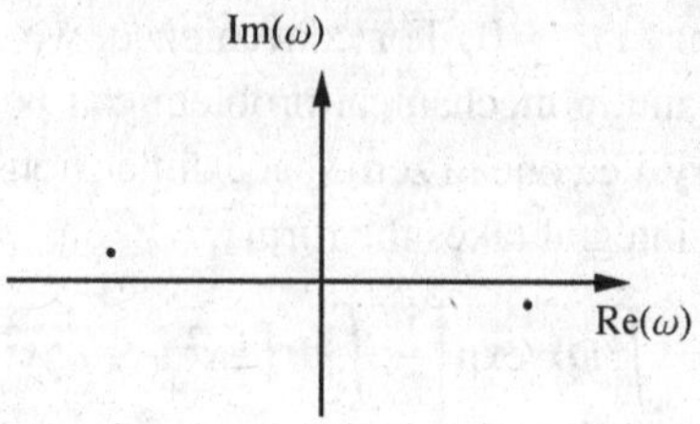

Fig. 2.1 Poles in the complex ω-plane of the Feynman propagator with quadratic dependence on the frequency ω.

In the canonical quantization approach, the correlation function $G(t)$ of the harmonic oscillator is defined by the time-ordered product

$$G(t) = \langle 0|\mathrm{T}\,\hat{x}(t)\hat{x}(0)|0\rangle \tag{2.95}$$

where T is the time-ordering operator

$$\mathrm{T}\,\hat{x}(t)\hat{x}(0) \equiv \begin{cases} \hat{x}(t)\hat{x}(0) & \text{for } t > 0 \\ \hat{x}(0)\hat{x}(t) & \text{for } 0 > t, \end{cases} \tag{2.96}$$

and

$$\hat{x}(t) = \mathrm{e}^{(\mathrm{i}/\hbar)Ht}\,\hat{x}(0)\,\mathrm{e}^{-(\mathrm{i}/\hbar)Ht}, \tag{2.97}$$

with H the Hamiltonian, is the position operator in the Heisenberg picture where operators are time dependent. In Eq. (2.95), use is made of Dirac's bra-ket notation of quantum mechanics, with $|0\rangle$ denoting the ground state. The correlation function (2.95) is related to the propagator through $G(t) = \mathrm{i}\hbar D_{\mathrm{F}}(t)$, cf. Eq. (2.82). It is readily evaluated using the algebraic method where the position operator is written in terms of creation ($a^\dagger$) and annihilation (a) operators as ($m = 1$)

$$\hat{x} = \sqrt{\frac{\hbar}{2\omega_{\mathrm{c}}}}(a + a^\dagger). \tag{2.98}$$

These operators satisfy the fundamental commutation relation

$$[a, a^\dagger] = 1. \tag{2.99}$$

In terms of these operators, the Hamiltonian takes the simple form

$$H = \hbar\omega_{\mathrm{c}}\left(\hat{n} + \tfrac{1}{2}\right) \tag{2.100}$$

with $\hat{n} \equiv a^\dagger a$. If $|n\rangle$ denotes the eigenstate of this operator, $\hat{n}|n\rangle = n|n\rangle$ with eigenvalue n, then

$$\begin{aligned} a|n\rangle &= \sqrt{n}|n\rangle, \\ a^\dagger|n\rangle &= \sqrt{n+1}|n+1\rangle, \\ H|n\rangle &= \hbar\omega_{\mathrm{c}}\left(n + \tfrac{1}{2}\right). \end{aligned} \tag{2.101}$$

It is readily established that n is an integer, $n = 0, 1, 2, \cdots$. This eigenvalue physically denotes the number of quanta of vibration present in the state $|n\rangle$. The creation operator $a^\dagger$ acting on this state adds a quantum of vibration, while the annihilation operator a subtracts such a quantum from that state. The ground state $|0\rangle$ contains no quanta of vibration so that $a|0\rangle = 0$. With the rules given, the correlation function can be easily evaluated with the expected result

$$G(t) = \frac{\hbar}{2\omega_c} \mathrm{e}^{-\mathrm{i}\omega_c|t|} = \mathrm{i}\hbar D_F(t). \tag{2.102}$$

2.5 Feynman Diagrams

For interacting theories, the functional integral (2.48) can usually no longer be evaluated in closed form as is the case for noninteracting theories. One then has to resort to perturbation theory, where the partition function and the correlation or Green functions are calculated approximately by expanding in the coupling constant. Such an expansion, which can be conveniently visualized by Feynman diagrams, is tantamount to perturbatively including quantum corrections, with each successive order carrying an additional power of $\hbar$.

Consider the interacting bosonic theory

$$\mathcal{L} = \frac{1}{2}\phi(x)K(\mathrm{i}\partial)\phi(x) + \mathcal{L}_\mathrm{I}(\phi) + J(x)\phi(x) \tag{2.103}$$

featuring a real field $\phi(x)$ for simplicity, and with a source term included. The first term, with $K(\mathrm{i}\partial)$ some differential operator, is quadratic in the field. This free part defines the Feynman propagator $\Delta_F(x)$ of the theory through

$$K(\mathrm{i}\partial)\Delta_F(x) = \delta(x). \tag{2.104}$$

The interaction term $\mathcal{L}_\mathrm{I}(\phi)$ is assumed to be a polynomial in $\phi(x)$. The partition function $Z[J]$ in Minkowski spacetime,

$$Z[J] = \int \mathrm{D}\phi \exp\left(\frac{\mathrm{i}}{\hbar}\int \mathrm{d}^d x\, \mathcal{L}\right), \tag{2.105}$$

can be rewritten as

$$\begin{aligned} Z[J] &= \int \mathrm{D}\phi \exp\left[\frac{\mathrm{i}}{\hbar}\int \mathrm{d}^d x\, \mathcal{L}_\mathrm{I}(\phi)\right] \exp\left[\frac{\mathrm{i}}{\hbar}\int \mathrm{d}^d x \left(\tfrac{1}{2}\phi K\phi + J\phi\right)\right] \\ &= \left\langle \sum_{n=0}^{\infty} \frac{1}{n!}\left[\frac{\mathrm{i}}{\hbar}\int \mathrm{d}^d x\, \mathcal{L}_\mathrm{I}(\phi)\right]^n \right\rangle_0 \\ &= \exp\left[\frac{\mathrm{i}}{\hbar}\int \mathrm{d}^d x\, \mathcal{L}_\mathrm{I}\left(-\mathrm{i}\hbar \frac{\delta}{\delta J(x)}\right)\right] Z_0[J] \end{aligned} \tag{2.106}$$

where the ensemble average $\langle\cdots\rangle_0$ is to be taken with respect to the free theory with the partition function

$$
\begin{aligned}
Z_0[J] &= \int \mathrm{D}\phi \, \exp\left\{\frac{\mathrm{i}}{\hbar}\int \mathrm{d}^d x\left[\tfrac{1}{2}\phi(x)K(\mathrm{i}\partial)\phi(x) + J(x)\phi(x)\right]\right\} \\
&= Z_0[0]\exp\left[-\frac{\mathrm{i}}{2\hbar}\int \mathrm{d}^d x\,\mathrm{d}^d x'\,J(x)\Delta_\mathrm{F}(x-x')J(x')\right], \qquad (2.107)
\end{aligned}
$$

where $Z_0[0]$ is the functional determinant

$$
Z_0[0] = \mathrm{Det}^{-1/2}\left[K(\mathrm{i}\partial)\right]. \qquad (2.108)
$$

In the last step in Eq. (2.106), it is used that any $\phi(x)$ can be brought down by taking the functional derivative with respect to the source $J(x)$. The partition function, which is represented by the diagram

$$Z[J] =$$ 

, (2.109)

can formally be expanded as

$$
\frac{Z[J]}{Z[0]} = \sum_{n=0}^{\infty}\frac{1}{n!}\left(\frac{\mathrm{i}}{\hbar}\right)^n \int \mathrm{d}^d x_1\cdots \mathrm{d}^d x_n G^{(n)}(x_1,\ldots,x_n)J(x_1)\cdots J(x_n) \qquad (2.110)
$$

with the expansion coefficients given by the correlation or Green functions:

$$
\begin{aligned}
G^{(n)}(x_1,\ldots,x_n) &\equiv \langle\phi(x_1)\cdots\phi(x_n)\rangle \\
&= \frac{1}{Z[0]}\left.\frac{(-\mathrm{i}\hbar)^n\delta^n Z[J]}{\delta J(x_1)\cdots\delta J(x_n)}\right|_{J=0}. \qquad (2.111)
\end{aligned}
$$

They are represented diagrammatically as:

$$G^{(n)}(x_1,\ldots,x_n) =$$ 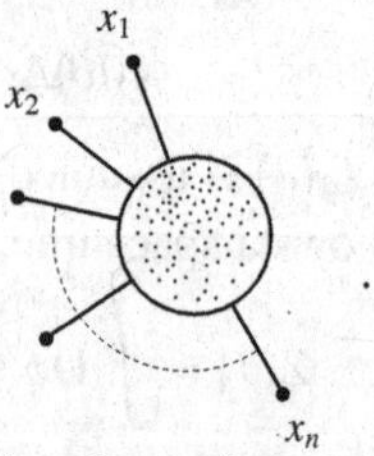

. (2.112)

That is, $Z[J]$ is the generating functional for Green functions. The first term in the series (2.110), corresponding to $n = 0$, is understood to be unity. The series is represented diagrammatically as

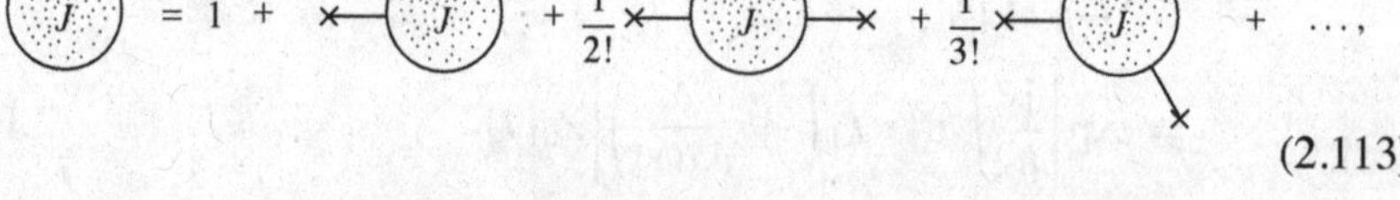

(2.113)

with the rule

$$\frac{\mathrm{i}}{\hbar}\int \mathrm{d}^d x\, J(x) \quad \underset{x}{\bullet}\!\!-\!\!-\!\!-\; = \;\times\!\!-\!\!-\!\!- \tag{2.114}$$

and where a normalization factor $Z[0]$ in each term at the right side of Eq. (2.113) is understood. This factor will be suppressed in all diagrams below.

To illustrate how to compute Green functions, we calculate the four-point Green function $G_0^{(4)}(x_1, x_2, x_3, x_4)$ of the free theory, for which the generating functional (2.107) can be represented as

$$Z_0[J] = \exp(\tfrac{1}{2}\,\times\!\!-\!\!-\!\!-\!\!\times)\,, \tag{2.115}$$

where

$$\times\!\!-\!\!-\!\!-\!\!\times \;\equiv -\frac{\mathrm{i}}{\hbar}\int \mathrm{d}^d x\, \mathrm{d}^d x'\, J(x)\Delta_{\mathrm{F}}(x-x')J(x'). \tag{2.116}$$

With the rules

$$\begin{aligned} -\mathrm{i}\hbar\frac{\delta}{\delta J(x_2)}\,\frac{1}{2}\;\times\!\!-\!\!-\!\!-\!\!\times &= -\int \mathrm{d}^d x\, J(x)\Delta_{\mathrm{F}}(x-x_2) \\ &\equiv \;\times\!\!-\!\!-\!\!-\underset{x_2}{\bullet} \end{aligned} \tag{2.117}$$

and

$$\begin{aligned} -\mathrm{i}\hbar\frac{\delta}{\delta J(x_1)}\;\times\!\!-\!\!-\!\!-\underset{x_2}{\bullet} &= \mathrm{i}\hbar\Delta_{\mathrm{F}}(x_1-x_2) \\ &\equiv \;\underset{x_1}{\bullet}\!\!-\!\!-\!\!-\underset{x_2}{\bullet}\;, \end{aligned} \tag{2.118}$$

the functional derivatives can be carried out one after the other in a straightforward manner to yield

$G_0^{(4)}(x_1, x_2, x_3, x_4) =$ 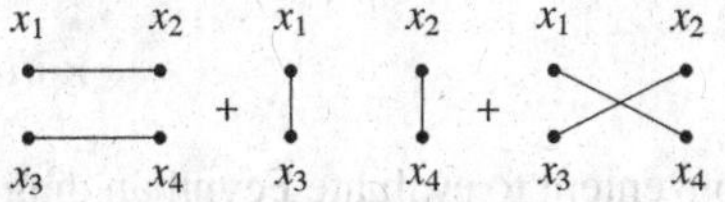 , (2.119)

or in algebraic form

$$\begin{aligned} G_0^{(4)}(x_1, x_2, x_3, x_4) &= \mathrm{i}\hbar\Delta_{\mathrm{F}}(x_1-x_2)\mathrm{i}\hbar\Delta_{\mathrm{F}}(x_3-x_4) \\ &\quad + \mathrm{i}\hbar\Delta_{\mathrm{F}}(x_1-x_4)\mathrm{i}\hbar\Delta_{\mathrm{F}}(x_2-x_3) \\ &\quad + \mathrm{i}\hbar\Delta_{\mathrm{F}}(x_1-x_3)\mathrm{i}\hbar\Delta_{\mathrm{F}}(x_2-x_4). \end{aligned} \tag{2.120}$$

For an example involving an interacting theory, consider the interaction term

$$\mathcal{L}_{\mathrm{I}} = -\frac{\lambda}{4!}\phi^4(x). \tag{2.121}$$

To first order in the self-coupling λ, the partition function assumes the form

$$Z[J] = Z_0[J] - \frac{\mathrm{i}}{\hbar}\frac{\lambda}{4!}\int \mathrm{d}^d x\,(-\mathrm{i}\hbar)^4\frac{\delta^4}{\delta J^4(x)}Z_0[J]. \tag{2.122}$$

Applying the above rules, we obtain for the second term

$$\frac{(-i\hbar)^4}{Z[0]}\frac{\delta^4}{\delta^4 J(x)}Z_0[J] = -3\hbar^2\Delta_F^2(0) + 6i\hbar\Delta_F(0)\left[\int d^d x'\,\Delta_F(x-x')J(x')\right]^2 + \left[\int d^d x'\,\Delta_F(x-x')J(x')\right]^4, \tag{2.123}$$

where the first term on the right corresponds to the three terms in Eq. (2.120) with coinciding arguments, $\Delta_F(x-x) = \Delta_F(0)$. In this way, the generating functional becomes to this order

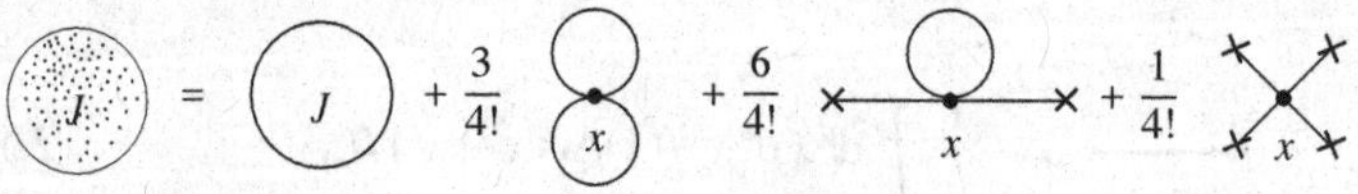

(2.124)

where a vertex stands for

$$\equiv -i\frac{\lambda}{\hbar}\int d^d x. \tag{2.125}$$

It gives rise to the two-point Green function $G^{(2)}(x_1, x_2)$

$G^{(2)}(x_1, x_2) =$ 

$+\ \frac{12}{4!}$ 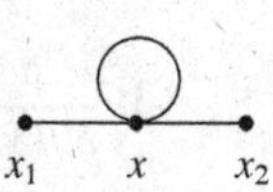

(2.126)

or in formula

$$G^{(2)}(x_1, x_2) = i\hbar\Delta_F(x_1 - x_2) - i\frac{\lambda}{\hbar}\frac{12}{4!}\int d^d x\, i\hbar\Delta_F(x_1 - x) \times i\hbar\Delta_F(0)\, i\hbar\Delta_F(x - x_2). \tag{2.127}$$

It is usually convenient to evaluate Feynman diagrams, by which we mean any diagram obtained by connecting vertices and sources by propagators, in momentum space. For the example at hand, where

$$\Delta_F(x) = \int \frac{d^4 k}{(2\pi)^4}\frac{e^{-ik\cdot x}}{K(k)} \tag{2.128}$$

by Eq. (2.104), the two-point Green function reads in momentum space

$$G^{(2)}(k) = \frac{i\hbar}{K(k)} - i\frac{\lambda}{\hbar}\frac{12}{4!}\int \frac{d^d l}{(2\pi)^d}\frac{i\hbar}{K(k)}\frac{i\hbar}{K(l)}\frac{i\hbar}{K(k)}. \tag{2.129}$$

By going through the steps leading to this result, one readily recognizes the following Feynman rules in momentum space for internal and external lines, and for

vertices, respectively

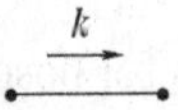

$$\equiv \int \frac{\mathrm{d}^d k}{(2\pi)^d} \frac{\mathrm{i}\hbar}{K(k)}, \tag{2.130}$$

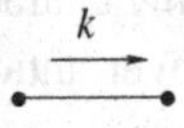

$$\equiv \int \frac{\mathrm{d}^d k}{(2\pi)^d} \frac{\mathrm{i}\hbar}{K(k)} \mathrm{e}^{+\mathrm{i}k\cdot x}, \tag{2.131}$$

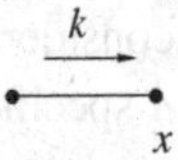

$$\equiv \int \frac{\mathrm{d}^d k}{(2\pi)^d} \frac{\mathrm{i}\hbar}{K(k)} \mathrm{e}^{-\mathrm{i}k\cdot x}, \tag{2.132}$$

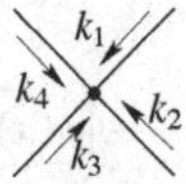

$$\equiv -\mathrm{i}\frac{\lambda}{\hbar}(2\pi)^d \delta(k_1 + k_2 + k_3 + k_4), \tag{2.133}$$

where by an internal line is meant a line not connected to a source. The momenta in these diagrams flow in the direction of the arrows. External lines carry an additional exponential: $\mathrm{e}^{+\mathrm{i}k\cdot x}$ for an outgoing line and $\mathrm{e}^{-\mathrm{i}k\cdot x}$ for an incoming line. The frequencies and wave vectors in the last diagram, and for vertices in general, are taken to flow inwards.

Note that these Feynman rules can be directly read off from the Lagrangian density (2.103) with the interaction term (2.121) without going through the intermediate steps. In this *diagrammatic approach*, which we will adopt in the following, the Lagrangian density is taken as defining the Feynman rules: the propagators are defined by the quadratic part, and the vertices by the interaction terms by means of taking repeated derivatives with respect to the fields until none are left.

Translational invariance, by which, for example, the two-point Green function $G(x, x')$ depends only on the difference of its arguments, $G(x, x') = G(x - x')$, implies overall momentum conversation. Indeed, without translational invariance

$$G(x, x') = \int \frac{\mathrm{d}^d k}{(2\pi)^d} \frac{\mathrm{d}^d k'}{(2\pi)^d} G(k, k') \mathrm{e}^{-\mathrm{i}(k\cdot x + k'\cdot x')}, \tag{2.134}$$

while with it

$$G(x - x') = \int \frac{\mathrm{d}^d k}{(2\pi)^d} G(k) \mathrm{e}^{-\mathrm{i}k\cdot(x - x')} \tag{2.135}$$

so that $G(k, k') = (2\pi)^d \delta(k + k') G(k)$ here. Overall momentum conservation is a property of all Feynman diagrams considered in the following ("what comes in goes out").

The combinatorial factor $12/4! = \frac{1}{2}$ in front of the second diagram of Eq. (2.127) entered in a straightforward manner. Such combinatorial factors can also be determined directly from the diagrams by counting the number of ways a given diagram can be drawn without changing its topology. There is, however, no need to go through the recipe in detail because most of the diagrams we will encounter below have simple symmetry factors.

For fermionic theories, the same Feynman rules as for bosonic theories apply with one exception which is related to the anticommuting character of Grassmann fields. According to Eq. (2.71), a Gaussian integral over anticommuting Grassmann variables produces a determinant and not its inverse as for commuting variables. To see what this implies for the Feynman rules, consider the free fermionic theory in the presence of a constant background field A specified by the partition function

$$Z[A] = \int \mathrm{D}\psi^* \mathrm{D}\psi \exp\left\{\frac{\mathrm{i}}{\hbar}\int \mathrm{d}^d x\, \psi^*(x)[K(\mathrm{i}\partial) + A]\psi(x)\right\}, \tag{2.136}$$

where, as before, $K(\mathrm{i}\partial)$ is some differential operator. Being quadratic in the Grassmann fields, the integrals can be carried out in closed form to give $Z[A] = \mathrm{Det}(K + A)$, cf. Eq. (2.72). This can be alternatively written as

$$Z[A] = \mathrm{Det}(K)\, \mathrm{e}^{(\mathrm{i}/\hbar)\Gamma[A]}, \tag{2.137}$$

with

$$\frac{\mathrm{i}}{\hbar}\Gamma[A] \equiv \mathrm{Tr}\ln(1 + S_\mathrm{F} A) = -\sum_{n=1}^{\infty}\frac{(-1)^n}{n}\mathrm{Tr}\,(S_\mathrm{F} A)^n . \tag{2.138}$$

Here, S_F is the fermion propagator

$$S_\mathrm{F}(x) = \int \frac{\mathrm{d}^d k}{(2\pi)^d}\frac{\mathrm{e}^{-\mathrm{i}k\cdot x}}{K(k)}, \tag{2.139}$$

which is related to the two-point Green function through, cf. (2.118)

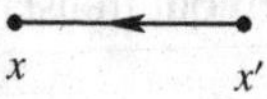

$$\equiv \langle\psi(x)\psi^*(x')\rangle = \mathrm{i}\hbar S_\mathrm{F}(x - x'), \tag{2.140}$$

and the trace Tr in Eq. (2.138) stands for the integral over spacetime and momentum, cf. Eq. (2.62)

$$\mathrm{Tr} \equiv \int \mathrm{d}^d x \int \frac{\mathrm{d}^d k}{(2\pi)^d}. \tag{2.141}$$

The right side of Eq. (2.138) is an infinite sum over one-loop diagrams of the form

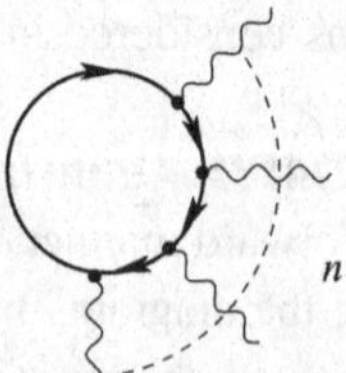

where a wiggly line denotes the external field A, while the vertex stands for $i/\hbar$. Because the field ψ is complex, the lines now carry an additional arrow. Specifically, a field ψ^* is represented by a line with an arrow pointing outwards, while for a field ψ the direction of the arrow is reversed, see Eq. (2.140). In comparison to its bosonic counterpart, the right side of Eq. (2.138) involves an additional minus sign. This can be accounted for by giving each closed ψ-loop a minus sign. This rule is general and also applies to diagrams involving more than one fermion loop.

2.6 Connected Green Functions

The Green functions (2.111) usually consist of disconnected pieces, each of which is in itself a connected diagram. Let

$$\frac{i}{\hbar} W[J] = \text{(2.142)}$$

denote the generating functional for the connected Green functions $G_c^{(n)}(x_1, \ldots, x_n)$:

$$\frac{i}{\hbar} W[J] = \sum_{n=1}^{\infty} \frac{1}{n!} \left(\frac{i}{\hbar}\right)^n \int d^d x_1 \cdots d^d x_n G_c^{(n)}(x_1, \ldots, x_n) J(x_1) \cdots J(x_n), \quad (2.143)$$

with

$$G_c^{(n)}(x_1, \ldots, x_n) \equiv \langle \phi(x_1) \cdots \phi(x_n) \rangle_c$$

$$= (-i\hbar)^{n-1} \left. \frac{\delta^n W[J]}{\delta J(x_1) \cdots \delta J(x_n)} \right|_{J=0}. \quad (2.144)$$

By analogy with Eq. (2.113), the infinite series (2.143) is depicted as

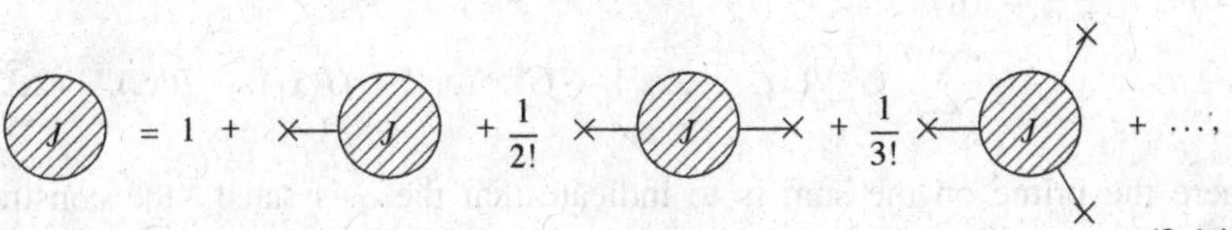

$$= 1 + \times\!\!-\!\!\bigcirc + \frac{1}{2!} \times\!\!-\!\!\bigcirc\!\!-\!\!\times + \frac{1}{3!} \times\!\!-\!\!\bigcirc\!\!\!\!\begin{array}{c}\times\\\\\times\end{array} + \cdots; \quad (2.145)$$

while the connected Green functions are represented diagrammatically in analogy with (2.112) as

$$G_c^{(n)}(1, \ldots, n) = \qquad . \qquad (2.146)$$

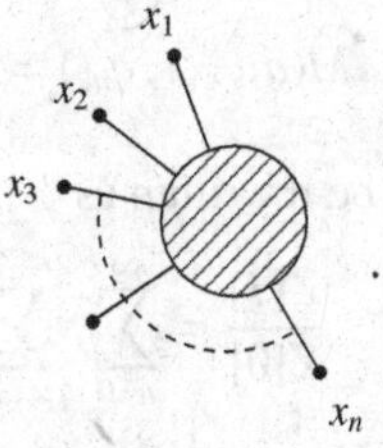

To obtain the relation between the generators $Z[J]$ and $iW[J]/\hbar$ of disconnected and connected Green functions, pull out a leg from $Z[J]$ by taking the
functional derivative with respect to the source $J(x)$:

$$-i\hbar\frac{\delta}{\delta J(x)} \left(\begin{array}{c} J \end{array}\right) = \underset{x}{\bullet}\!-\!\!\left(\begin{array}{c} J \end{array}\right) = \underset{x}{\bullet}\!-\!\!\left(\begin{array}{c} J \end{array}\right)\left(\begin{array}{c} J \end{array}\right). \tag{2.147}$$

Each term in the infinite series represented by the second diagram splits into two
parts: the part that is connected to the external leg and the remainder, which in
each term consists of the sum of all possible diagrams, which is precisely $Z[J]$.
In algebraic form, the diagrammatic relation (2.147) translates into the differential
equation

$$-i\hbar\frac{\delta Z[J]}{\delta J(x)} = \frac{\delta W[J]}{\delta J(x)}Z[J], \tag{2.148}$$

which has the solution

$$Z[J] = Z[0]\,e^{(i/\hbar)W[J]}. \tag{2.149}$$

To establish this relation more explicitly, consider the n-point Green function
$G^{(n)}(x_1,\ldots,x_n)$. It splits into, say, q_1 connected n_1-point Green functions $G^{(n_1)}$,
q_2 connected n_2-point Green functions $G^{(n_2)}, \ldots, q_p$ connected n_p-point Green
functions $G^{(n_p)}$ with

$$\sum_{\nu=1}^{p} q_\nu n_\nu = n. \tag{2.150}$$

With this splitting, the generating functional (2.110) takes the form

$$\frac{Z[J]}{Z[0]} = \sum_{n=0}^{\infty}\frac{1}{n!}\left(\frac{i}{\hbar}\right)^n \int d^d x_1 \cdots d^d x_n$$
$$\times \sum_{q_1,\ldots,q_p}{}' G^{(n_1)}(x_1,\ldots,x_{n_1})\cdots G^{(n_p)}(\ldots,x_n)J(x_1)\cdots J(x_n), \tag{2.151}$$

where the prime on the sum is to indicate that the q_ν's satisfy the constraint
(2.150). Since the connected Green functions are symmetric with respect to per-
mutations of their arguments, the number $M(q_1,\ldots,q_{n_\nu})$ of independent terms in
the summand for given q_ν's satisfying the constraint (2.150) is

$$M(q_1,\ldots,q_{n_\nu}) = \frac{n!}{\prod_{\nu=1}^{p}(n_\nu!)^{q_\nu}q_\nu!} \tag{2.152}$$

so that Eq. (2.151) can be rewritten as

$$\frac{Z[J]}{Z[0]} = \sum_{n=0}^{\infty}\sum_{q_1,\ldots,q_p}{}'\prod_{\nu=1}^{p}\frac{(A_{n_\nu})^{q_\nu}}{q_\nu!}, \tag{2.153}$$

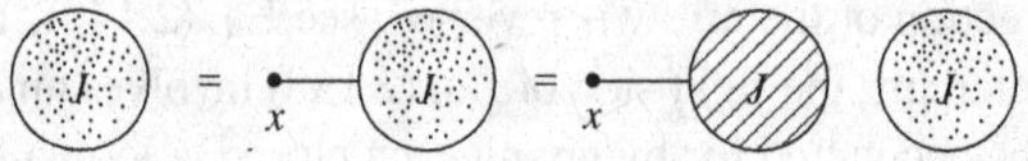

with

$$A_{n_\nu} = \frac{1}{n_\nu!}\left(\frac{\mathrm{i}}{\hbar}\right)^{n_\nu} \int \mathrm{d}^d x_1 \cdots \mathrm{d}^d x_{n_\nu} G^{(n_\nu)}(x_1, \ldots, x_{n_\nu}) J(x_1) \cdots J(x_{n_\nu}). \quad (2.154)$$

The sum over n lifts the constraint (2.150) so that the q_ν's become independent, and

$$\frac{Z[J]}{Z[0]} = \prod_{\nu=1}^{\infty} \sum_{q_\nu=0}^{\infty} \frac{(A_{n_\nu})^{q_\nu}}{q_\nu!} = \prod_{\nu=1}^{\infty} \mathrm{e}^{A_{n_\nu}}. \quad (2.155)$$

The remaining infinite product of exponentials can be written as an exponential of an infinite sum

$$\frac{Z[J]}{Z[0]} = \exp\left[\sum_{n=1}^{\infty} \frac{1}{n!}\left(\frac{\mathrm{i}}{\hbar}\right)^{n} \int \mathrm{d}^d x_1 \cdots \mathrm{d}^d x_n G_{\mathrm{c}}^{(n)}(x_1, \ldots, x_n) J(x_1) \cdots J(x_n)\right], \quad (2.156)$$

which is the expected form (2.149), with $W[J]$ given in Eq. (2.143).

2.7 Effective Action

An arbitrary connected diagram usually falls apart into two disconnected pieces when a single internal line is cut. The special class of connected diagrams which cannot be separated into two pieces by cutting a single internal line is called *one-particle irreducible* (1PI). The generating functional $\Gamma[\phi_{\mathrm{c}}]$ of 1PI or *proper* Green functions $\Gamma^{(n)}(x_1, \ldots, x_n)$ is given by the Legendre transform of $W[J]$:

$$\Gamma[\phi_{\mathrm{c}}] = W[J] - \int \mathrm{d}^d x\, J(x) \phi_{\mathrm{c}}(x), \quad (2.157)$$

where $\phi_{\mathrm{c}}(x)$ is the average of the fluctuating field $\phi(x)$ in the presence of an external source $J(x)$,

$$\phi_{\mathrm{c}}(x) = \frac{\delta W[J]}{\delta J(x)} = $$ 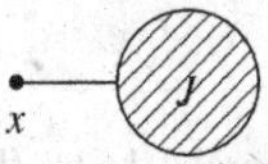 . (2.158)

Specifically,

$$\Gamma(\phi_{\mathrm{c}}) = \sum_{n=1}^{\infty} \frac{1}{n!} \int \mathrm{d}^d x_1 \cdots \mathrm{d}^d x_n \Gamma^{(n)}(x_1, \ldots, x_n)\, \phi_{\mathrm{c}}(x_1) \cdots \phi_{\mathrm{c}}(x_n), \quad (2.159)$$

and diagrammatically

$$\Gamma^{(n)}(x_1, \ldots, x_n) = $$ 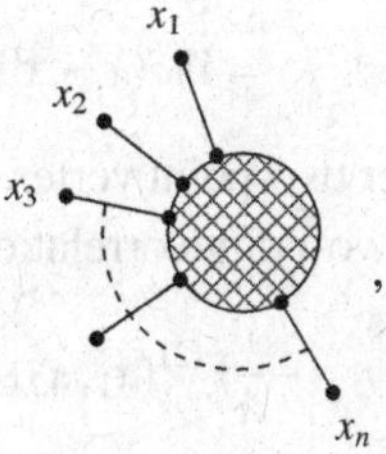 , (2.160)

where an external leg connects always to a vertex, and, by definition, never to a propagator. This has been indicated by the dots on the edge of the 1PI diagram. The generating functional $\Gamma[\phi_c]$, which is a functional of the *classical field* $\phi_c(x)$, is called the *effective action*. This is because it satisfies the classical field equation

$$\begin{aligned}\frac{\delta\Gamma[\phi_c]}{\delta\phi_c(x)} =& \frac{\delta W[J]}{\delta\phi_c(x)} - J(x) - \int d^d x' \, \frac{\delta J(x')}{\delta\phi_c(x)}\phi_c(x') \\ =& - J(x),\end{aligned} \tag{2.161}$$

and, therefore, plays a similar role as a classical action. In deriving Eq. (2.161), it is used that

$$\begin{aligned}\frac{\delta W[J]}{\delta\phi_c(x)} &= \int d^d x' \, \frac{\delta W[J]}{\delta J(x')}\frac{\delta J(x')}{\delta\phi_c(x)} \\ &= \int d^d x' \, \phi_c(x')\,\frac{\delta J(x')}{\delta\phi_c(x)}.\end{aligned} \tag{2.162}$$

The functional derivative of the classical field $\phi_c(x)$ with respect to the source yields by Eq. (2.158) the connected (two-point) Green function

$$-i\hbar\frac{\delta\phi_c(x)}{\delta J(x')} = -i\hbar\frac{\delta^2 W[J]}{\delta J(x')\delta J(x)} = G_c(x, x'). \tag{2.163}$$

This equation can be alternatively stated as

$$-i\hbar\frac{\delta}{\delta J(x)} = \int d^d x' G_c(x, x')\frac{\delta}{\delta\phi_c(x')}, \tag{2.164}$$

or as diagram as

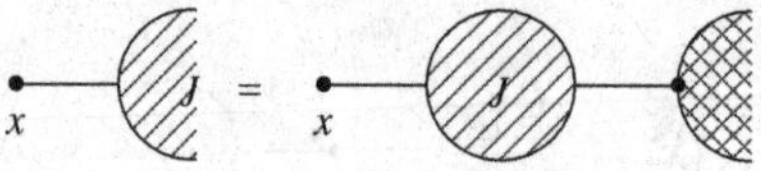

. (2.165)

Applied to the field equation (2.161), this rule gives

$$\begin{aligned}-i\hbar\frac{\delta^2\Gamma[\phi_c]}{\delta J(x')\delta\phi_c(x)} &= \int d^d x'' G_c(x', x'')\frac{\delta^2\Gamma[\phi_c]}{\delta\phi_c(x'')\delta\phi_c(x)} \\ &= \int d^d x'' G_c(x', x'')\Gamma^{(2)}(x'', x) \\ &= i\hbar\delta(x - x'),\end{aligned} \tag{2.166}$$

showing that the proper two-point vertex $\Gamma^{(2)}$ is, apart from a prefactor, the inverse of the full connected two-point correlation function,

$$-\frac{i}{\hbar}\Gamma^{(2)}(x_1, x_2) = G_c^{-1}(x_1, x_2). \tag{2.167}$$

It can be alternatively stated as

$$G_c(x,y) = -\,\mathrm{i}\hbar\frac{\delta^2 W[J]}{\delta J(x)\delta J(y)} = -\int \mathrm{d}^d x' \int \mathrm{d}^d y' G_c(x,x')\frac{\mathrm{i}}{\hbar}\Gamma^{(2)}(x',y')G_c(y',y), \tag{2.168}$$

or diagrammatically as

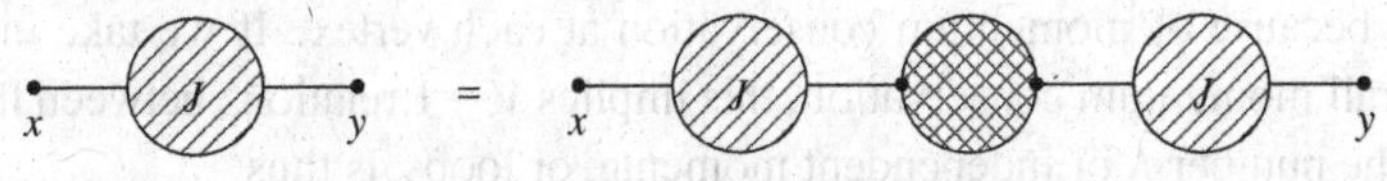

(2.169)

By taking successive functional derivatives $\delta/\delta J$ of the last equation in Eq. (2.168), one obtains by Eq. (2.164) the decomposition of the connected Green functions in terms of 1PI Green functions. For example,

$$G_c^{(3)}(x,y,z) = (-\mathrm{i}\hbar)^2\frac{\delta^3 W[J]}{\delta J(x)\delta J(y)\delta J(z)} = \int \mathrm{d}^d x' \mathrm{d}^d y' \mathrm{d}^d z' G_c(x,x')G_c(y,y')G_c(z,z')\frac{\mathrm{i}}{\hbar}\Gamma^{(3)}(x',y',z'), \tag{2.170}$$

or

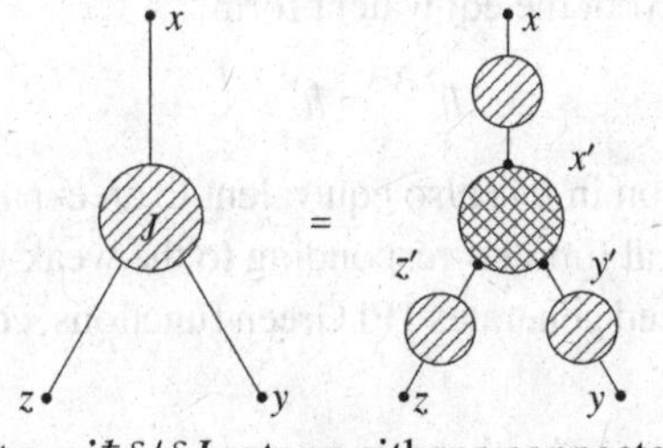

(2.171)

The differential operator $-\mathrm{i}\hbar\delta/\delta J$ acts on either a connected two-point Green function, or on some proper vertex $\Gamma^{(n)}$. In the former case, it produces by Eq. (2.170) a three-point vertex with a new external line added, i.e., diagrammatically,

$$-\mathrm{i}\hbar\frac{\delta}{\delta J(x)}$$ 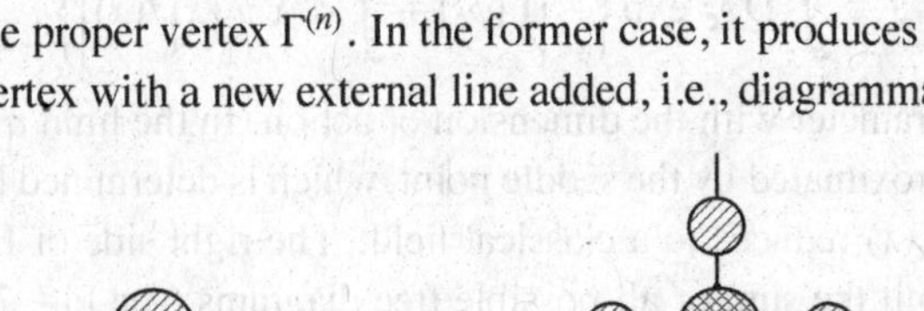 , (2.172)

while in the latter case, it adds by Eq. (2.164) an extra leg to the vertex,

$$-\mathrm{i}\hbar\frac{\delta}{\delta J(x)}$$ 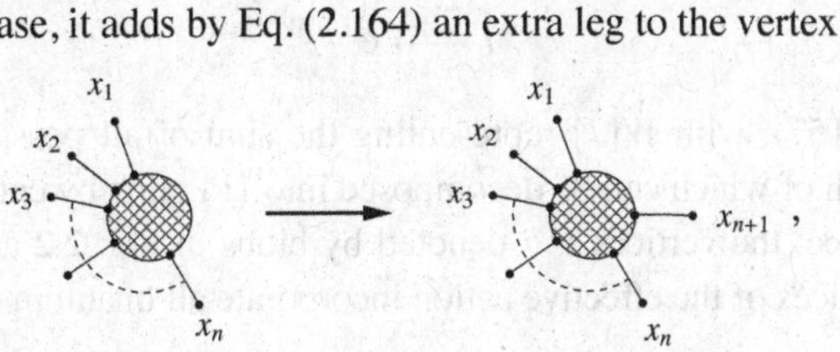

, (2.173)

As stated in the introduction of Sec. 2.5, quantum effects are perturbatively included by expanding in powers of $\hbar$. Diagrammatically, this corresponds to an expansion in *loops*. To see this, consider a proper vertex diagram with V vertices and I internal lines. According to Eq. (2.130), each propagator contributes a factor of $\hbar$ to the diagram, while each vertex comes with a factor of $\hbar^{-1}$, see, for example, Eq. (2.133). The diagram therefore carries a factor of $\hbar^{I-V}$. Since momentum flows through each line, there are I internal momenta. They are, however, not all independent because of momentum conservation at each vertex. If we take into account overall momentum conservation, this implies $V - 1$ relations between the momenta. The number L of independent momenta, or loops, is thus

$$L = I - V + 1, \tag{2.174}$$

and

$$\hbar^{I-V} = \hbar^{L-1}, \tag{2.175}$$

showing that an expansion in powers of $\hbar$ is tantamount to an expansion in loops. Tree diagrams are diagrams without loops ($L = 0$). They are, consequently, the only ones surviving the classical limit ($\hbar \to 0$). Using that $I = 2V$, we can cast the left side of Eq. (2.175) in the equivalent form

$$\hbar^{I-V} = \hbar^{V}, \tag{2.176}$$

showing that the expansion in $\hbar$ is also equivalent to an expansion in the coupling constant, with the classical limit corresponding to the weak-coupling limit.

To see that $\Gamma[\phi_\mathrm{c}]$ indeed generates 1PI Green functions, consider the functional integral

$$I \equiv \int \mathrm{D}\phi_\mathrm{c} \exp\left\{\frac{\mathrm{i}}{a}\left[\Gamma[\phi_\mathrm{c}] + \int \mathrm{d}^d x\, \phi_\mathrm{c}(x) J(x)\right]\right\}, \tag{2.177}$$

where a is a parameter with the dimension of action. In the limit $a \to 0$, the integral can be approximated by the saddle point, which is determined by Eq. (2.161), and the field $\phi_\mathrm{c}(x)$ reduces to a classical field. The right side of Eq. (2.177) denotes in this limit the sum of all possible tree diagrams (see Fig. 2.2) defined by the action $\Gamma[\phi_\mathrm{c}]$ with vertices $\Gamma^{(n)}$. On the other hand,

$$I \xrightarrow[a \to 0]{} \mathrm{e}^{(\mathrm{i}/a)W[J]}, \tag{2.178}$$

by Eq. (2.157), with $W[J]$ representing the sum of all possible connected diagrams, each of which can be decomposed into 1PI parts by cutting single internal lines. Hence, the vertices $\Gamma^{(n)}$ denoted by blobs in Fig. 2.2 are indeed 1PI. The proper vertices of the effective action incorporate all quantum effects.

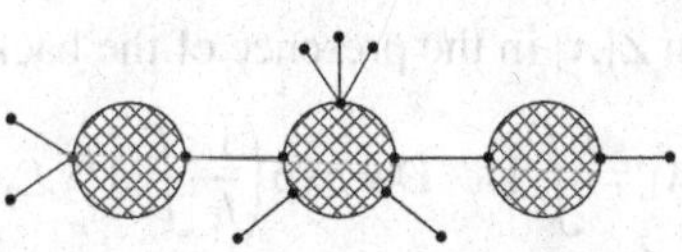

Fig. 2.2 An example of a tree diagram made up of three 1PI parts.

2.8 Linear Response Theory

In this section, we study the response of a system to some externally applied field. In linear response theory, the changes in observables are calculated to first order in the external field.

Consider the theory (1.44) and minimally couple it to an external gauge field by letting

$$\partial^\mu \to \partial^\mu + \mathrm{i}(q/\hbar c)A^\mu. \tag{2.179}$$

Here, we introduced the four-vector

$$A^\mu \equiv (c\Phi, \mathbf{A}), \tag{2.180}$$

with Φ the scalar and $\mathbf{A}$ the vector potential from which the magnetic induction $\mathbf{B}$ and the electric field $\mathbf{E}$ follow as

$$\mathbf{B} = \nabla \times \mathbf{A}, \quad \mathbf{E} = -\frac{1}{c}\frac{\partial}{\partial t}\mathbf{A} - \nabla\Phi. \tag{2.181}$$

The extra factor of c in the time component of A^μ is included so that with our convention $x^\mu = (t, \mathbf{x})$ the minimal coupling takes a simple form. After coupling to the external gauge field, the Lagrangian density becomes

$$\mathcal{L}_\mathrm{e} = \psi^*\left(\mathrm{i}\hbar\frac{\partial}{\partial t} - \frac{q}{c}A^0\right)\psi - \frac{1}{2m}\left(\mathrm{i}\hbar\nabla - \frac{q}{c}\mathbf{A}\right)\psi^* \cdot \left(-\mathrm{i}\hbar\nabla - \frac{q}{c}\mathbf{A}\right)\psi - \mathcal{V}. \tag{2.182}$$

It describes particles of mass m and electric charge q. For electrons, the charge is negative $q = -e < 0$. The electric current density j_e^μ can be obtained by taking a functional derivative of the action,

$$\frac{q}{c}j_\mathrm{e}^\mu(x) = -\frac{\delta S_\mathrm{e}}{\delta A_\mu(x)}. \tag{2.183}$$

Explicitly, cf. Eq. (1.58),

$$j_\mathrm{e}^0 = \psi^*\psi, \quad \mathbf{j}_\mathrm{e} = -\mathrm{i}\frac{\hbar}{2m}\psi^*\overleftrightarrow{\nabla}\psi - \frac{q}{mc}\mathbf{A}\psi^*\psi, \tag{2.184}$$

where the last term in the current, known as the *diamagnetic* contribution, is required for gauge invariance.

The partition function $Z[A]$ in the presence of the background field,

$$Z[A] = \int \mathrm{D}\psi^* \mathrm{D}\psi \, \exp\left(\frac{\mathrm{i}}{\hbar}\int \mathrm{d}^d x \mathcal{L}_\mathrm{e}\right), \tag{2.185}$$

can be expanded around the partition function $Z[0]$ in the absence of the external field in powers of A^μ as

$$Z[A] = Z[0] \exp\left[\frac{\mathrm{i}}{2\hbar}\frac{q^2}{c^2}\int \mathrm{d}^d x \, \mathrm{d}^d x' \, A^\mu(x)\, \Pi_{\mu\nu}(x,x')\, A^\nu(x') + \cdots\right], \tag{2.186}$$

where higher orders in the gauge field are omitted. The leading term in the expansion, with $\Pi_{\mu\nu}(x,x')$ the so-called *polarization tensor*, comprises linear response theory. The definition adopted here is the standard convention used in particle physics, it differs a minus sign from what is often used in condensed matter theory.

Since the theory is invariant under the gauge transformations

$$A^\mu(x) \to A^\mu(x) + \partial^\mu \alpha(x), \quad \psi(x) \to \mathrm{e}^{-\mathrm{i}(q/\hbar c)\alpha(x)}\psi(x), \tag{2.187}$$

the polarization tensor is transverse

$$\partial^\mu \Pi_{\mu\nu}(x,x') = 0 \tag{2.188}$$

as can be checked explicitly by applying a gauge transformation to the gauge fields in the partition function (2.186) and integrating by parts. For a translational invariant theory $\Pi_{\mu\nu}(x,x') = \Pi_{\mu\nu}(x-x')$, and

$$\int \mathrm{d}^d x \, \mathrm{d}^d x' \, A^\mu(x)\, \Pi_{\mu\nu}(x,x')\, A^\nu(x') = \int \frac{\mathrm{d}^d k}{(2\pi)^d} A^\mu(-k)\, \Pi_{\mu\nu}(k)\, A^\nu(k). \tag{2.189}$$

The induced current density $\langle j_\mathrm{e}^\mu(x)\rangle$, where the ensemble average is with respect to the partition function (2.185), is obtained by taking the functional derivative of the partition function,

$$\frac{q}{c}\langle j_\mathrm{e}^\mu(x)\rangle = \mathrm{i}\hbar \frac{1}{Z[A]}\frac{\delta Z[A]}{\delta A_\mu(x)} = \mathrm{i}\hbar \frac{\delta \ln Z[A]}{\delta A_\mu(x)}, \tag{2.190}$$

in the same way as in Eq. (2.183) the classical current is obtained by taking the functional derivative of the classical action. In terms of the polarization tensor, this relation assumes the form

$$\langle j_\mathrm{e}^\mu(x)\rangle = -\frac{q}{c}\int \mathrm{d}^d x' \, \Pi^{\mu\nu}(x,x') A_\nu(x'), \tag{2.191}$$

or, after Fourier transforming,

$$\langle j_\mathrm{e}^\mu(k)\rangle = -\frac{q}{c}\Pi^{\mu\nu}(k) A_\nu(k). \tag{2.192}$$

Similarly,

$$\frac{q^2}{c^2}\Pi^{\mu\nu}(x,x') = -\mathrm{i}\hbar\frac{\delta^2 \ln Z[A]}{\delta A_\mu(x)\delta A_\nu(x')}, \tag{2.193}$$

or

$$\frac{q^2}{c^2}\Pi^{\mu\nu}(x,x') = -\mathrm{i}\hbar\frac{\delta}{\delta A_\mu(x)}\left(\frac{1}{Z[A]}\frac{\delta Z[A]}{\delta A_\nu(x')}\right)$$
$$= -\frac{\mathrm{i}q^2}{\hbar c^2}\langle j_\mathrm{e}^\mu(x)\rangle\langle j_\mathrm{e}^\nu(x')\rangle - \mathrm{i}\hbar\frac{1}{Z[A]}\frac{\delta^2 Z[A]}{\delta A_\mu(x)\delta A_\nu(x')}.$$

For the partition function (2.185), this leads to

$$\Pi^{\mu\nu}(x,x') = \frac{\mathrm{i}}{\hbar}\Lambda^{\mu\nu}(x,x') - \frac{c}{q}\left\langle\frac{\delta j_\mathrm{e}^\mu(x)}{\delta A_\nu(x')}\right\rangle, \tag{2.194}$$

with

$$\Lambda^{\mu\nu}(x,x') \equiv \langle j_\mathrm{e}^\mu(x) j_\mathrm{e}^\nu(x')\rangle - \langle j_\mathrm{e}^\mu(x)\rangle\langle j_\mathrm{e}^\nu(x')\rangle \tag{2.195}$$

the connected *current-current* correlation function. The last term in Eq. (2.194), with $\langle\delta j_\mathrm{e}^\mu(x)/\delta A_\nu(x')\rangle = \langle\delta j_\mathrm{e}^\nu(x')/\delta A_\mu(x)\rangle$, survives when the current has a diamagnetic contribution as in Eq. (2.184). Equation (2.194) shows that apart from this term, the polarization tensor is given by the current-current correlation function.

The polarization tensor satisfies various sum rules. The first one is a direct consequence of Eq. (2.188), which implies

$$\frac{\mathrm{i}}{\hbar}\partial_\mu\Lambda^{\mu i}(x,x') = -\frac{1}{m}\partial^i\left[n(x)\delta(x-x')\right],$$
$$\partial_\mu\Lambda^{\mu 0}(x,x') = \partial_\mu\Lambda^{0\mu}(x,x') = 0. \tag{2.196}$$

If the average particle number density is constant, $n(x) = \bar{n}$, these relations can be combined into

$$\frac{\mathrm{i}}{\hbar}\omega^2\Lambda^{00}(k) - \frac{\mathrm{i}}{\hbar}k_i k_j\Lambda^{ij}(k) = -\frac{\bar{n}}{m}\mathbf{k}^2, \tag{2.197}$$

after Fourier transforming. Assuming that both $\Lambda^{00}(k)$ and $\Lambda^{ij}(k)$ behave asymptotically as $1/\omega^2$, we obtain what is called the *f sum rule*,

$$\lim_{\omega\to\infty}\Pi^{00}(k) = -\frac{\bar{n}}{m}\frac{\mathbf{k}^2}{\omega^2}. \tag{2.198}$$

A second sum rule follows from the observation that because the theory is linear in the time derivatives, the zeroth component of the external gauge field A^μ couples to the theory in the same way as the chemical potential μ. From this, the following relation derives from Eq. (2.193):

$$\lim_{\mathbf{k}\to 0}\Pi^{00}(0,\mathbf{k}) = \int \mathrm{d}^d x\,\Pi^{00}(t,\mathbf{x}) = -\frac{1}{V}\frac{\partial^2\Omega}{\partial\mu^2}, \tag{2.199}$$

where

$$\Omega \equiv i\hbar \mathcal{T}^{-1} \ln Z, \tag{2.200}$$

with $\mathcal{T}$ the extent of the time dimension, is the zero-temperature Minkowski counterpart of Eq. (2.8). As will be demonstrated explicitly in Sec. 6.5 below, the order in which the limits are taken is usually important. The notation in Eq. (2.199) is short for the order

$$\lim_{\mathbf{k}\to 0} \Pi^{00}(0, \mathbf{k}) = \lim_{\mathbf{k}\to 0} \lim_{\omega\to 0} \Pi^{00}(\omega, \mathbf{k}), \tag{2.201}$$

describing the case of a static potential in the limit of long wave lengths. The reversed order $\lim_{\omega\to 0} \lim_{\mathbf{k}\to 0} \Pi^{00}(\omega, \mathbf{k})$ describes the case of a uniform potential in the limit of zero frequency. The right side of Eq. (2.199), which is known in statistical physics as the *compressibility sum rule*, yields the (isothermal) compressibility κ_T,

$$-\frac{1}{V}\frac{\partial^2 \Omega}{\partial \mu^2} = \left.\frac{\partial \bar{n}}{\partial \mu}\right|_T = \bar{n}^2 \kappa_T. \tag{2.202}$$

The compressibility is in turn related to the speed of sound c_s via

$$\kappa_T = \frac{1}{m\bar{n}c_s^2} \tag{2.203}$$

so that

$$\lim_{\mathbf{k}\to 0} \Pi^{00}(0, \mathbf{k}) = \bar{n}^2 \kappa_T = \frac{\bar{n}}{mc_s^2}. \tag{2.204}$$

The speed of sound has been given a subscript here to distinguish it from the speed of light. Although the derivation is given for zero temperature, the sum rule also applies at finite temperature. In that case, the temperature is to be taken fixed as indicated by the subscript T in Eq. (2.202). For example, for an ideal classical gas, whose grand canonical thermodynamic potential is given in Eq. (2.31), the expressions (2.202) and (2.203) lead to the standard result for the speed of sound

$$c_s^2 = k_B T/m. \tag{2.205}$$

The sum rules (2.198) and (2.204) are often represented in terms of the so-called *dynamic structure factor* $S(k)$, which is related to $\Pi^{00}(k)$ through

$$\Pi^{00}(k) = -\frac{1}{\hbar}\int_0^\infty d\omega' \left(\frac{1}{\omega - \omega' + i\eta} - \frac{1}{\omega + \omega' + i\eta}\right) S(\omega', \mathbf{k}). \tag{2.206}$$

At the absolute zero of temperature, the dynamic structure factor is real and positive, and vanishes for negative frequencies $\omega < 0$. Note that in contrast to

Eq. (2.94), the frequency ω is replaced with $\omega + i\eta$ in both terms in the decomposition (2.206), as is appropriate for response theory. By the identity

$$\frac{1}{\omega - \omega' + i\eta} = \frac{\mathrm{P}}{\omega - \omega'} - i\pi\delta(\omega - \omega'), \tag{2.207}$$

where P stands for principal part, the imaginary part of Eq. (2.206) can be written as

$$\mathrm{Im}\,\Pi^{00}(k) = \frac{\pi}{\hbar}[S(\omega, \mathbf{k}) - S(-\omega, \mathbf{k})]. \tag{2.208}$$

In terms of $S(k)$, the sum rules read,

$$\int_0^\infty d\omega \hbar\omega S(k) = \frac{\bar{n}\hbar^2\mathbf{k}^2}{2m}, \qquad \lim_{\mathbf{k}\to 0} \int_0^\infty d\omega \frac{S(k)}{\hbar\omega} = \frac{\bar{n}}{2mc_s^2}, \tag{2.209}$$

respectively, showing that they are exhausted by the excitation spectrum $\omega(\mathbf{k}) = c_s|\mathbf{k}|$,

$$\begin{aligned} \lim_{\mathbf{k}\to 0} S(k) &= \frac{\bar{n}\hbar|\mathbf{k}|}{2mc_s}\delta[\omega - \omega(\mathbf{k})] \\ &= \frac{\bar{n}\hbar\mathbf{k}^2}{2m\omega(\mathbf{k})}\delta[\omega - \omega(\mathbf{k})]. \end{aligned} \tag{2.210}$$

This result is valid for any compressible system boasting a gapless sound mode.

Equation (2.210) is usually expressed in the equivalent form

$$E(\mathbf{k}) = \frac{\hbar^2\mathbf{k}^2}{2mS(\mathbf{k})} \tag{2.211}$$

due to Feynman, where $E(\mathbf{k}) = \hbar\omega(\mathbf{k}) = \hbar c_s|\mathbf{k}|$ is the excitation energy, and $S(\mathbf{k})$ the *static structure factor* defined through

$$\begin{aligned} S(\mathbf{k}) &\equiv \frac{1}{\bar{n}}\int_0^\infty d\omega S(k) \\ &= \frac{\hbar|\mathbf{k}|}{2mc_s}, \end{aligned} \tag{2.212}$$

which is dimensionless. The integral denotes, apart from a prefactor, the Fourier transform of $S(t, \mathbf{k})$ taken at $t = 0$. By Eqs. (2.208) and (2.194), the static structure factor is related to the instantaneous density-density correlation function through

$$\begin{aligned} \bar{n}S(\mathbf{k}) &= \hbar\int_0^\infty \frac{d\omega}{\pi}\mathrm{Im}\,\Pi^{00}(k) \\ &= \langle \delta n(\mathbf{k})\delta n(-\mathbf{k})\rangle, \end{aligned} \tag{2.213}$$

where $\delta n(\mathbf{k}) \equiv j_e^0(\mathbf{k}) - \langle j_e^0(\mathbf{k})\rangle$. That is, the static structure factor gives a measure of the average square density fluctuations.

2.9 Ward Identities

At the classical level, Noether's theorem asserts that symmetries of the action imply the conservation of currents. Since in deriving these conservation laws use is made of the field equations, the question arises what happens at the quantum level, where the functional integral includes all possible field configurations which do not in general satisfy the field equations.

To address this question, consider the theory (1.44) which, by invariance under global U(1) phase transformations (1.54), possesses the conserved current (1.56). A quick way to arrive at this current is to let the transformation parameter α be a function of x. Under such a spacetime-dependent phase transformation, the action namely changes as

$$\delta S = -\hbar \int \mathrm{d}^d x\, j^\mu(x) \partial_\mu \alpha(x). \tag{2.214}$$

For the following, it suffices to restrict ourselves to infinitesimal transformations

$$\psi(x) \to \psi'(x) = [1 + \mathrm{i}\alpha(x)]\,\psi(x), \quad \psi^*(x) \to \psi'^*(x) = [1 - \mathrm{i}\alpha(x)]\,\psi^*(x), \tag{2.215}$$

and study the behavior of the partition function,

$$Z[J] = \int \mathrm{D}\psi^* \mathrm{D}\psi \, \exp\left\{\frac{\mathrm{i}}{\hbar} \int \mathrm{d}^d x \,[\mathcal{L}(\psi, \psi^*) + J^*(x)\psi(x) + J(x)\psi^*(x)]\right\}, \tag{2.216}$$

under these transformations. Since the measure is invariant under these transformations, i.e., $\mathrm{D}\psi'^*\mathrm{D}\psi' = \mathrm{D}\psi^*\mathrm{D}\psi$, the partition function changes as

$$\begin{aligned} Z[J] \to \int \mathrm{D}\psi^* \mathrm{D}\psi \, \exp\Big(\frac{\mathrm{i}}{\hbar} \int \mathrm{d}^d x \Big\{\mathcal{L}(\psi, \psi^*) - \hbar j^\mu(x)\partial_\mu \alpha(x) \\ + J^*(x)\,[1 + \mathrm{i}\alpha(x)]\,\psi(x) + J(x)\,[1 - \mathrm{i}\alpha(x)]\,\psi^*(x)\Big\}\Big). \end{aligned}$$

By invariance, the right side equals $Z[J]$. Expanding to first order in $\alpha(x)$ and using that this parameter is arbitrary, we arrive at the equation

$$\begin{aligned} \int \mathrm{D}\psi^* \mathrm{D}\psi \, \exp\left\{\frac{\mathrm{i}}{\hbar} \int \mathrm{d}^d x [\mathcal{L}(\psi, \psi^*) + J^*(x)\psi(x) + J(x)\psi^*(x)]\right\} \\ \times [\hbar \partial_\mu j^\mu(y) + \mathrm{i}J^*(y)\psi(y) - \mathrm{i}J(y)\psi^*(y)] = 0, \end{aligned} \tag{2.217}$$

which, by taking functional derivatives with respect to the sources, generates relations between Green functions

$$\begin{aligned} \left\langle \frac{\partial}{\partial y^\mu} j^\mu(y)\psi(x_1)\cdots\psi(x_m)\psi^*(x_1')\cdots\psi^*(x_n')\right\rangle = \\ -\left[\sum_{i=1}^m \delta(y - x_i) - \sum_{i=1}^n \delta(y - x_i')\right] \langle \psi(x_1)\cdots\psi(x_m)\psi^*(x_1')\cdots\psi^*(x_n')\rangle. \end{aligned} \tag{2.218}$$

These so-called *Ward identities* reflect the symmetries of the classical theory at the quantum level.

As an application, consider the connected three-point Green function

$$G_{\rm c}^{\mu}(y,x,x') \equiv \langle j^{\mu}(y)\psi(x)\psi^{*}(x')\rangle. \tag{2.219}$$

After introducing the source term $\mathcal{L}_{\rm I} = -A^{\mu} j_{\mu}$, this correlation function can be obtained from the generating functional $W[J, J^{*}, A^{\mu}]$ for connected Green functions by taking derivatives with respect to the sources

$$\begin{aligned} G_{\rm c}^{\mu}(y,x,x') &= -(-{\rm i}\hbar)^{2}\frac{\delta^{3}W[J]}{\delta A_{\mu}(y)\delta J^{*}(x)\delta J(x')} \\ &= \int {\rm d}^{d}\bar{x}\int {\rm d}^{d}\bar{x}' G_{\rm c}(x,\bar{x})\Gamma^{\mu}(y,\bar{x},\bar{x}')G_{\rm c}(\bar{x}',x'). \end{aligned} \tag{2.220}$$

The last step, with

$$\Gamma^{\mu}(y,\bar{x},\bar{x}') \equiv \frac{\delta^{3}\Gamma[\psi_{\rm c},\psi_{\rm c}^{*},A^{\mu}]}{\delta A_{\mu}(y)\delta\psi_{\rm c}^{*}(\bar{x})\delta\psi_{\rm c}(\bar{x}')} \tag{2.221}$$

the proper three-point vertex, follows from the complex generalization of Eq. (2.168). The correlation function satisfies the Ward identity

$$\frac{\partial}{\partial y^{\mu}}G_{\rm c}^{\mu}(y,x,x') = -\left[\delta(y-x)-\delta(y-x')\right]G_{\rm c}(x,x'). \tag{2.222}$$

With the Fourier transform

$$(2\pi)^{4}\delta(q+k+k')G_{\rm c}^{\mu}(q,k,k') = \int {\rm d}^{d}y\,{\rm d}^{d}x\,{\rm d}^{d}x'\,{\rm e}^{{\rm i}q\cdot y+{\rm i}k\cdot x+{\rm i}k'\cdot x'}G_{\rm c}^{\mu}(y,x,x'), \tag{2.223}$$

where the delta function on the left imposes overall momentum conservation, the Ward identity becomes in momentum space

$${\rm i}q_{\mu}G_{\rm c}^{\mu}(q,k,-k-q) = G_{\rm c}(k) - G_{\rm c}(k+q). \tag{2.224}$$

In terms of the proper three-point vertex Γ^{μ} introduced in Eq. (2.220), whose Fourier transform reads

$$G_{\rm c}^{\mu}(q,k,-k-q) = G_{\rm c}(k)\Gamma^{\mu}(q,k,-k-q)G_{\rm c}(k+q), \tag{2.225}$$

it takes the form

$${\rm i}q_{\mu}\Gamma^{\mu}(q,k,-k-q) = G_{\rm c}^{-1}(k+q) - G_{\rm c}^{-1}(k). \tag{2.226}$$

In the limit $q \to 0$, this reduces to the differential form

$${\rm i}\Gamma^{\mu}(0,k,-k) = \frac{\partial G_{\rm c}^{-1}(k)}{\partial k_{\mu}} \tag{2.227}$$

we will use in Sec. 4.9 below.

Notes

(i) Quantum statistical mechanics is a topic discussed in most textbooks on statistical mechanics, see, for example, [Kubo (1965)] or [Pathria (1996)].

(ii) Introductions to finite-temperature field theory can be found in [Abrikosov *et al.* (1963)] and [Fetter and Walecka (1971)], where canonical quantization is used, and, more in line with our functional integral approach, in [Popov (1983, 1987)] and in [Kapusta (1989)]. Classic references to relativistic finite-temperature field theory using the functional integral approach are [Bernard (1974)] and [Dolan and Jackiw (1974)]. For a review of the various finite-temperature formalisms in quantum field theory, see [Landsman and van Weert (1987)].

(iii) The formal connection between inverse temperature and imaginary time was pointed out by Bloch (1932).

(iv) The functional integral approach to field quantization is covered in most modern textbooks on quantum field theory, see, for example, [Ramond (1981)]. Classic references are [Abers and Lee (1973)] and [Coleman (1975)]. This Erice lecture and others by Coleman are reprinted in [Coleman (1988)]. The approach originates from Feynman's spacetime approach to quantum mechanics [Feynman (1948)].

(v) Feynman diagrams are introduced in any modern textbook on quantum field theory, see, for example, [Ramond (1981)]. Classic references to the diagrammatic approach to quantum field theory are ['t Hooft and Veltman (1974)] and [Cvitanović (1983)].

(vi) Effective actions are covered in the references cited in the previous two items. See also [Weinberg (1996)] for an in-depth discussion.

(vii) Linear response theory is presented in, for example, [Fetter and Walecka (1971)]. Our presentation is based on [Fradkin (1967); Kapusta (1989)], and [Fradkin (1991)].

(viii) Ward introduced the identities now named after him in [Ward (1950)]. They were generalized by Takahashi (1957). The identities are discussed in most modern textbooks on quantum field theory, see, for example, [Ramond (1981)]. The role of (generalized) Ward identities in canonical quantization is elucidated in [Takahashi (1986)].

Chapter 3

Calculation Tools

This chapter introduces various calculation tools which will be put to use in later chapters. The tools are introduced in the context of simple physical systems to demonstrate how they work. This practical chapter also introduces concepts that will repeatedly surface when we proceed, such as the geometric phase and renormalization.

3.1 Derivative Expansion

One of the calculation tools which will be repeatedly used in the following is the *derivative expansion method* to systematically compute (one-loop) quantum corrections as a series expansion in powers of derivatives. With the exception of certain two-dimensional models, the series usually cannot be summed exactly. The derivative expansion method we adopt is closely related to evaluating Feynman diagrams. The advantage of this algebraic procedure over evaluating diagrams is that it is fairly foolproof in that no symmetry factors have to be included by hand, no double counting is possible, and all relevant diagrams are automatically accounted for—matters that usually make *diagrammar* prone to errors. Because the derivative expansion method is straightforward, many computations can be carried out by using a symbolic manipulation program such as REDUCE, SCHOONSCHIP, or FORM.

Consider a Lagrangian density of a general form:

$$\mathcal{L} = \psi^\dagger \left[K(\mathrm{i}\partial) - M(A, J)\right] \psi. \tag{3.1}$$

Here, ψ is a Grassmann field describing fermions interacting bilinearly with external scalar and gauge fields, collectively denoted by A, and sources J (internal symmetry labels are suppressed), and $K(\mathrm{i}\partial)$ is some differential operator. All fields except ψ will be treated classically. The function M depends on the sources and the external fields and may include a constant term. The object is to integrate out the fermionic degrees of freedom so as to obtain an effective action which depends only on the background fields. We start with the generating functional

$$Z[A, J] = \int \mathrm{D}\psi^\dagger \, \mathrm{D}\psi \, \mathrm{e}^{(\mathrm{i}/\hbar)S}, \tag{3.2}$$

with S the classical action $S = \int \mathrm{d}^d x \, \mathcal{L}$. The integral over the Grassmann fields is Gaussian and yields a determinant

$$Z[A, J] = \mathrm{Det}\,[K(p) - M(A, J)], \tag{3.3}$$

which summarizes the one-loop effects. Here, $p^\mu = \mathrm{i}\hbar\partial^\mu = (\hbar\partial_t, -\mathrm{i}\hbar\nabla)$. By the identity (2.55), this can be rewritten as

$$Z[A, J] = \mathrm{e}^{(\mathrm{i}/\hbar)\Gamma_1[A,J]}, \tag{3.4}$$

with $\Gamma_1[A, J]$ the one-loop effective action,

$$\Gamma_1[A, J] = -\mathrm{i}\hbar \,\mathrm{Tr}\ln\,[\,K(p) - M(A, J)], \tag{3.5}$$

or more explicitly, cf. Eq. (2.62)

$$\Gamma_1[A, J] = -\mathrm{i}\hbar \,\mathrm{tr} \int \mathrm{d}^d x \int \frac{\mathrm{d}^d k}{(2\pi)^d} \, \mathrm{e}^{\mathrm{i}k\cdot x} \ln\,[K(p) - M(A, J)]\, \mathrm{e}^{-\mathrm{i}k\cdot x}. \tag{3.6}$$

This formal expression is treated as follows by the derivative expansion method. First, the logarithm is expanded in a Taylor series. Each term in the series contains powers of the derivative p^μ which operates on *every* spacetime-dependent field which appears to the right. All these operators are shifted to the left by repeatedly applying the identity

$$f(x)p^\mu g(x) = (p^\mu - \mathrm{i}\hbar\partial^\mu) f(x) g(x), \tag{3.7}$$

where $f(x)$ and $g(x)$ are arbitrary functions of spacetime, and the derivative $\partial^\mu = (\partial_t, -\nabla)$ is defined as acting *only* on the next function to the right. One then integrates by parts so that all the p^μ's act to the left where only a factor $\exp(\mathrm{i}k \cdot x)$ stands. Ignoring total derivatives and taking into account the minus signs which arise when integrating by parts, one recognizes that all occurrences of p^μ (an operator) are replaced with ($\hbar$ time) k^μ (an integration variable). The exponential function $\exp(-\mathrm{i}k \cdot x)$ can at this stage be moved to the left where it it is multiplied with $\exp(\mathrm{i}k \cdot x)$ to give unity. The momentum integration can now in principle be carried out so that the effective action assumes the form of an integral over a local density, $\Gamma_1 = \int \mathrm{d}^d x \, \mathcal{L}_1$, or more generally

$$\Gamma = \int \mathrm{d}^d x \, \mathcal{L}_{\mathrm{eff}} \tag{3.8}$$

if all quantum effects are included. In the limit $\hbar \to 0$, the effective Lagrangian $\mathcal{L}_{\mathrm{eff}}$ reduces to the classical Lagrangian.

3.2 Free Electron Gas in an External Magnetic Field

To demonstrate the use of the derivative expansion method, we consider a free electron gas at the absolute zero of temperature in the presence of a uniform and static magnetic field **H**. We assume that the induction field $\mathbf{B} = \nabla \times \mathbf{A}(\mathbf{x})$ acting on the individual electrons coincides with the applied magnetic field **H**. In addition to assuming that the electrons have no mutual interactions, we also ignore their spin for the moment. The system is described by the Lagrangian density

$$\mathcal{L} = \psi^* \mathrm{i}\hbar \frac{\partial}{\partial t} \psi - \frac{1}{2m} \psi^* \left(-\mathrm{i}\hbar \nabla + \frac{e}{c} \mathbf{A} \right)^2 \psi + \mu \psi^* \psi, \tag{3.9}$$

with $-e < 0$ the charge of the electrons and μ the chemical potential, which for a free electron gas coincides with the Fermi energy, $\mu = \hbar^2 k_\mathrm{F}^2/2m$. The one-component Grassmann field ψ represents the spinless electrons. When the spin of the electrons is taken into account, as in Sec. 3.9 below, ψ must be replaced with a two-component field ψ_σ with $\sigma = \uparrow, \downarrow$, and the Lagrangian must be augmented by the Pauli term. After integrating out the Grassmann fields, we arrive at the one-loop effective action

$$\begin{aligned} \Gamma_1 &= -\mathrm{i}\hbar \,\mathrm{Tr} \ln \left[p_0 - \frac{1}{2m} \mathbf{p}^2 + \mu - \Lambda(\mathbf{A}) \right] \\ &= \mathrm{i}\hbar \,\mathrm{Tr} \sum_{\ell=1}^{\infty} \frac{1}{\ell} \left[S_\mathrm{F}(p) \Lambda(\mathbf{A}) \right]^\ell \end{aligned} \tag{3.10}$$

up to an irrelevant additive constant corresponding to setting the vector potential to zero in the first line. Here,

$$\Lambda(\mathbf{A}) \equiv \frac{e}{2mc} (\mathbf{p} \cdot \mathbf{A} + \mathbf{A} \cdot \mathbf{p}) + \frac{e^2}{2mc^2} \mathbf{A}^2, \tag{3.11}$$

and S_F is the electron propagator

$$S_\mathrm{F}(k) = \frac{1}{\hbar\omega - \xi(\mathbf{k}) + \mathrm{i}\,\mathrm{sgn}(\omega)\eta}, \tag{3.12}$$

with $\xi(\mathbf{k}) \equiv \epsilon(\mathbf{k}) - \mu$ the kinetic energy measured relative to the Fermi surface $\mu = \hbar^2 k_\mathrm{F}^2/2m$. The propagator is related to the average particle number density n through

$$\mathrm{i}\hbar \int \frac{\mathrm{d}^d k}{(2\pi)^d} S_\mathrm{F}(k) = -n, \tag{3.13}$$

where the minus sign arises because the fields anticommute. For the case at hand, this formula can be checked explicitly. With the required convergence factor $\exp(\mathrm{i}\omega 0^+)$ included [see below Eq. (2.69)], the frequency integral yields

$$\hbar \int \frac{\mathrm{d}\omega}{2\pi} \frac{\mathrm{e}^{\mathrm{i}\omega 0^+}}{\hbar\omega - \xi(\mathbf{k}) + \mathrm{i}\,\mathrm{sgn}(\omega)\eta} = \mathrm{i}\theta[-\xi(\mathbf{k})]. \tag{3.14}$$

The integral over the wave vectors then becomes simple, with the result

$$\mathrm{i}\hbar \int \frac{\mathrm{d}^d k}{(2\pi)^d} S_{\mathrm{F}}(k) = -\int \frac{\mathrm{d}^D k}{(2\pi)^D} \theta[-\xi(\mathbf{k})] = -\frac{\Omega_D}{(2\pi)^D} \frac{1}{D} k_{\mathrm{F}}^D, \tag{3.15}$$

where the right sides give minus the particle number density (per spin degree of freedom), which is $n = k_F^3/6\pi^2$ in three and $n = k_F^2/4\pi$ in two space dimensions.

We wish to calculate the magnetic susceptibility χ. For an isotropic system and small enough applied field, χ is defined through $V\mathbf{M} = \chi \mathbf{H}$, with $\mathbf{M}$ the magnetization, $\mathbf{M} \equiv -\partial\Omega/\partial\mathbf{H}$ and Ω the zero-temperature thermodynamic potential (2.200). Systems with $\chi > 0$ polarize and in this way enhance the applied field, i.e., are *paramagnetic*, while those with $\chi < 0$ oppose the applied field and are *diamagnetic*. With $\mathbf{H} = \mathbf{B}$, it follows that in the linear isotropic approximation

$$\chi = -\frac{1}{V} \frac{\partial^2 \Omega}{\partial B^2}, \tag{3.16}$$

with $B = |\mathbf{B}|$. For the case at hand, Ω assumes the form $\Omega = -\int \mathrm{d}^D x \mathcal{L}_{\mathrm{eff}}$, where the effective Lagrangian $\mathcal{L}_{\mathrm{eff}}$ also includes the classical Maxwell term $-(1/8\pi)B^2$, see Eq. (3.8). The terms in the expansion (3.10) of interest to us are quadratic in the vector potential $\mathbf{A}$. Applying the shift rule (3.7), we obtain for the second term ($\ell = 2$) in the expansion

$$\begin{aligned} \Gamma_1^{(\ell=2)} &\equiv \mathrm{i}\hbar \frac{e^2}{8m^2c^2} \operatorname{Tr} S_{\mathrm{F}}(p) (\mathbf{p}\cdot\mathbf{A} + \mathbf{A}\cdot\mathbf{p}) S_{\mathrm{F}}(p) (\mathbf{p}\cdot\mathbf{A} + \mathbf{A}\cdot\mathbf{p}) \\ &= \mathrm{i}\hbar \frac{e^2}{8m^2c^2} \operatorname{Tr} S_{\mathrm{F}}(p) S_{\mathrm{F}}(p_0, \mathbf{p} + \mathrm{i}\hbar\nabla) \left(2p^i + \mathrm{i}\hbar\partial_i\right)\left(2p^j + \mathrm{i}\hbar\partial_j\right) A^i A^j, \end{aligned} \tag{3.17}$$

where the derivatives $\nabla = (\partial_1, \partial_2, \cdots, \partial_D)$ only operate on the first $\mathbf{A}(\mathbf{x})$ to their right. The superscript on Γ indicates the term of the expansion considered. An integral of the type appearing in Eq. (3.17) will be studied in closed form in a later chapter, see Sec. 6.5. To illustrate the derivative expansion method, we here expand the expression up to second order in derivatives using

$$\begin{aligned} S_{\mathrm{F}}(\omega, \mathbf{k} + \mathbf{q}) &= \frac{1}{\hbar\omega - \xi(\mathbf{k}+\mathbf{q}) + \mathrm{i}\,\mathrm{sgn}(\omega)\eta} \\ &= S_{\mathrm{F}}(k) \left[1 + \frac{\hbar^2}{2m} S_{\mathrm{F}}(k) \left(2\mathbf{k}\cdot\mathbf{q} + \mathbf{q}^2\right) + \frac{\hbar^4}{m^2} S_{\mathrm{F}}^2(k) (\mathbf{k}\cdot\mathbf{q})^2\right]. \end{aligned} \tag{3.18}$$

To evaluate the various integrals appearing in the expansion, we use the result (3.15) as a master equation to obtain, for example,

$$\begin{aligned} \mathrm{i}\hbar \int \frac{\mathrm{d}^4 k}{(2\pi)^4} S_{\mathrm{F}}^2(k) &= -\mathrm{i}\hbar \frac{\partial}{\partial\mu} \int \frac{\mathrm{d}^4 k}{(2\pi)^4} S_{\mathrm{F}}(k) \\ &= \int \frac{\mathrm{d}^3 k}{(2\pi)^3} \delta[-\xi(\mathbf{k})] = \frac{1}{2\pi^2} \frac{m k_{\mathrm{F}}}{\hbar^2}, \end{aligned} \tag{3.19}$$

where we specialized to $d = 4$, and

$$
\begin{aligned}
\mathrm{i}\hbar \int \frac{\mathrm{d}^4 k}{(2\pi)^4} k^i k^j S_{\mathrm{F}}^3(k) &= \mathrm{i}\frac{\hbar}{2}\frac{\partial}{\partial \mu} \int \frac{\mathrm{d}^4 k}{(2\pi)^4} k^i k^j S_{\mathrm{F}}^2(k) \\
&= \frac{1}{6}\delta^{ij} \int \frac{\mathrm{d}^3 k}{(2\pi)^3} \mathbf{k}^2 \frac{\partial}{\partial \epsilon(\mathbf{k})} \delta[-\xi(\mathbf{k})] \qquad (3.20) \\
&= -\frac{1}{4\pi^2}\frac{m^2 k_{\mathrm{F}}}{\hbar^4}\delta^{ij}.
\end{aligned}
$$

Pasting the pieces together, we arrive at the the one-loop effective Lagrangian, which is related to Γ_1 through $\Gamma_1 = \int \mathrm{d}^d x \, \mathcal{L}_1$,

$$
\mathcal{L}_1^{(\ell=2)} = \frac{e^2 n}{2mc^2} A^i \left[\delta_{ij} - \frac{1}{4k_{\mathrm{F}}^2} \left(\partial_i \partial_j - \nabla^2 \delta_{ij} \right) \right] A^j, \tag{3.21}
$$

with $n = k_{\mathrm{F}}^3/6\pi^2$ the particle number density. The first term on the right cancels the diamagnetic contribution in the first term ($\ell = 1$) of the expansion (3.10),

$$
\Gamma_1^{(\ell=1)} \equiv \mathrm{i}\hbar \frac{e^2}{2mc^2} \operatorname{Tr} S_{\mathrm{F}}(p) \mathbf{A}^2. \tag{3.22}
$$

The remaining two terms in Eq. (3.21) yield after partial integration

$$
\mathcal{L}_1^{(\ell=1,2)} = -\frac{e^2}{48\pi^2}\frac{k_{\mathrm{F}}}{mc^2} B^2. \tag{3.23}
$$

This one-fermion-loop contribution, which modifies the Maxwell term $-(1/8\pi)B^2$ will be discussed in Sec. 3.9 below after also the spin of the electrons is accounted for.

3.3 Goldstone-Wilczek Currents

As a second example of the derivative expansion method, we consider gapless fermions moving in one space dimension described by a massless Dirac-like theory. As will be shown in Sec. 9.2 below, relativisticlike fermions naturally emerge in the context of quasi one-dimensional condensed matter systems.

We couple the fermions to two external real scalar fields ϕ_1 and ϕ_2 through a Yukawa interaction

$$
\mathcal{L} = \bar{\psi} \mathrm{i} \partial\!\!\!/ \psi - g \bar{\psi} \left(\phi_1 + \mathrm{i}\gamma^5 \phi_2 \right) \psi. \tag{3.24}
$$

Here, $A\!\!\!/ \equiv A_\mu \gamma^\mu$, with γ^μ the Dirac matrices, denotes the Feynman slash notation. The Dirac matrices obey the algebra

$$
\{\gamma^\mu, \gamma^\nu\} \equiv \gamma^\mu \gamma^\nu + \gamma^\nu \gamma^\mu = 2\eta^{\mu\nu}. \tag{3.25}
$$

Although not needed in the following, they can in $d = 2$ be represented in terms of the Pauli matrices $\boldsymbol{\tau}$ as, for example,

$$\gamma^0 = \tau^1, \quad \gamma^1 = \mathrm{i}\tau^2. \tag{3.26}$$

Moreover, in the Lagrangian (3.24), $\bar{\psi} \equiv \psi^\dagger \gamma^0$ and $\gamma^5 \equiv \gamma^0\gamma^1 = -\tau^3$. The definition of γ^5 is such that it anticommutes with the other gamma matrices,

$$\left\{\gamma^5, \gamma^\mu\right\} = 0, \tag{3.27}$$

and $(\gamma^5)^2 = 1$. The matrices γ^0 and γ^5 are Hermitian, $\gamma^{0\dagger} = \gamma^0$, while γ^1 is anti-Hermitian, $\gamma^{1\dagger} = -\gamma^1$. The first Yukawa coupling in the Lagrangian (3.24) physically denotes the electron-phonon interaction in one space dimension, with ϕ_1 measuring the displacement of the atoms from their equilibrium positions, see Sec. 1.1. The second coupling, involving the γ^5 matrix, is a natural generalization of the first. For notational convenience, we in this section adopt natural units where $\hbar = c = 1$. The theory (3.24) is invariant under the chiral transformations

$$\begin{aligned} \phi_1 &\to \phi_1 \cos\theta + \phi_2 \sin\theta \\ \phi_2 &\to -\phi_1 \sin\theta + \phi_2 \cos\theta \\ \psi &\to \mathrm{e}^{\mathrm{i}\gamma^5\theta/2}\psi. \end{aligned} \tag{3.28}$$

To bring out this invariance more explicitly, we write

$$\phi_1 + \mathrm{i}\gamma^5\phi_2 = \rho\, \mathrm{e}^{\mathrm{i}\gamma^5\theta}, \tag{3.29}$$

where $\rho \equiv (\phi_1^2 + \phi_2^2)^{1/2}$ and

$$\theta \equiv \arctan(\phi_2/\phi_1). \tag{3.30}$$

With the decomposition

$$\psi = s\chi, \quad s \equiv \mathrm{e}^{-\mathrm{i}\gamma^5\theta/2}, \tag{3.31}$$

the Lagrangian density assumes the form

$$\mathcal{L} = \bar{\chi}\left(\mathrm{i}\partial\!\!\!/ - g\rho - V\!\!\!/\gamma^5 - J\!\!\!/\right)\chi, \tag{3.32}$$

with

$$V_\mu\gamma^5 \equiv -\mathrm{i}s^\dagger\partial_\mu s = -\frac{1}{2}\partial_\mu\theta\gamma^5, \tag{3.33}$$

and where also a source term $\mathcal{L}_{\mathrm{I}} = -\bar{\psi}\gamma^\mu\psi J_\mu$ is included.

In the following, we assume that the background fields change slowly in space and time. We are interested in the current density $\langle j^\mu\rangle \equiv \langle\bar{\psi}\gamma^\mu\psi\rangle$ induced by these slowly varying fields. Such currents, which are known as *Goldstone-Wilczek currents*, can be computed in an expansion in derivatives of the background fields.

To lowest order, ρ can be taken to be constant, $\rho = \bar{\rho}$, so that the one-loop effective action obtained after integrating out the fermionic degrees of freedom can be written as

$$\Gamma_1 = \mathrm{i}\,\mathrm{Tr} \sum_{\ell=1}^{\infty} \frac{1}{\ell} \left[S_{\mathrm{F}}(\not{V}\gamma^5 + \not{J}) \right]^{\ell}, \tag{3.34}$$

apart from an irrelevant additive constant. Here, S_{F} is the fermion propagator

$$S_{\mathrm{F}}(k) = \frac{1}{\not{k} - \bar{\rho}} = \frac{\not{k} + \bar{\rho}}{k^2 - \bar{\rho}^2 + \mathrm{i}\eta}, \tag{3.35}$$

where in the last step a small positive switching parameter $\eta = 0^+$ is introduced for causality, see Sec. 2.4.

The part of interest to us is the contribution linear in the source contained in the $\ell = 2$ term in the expansion (3.34):

$$\begin{aligned} \Gamma_1^{(\ell=2)} &= \mathrm{i}\,\mathrm{Tr} \frac{1}{\not{p} - \bar{\rho}} \not{V}\gamma^5 \frac{1}{\not{p} - \bar{\rho}} \not{J} \\ &= \mathrm{i}\,\mathrm{Tr} \left(-\frac{\bar{\rho}}{p^2 - \bar{\rho}^2} \not{V} \frac{\bar{\rho}}{p^2 - \bar{\rho}^2} \not{J}\gamma^5 + \frac{\not{p}}{p^2 - \bar{\rho}^2} \not{V} \frac{\not{p}}{p^2 - \bar{\rho}^2} \not{J}\gamma^5 \right), \end{aligned} \tag{3.36}$$

where we suppressed the η dependence, and where, as before, the superscript on Γ indicates the term of the expansion considered. In deriving this, the cyclic property of the trace is used to combine two contributions into one. By the trace identities

$$\mathrm{tr}\, \gamma^\mu \gamma^\nu \gamma^5 = -2\epsilon^{\mu\nu} \tag{3.37}$$

$$\mathrm{tr}\, \gamma^\mu \gamma^\nu \gamma^\lambda \gamma^\sigma \gamma^5 = 2\left(-\eta^{\mu\nu}\epsilon^{\lambda\sigma} + \eta^{\mu\lambda}\epsilon^{\nu\sigma} + \eta^{\nu\lambda}\epsilon^{\sigma\mu} \right), \tag{3.38}$$

which follow directly from the Dirac algebra (3.25), and the relation

$$\gamma^\mu \gamma^5 = \epsilon^{\mu\nu} \gamma_\nu, \tag{3.39}$$

where $\epsilon^{\mu\nu}$ is the antisymmetric Levi-Civita tensor in two spacetime dimensions with $\epsilon_{01} = 1$ and $\epsilon_{\mu\nu}\epsilon^{\mu\nu} = -2$, the contribution gives

$$\begin{aligned} \mathcal{L}_1^{(\ell=2)} = 2\mathrm{i}\epsilon^{\mu\nu} \int \frac{\mathrm{d}^2 k}{(2\pi)^2} & \frac{1}{(k^2 - \bar{\rho}^2 + \mathrm{i}\eta)^2} \\ & \times \left(\bar{\rho}^2 V_\mu J_\nu - k \cdot V k_\mu J_\nu + k^2 V_\mu J_\nu + k \cdot V J_\mu k_\nu \right) \end{aligned} \tag{3.40}$$

to lowest order in derivatives. The last three integrals diverge logarithmically in the ultraviolet as can be checked by power counting. To regularize them, we evaluate the integrals in $d = 2 - \varepsilon$ dimensions with ε small, and use

$$\int \frac{\mathrm{d}^d k}{(2\pi)^d} \frac{k^\mu k^\nu}{(k^2 - \bar{\rho}^2 + \mathrm{i}\eta)^2} = \frac{\eta^{\mu\nu}}{d} \int \frac{\mathrm{d}^d k}{(2\pi)^d} \frac{k^2}{(k^2 - \bar{\rho}^2 + \mathrm{i}\eta)^2} \tag{3.41}$$

to obtain

$$\mathcal{L}_1^{(\ell=2)} = 2\mathrm{i}\epsilon^{\mu\nu} V_\mu J_\nu \lim_{d\to 2} \int \frac{\mathrm{d}^d k}{(2\pi)^d} \frac{1}{(k^2 - \bar{\rho}^2 + \mathrm{i}\eta)^2} \left[\bar{\rho}^2 + (1 - 2/d)\, k^2\right]. \tag{3.42}$$

The factor $(1 - 2/d)$, which we would have missed had we remained in $d = 2$, renders the corresponding integral finite in the limit $d \to 2$. Using the basic formulas of dimensional regularization

$$\int \frac{\mathrm{d}^d k}{(2\pi)^d} \frac{1}{(k^2 - M^2 + \mathrm{i}\eta)^n} = \mathrm{i} \frac{(-1)^n}{(4\pi)^{d/2}} \frac{\Gamma(n - d/2)}{\Gamma(n)} \frac{1}{(M^2 - \mathrm{i}\eta)^{n-d/2}}, \tag{3.43}$$

with $\Gamma(n)$ the gamma function, and

$$\int \frac{\mathrm{d}^d k}{(2\pi)^d} \frac{k^\mu k^\nu}{(k^2 - M^2 + \mathrm{i}\eta)^n} = \frac{\mathrm{i}}{2} \frac{(-1)^{n-1}}{(4\pi)^{d/2}} \frac{\Gamma(n - 1 - d/2)}{\Gamma(n)} \frac{\eta^{\mu\nu}}{(M^2 - \mathrm{i}\eta)^{n-1-d/2}}, \tag{3.44}$$

we find for $d \to 2$

$$\mathcal{L}_1^{(\ell=2)} = \frac{1}{\pi} \epsilon^{\mu\nu} J_\mu V_\nu \tag{3.45}$$

and

$$\langle j^\mu \rangle = -\frac{1}{\pi} \epsilon^{\mu\nu} V_\nu = \frac{1}{2\pi} \epsilon^{\mu\nu} \partial_\nu \theta \tag{3.46}$$

by Eq. (3.33). More explicitly, the Goldstone-Wilczek current induced by the slowly varying background fields reads

$$\langle j^\mu \rangle = \frac{1}{2\pi} \epsilon^{\mu\nu} \epsilon_{ab} \frac{\phi_a \partial_\nu \phi_b}{\phi_1^2 + \phi_2^2}, \tag{3.47}$$

with ϵ_{ab} $(a, b = 1, 2)$ defined such that $\epsilon_{12} = 1$. The induced current physically describes the polarization of the fermion vacuum by the varying background fields. Note that this lowest-order expression in derivatives is independent of the parameters of the theory (3.24). The current, which is conserved $\partial_\mu \langle j^\mu \rangle = 0$, becomes singular when the two fields vanish simultaneously. Higher-order contributions in derivatives can be calculated to arbitrary order by applying the derivative expansion method to the expression (3.36).

In Sec. 9.9 below, it will be shown that when the background fields configuration is topologically nontrivial, the defects contained in the configuration may acquire unusual quantum numbers through polarization effects.

3.4 Optical Phonons

To demonstrate the use of the derivative expansion method in the context of a purely bosonic theory, we consider the ϕ^4 theory, defined by the Lagrangian density

$$\mathcal{L} = \frac{1}{2}\left[(\partial_t\phi)^2 - c_s^2(\partial_1\phi)^2 - \mu^2\phi^2\right] - \frac{1}{4!}\lambda\phi^4 \tag{3.48}$$

with ϕ a real scalar field in two spacetime dimensions. The first two terms, with c_s the speed of sound, describe acoustic sound waves, or, after quantizing, *acoustic* phonons characterized by a gapless excitation spectrum, see Eqs. (1.10) and (1.9). The two added terms in the Lagrangian density (3.48), with the parameters μ and λ, result in an energy gap. The theory therefore describes *optical* phonons in one space dimension. For notational convenience, we set $c_s = 1$ in this example. To obtain the one-loop effective action, we use the *semiclassical expansion* and set $\phi = \phi_c + \tilde{\phi}$, with $\phi_c(x)$ a solution of the field equation, and expand the Lagrangian density about this solution up to second order in $\tilde{\phi}$:

$$\mathcal{L} = \mathcal{L}_c + \frac{1}{2}\left[(\partial_\mu\tilde{\phi})^2 - \left(\mu^2 + \tfrac{1}{2}\lambda\phi_c^2\right)\tilde{\phi}^2\right] + O(\tilde{\phi}^3), \tag{3.49}$$

where $\mathcal{L}_c$ denotes the Lagrangian density with $\phi = \phi_c$, i.e., $\mathcal{L}_c = \mathcal{L}(\phi_c)$. With the higher-order terms neglected, the partition function

$$Z = \int \mathrm{D}\phi\, \mathrm{e}^{(\mathrm{i}/\hbar)S} \tag{3.50}$$

reduces to a Gaussian functional integral

$$\begin{aligned} Z &= \mathrm{e}^{(\mathrm{i}/\hbar)S_c} \int \mathrm{D}\tilde{\phi}\, \exp\left\{\mathrm{i}\frac{\hbar}{2}\int \mathrm{d}^d x\left[(\partial_\mu\tilde{\phi})^2 - \left(\mu^2 + \tfrac{1}{2}\lambda\phi_c^2\right)\tilde{\phi}^2\right]\right\} \\ &= \mathrm{e}^{(\mathrm{i}/\hbar)S_c}\, \mathrm{Det}^{-1/2}\left[p^2 - \mu^2 - \tfrac{1}{2}\lambda\phi_c^2(x)\right], \end{aligned} \tag{3.51}$$

where $p^2 = p^\mu p_\mu = p_0^2 - \mathbf{p}^2$ is the derivative introduced in Sec. 3.1 with $\hbar = 1$. The semiclassical expansion corresponds to the *stationary-phase*, or saddle-point approximation of the functional integral (3.50) with the Gaussian fluctuations included. Because the exponential $\mathrm{e}^{(\mathrm{i}/\hbar)S_c}$ oscillates rapidly, the major contribution to the integral comes from the vicinity of the saddle point where $\delta S/\delta\phi = 0$. In passing we note that the so-called *mean-field approximation* is obtained by including *only* the contribution of the saddle point and ignoring all fluctuations. This amounts to omitting the determinant in Eq. (3.51).

That determinant leads to the one-loop effective action

$$\Gamma_1 = \mathrm{i}\frac{\hbar}{2}\,\mathrm{Tr}\ln\left[p^2 - \mu^2 - \tfrac{1}{2}\lambda\phi_c^2(x)\right]. \tag{3.52}$$

For a constant field $\phi_c(x) = v$, the effective action reduces to the effective potential density

$$\mathcal{V}_1(v) = -\mathrm{i}\frac{\hbar}{2}\int \frac{\mathrm{d}^d k}{(2\pi)^d} \ln\left(k^2 - \mu^2 - \tfrac{1}{2}\lambda v^2\right) \tag{3.53}$$

whose explicit form does not concern us here.

If the classical field $\phi_c(x)$ depends on spacetime, the operator $p^2 - \mu^2 - \frac{1}{2}\lambda\phi_c^2(x)$ in Eq. (3.52) is no longer diagonal in momentum space and one has to resort to a derivative expansion. This expansion yields a result of the form

$$\Gamma_1 = \int \mathrm{d}^d x\left[-\mathcal{V}_1(\phi) + \tfrac{1}{2}\mathcal{Z}_1(\phi)(\partial\phi)^2 + \cdots\right], \tag{3.54}$$

with $\mathcal{V}_1$ the one-loop effective potential (3.53). To compute $\mathcal{Z}_1(\phi)$, we write $\phi_c(x) = v + \eta(x)$, and expand the effective action to second order in η. Equation (3.54) then gives

$$\begin{aligned}\delta\Gamma_1 &\equiv \Gamma_1[v+\eta] - \Gamma_1[v] \\ &= \int \mathrm{d}^d x\left[-\frac{\partial\mathcal{V}_1}{\partial v}\eta - \frac{1}{2}\frac{\partial^2\mathcal{V}_1}{\partial v^2}\eta^2 + \frac{1}{2}\mathcal{Z}_1(v)(\partial_\mu\eta)^2 + \cdots\right]\end{aligned} \tag{3.55}$$

while Eq. (3.52) leads to

$$\begin{aligned}\delta\Gamma_1 =& -\mathrm{i}\hbar\frac{\lambda}{2}\,\mathrm{Tr}\,\frac{1}{p^2 - M^2}\left(v\eta + \tfrac{1}{2}\eta^2\right) \\ & -\mathrm{i}\hbar\frac{\lambda^2 v^2}{4}\,\mathrm{Tr}\,\frac{1}{p^2 - M^2}\eta\frac{1}{p^2 - M^2}\eta,\end{aligned} \tag{3.56}$$

with $M^2 \equiv \mu^2 + \frac{1}{2}\lambda v^2$. Moving the derivatives p^μ to the left by using the shift rule (3.7) with $\hbar = 1$, one obtains

$$\begin{aligned}\delta\Gamma_1 =& -\mathrm{i}\hbar\frac{\lambda}{4}\,\mathrm{Tr}\,\frac{1}{p^2 - M^2}(2v\eta + \eta^2) \\ & -\mathrm{i}\hbar\frac{\lambda^2 v^2}{4}\,\mathrm{Tr}\frac{1}{p^2 - M^2}\frac{1}{(p - \mathrm{i}\partial)^2 - M^2}\eta\eta,\end{aligned} \tag{3.57}$$

where we recall the definition of the derivative ∂_μ as operating only on the first field to its right. The second denominator in the last term of Eq. (3.57) can be expanded in powers of derivatives as

$$\frac{1}{(p - \mathrm{i}\partial)^2 - M^2} = \frac{1}{p^2 - M^2}\sum_{\ell=0}^{\infty}\left(\frac{2\mathrm{i}p\cdot\partial + \partial^2}{p^2 - M^2}\right)^\ell. \tag{3.58}$$

The three terms in Eq. (3.57) without a derivative precisely yield the first two terms at the right side of Eq. (3.55) as can be verified by using the basic formula (3.43) of dimensional regularization. The coefficient $\mathcal{Z}_1(v)$ is determined by the terms quadratic in ∂_μ in the expansion (3.58). By Eq. (3.44) this gives after an integration by parts

$$\mathcal{Z}_1(v) = \frac{1}{12}\frac{\Gamma(3-d/2)}{(4\pi)^{d/2}}\frac{\lambda^2 v^2}{(M^2)^{3-d/2}}, \tag{3.59}$$

which reduces to

$$\mathcal{Z}_1(v) = \frac{1}{48\pi}\frac{\lambda^2 v^2}{\left(\mu^2 + \frac{1}{2}\lambda v^2\right)^2} \tag{3.60}$$

in $d = 2$. The coefficient $\mathcal{Z}_1(\phi)$ in Eq. (3.54) is given by the same expression with v replaced with ϕ. Higher-order terms can be computed in a similar fashion.

3.5 Geometric Phase

Consider the quantum mechanical problem of a single particle described by a Hamiltonian $H(\boldsymbol{\lambda})$ which depends on a set of external parameters $\boldsymbol{\lambda} = (\lambda_1, \lambda_2, \cdots)$. The time evolution of the state $|\psi(t)\rangle$ of the system is determined by the Schrödinger equation

$$H(\boldsymbol{\lambda})|\psi(t)\rangle = \mathrm{i}\hbar\frac{\partial}{\partial t}|\psi(t)\rangle. \tag{3.61}$$

Assume that $H(\boldsymbol{\lambda})$ depends on time only through the parameters $\boldsymbol{\lambda}(t)$ and let $H(t) \equiv H[\boldsymbol{\lambda}(t)]$. At any instant, a complete set of energy eigenstates $|n(t)\rangle$, with $|n(t)\rangle \equiv |n[\boldsymbol{\lambda}(t)]\rangle$ can be introduced, satisfying

$$H(t)|n(t)\rangle = E_n(t)|n(t)\rangle, \tag{3.62}$$

with instantaneous energies $E_n(t)$. Assume further that the time evolution is sufficiently slow so that the Schrödinger equation can be solved in the adiabatic limit, meaning that the system remains in the eigenstate $|n(t)\rangle$ of $H(t)$ which smoothly evolves from the initial state $|n(0)\rangle$ at $t = 0$. Here, it is assumed that the eigenstate is nondegenerate. At time t, the state $|\psi_n(t)\rangle$ is then given by

$$|\psi_n(t)\rangle = \exp\left[-\frac{\mathrm{i}}{\hbar}\int_0^t \mathrm{d}t'\, E_n(t') + \mathrm{i}\gamma_n(t)\right]|n(t)\rangle, \tag{3.63}$$

where the first factor on the right is the standard dynamic phase factor. The phase $\gamma_n(t)$ reads

$$\begin{aligned}\gamma_n(t) &= \int_0^t \mathrm{d}t' \langle n(t')|\mathrm{i}\frac{\partial}{\partial t'}|n(t')\rangle \\ &= \mathrm{i}\int_0^t \mathrm{d}t' \frac{\mathrm{d}\boldsymbol{\lambda}}{\mathrm{d}t'}\cdot\langle n(t')|\nabla_{\boldsymbol{\lambda}}|n(t')\rangle \\ &= \int_{\boldsymbol{\lambda}(0)}^{\boldsymbol{\lambda}(t)} \mathrm{d}\boldsymbol{\lambda}\cdot\mathbf{A}_n,\end{aligned} \tag{3.64}$$

where we introduced the notation $\nabla_\lambda \equiv (\partial/\partial\lambda_1, \partial/\partial\lambda_2, \cdots)$ and

$$\mathbf{A}_n \equiv \mathrm{i}\langle n|\nabla_\lambda|n\rangle. \tag{3.65}$$

If the time evolution is such that at $t = \mathcal{T}$ the system returns to its initial configuration, a closed path Γ is traced out in parameter space. The *Berry* or *geometric phase*

$$e^{\mathrm{i}\gamma_n(\Gamma)} \equiv \exp\left(\mathrm{i}\oint_\Gamma \mathrm{d}\boldsymbol{\lambda}\cdot\mathbf{A}_n\right), \tag{3.66}$$

acquired by the state is independent of the precise location of Γ, provided no singularities are encountered when smoothly deforming the loop in parameter space. In the absence of any singularity in parameter space, i.e., when this space is topologically trivial, the loop can be shrunk to zero and the geometric phase vanishes.

By Stoke's theorem, the line integral can be converted into a surface integral, giving

$$\begin{aligned}\gamma_n(\Gamma) &= \int_S \mathrm{d}\mathbf{S}\cdot\nabla_\lambda\times\mathbf{A}_n \\ &= \int_S \mathrm{d}\mathbf{S}\cdot\mathbf{B}_n,\end{aligned} \tag{3.67}$$

with $\mathrm{d}\mathbf{S}$ denoting a surface element of the surface S in parameter space subtending the loop Γ, and $\mathbf{B}_n \equiv \nabla_\lambda\times\mathbf{A}_n$. Insertion of a complete set of eigenstates yields

$$\mathbf{B}_n = \mathrm{i}\sum_{m\neq n}\langle\nabla_\lambda n|m\rangle\times\langle m|\nabla_\lambda|n\rangle. \tag{3.68}$$

By the identity ($E_m \neq E_n$)

$$\langle m|\nabla_\lambda|n\rangle = \frac{\langle m|\nabla_\lambda H|n\rangle}{E_n - E_m}, \tag{3.69}$$

one finds

$$\mathbf{B}_n = \mathrm{i}\sum_{m\neq n}\frac{\langle n|\nabla_\lambda H|m\rangle\times\langle m|\nabla_\lambda H|n\rangle}{(E_n - E_m)^2}. \tag{3.70}$$

This expression identifies the singularities in parameter space as *diabolic points* where two energy levels cross and the Hamiltonian has a two-fold degeneracy.

It is instructive to consider the generic two-level Hamiltonian,

$$H(t) = \frac{1}{2}\boldsymbol{\lambda}(t)\cdot\boldsymbol{\sigma}, \tag{3.71}$$

with $\boldsymbol{\sigma}$ the Pauli matrices, as an example. The 2×2 Pauli matrices, which are generally not related to spin, naturally arise in perturbation theory when energy levels cross in parameter space. For a complex Hamiltonian, energy levels generically cross two at a time, and it requires adjusting three parameters to hit the diabolic point. In the language used to characterize topological defects, such a point is detected by surrounding it with an ordinary sphere S^2 in parameter space. In the vicinity of such a point, all other energy levels can be neglected, and only the two-level subsystem, described by the Hamiltonian (3.71) up to an irrelevant additive constant, need be considered. The two eigenstates $|n\rangle$ of the Hamiltonian (3.71), denoted by $|\pm\rangle$, are two-component "spinors" and eigenstates of the component of $\boldsymbol{\sigma}$ along $\boldsymbol{\lambda}$,

$$H|\pm\rangle = \pm\frac{1}{2}\lambda|\pm\rangle, \tag{3.72}$$

with eigenvalues $E_\pm = \pm\frac{1}{2}\lambda$ and $\lambda \equiv |\boldsymbol{\lambda}|$. At the origin of three-dimensional parameter space, $\boldsymbol{\lambda} = 0$, the two eigenvalues cross, identifying the origin as a diabolic point. For this two-level system, Eq. (3.70) gives for the $|+\rangle$ state

$$\mathbf{B}_+ = \frac{\mathrm{i}}{4}\frac{1}{\lambda^2}\langle+|\boldsymbol{\sigma}|-\rangle \times \langle-|\boldsymbol{\sigma}|+\rangle \tag{3.73}$$

and for the $|-\rangle$ state, $\mathbf{B}_- = -\mathbf{B}_+$. With the complete set of eigenstates taken out again, Eq. (3.73) reduces to the form

$$\mathbf{B}_+ = \frac{\mathrm{i}}{4}\frac{1}{\lambda^2}\langle+|\boldsymbol{\sigma}\times\boldsymbol{\sigma}|+\rangle = -\frac{1}{2}\frac{1}{\lambda^2}\langle+|\boldsymbol{\sigma}|+\rangle = -\frac{1}{2}\frac{\boldsymbol{\lambda}}{\lambda^3} \tag{3.74}$$

by the identity $\boldsymbol{\sigma}\times\boldsymbol{\sigma} = 2\mathrm{i}\boldsymbol{\sigma}$ and by Eq. (3.72). Equation (3.74) describes the "magnetic induction" field produced by a monopole with magnetic charge $-\frac{1}{2}$ residing at the location of the degeneracy at the origin of parameter space. The induction field satisfies the magnetic counterpart of Gauss's law

$$\nabla_\lambda \cdot \mathbf{B}_+ = -2\pi\delta(\boldsymbol{\lambda}). \tag{3.75}$$

The corresponding Berry phase reads,

$$\gamma_+(\Gamma) = -\frac{1}{2}\int_S \mathrm{d}\mathbf{S}\cdot\frac{\boldsymbol{\lambda}}{\lambda^3} = -\frac{1}{2}\Omega(\Gamma), \tag{3.76}$$

with $\Omega(\Gamma)$ the solid angle that the loop Γ subtends at the degeneracy. The phase change of the state which is adiabatically transported along Γ once equals the flux through the loop emanating from the monopole, see Fig. 3.1.

An associated vector potential (3.65) can be constructed by parametrizing the vector $\boldsymbol{\lambda}$ in terms of spherical coordinates as

$$\boldsymbol{\lambda} = \lambda(\sin\theta\cos\varphi, \sin\theta\sin\varphi, \cos\theta)^T, \tag{3.77}$$

where "T" denotes the transpose of the vector. The angles θ and φ depend on time. Then,

$$H = \frac{\lambda}{2}\begin{pmatrix}\cos\theta & \mathrm{e}^{-\mathrm{i}\varphi}\sin\theta \\ \mathrm{e}^{\mathrm{i}\varphi}\sin\theta & -\cos\theta\end{pmatrix} \tag{3.78}$$

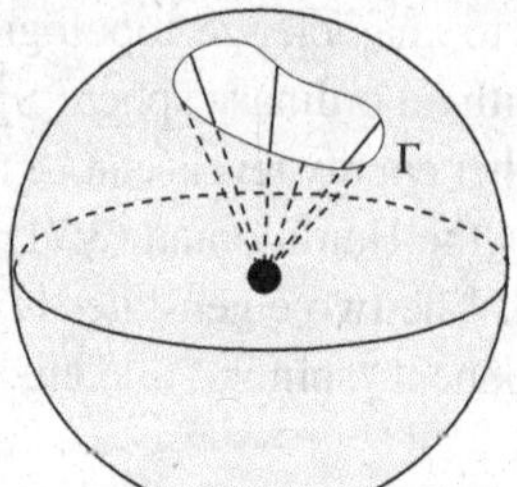

Fig. 3.1 Flux through the loop Γ emanated by the monopole at the origin of the sphere.

with the eigenstate

$$|+\rangle = \begin{pmatrix} \cos\theta/2 \\ e^{i\varphi}\sin\theta/2 \end{pmatrix}, \tag{3.79}$$

corresponding to the eigenvalue $+\frac{1}{2}\lambda$. Inserting this into Eq. (3.65), we find

$$\mathbf{A}_+ = -\frac{1}{2}\frac{1-\cos\theta}{\lambda\sin\theta}\mathbf{e}_\varphi, \tag{3.80}$$

where $\mathbf{e}_\varphi = (-\sin\varphi, \cos\varphi, 0)$ is the unit vector in the λ^1–λ^2 plane of parameter space. In deriving this, use is made of the trigonometric identity $\sin^2(\frac{1}{2}\theta) = \frac{1}{2}(1-\cos\theta)$ as well as of the explicit form of the Laplace operator in spherical coordinates

$$\nabla = \left(\mathbf{e}_\lambda\frac{\partial}{\partial\lambda}, \mathbf{e}_\theta\frac{1}{\lambda}\frac{\partial}{\partial\theta}, \mathbf{e}_\varphi\frac{1}{\lambda\sin\theta}\frac{\partial}{\partial\varphi}\right), \tag{3.81}$$

with $\mathbf{e}_\lambda = \boldsymbol{\lambda}/\lambda$ and $\mathbf{e}_\theta$ the two other unit vectors spanning the three-dimensional parameter space. Equation (3.80) gives for the phase (3.64) acquired by the state which is adiabatically transported along an open path in parameter space

$$\gamma_+(t) = \int_0^t dt'\frac{d\boldsymbol{\lambda}}{dt'}\cdot\mathbf{A}_+ = g\int_0^t dt'\frac{d\varphi}{dt'}(1-\cos\theta), \tag{3.82}$$

with $g = -\frac{1}{2}$ the monopole charge. For a closed path, this expression reduces to Eq. (3.76).

The vector potential (3.80), and therefore Eq. (3.82), is ill-defined at $\theta = \pi$. From Eq. (3.76) it follows that the flux through the loop Γ approaches zero as $\theta \to 0$ where the loop shrinks to a point. As Γ is lowered over a sphere centered at the location of the monopole, the flux increases in absolute value until at $\theta = \pi$ it takes the value $4\pi g$. Since the loop has now again shrunk to a point, the vector potential (3.80) must be singular here in order to satisfy

$$\gamma_+(\Gamma) = \int_\Gamma d\boldsymbol{\lambda}\cdot\mathbf{A}_+ = 4\pi g, \tag{3.83}$$

where Γ is an infinitesimal loop around the negative λ^3 axis at the south pole of the sphere. In order for the wave function to be single-valued there, the phase factor $e^{i\gamma_+(\Gamma)}$ must be unity. This is satisfied if and only if g is an integer or half-integer, which is Dirac's charge quantization condition. As in the case of topological defects, the quantization is for topological reasons. For the two-level system (3.71), the monopole charge is $g = -\frac{1}{2}$ and the single-valuedness condition is obeyed.

The above argument holds for any sphere of arbitrary radius and so the vector potential must be singular along the entire negative λ^3 axis. This constitutes the famous Dirac string. The string carries the magnetic flux from infinity to the monopole which then spreads it out radially. The location of the Dirac string is unphysical and can be shifted by a gauge transformation. For example, under the transformation $|+\rangle \to |+'\rangle = e^{-i\varphi}|+\rangle$, $\mathbf{A}_+ \to \mathbf{A}'_+ = \mathbf{A}_+ + \nabla\varphi$, and the vector potential assumes the form

$$\mathbf{A}'_+ = \frac{1}{2}\frac{1+\cos\theta}{\lambda\sin\theta}\mathbf{e}_\varphi, \tag{3.84}$$

which is singular along the positive rather than the negative λ^3 axis.

If the Hamiltonian is (pseudo)real, i.e., $H = \pm H^*$, then

$$\langle n|\nabla_\lambda H|m\rangle = \langle m|\nabla_\lambda H^*|n\rangle = \pm\langle m|\nabla_\lambda H|n\rangle, \tag{3.85}$$

and the "magnetic induction" field $\mathbf{B}_n$ in Eq. (3.70) vanishes. For such Hamiltonians only two parameters need be adjusted to make two energy levels cross. Such a diabolic point is detected by surrounding it with a circle S^1 in parameter space. An example of a real Hamiltonian obtains by setting $\lambda_2 = 0$ in Eq. (3.71). A closed loop Γ in parameter space then lies entirely in the λ^1–λ^3 plane. If it encloses a diabolic point, $\Omega = \pm 2\pi$, and by Eq. (3.76)

$$e^{i\gamma_\pm(\Gamma)} = -1, \tag{3.86}$$

implying that the real wave function changes sign when the state it describes is transported around this loop in parameter space.

Whereas diabolic points of a complex Hamiltonian are point defects in three-dimensional parameter space analogous to magnetic monopoles, diabolic points of a (pseudo)real Hamiltonian are point defects in two-dimensional parameter space analogous to point vortices.

3.6 Monopole Action

As a warm-up exercise for the next section, we recast the above quantum-mechanical findings concerning the single-particle state in a field-theoretical context. Consider the $(0+1)$-dimensional Lagrangian

$$L = \psi^\dagger(t)\,[i\hbar\partial_t - H(t)]\,\psi(t), \tag{3.87}$$

with $H(t)$ the time-dependent two-level Hamiltonian (3.71) and $\psi(t)$ a two-component Grassmann field. In this fermionic theory, the two energy levels $\pm\frac{1}{2}\lambda$ correspond to the single-particle state being empty or occupied. The associated zero-temperature partition function reads

$$Z = \int \mathrm{D}\psi^\dagger \mathrm{D}\psi \, \exp\left(\frac{\mathrm{i}}{\hbar}\int \mathrm{d}t \, L\right). \tag{3.88}$$

We decompose the fermion field as

$$\psi(t) = s(t)\chi(t) \tag{3.89}$$

with $s(t)$ a time-dependent SU(2) matrix, satisfying

$$s^\dagger \mathbf{e}_\lambda \cdot \boldsymbol{\sigma} s = \sigma^3. \tag{3.90}$$

where the right side is independent of time—that is, this unitary transformation rotates the time-dependent unit vector $\mathbf{e}_\lambda$ in a fixed direction. Without loss of generality, this direction can be taken to be the 3-direction. Explicitly, the SU(2) matrix is given by

$$s = \begin{pmatrix} \cos\theta/2 & -\mathrm{e}^{-\mathrm{i}\varphi}\sin\theta/2 \\ \mathrm{e}^{\mathrm{i}\varphi}\sin\theta/2 & \cos\theta/2 \end{pmatrix} \tag{3.91}$$

in spherical coordinates. The Lagrangian now becomes

$$L = \chi^\dagger\left(\mathrm{i}\hbar\partial_t - V_0 - \tfrac{1}{2}\lambda\sigma^3\right)\chi, \tag{3.92}$$

with $V_0 = V^0 \equiv -\mathrm{i}\hbar s^\dagger \partial_t s$, and where we recall that λ depends on time only through the angles θ and φ. The Feynman propagator of the theory is given by

$$S_\mathrm{F}(\omega) = \frac{\hbar\omega + \frac{1}{2}\lambda\sigma^3}{\hbar^2\omega^2 - \frac{1}{4}\lambda^2 + \mathrm{i}\eta}. \tag{3.93}$$

After integrating out the fermion field in Eq. (3.88), we arrive at the one-loop effective action

$$\Gamma_1 = -\mathrm{i}\hbar \,\mathrm{Tr}\ln\left(p_0 - V_0 - \tfrac{1}{2}\lambda\sigma^3\right), \tag{3.94}$$

with $p_0 = p^0 = \mathrm{i}\hbar\partial_t$ the derivative introduced in Sec. 3.1. The effective action possesses a term first order in time derivatives,

$$\begin{aligned} \Gamma_\mathrm{WZ} &\equiv \mathrm{i}\hbar \,\mathrm{Tr}\, S_\mathrm{F} V_0 \\ &= \mathrm{i}\hbar \,\mathrm{tr}\int \mathrm{d}t \int \frac{\mathrm{d}\omega}{2\pi}\frac{\frac{1}{2}\lambda\sigma^3}{\hbar^2\omega^2 - \frac{1}{4}\lambda^2 + \mathrm{i}\eta} V_0 \\ &= -\frac{\mathrm{i}}{2}\mathrm{sgn}(\lambda)\hbar \,\mathrm{tr}\int \mathrm{d}t \, \sigma^3 s^\dagger \partial_t s \end{aligned} \tag{3.95}$$

which, with the explicit expression (3.91), reproduces the geometric phase (3.82)

$$\Gamma_{\rm WZ} = \frac{1}{2}\mathrm{sgn}(\lambda)\hbar \int \mathrm{d}t \frac{\mathrm{d}\varphi}{\mathrm{d}t}(1-\cos\theta), \tag{3.96}$$

apart from a minus sign. The extra minus sign in Eq. (3.96) derives from the fermion loop which gave rise to this effective action. Equation (3.96) can be cast in the equivalent form

$$\Gamma_{\rm WZ} = -\frac{1}{2}\hbar \sum_n \mathrm{sgn}(E_n)\gamma_n(t) \tag{3.97}$$

involving both energy eigenstates $n = \pm$ with eigenvalues $E_\pm = \pm\frac{1}{2}\lambda$ and Berry phases $\gamma_- = -\gamma_+$. This example illustrates that the adiabatic approximation used to derive the geometric phase corresponds in the framework of effective actions to the term first order in time derivatives. Such terms, which persist in the adiabatic limit, and which have coefficients that are quantized for topological reasons are known as *Wess-Zumino* terms. Another important property of Wess-Zumino terms is that they have a singular, gauge non-invariant form when written in a local fashion, as in Eq. (3.96).

A regular form of the Berry phase and the Wess-Zumino term can be obtained at the expense of introducing a new variable $0 \leq u \leq 1$ and a new function $\lambda(t,u)$ such that

$$\lambda(t,1) = \lambda(t), \quad \lambda(t,0) = (0,0,1). \tag{3.98}$$

In other words, the new function, which is assumed to depend on t and u only through the angles θ and φ, smoothly interpolates between the constant vector $(0,0,1)$ and $\lambda(t)$, i.e., $\lambda(t,u)$ is a homotopy as defined in Eq. (1.129). The Berry phase (3.82) can then be equivalently written as

$$\begin{aligned}
\gamma_+ &= \int_0^1 \mathrm{d}u \int \mathrm{d}t \left[\frac{\mathrm{d}}{\mathrm{d}u}\left(\frac{\mathrm{d}\lambda}{\mathrm{d}t}\cdot \mathbf{A}_+\right) - (t \leftrightarrow u)\right] \\
&= \int_0^1 \mathrm{d}u \int \mathrm{d}t \left[\frac{\partial \mathbf{A}_+}{\partial u}\cdot \frac{\partial \lambda}{\partial t} - (t \leftrightarrow u)\right] \\
&= \int_0^1 \mathrm{d}u \int \mathrm{d}t\, (\nabla_\lambda \times \mathbf{A}_+)\cdot(\partial_u \lambda \times \partial_t \lambda) \\
&= g \int_0^1 \mathrm{d}u \int \mathrm{d}t\, \mathbf{e}_\lambda \cdot (\partial_u \mathbf{e}_\lambda \times \partial_t \mathbf{e}_\lambda),
\end{aligned} \tag{3.99}$$

where we subtracted a total time derivative in the first line, and used the chain rule together with the identity

$$\mathbf{a}\cdot\mathbf{c}\,\mathbf{b}\cdot\mathbf{d} - \mathbf{a}\cdot\mathbf{d}\,\mathbf{b}\cdot\mathbf{c} = (\mathbf{a}\times\mathbf{b})\cdot(\mathbf{c}\times\mathbf{d}) \tag{3.100}$$

in the third step to arrive at the gradient operator ∇_λ. More explicitly,

$$\gamma_+ = g \int_0^1 \mathrm{d}u \int \mathrm{d}t\, [\partial_t \cos\theta\, \partial_u \varphi - (t \leftrightarrow u)], \tag{3.101}$$

with $g = -\frac{1}{2}$.

3.7 Wess-Zumino Terms

Consider a field theory specified by the Lagrangian density

$$\mathcal{L} = \psi^\dagger(x)[\mathrm{i}\hbar\partial_t - H(x)]\psi(x), \tag{3.102}$$

with ψ an anticommuting Grassmann field describing fermions, and H a single-particle Hamiltonian which is assumed to depend on time only through the time dependence of the background fields. The zero-temperature partition function,

$$\begin{aligned} Z(\mathcal{T}) &= \int \mathrm{D}\psi^\dagger \mathrm{D}\psi \exp\left[\frac{\mathrm{i}}{\hbar}\int_0^{\mathcal{T}} \mathrm{d}t \int \mathrm{d}^D x\, \psi^\dagger(\mathrm{i}\hbar\partial_t - H)\psi\right] \\ &= \mathrm{Det}(p_0 - H), \end{aligned} \tag{3.103}$$

with $\mathcal{T} \to \infty$, gives rise to the one-loop effective action

$$\Gamma_1(\mathcal{T}) = -\mathrm{i}\hbar \ln Z(\mathcal{T}) = -\mathrm{i}\hbar\, \mathrm{Tr} \ln(p_0 - H) \tag{3.104}$$

which we wish to evaluate in the adiabatic limit, where the background fields vary slowly in time. It is assumed that these fields return to their initial configurations at time $t = \mathcal{T}$.

Consider the eigenvalue equation (in bra-ket notation) for the operator $\mathrm{i}\hbar\partial_t - H$

$$(\mathrm{i}\hbar\partial_t - H)\,|\chi\rangle = \lambda|\chi\rangle. \tag{3.105}$$

Unlike the Grassmann field ψ in Eq. (3.102), the function $\langle x|\chi\rangle \equiv \chi(x)$ is an ordinary, albeit multicomponent, function subject to antiperiodic boundary conditions in time,

$$\chi(t + \mathcal{T}, \mathbf{x}) = -\chi(t, \mathbf{x}) \tag{3.106}$$

for all t and $\mathbf{x}$. A standard method to solve the eigenvalue equation (3.105) is to write the solutions in terms of zero modes, i.e., in terms of the solutions f_n of the corresponding homogeneous equation

$$(\mathrm{i}\hbar\partial_t - H)\, f(x) = 0, \tag{3.107}$$

satisfying

$$f_n(t + \mathcal{T}, \mathbf{x}) = \mathrm{e}^{-\mathrm{i}\alpha_n} f_n(t, \mathbf{x}) \tag{3.108}$$

for all t and $\mathbf{x}$. The α_n's are known as *Floquet indices*. Given these zero modes, it can be checked by direct substitution that the functions

$$\langle x|\chi_{m,n}\rangle = \mathrm{e}^{-\mathrm{i}(\nu_m - \alpha_n/\mathcal{T})t} f_n(x), \tag{3.109}$$

with $\nu_m \equiv (\pi/\mathcal{T})(2m + 1)$ the real-time counterparts of the fermionic Matsubara frequencies (2.47), are eigenfunctions of Eq. (3.105) with eigenvalues

$$\lambda_{m,n} = \hbar\nu_m - \hbar\alpha_n/\mathcal{T}. \tag{3.110}$$

Comparison of Eq. (3.107) to Eq. (3.61) and its solution (3.63) shows that for adiabatically changing background fields

$$|f_n(\mathcal{T})\rangle = \exp\left[-\frac{\mathrm{i}}{\hbar}\int_0^{\mathcal{T}} \mathrm{d}t E_n(t) + \mathrm{i}\gamma_n(\mathcal{T})\right]|n(\mathcal{T})\rangle, \qquad (3.111)$$

where $|n(t)\rangle$ is an eigenstate of the instantaneous Hamiltonian $H(t)$ with energy $E_n(t)$, and $\gamma_n(\mathcal{T})$ is the geometric phase picked up by this state during the time evolution from $t = 0$ to $t = \mathcal{T}$ in which the adiabatically changing background fields return to their initial configurations. It follows that the Floquet indices read in this case

$$\alpha_n = \frac{1}{\hbar}\int_0^{\mathcal{T}} \mathrm{d}t E_n(t) - \gamma_n(\mathcal{T}). \qquad (3.112)$$

Using the $\chi_{m,n}$'s as basis to evaluate the trace in Eq. (3.104), we obtain, cf. Eq. (2.62)

$$\begin{aligned}
\Gamma_1 &= -\mathrm{i}\frac{\hbar}{\mathcal{T}}\sum_{m,n}\langle\chi_{m,n}|\ln(\mathrm{i}\hbar\partial_t - H)|\chi_{m,n}\rangle \\
&= -\mathrm{i}\hbar\int_0^{\mathcal{T}}\mathrm{d}t\int \mathrm{d}^D x\,\frac{1}{\mathcal{T}}\sum_{m,n}\langle\chi_{m,n}|x\rangle\ln(\hbar\nu_m - \hbar\alpha_n/\mathcal{T})\langle x|\chi_{m,n}\rangle \qquad (3.113)\\
&= -\mathrm{i}\hbar\int_0^{\mathcal{T}}\mathrm{d}t\sum_n\int\frac{\mathrm{d}\omega}{2\pi}\ln(\hbar\omega - \hbar\alpha_n/\mathcal{T}),
\end{aligned}$$

where in the second step the unit operator $\int \mathrm{d}^d x|x\rangle\langle x|$ is inserted, while in the last step the sum over the frequency is replaced with an integral, $(1/\mathcal{T})\sum_m \to (1/2\pi)\int \mathrm{d}\omega$, which is justified in the limit $\mathcal{T} \to \infty$. The prefactors in this transformation follow from the definition of ν_m below Eq. (3.109), which implies that $\Delta\nu_m = (2\pi/\mathcal{T})\Delta m$, with $\Delta m = 1$ the minimal (positive) change in the summation variable. The $\chi_{m,n}$'s are assumed to be normalized to unity,

$$\int \mathrm{d}^D x\,\langle\chi_{m,n}|t,\mathbf{x}\rangle\langle t,\mathbf{x}|\chi_{m,n}\rangle = \langle\chi_{m,n}|\chi_{m,n}\rangle = 1. \qquad (3.114)$$

We now specialize to a relativisticlike theory by adding the small positive switching parameter $\eta = 0^+$ to the argument of the logarithm in Eq. (3.104), see Sec. 2.4. With the help of the integral

$$\hbar\int\frac{\mathrm{d}\omega}{2\pi}\frac{1}{\hbar\omega - x + \mathrm{i}\,\mathrm{sgn}(\omega)\eta} = -\frac{\mathrm{i}}{2}\mathrm{sgn}(x), \qquad (3.115)$$

the integration over the frequency in Eq. (3.113) is readily carried out with the result

$$\Gamma_1 = \frac{1}{2}\frac{\hbar}{\mathcal{T}}\int_0^{\mathcal{T}}\mathrm{d}t\sum_n|\alpha_n|. \qquad (3.116)$$

For time-independent background fields $\alpha_n = E_n \mathcal{T}/\hbar$. If the eigenvalues $E_n(t)$ do not cross zero, the first term at the right side of Eq. (3.112) is of order $O(\mathcal{T})$ while the second is of order $O(1)$. The effective action thus becomes in the limit $\mathcal{T} \to \infty$

$$\Gamma_1 = \frac{1}{2}\int_0^{\mathcal{T}} dt \sum_n |E_n| - \frac{1}{2}\hbar \sum_n \mathrm{sgn}(E_n)\gamma_n. \tag{3.117}$$

The first term on the right simply denotes minus the ground state energy, or zero-point energy of the system,

$$V_1 = -\frac{1}{2}\sum_n |E_n|, \tag{3.118}$$

while the second term is the exact analog of the Wess-Zumino term (3.97). The minus sign at the right of Eq. (3.118) is characteristic of Grassmann fields and shows that fermions contribute with a negative sign to the zero-point energy .

3.8 Landau-Levels Technique

For charged systems, one-loop quantum corrections can often be evaluated in a physically intuitive way by using that a charged particle in an external magnetic field has a discrete energy spectrum.

Consider the Dirac Hamiltonian in three space dimensions

$$H = c\boldsymbol{\alpha}\cdot\left(\mathbf{p} + \frac{e}{c}\mathbf{A}\right) + \beta mc^2, \tag{3.119}$$

with $p^i = i\hbar\partial^i = -i\hbar\partial_i$, describing a relativistic electron of mass m and charge $-e < 0$. Here, $\boldsymbol{\alpha}$ and β denote 4×4 matrices which can be represented as

$$\boldsymbol{\alpha} = \begin{pmatrix} 0 & \boldsymbol{\sigma} \\ \boldsymbol{\sigma} & 0 \end{pmatrix}, \quad \beta = \begin{pmatrix} 1 & 0 \\ 0 & -1 \end{pmatrix}, \tag{3.120}$$

with $\boldsymbol{\sigma}$ the Pauli matrices. These matrices satisfy the Clifford algebra

$$\left\{\alpha^i, \alpha^j\right\} \equiv \alpha^i\alpha^j + \alpha^j\alpha^i = 2\delta^{ij}, \tag{3.121}$$

and in addition the relations

$$\left[\alpha^i, \alpha^j\right] = 2i\epsilon^{ijk}\Sigma^k, \quad \text{with} \quad \Sigma^k \equiv \begin{pmatrix} \sigma^k & 0 \\ 0 & \sigma^k \end{pmatrix}, \tag{3.122}$$

$$\left\{\alpha^i, \beta\right\} = 0, \quad \beta^2 = 1. \tag{3.123}$$

For a constant external magnetic field in the 3-direction, the vector potential may be chosen as

$$\mathbf{A} = \left(0, Bx^1, 0\right), \tag{3.124}$$

where it is used that for a charged particle in vacuum, the induction field B, which is the field felt by the charge, equals the applied magnetic field H, $B = H$. The spectrum of the one-particle Dirac Hamiltonian is readily determined by considering the square of the operator and using that

$$
\begin{aligned}
\left[\alpha \cdot \left(\mathbf{p} + \frac{e}{c}\mathbf{A}\right)\right]^2 &= \left(\tfrac{1}{2}\left\{\alpha^i, \alpha^j\right\} + \tfrac{1}{2}\left[\alpha^i, \alpha^j\right]\right)\left(\mathbf{p} + \frac{e}{c}\mathbf{A}\right)^i \left(\mathbf{p} + \frac{e}{c}\mathbf{A}\right)^j \\
&= \left(\mathbf{p} + \frac{e}{c}\mathbf{A}\right)^2 + \frac{\mathrm{i}}{2}\frac{e\hbar}{c}\left[\alpha^i, \alpha^j\right]\partial^i A^j \\
&= \left(\mathbf{p} + \frac{e}{c}\mathbf{A}\right)^2 + \frac{e\hbar}{c}\mathbf{B} \cdot \mathbf{\Sigma}. \qquad (3.125)
\end{aligned}
$$

This gives

$$
H^2 = c^2\left(\mathbf{p} + \frac{e}{c}\mathbf{A}\right)^2 + m^2c^4 + e\hbar c\mathbf{B} \cdot \mathbf{\Sigma}, \qquad (3.126)
$$

where we omitted the 4×4 unit matrix in the first and second term on the right. It is understood that each term constitutes a 4×4 matrix. The matrix H^2 is seen to be block diagonal with two identical 2×2 parts. The first term on the right is formally (c^2 times) the Hamiltonian of a *nonrelativistic* electron of mass $M = \frac{1}{2}$ in a uniform constant magnetic field. The eigenvalues λ_n of that problem, first solved by Landau, are given by

$$
\lambda_n = \frac{\hbar}{Mc}|eB|\left(n + \tfrac{1}{2}\right) + \hbar^2 k_3^2 \qquad (3.127)
$$

where $n = 0, 1, 2 \cdots$ labels the Landau levels, and the last term corresponds to free linear motion parallel to the applied field. Given this result, the energy eigenvalues $E_{n,\sigma}$ of the one-particle Dirac Hamiltonian H now follow as

$$
E_{n,\sigma} = \pm\sqrt{2\hbar c|eB|\left(n + \tfrac{1}{2}\right) + \hbar^2c^2k_3^2 + m^2c^4 + \hbar ceB\sigma}, \qquad (3.128)
$$

which can be positive or negative. Here, $\sigma = \pm 1$ indicates whether the electron spin has a component along the applied field ($\sigma = 1$) or in the opposite direction ($\sigma = -1$). As an aside, we note that the eigenvalues of a charged relativistic spin-0 particle are given by Eq. (3.128) with the spin term set to zero.

In the nonrelativistic limit, which corresponds to taking m large and disregarding the negative energy levels, the relativistic spectrum reduces to

$$
E_{n,\sigma} = \frac{\hbar}{mc}|eB|\left(n + \tfrac{1}{2}\right) + \frac{\hbar^2 k_3^2}{2m} + \frac{\hbar e}{2mc}B\sigma, \qquad (3.129)
$$

up to an irrelevant additive constant. Note that electrons with their spin aligned with the applied field have a larger energy than when their spin points in the opposite direction.

In both the relativistic and the nonrelativistic case, each Landau level has a degeneracy of $|eB|/2\pi\hbar c$ per unit area perpendicular to the applied field. This degeneracy is due to (generalized) translational invariance in the x^1–x^2 plane, as can be seen as follows. In the gauge (3.124), the problem is manifestly invariant under translations in the 2-direction. Under translations in the 1-direction $x^1 \to x^1 + a$, the vector potential however changes, $A^2 \to A^2 + Ba$. But this change can be canceled by a gauge transformation $A^\mu \to A^\mu + \partial^\mu \Lambda$ with gauge parameter $\Lambda = Bax^2$, implying that the problem is invariant under the combined action of a translation and a gauge transformation. Let L be the extent of the system in the 2-direction. The wave vector in this direction is then quantized as

$$\Delta k^2 = 2\pi l / L \tag{3.130}$$

with l an integer. Since k^2 appears in the Hamiltonian in the combination $\hbar k^2 + (e/c)A^2$ it follows that in the gauge (3.124)

$$\hbar \Delta k^2 = (|eB|/c)\Delta x^1, \tag{3.131}$$

where both Δk^2 and Δx^1 are defined to be positive. On combining the last two equations, we deduce that a single state, corresponding to $l = 1$ in Eq. (3.130), occupies a spatial strip with area

$$L \Delta x^1 = 2\pi \frac{\hbar c}{|eB|}. \tag{3.132}$$

And, hence, the number of states per unit area in a single Landau level is $|eB|/2\pi\hbar c$. Since in deriving this result we merely used the specific form of the minimal coupling, the result is valid for any gauge-invariant theory, be it relativistic or nonrelativistic.

To demonstrate the use of Landau levels to efficiently calculate one-loop quantum corrections, we turn to a problem in quantum electrodynamics (QED) and consider the zero-temperature partition function of relativistic electrons in a constant magnetic background field $H = B$ pointing in the positive 3-direction

$$Z = \int \mathrm{D}\psi^\dagger \mathrm{D}\psi \, \exp\left\{ \frac{\mathrm{i}}{\hbar} \int \mathrm{d}^4 x \left[\psi^\dagger (\mathrm{i}\hbar\partial_t - H)\psi - \frac{1}{8\pi} B^2 \right] \right\} \tag{3.133}$$

with H the one-particle Dirac Hamiltonian (3.119). Matching the 4×4 structure of H, the Grassmann field ψ has four components, describing electrons and their antiparticles, positrons, each with two spin states. The effective action $\Gamma = -\mathrm{i}\hbar \ln Z$ up to one-loop order obtained after integrating out the electrons reads

$$\begin{aligned} \Gamma[B] &= -\frac{1}{8\pi} \int \mathrm{d}^4 x \, B^2 - \mathrm{i}\hbar \, \mathrm{Tr} \ln(p_0 - H) \\ &= -\frac{1}{8\pi} \int \mathrm{d}^4 x \, B^2 - \mathrm{i}\hbar \int \mathrm{d}^4 x \frac{|eB|}{2\pi\hbar c} \sum_{n,\sigma} \int \frac{\mathrm{d}\omega}{2\pi} \frac{\mathrm{d}k^3}{2\pi} \ln(\hbar\omega - E_{n,\sigma}), \end{aligned} \tag{3.134}$$

where the sum extends over all Landau levels labeled by $n = 0, 1, 2 \cdots$ and over the two spin projections. Both the positive ($E_{n,\sigma} > 0$) and negative ($E_{n,\sigma} < 0$) energy eigenvalues are to be included in Eq. (3.134). The degeneracy per Landau level has been accounted for by including the factor $|eB|/2\pi\hbar c$. To assure causality, the infinitesimal switching parameter $\eta = 0^+$ is to be introduced in Eq. 3.134) by letting $\hbar\omega \to \hbar\omega + \mathrm{i}\,\mathrm{sgn}(\omega)\eta$. With the help of the result Eq. (3.115), the frequency integral in Eq. (3.134) can be evaluated to give for the corresponding Lagrangian density

$$\mathcal{L}_{\rm eff} = -\frac{1}{8\pi}B^2 + \frac{|eB|}{4\pi\hbar c}\int\frac{\mathrm{d}k^3}{2\pi}\sum_\sigma\sum_{n=0}^{\infty}|E_{n,\sigma}|, \tag{3.135}$$

where we repeat that both the positive and negative energy eigenvalues are to be included. The sum over the Landau levels is best evaluated by use of the Schwinger proper-time representation of the square root,

$$\frac{1}{a^z} = \frac{1}{\Gamma(z)}\int_0^\infty\frac{\mathrm{d}s}{s}\,s^z\,\mathrm{e}^{-as}, \tag{3.136}$$

with $z = -\frac{1}{2}$ and $\Gamma(-\frac{1}{2}) = -2\sqrt{\pi}$. The remaining integral over k^3 is Gaussian and readily evaluated. After subtracting the B-independent part, which represents the free-particle contribution to the partition function in the absence of an external field, we obtain in this way

$$\mathcal{L}_{\rm eff} = -\frac{1}{8\pi}B^2 - \frac{1}{8\pi^2}\frac{1}{(\hbar c)^3}\int_0^\infty\frac{\mathrm{d}s}{s^3}\mathrm{e}^{-m^2c^4s}\left(\hbar c|eB|s\frac{\cosh(\hbar ceBs)}{\sinh(\hbar c|eB|s)} - 1\right). \tag{3.137}$$

The sinh factor arises from the orbital motion of the charged particles, while the cosh factor stems from their spin.

Expanding the effective Lagrangian in a power series in the coupling constant e, we note that the first term diverges. To regularize this divergence, we apply dimensional regularization and consider the system in arbitrary $d = 4-\varepsilon$ dimensions with ε small by replacing the momentum integral $\int \mathrm{d}k^3/2\pi$ with

$$\int_{-\infty}^\infty\frac{\mathrm{d}k^3}{2\pi} \to \frac{\Omega_{1-\varepsilon}}{(2\pi)^{1-\varepsilon}}\int_0^\infty \mathrm{d}k k^{-\varepsilon}, \tag{3.138}$$

where $\Omega_d = 2\pi^{d/2}/\Gamma(d/2)$ is the surface of a unit hypersphere embedded in d space dimensions. We proceed by using that

$$(\hbar c)^\varepsilon\int_0^\infty\frac{\mathrm{d}s}{s^{1-\varepsilon/2}}\mathrm{e}^{-m^2c^4s} = \Gamma\left(\frac{\varepsilon}{2}\right)\kappa^{-\varepsilon}\left(\frac{\hbar\kappa}{mc}\right)^\varepsilon \approx \frac{2}{\varepsilon}\kappa^{-\varepsilon} \tag{3.139}$$

to logarithmic accuracy. Here, κ is an arbitrary parameter of dimension inverse length introduced so that the dimensionless ratio $\hbar\kappa/mc$ appears as argument of the function $f(x) \equiv x^\varepsilon$, which can then be expanded about $\varepsilon = 0$. The first term in the expansion of the effective Lagrangian then reads

$$\mathcal{L}_{\text{eff}} = -\frac{1}{8\pi}\left(1 + \frac{2}{3\pi\varepsilon}\frac{e^2\kappa^{-\varepsilon}}{\hbar c}\right)B^2 + \cdots , \tag{3.140}$$

where the combination $e^2\kappa^{-\varepsilon}/\hbar c$ is dimensionless in $4 - \varepsilon$ dimensions. The divergence, which arises when taking $\varepsilon \to 0$, appears in a term of a form present in the original Lagrangian. It can therefore be absorbed by introducing a renormalized coupling constant e_r and a renormalized field B_r as follows. Since the magnetic vector potential $\mathbf{A}$ is coupled to the electron field in a gauge-invariant way through minimal substitution $\nabla \to \nabla + i(e/\hbar c)\mathbf{A}$, the combination $e\mathbf{A}$ and also eB is invariant under renormalization. That is,

$$eB = e_r\kappa^{\varepsilon/2}B_r, \tag{3.141}$$

where the arbitrary inverse length scale κ introduced in Eq. (3.140) is used to remove the dimension of the renormalized coupling constant when measured in natural units where $\hbar = c = 1$. More specifically, $e_r^2/\hbar c$ is dimensionless. Equation (3.141) can be used to rewrite Eq. (3.140) as

$$\begin{aligned}\mathcal{L}_{\text{eff}} &= -\frac{1}{8\pi}\frac{\kappa^\varepsilon}{e^2}\left(1 + \frac{2}{3\pi\varepsilon}\frac{e^2\kappa^{-\varepsilon}}{\hbar c}\right)(e_rB_r)^2 + \cdots \\ &\equiv -\frac{1}{8\pi}B_r^2 + \cdots ,\end{aligned} \tag{3.142}$$

where the relation between the bare and renormalized coupling constants reads to this order

$$\frac{1}{e_r^2} = \frac{\kappa^\varepsilon}{e^2}\left(1 + \frac{2}{3\pi\varepsilon}\frac{e^2\kappa^{-\varepsilon}}{\hbar c}\right), \tag{3.143}$$

or

$$\frac{1}{e^2} = \frac{\kappa^{-\varepsilon}}{e_r^2}\left(1 - \frac{2}{3\pi\varepsilon}\frac{e_r^2}{\hbar c}\right). \tag{3.144}$$

For so-called *renormalizable theories*, all divergences can be absorbed in a redefinition of the fields and parameters appearing in the original theory. The renormalized coupling constant is the physical parameter that can be measured in experiment. The parameter κ is interpreted as the scale at which the charge is probed, and $e_r = e_r(\kappa)$. Since the bare coupling constant e is independent of κ, it follows that

$$\beta(e_r^2) \equiv \kappa\frac{de_r^2}{d\kappa} = -\varepsilon e_r^2 + \frac{2}{3\pi}\frac{e_r^4}{\hbar c} + O(e_r^6), \tag{3.145}$$

where on the left we introduced the so-called *beta* function. In the physical limit $\varepsilon \to 0$, this differential, or flow equation gives

$$\frac{\hbar c}{e_r^2(\kappa')} - \frac{\hbar c}{e_r^2(\kappa)} = -\frac{2}{3\pi}\ln\left(\frac{\kappa'}{\kappa}\right), \tag{3.146}$$

showing that for small e_r, the physical charge increases with increasing scale $\kappa' > \kappa$. *Vice versa*, the electric charge becomes weaker at large distances. This continues until the electron Compton wave length $\lambda_C \equiv 2\pi\hbar/mc$ is reached, beyond which the electric charge remains constant—that is, λ_C acts as an infrared cutoff. The scale dependence of the physical charge shows that the QED vacuum can be thought of as a polarizable medium that screens charges. The medium is comprised of virtual pairs of electrons and positrons, the antiparticles of electrons, that constantly pop up from the vacuum and quickly disappear again.

The higher-order terms in the expansion of the one-loop expression (3.137) are all finite. Expanding it to fourth order in B, one finds

$$\mathcal{L}_{\text{eff}} = -\frac{1}{8\pi}B_r^2 - \frac{1}{360\pi^2}\frac{\hbar e_r^4}{c^7 m^4}B_r^4 + O\left(e_r^6\right), \tag{3.147}$$

where in the second term on the right, Eq. (3.141) with $\varepsilon \to 0$ is used.

For a charged scalar field in a background magnetic field, the expression for the one-loop effective action is obtained from Eq. (3.137) by omitting the spin part, and multiplying it by a factor of $-\frac{1}{2}$. This factor arises because, in contrast to a boson loop, a fermion loop carries a minus sign, and because spin-$\frac{1}{2}$ particles have two spin orientations.

3.9 Magnetic Susceptibility of an Electron Gas

As an application of the Landau-levels technique in the context of condensed matter, we consider a free electron gas in a background magnetic field. As detailed in Sec. 2.3, integrands as in Eq. (3.115) carry in nonrelativistic theories an additional convergence factor so that, cf. (3.14)

$$\hbar\int\frac{d\omega}{2\pi}\frac{e^{i\omega 0^+}}{\hbar\omega - x + i\,\text{sgn}(\omega)\eta} = i\theta(-x). \tag{3.148}$$

Given the eigenvalues (3.129) with a chemical potential μ included by letting $E_{n,\sigma} \to E_{n,\sigma} - \mu$, we obtain as nonrelativistic counterpart of the effective Lagrangian (3.135)

$$\mathcal{L}_1 = \frac{|eB|}{2\pi\hbar c}\int\frac{dk^3}{2\pi}\sum_{n=0}^{\infty}\sum_{\sigma=\pm 1}(\mu - E_{n,\sigma})\theta(\mu - E_{n,\sigma}), \tag{3.149}$$

which only includes positive-energy Landau levels. In Eq. (3.149), the magnetic energy density $(1/8\pi)B^2$ has been omitted for brevity. We evaluate this expression in two stages by first omitting the spin term in Eq. (3.129). That term is easily dealt with at the end for it merely shifts the Landau levels by $\pm\hbar eB/2mc$. The integral over k^3 is readily evaluated with the result for the orbital contribution to the effective Lagrangian

$$\mathcal{L}_{1,\mathrm{orbi}} = \frac{|eB|}{3\pi^2\hbar c}\left(\frac{2m}{\hbar^2}\right)^{1/2}\sum_{n=0}^{i}\left[\mu - \frac{\hbar}{mc}|eB|(n+\tfrac{1}{2})\right]^{3/2}, \tag{3.150}$$

where the sum over n is up to the integer i defined by

$$i \equiv \left[\frac{\mu mc}{\hbar|eB|} - \frac{1}{2}\right], \tag{3.151}$$

and $[x]$ is the integer-part function, denoting the largest integer less than x. For sufficiently small magnetic fields, the summation in Eq. (3.150) may be carried out by using Euler's summation formula

$$\sum_{n=0}^{i} f(n) = \int_{-\frac{1}{2}}^{i+\frac{1}{2}} \mathrm{d}x\, f(x) - \frac{1}{24}\left[f'\left(i+\tfrac{1}{2}\right) - f'\left(-\tfrac{1}{2}\right)\right] + \cdots, \tag{3.152}$$

giving

$$\mathcal{L}_{1,\mathrm{orbi}} = \frac{g}{15\pi^2}\left(\frac{2m}{\hbar^2}\right)^{3/2}\mu^{5/2} - \frac{g}{24\pi^2}\left(\frac{2m}{\hbar^2}\right)^{-1/2}\mu^{1/2}\left(\frac{eB}{\hbar c}\right)^2 + \cdots, \tag{3.153}$$

where we replaced $i+1$ with i, which is justified for small B. The first term at the right side of Eq. (3.153), which is independent of the magnetic field, is the free particle contribution in the absence of an applied field, while the second term reproduces, apart form the spin factor $g = 2$, the result (3.23) which was obtained by means of the derivative expansion method. Using the definition (3.16) of the magnetic susceptibility χ, we obtain from Eq. (3.153) a negative contribution to the susceptibility

$$\chi_{\mathrm{orbi}} = -\frac{1}{6\pi^2}\frac{e^2}{\hbar^2 c^2}\left(\frac{\hbar^2\mu}{2m}\right)^{1/2}, \tag{3.154}$$

known as the Landau *diamagnetic* susceptibility. It represents the effect of orbital motion, which produces a magnetic field that tends to oppose the applied field.

The spin term in Eq. (3.129) can be accounted for in the expression (3.153) for the effective Lagrangian by letting

$$\mu \to \mu_{\uparrow,\downarrow} = \mu \mp \frac{\hbar}{2mc}eB \tag{3.155}$$

for spin up and spin down electrons, respectively. This leads to a positive contribution to the susceptibility

$$\chi_{\rm spin} = \mu_{\rm B}^2 g\nu(0), \tag{3.156}$$

known as the Pauli *paramagnetic* susceptibility. By aligning with the applied field, the magnetic spin moments enhance rather than oppose the applied field, where it is recalled that the spin and the magnetic moment of an electron point in opposite directions. The sum of the two susceptibilities,

$$\chi = \chi_{\rm spin} + \chi_{\rm orbi} = \mu_{\rm B}^2 g\nu(0)\left(1 - \tfrac{1}{3}\right), \tag{3.157}$$

shows that a free electron gas in three space dimensions is paramagnetic as the Landau diamagnetic contribution only compensates part of the Pauli paramagnetic contribution. In Eq. (3.157), $\mu_{\rm B} \equiv e\hbar/2mc$ is the Bohr magneton, and $\nu(0) = mk_{\rm F}/2\pi^2\hbar^2$, with $k_{\rm F}$ the Fermi wave number, is the density of states at the Fermi surface per spin degree of freedom. The ratio of the two susceptibilities is

$$\chi_{\rm spin}/\chi_{\rm orbi} = -\frac{1}{3}. \tag{3.158}$$

Exactly the same ratio is found in the relativistic case (3.137) as the Paul paramagnetic contribution is obtained by expanding $\cosh(x) = 1 + \frac{1}{2}x^2 + \cdots$, while the Landau diamagnetic contribution is obtained by expanding $x\sinh^{-1}(x) = 1 - \frac{1}{6}x^2 + \cdots$. However, the overall sign is opposite, i.e., $\chi_{\rm QED} < 0$ as follows from Eq. (3.140) and the definition (3.16), implying that the QED vacuum is diamagnetic.

Notes

(i) REDUCE, developed by Hearn and others, is one of the most widely used general computer algebra systems. It is available for most common computing platforms at a moderate price, see [Hearn (2004)].

(ii) SCHOONSCHIP is an algebraic manipulation program developed by Veltman to handle large problems in particle physics. Although copyrighted by its author, the executable code is freely available for various (older) computer platforms, see [Veltman and Williams (1993)].

(iii) FORM is an intellectual descendant of SCHOONSCHIP developed by Vermaseren to handle very large expressions of millions of terms in an efficient way. The program runs on various computing systems and is available free of charge, see [Vermaseren (2007)].

(iv) The derivative expansion method was developed by Fraser (1985) where it was applied to ϕ^4 theory. Section 3.4 closely follows this reference. The application of this algebraic procedure to fermionic theories was put forward in [Aitchison and Fraser (1984)] and [Aitchison and Fraser (1985)]. For applications to exactly solvable two-dimensional models and finite-temperature field theory, see [Das (1999)].

(v) The semiclassical method is well covered in the textbook [Rajaraman (1982)].

(vi) The method to calculate charges and currents induced by solitons in quantum field theory was introduced by Goldstone and Wilczek (1981). See [Wilczek (2002)] for an overview and [MacKenzie (1984)] for detailed calculations.

(vii) The dimensional regularization technique to render diverging integrals finite by continuation from a fixed to an arbitrary number d of spacetime dimensions was introduced by 't Hooft and Veltman (1972).

(viii) Our presentation of the geometric phase closely follows the original paper [Berry (1984)]. Together with about 40 other papers on the geometric phase, including some predating Berry's, it is reprinted in [Shapere and Wilczek (1989)]. Prior to Berry's seminal paper, this phase was generally thought to be inconsequential, see, for example [Schiff (1968)].

(ix) The Wess-Zumino term is used here in the sense developed by Witten (1983).

(x) The connection between the Berry phase and Wess-Zumino terms was established independently in [Niemi and Semenoff (1985)] and in [Stone (1986)]. It is further detailed in [Aitchison (1987); Niemi (1987)], and in [Zuk (1987)].

(xi) The use of Floquet indices to calculate fermionic determinants was first demonstrated by Dashen *et al.* (1975). For further details, see [Rajaraman (1982)].

(xii) The problem of a nonrelativistic charged particle moving in a uniform magnetic field was solved by Landau (1930).

(xiii) The Landau-levels technique to calculate one-fermion-loop quantum corrections in QED was introduced by Nielsen (1981). The technique was applied in that paper primarily to QCD (quantum chromodynamics) to establish asymptotic freedom.

(xiv) The powerful calculation technique, known as the Schwinger proper-time method, which we repeatedly use throughout the book, was introduced in [Schwinger (1951)].

(xv) The effective Lagrangian (3.147) for photons to fourth order in an external magnetic field with the contributions due to an external electric field included was first obtained by Euler and Kockel (1935), while the expression (3.137) to all orders was first calculated by Heisenberg and Euler (1936).

Chapter 4

Bose-Einstein Condensation

In 1995, Cornell and Wieman (JILA, University of Colorado at Boulder) succeeded for the first time in creating an atomic Bose-Einstein condensate by laser-cooling a magnetically trapped gas of ^{87}Rb atoms. Such Bose gases are now routinely produced by experimental groups around the world. A Bose-Einstein condensate, involving identical particles sharing the same quantum state, constitutes a new quantum state of matter that is sometimes referred to as a "fifth state of matter". It is one of the most prevalent topics in contemporary condensed matter physics.

4.1 Ideal Bose Gas

We first study Bose-Einstein condensation (BEC) in an ideal uniform Bose gas. As starting point we return to the expression (2.63) for the thermodynamic potential

$$\frac{\Omega}{V} = \frac{\Omega_{\mathbf{k}=0}}{V} + \frac{1}{\beta}\sum_{n=-\infty}^{\infty}\int \frac{\mathrm{d}^D k}{(2\pi)^D} \ln\left[-\mathrm{i}\hbar\omega_n + \epsilon(\mathbf{k}) - \mu\right], \tag{4.1}$$

with $\epsilon(\mathbf{k}) = \hbar^2\mathbf{k}^2/2m$ the kinetic energy of the free atoms. A possible contribution of the zero wave-vector $\mathbf{k} = 0$ state is included here. This contribution is not captured by the second term because the density of states, behaving as $\nu(\epsilon) \sim \epsilon^{D/2-1}$, is zero for the $\mathbf{k} = 0$ state in $D > 2$. It therefore has to be treated separately. Explicitly,

$$\begin{aligned}\Omega_{\mathbf{k}=0} &= \frac{1}{\beta}\sum_{n=-\infty}^{\infty} \ln(-\mathrm{i}\hbar\omega_n - \mu) \\ &= \frac{1}{\beta}\ln(1 - \mathrm{e}^{-\alpha}).\end{aligned} \tag{4.2}$$

where we repeated the steps leading to Eq. (2.67). Formally, the sum diverges, but the regularization method we use to evaluate it produces a finite result. The special status of the zero wave-vector state follows from considering the pressure $P = -\partial\Omega/\partial V$, showing that this ground state does not contribute. It also does not contribute to the average energy $\langle E \rangle = \partial(\beta\Omega)/\partial\beta$. At sufficiently high temperatures, the $\mathbf{k} = 0$ state can be ignored. But at a certain temperature $T_{\rm BEC}$, particles start to condense in this ground state. By further lowering the temperature, the ground state fills up with a finite fraction of the total number of particles present in the system.

The integration over wave vectors in Eq. (4.1) can be conveniently carried out by introducing the Schwinger proper-time representation (2.64) of the logarithm. Together with the Poisson summation formula (2.66) to carry out the sum over the Matsubara frequencies, this yields the fugacity series

$$\begin{aligned} \frac{\Omega}{V} &= \frac{\Omega_{\mathbf{k}=0}}{V} - \frac{1}{\beta\lambda_T^D}\sum_{w=1}^{\infty} w^{-D/2-1}\mathrm{e}^{-\alpha w} \\ &= \frac{\Omega_{\mathbf{k}=0}}{V} - \frac{1}{\beta\lambda_T^D} g_{D/2+1}(z), \end{aligned} \tag{4.3}$$

where

$$g_s(z) \equiv \sum_{w=1}^{\infty} \frac{z^w}{w^s}, \tag{4.4}$$

with $z = \exp(-\alpha) = \exp(\beta\mu)$, is the so-called *Bose function*. For $z = 1$ it reduces to the Riemann ζ function $g_s(1) = \zeta(s)$,

$$\zeta(s) \equiv \sum_{w=1}^{\infty} \frac{1}{w^s}. \tag{4.5}$$

Finally, λ_T in Eq. (4.3) is the de Broglie thermal wave length (2.33). If the mass of the particles is large or if the temperature is high (β small), the thermal wave length is small. By Eq. (2.12), the average particle number density $n \equiv \langle N \rangle / V$ can be obtained by taking the derivative of Ω with respect to α, or equivalently,

$$n = -\frac{1}{V}\frac{\partial\Omega}{\partial\mu}. \tag{4.6}$$

In the normal gas state, the chemical potential is negative so that the energy spectrum,

$$\xi(\mathbf{k}) \equiv \epsilon(\mathbf{k}) - \mu \tag{4.7}$$

is gapped with $-\mu > 0$ denoting the minimum of energy. In this state, the occupancy of the zero wave-vector state can be ignored.

Since the lowest energy level of a free boson is zero, $\epsilon(0) = 0$, the critical value of the chemical potential, defined in Eq. (2.6), is also zero in the thermodynamic limit, where $\langle N \rangle, V \to \infty$, such that the average particle density n is fixed. For $\mu \to 0^-$, Eq. (4.6) gives

$$n = \frac{\zeta(D/2)}{\lambda_{T_{\rm BEC}}^D}, \tag{4.8}$$

which defines the condensation temperature

$$k_{\rm B} T_{\rm BEC} = \frac{2\pi\hbar^2}{m}\left(\frac{n}{\zeta(D/2)}\right)^{2/D}, \tag{4.9}$$

where it is assumed that the particle number density n is given. Equation (4.9) shows that the condensation temperature of a free uniform Bose gas is determined by the mass m of the atoms involved and the particle number density. If the mass of the particles is large or the particle number density small, the condensation temperature is low. Since $n\lambda_{T_{\rm BEC}}^D \sim 1$ by Eq. (4.8), the thermal wave length must be of order the average interparticle spacing for Bose-Einstein condensation to happen. This means either a high particle number density or ultracold temperatures. Below $T_{\rm BEC}$, the chemical potential remains locked at the critical value $\mu_{\rm c} = 0$ in the thermodynamic limit.

In a finite system, the chemical potential remains finite. The $\mathbf{k} = 0$ contribution (4.2) then yields the additional term to the particle number density

$$n_0 = \frac{1}{V}\frac{1}{{\rm e}^{\alpha} - 1} = \frac{1}{V}\sum_{w=1}^{\infty} {\rm e}^{-\alpha w}, \tag{4.10}$$

by Eq. (4.6). It physically represents the number density of particles residing in the condensate. The total particle number density becomes

$$n = n_0 + \frac{1}{\lambda_T^D} g_{D/2}(z). \tag{4.11}$$

When the ground state harbors a finite fraction of the particles, it follows from Eq. (4.10) that $\alpha \sim 1/\langle N \rangle$, showing that α and thus $\mu = -\alpha/\beta$ vanish in the thermodynamic limit. At the absolute zero of temperature, all the atoms of an ideal Bose gas share the same $\mathbf{k} = 0$ quantum state.

4.2 Critical Properties

We next consider the critical properties of an ideal uniform Bose gas as summarized by the critical exponents. Describing the system in equilibrium, these universal properties are exclusively determined by the zero-frequency mode in the

Fourier series (2.41). The contribution of this mode to the thermodynamic potential can be obtained from the fugacity series (4.3) by replacing the sum over w by an integral:

$$\sum_{w=1}^{\infty} w^{-D/2-1} \mathrm{e}^{-\alpha w} \to \int_0^{\infty} \mathrm{d}w\, w^{-D/2-1} \mathrm{e}^{-\alpha w} = \alpha^{D/2} \Gamma(-D/2), \tag{4.12}$$

as can be explicitly checked by directly evaluating the $n = 0$ term in Eq. (4.1) with the help of the Schwinger proper-time representation (2.64). This mode contributes the term $\Gamma(1 - D/2)\alpha^{D/2-1}/\lambda_T^D$ to the particle number density so that slightly above $T_{\rm BEC}$:

$$n \approx \frac{1}{\lambda_T^D} \left[\zeta(D/2) + \Gamma(1 - D/2)\alpha^{D/2-1} \right]. \tag{4.13}$$

With the expression (4.8) valid at $T = T_{\rm BEC}$, it then follows that for $T > T_{\rm BEC}$

$$\alpha^{D/2-1} \approx \frac{\zeta(D/2)}{\Gamma(-D/2)} \left(\frac{T}{T_{\rm BEC}} - 1 \right). \tag{4.14}$$

For fixed particle number density, the chemical potential therefore tends to zero in a D-dependent way as

$$\mu(T) \sim -(T - T_{\rm BEC})^{2/(D-2)}. \tag{4.15}$$

Below the condensation temperature, the chemical potential of the noninteracting theory remains zero in the thermodynamic limit.

As an aside, note that since the last term in Eq. (4.13) represents the contribution of the zero-frequency mode, the first term at the right, which determines the condensation temperature, represents the contribution of the dynamic modes. In other words, although the critical properties are determined exclusively by the zero-frequency mode, the condensation temperature $T_{\rm BEC}$ of an ideal Bose gas is determined by the dynamic modes only.

The derivative of Eq. (4.13) with respect to the chemical potential μ yields by Eq. (2.202) the compressibility. It is seen to scale as $\kappa \sim (-\mu)^{D/2-2} \sim (T - T_{\rm BEC})^{(D-4)/(D-2)}$ when the condensation temperature $T_{\rm BEC}$ is approached from above, giving

$$\alpha_{\rm x} = \frac{4-D}{D-2} \tag{4.16}$$

for the critical exponent defined through

$$\kappa \sim (T - T_{\rm BEC})^{-\alpha_{\rm x}}. \tag{4.17}$$

This exponent is negative for $2 < D < 4$, implying that the compressibility remains finite at the condensation temperature.

To compute the other exponents, we consider the static correlation function $G(\mathbf{x})$, which is obtained from the time-dependent function (2.80) by setting $\tau = 0$. Proceeding as in the previous section, we apply Eq. (3.136) with $z = 1$ to write the numerator in the resulting integrand as an integral over the Schwinger proper time, and use the Poisson summation formula (2.66) to replace the sum over the Matsubara frequencies with a chain of delta functions. The integrals are now readily evaluated with the result

$$G(\mathbf{x}) = \frac{1}{\lambda_T^D} \sum_{w=0}^{\infty} w^{-D/2} \, \mathrm{e}^{-\alpha w} \exp\left(-\frac{\pi r^2}{\lambda_T^2 w} \right) \tag{4.18}$$

with $r \equiv |\mathbf{x}|$. As for the thermodynamic potential, the contribution of the zero-frequency mode to the correlation function is obtained by replacing the sum over w with an integral, yielding

$$G(\mathbf{x}) = \frac{2}{\lambda_T^2 (2\pi \xi r)^{D/2-1}} K_{D/2-1}(r/\xi), \tag{4.19}$$

where $K_n(x)$ denotes the modified Bessel function of the second kind and ξ is the finite-temperature correlation length, cf. Eq. (1.74)

$$\xi \equiv \frac{\lambda_T}{\sqrt{4\pi\alpha}}. \tag{4.20}$$

Asymptotically, for $r \gg \xi$,

$$G(\mathbf{x}) \sim \frac{1}{\lambda_T^2 (2\pi\xi)^{(D-3)/2} r^{(D-1)/2}} \, \mathrm{e}^{-r/\xi}. \tag{4.21}$$

As the condensation temperature T_{BEC} is approached from above, the correlation length diverges as $\xi \sim (-\mu)^{-1/2} \sim (T - T_{\mathrm{BEC}})^{-1/(D-2)}$ so that the correlation length exponent ν is

$$\nu = \frac{1}{D-2}. \tag{4.22}$$

Since $K_n(x) \sim 2^{n-1}\Gamma(n)/x^n$ for $x \to 0$, Eq. (4.19) takes the form

$$G(\mathbf{x}) \sim \frac{\Gamma(D/2-1)}{\pi^{D/2-1} \lambda_{T_{\mathrm{BEC}}}^2} \frac{1}{r^{D-2}} \tag{4.23}$$

right at the condensation temperature. This shows that the critical exponent η, defined through

$$G(\mathbf{x}) \sim 1/r^{D-2+\eta}, \tag{4.24}$$

is zero, $\eta = 0$. Despite describing a noninteracting system, the critical exponents obtained here are, apart from η, nontrivial for $D < 4$. The cause for this can be traced back to the nontrivial behavior (4.15) of the chemical potential upon approaching the condensation temperature, which in turn is caused by the constraint that the particle number density is fixed.

4.3 Ideal Bose Gas in a Harmonic Trap

We next consider mutually noninteracting particles trapped in an external harmonic oscillator potential

$$U_{\rm ex}(\mathbf{x}) = \frac{1}{2} m \sum_{i=1}^{D} \omega_i x_i^2, \tag{4.25}$$

characterized by the angular frequencies ω_i. The single-particle energy levels in this confining potential are

$$\epsilon_{\{n_i\}} = \sum_{i=1}^{D} \left(n_i + \tfrac{1}{2}\right) \hbar\omega_i, \tag{4.26}$$

with $n_i = 0, 1, 2 \cdots$. The lowest energy level corresponds to setting all $n_i = 0$. The critical value of the chemical potential, defined in Eq. (2.6), then follows as

$$\mu_{\rm c} = \frac{D}{2} \hbar\bar{\omega}, \tag{4.27}$$

where $\bar{\omega} \equiv \sum_{i=1}^{D} \omega_i / D$ is the arithmetic average of the trapping frequencies. The atoms condensed in this lowest energy state, being mutually noninteracting, are all described by the single-particle ground-state wave function of the harmonic oscillator. Their coherent behavior therefore produces the classical field

$$\psi_{\rm c}(\mathbf{x}) \sim {\rm e}^{-r^2/2\ell^2}, \tag{4.28}$$

where for simplicity we consider an isotropic trap ($\omega_i = \omega$), and

$$\ell \equiv \sqrt{\hbar/m\omega} \tag{4.29}$$

defines the linear size of the condensate. The Gaussian profile of the condensate, with its sharp peak at the center of the trap, is an important experimental signature of BEC in a harmonic trap. The atoms in excited states of the harmonic oscillator form what is termed the *thermal cloud*. To arrive at an estimate of the spatial extent of this cloud, we apply the semiclassical approximation by writing

$$n(\mathbf{x}) - n_0(\mathbf{x}) \sim {\rm e}^{-\beta U_{\rm ex}(\mathbf{x})} = {\rm e}^{-r^2/2b^2}, \tag{4.30}$$

with $b \equiv \sqrt{k_{\rm B}T/m\omega^2}$. Since $b/\ell = \sqrt{k_{\rm B}T/\hbar\omega} \gg 1$, the thermal cloud is spread out much more than the condensate.

The general expression (2.9) of the thermodynamic potential becomes for an ideal Bose gas in a harmonic trap

$$\Omega = -\frac{1}{\beta} \sum_{w=1}^{\infty} \frac{1}{w} \frac{z^w}{\prod_{i=1}^{D} (1 - {\rm e}^{-\beta\hbar\omega_i w})} \tag{4.31}$$

where the logarithm is expanded in a fugacity series with the fugacity

$$z \equiv e^{\beta(\mu-\mu_c)} \tag{4.32}$$

defined with the zero-point energy included. The fugacity so defined tends to unity at the condensation temperature. The average particle number follows as

$$\langle N \rangle = \sum_{w=1}^{\infty} \frac{z^w}{\prod_{i=1}^{D} (1 - e^{-\beta\hbar\omega_i w})}, \tag{4.33}$$

where the contribution of the ground state is accounted for in this expression. This is in contrast to a uniform Bose gas, where this contribution is to be treated separately.

The condensation temperature $T_{\rm BEC}$ at fixed particle number can be obtained from Eq. (4.33) by expanding it as

$$\langle N \rangle = \frac{z}{1-z} + \frac{1}{(\beta\hbar\tilde{\omega})^D} \left[g_D(z) + \frac{D}{2}\beta\hbar\bar{\omega}\, g_{D-1}(z) + O\,(\beta\hbar\omega_i)^2 \right] \tag{4.34}$$

valid for $k_{\rm B}T \gg \hbar\omega_i$. Here, $\tilde{\omega} \equiv \left(\prod_{i=1}^{D} \omega_i\right)^{1/D}$ denotes the geometric average of the trapping frequencies. The lowest-order term in this expansion gives the fraction of particles residing in the ground state,

$$\langle N_0 \rangle = \frac{z}{1-z}. \tag{4.35}$$

Since $T_{\rm BEC}$ is defined as the lowest temperature where all the particles can still be accommodated in the excited states, Eq. (4.34) gives

$$\frac{T_{\rm BEC}}{T_{\rm BEC}^{(0)}} = 1 - \frac{1}{2}\frac{\zeta(D-1)}{\zeta^{1-1/D}(D)}\frac{\bar{\omega}}{\tilde{\omega}}\langle N \rangle^{-1/D}, \tag{4.36}$$

where

$$k_{\rm B}T_{\rm BEC}^{(0)} = \frac{\hbar\tilde{\omega}\langle N \rangle^{1/D}}{\zeta^{1/D}(D)} \tag{4.37}$$

is the condensation temperature in lowest order. The result (4.36) shows that finite-size effects, represented by the last term, lower the condensation temperature. The fraction of particles in the condensate can now be written as

$$\frac{\langle N_0 \rangle}{\langle N \rangle} = 1 - \left(\frac{T}{T_{\rm BEC}^{(0)}}\right)^D \left[1 + \frac{D}{2}\frac{\zeta(D-1)}{\zeta(D)}\frac{\hbar\bar{\omega}}{k_{\rm B}T}\right], \tag{4.38}$$

showing that the first-order correction reduces the condensate fraction.

Slightly above the condensation temperature, the fugacity can be expanded as $z = 1 + \beta(\mu - \mu_c) + \cdots$. It then follows from Eq. (4.34) that the chemical potential approaches its critical value as

$$\mu - \mu_c = -D\frac{\zeta(D)}{\zeta(D-1)}k_{\rm B}\left(T - T_{\rm BEC}^{(0)}\right) \tag{4.39}$$

This approach differs from that in a uniform Bose gas given in Eq. (4.15).

4.4 Gross-Pitaevskii Theory

We next turn to Bose-Einstein condensation in dilute weakly interacting Bose gases. The Lagrangian density describing such systems is obtained by adding a self-interaction term to the free theory (2.37):

$$\mathcal{L} = \mathrm{i}\hbar\psi^* \frac{\partial}{\partial t}\psi - \frac{\hbar^2}{2m}\nabla\psi^* \cdot \nabla\psi + \mu|\psi|^2 - \frac{g}{2}|\psi|^4. \tag{4.40}$$

The interaction term, characterized by the coupling constant $g\,(> 0)$, is a local term built from fields at the same spacetime point. In principle, higher orders in the fields such as a $|\psi|^6$ three-particle interaction term can be included, but for a dilute gas the two-particle interaction term suffices. The true interatomic potential $V(\mathbf{x} - \mathbf{x}')$ has typically a repulsive hard core with a radius less than one nanometer, and a weak long-range attractive tail. In Eq. (2.37), $V(\mathbf{x} - \mathbf{x}')$ has been approximated by a local, i.e., delta-function potential,

$$V(\mathbf{x}) = g\delta(\mathbf{x}), \tag{4.41}$$

parametrized by a single parameter so that the interaction term in the Lagrangian $L_\mathrm{I} = \int \mathrm{d}^D x \mathcal{L}_\mathrm{I}$ becomes

$$\begin{aligned} L_\mathrm{I} &= -\frac{1}{2}\int \mathrm{d}^D x \mathrm{d}^D x' |\psi(\mathbf{x})|^2 V(\mathbf{x} - \mathbf{x}')|\psi(\mathbf{x}')|^2 \\ &= -\frac{g}{2}\int \mathrm{d}^D x |\psi(\mathbf{x})|^4. \end{aligned} \tag{4.42}$$

The coupling constant g can be related to the s-wave *scattering length* a characterizing the scattering of two free particles in vacuum as follows:

$$g = 4\pi\hbar^2 a/m. \tag{4.43}$$

To motivate this relation, recall from standard quantum mechanics that the amplitude f for the scattering of two identical particles in vacuum is given in the Born approximation by

$$f_\mathrm{B}(\mathbf{k}, \mathbf{k}') = -\frac{m}{4\pi\hbar^2}\int \mathrm{d}^3 x\, V(\mathbf{x})\, \mathrm{e}^{-\mathrm{i}(\mathbf{k}-\mathbf{k}')\cdot\mathbf{x}}, \tag{4.44}$$

with $\hbar\mathbf{k}$ and $\hbar\mathbf{k}'$ the momenta before and after the scattering event, and where the subscript B indicates the Born approximation. In the limit $|\mathbf{k}| = |\mathbf{k}'| \to 0$, the scattering amplitude becomes a constant,

$$\lim_{|\mathbf{k}|\to 0} f_\mathrm{B}(\mathbf{k}, \mathbf{k}) \to -a_\mathrm{B}, \tag{4.45}$$

which defines the s-wave scattering length. For the potential (4.41), these relations give in the Born approximation $g = 4\pi a_\mathrm{B}\hbar^2/m$. In Eq. (4.43), which goes beyond the Born approximation, a_B is replaced with the true scattering length a.

Central to the quantum description is the functional integral

$$Z = \int \mathrm{D}\psi^* \mathrm{D}\psi \, \mathrm{e}^{(\mathrm{i}/\hbar)S}, \tag{4.46}$$

representing the sum over all possible field configurations. Unlike for an ideal gas, this functional integral, involving the action $S = \int \mathrm{d}^d x \mathcal{L}$ with the Lagrangian density given in Eq. (4.40), cannot be evaluated in closed form. The simplest approximation of the integral corresponds to the mean-field approximation presented in Sec. 3.4. It is obtained by including only the contribution of the saddle point which is defined by the field configuration for which the phase factor in the integrand is stationary, i.e., by

$$\frac{\delta S}{\delta \psi^*(x)} = 0. \tag{4.47}$$

This condition yields precisely the Euler-Lagrange equation. Explicitly,

$$\mathrm{i}\hbar \frac{\partial}{\partial t}\psi_\mathrm{c}(x) = -\frac{\hbar^2}{2m}\nabla^2 \psi_\mathrm{c}(x) - \mu\psi_\mathrm{c}(x) + g|\psi_\mathrm{c}(x)|^2 \psi_\mathrm{c}(x). \tag{4.48}$$

Since fluctuations are ignored, this equation, known as the *Gross-Pitaevskii equation*, provides a classical description of a weakly interacting Bose gas. As argued in Sec. 3.4, the saddle point gives the dominant contribution to the integral. The mean-field approximation becomes particularly useful when a condensate is present for a macroscopic fraction of the atoms then behaves coherently. The density n_0 of particles residing in the condensate is given by

$$n_0(x) = |\psi_\mathrm{c}(x)|^2. \tag{4.49}$$

The role of the Gross-Pitaevskii equation in the context of Bose-Einstein condensates is similar to that of the Maxwell equations in electrodynamics. When needed, a trapping potential $U_\mathrm{ex}(\mathbf{x})$ can be included in the Gross-Pitaevskii equation by shifting the chemical potential $\mu \to \mu - U_\mathrm{ex}(\mathbf{x})$.

The Gross-Pitaevskii equation can be viewed as describing the condensate in a self-consistent Hartree potential $V_\mathrm{H}(\mathbf{x})$ produced by the condensate itself,

$$\begin{aligned} V_\mathrm{H}(\mathbf{x}) &\equiv \int \mathrm{d}^D x' V(\mathbf{x} - \mathbf{x}')|\psi_\mathrm{c}(\mathbf{x}')|^2 \\ &= g n_0(\mathbf{x}). \end{aligned} \tag{4.50}$$

For $T \lesssim 0.4 T_\mathrm{BEC}$ the Gross-Pitaevskii equation describes the static as well as the dynamic properties of a weakly interacting Bose gas reasonably well, typically within a few percent.

For a static Bose gas and with the gradient term ignored, the Gross-Pitaevskii equation reduces to the simple form

$$\mu - U_\mathrm{ex}(\mathbf{x}) = g n_0(\mathbf{x}), \tag{4.51}$$

which gives a relation between the chemical potential and the number density of condensed particles. This approximation is known as the *Thomas-Fermi approximation*. It is valid when the density of particles is sufficiently high. For an isotropic harmonic trapping potential (4.25) with $\omega_i = \omega, \; (i = 1, \ldots, D)$, the condensate density profile $n_0(\mathbf{x})$ takes the form

$$n_0(\mathbf{x}) = \frac{1}{g}\left(\mu - \tfrac{1}{2}m\omega^2 r^2\right), \tag{4.52}$$

with $r = |\mathbf{x}|$. The linear size $r_{\rm TF}$ of the condensate is determined in this approximation by the condition $n_0(r_{\rm TF}) = 0$, giving

$$\mu = \frac{1}{2}m\omega^2 r_{\rm TF}^2. \tag{4.53}$$

In terms of the average number $\langle N_0 \rangle$ of condensed particles,

$$\begin{aligned} \langle N_0 \rangle &= \Omega_D \int_0^{r_{\rm TF}} \mathrm{d}r \, r^{D-1} \frac{1}{g}\left(\mu - \tfrac{1}{2}m\omega^2 r^2\right) \\ &= \frac{\Omega_D}{D(D+2)} \frac{2\mu}{g} \left(\frac{2\mu}{m\omega^2}\right)^{D/2}, \end{aligned} \tag{4.54}$$

the chemical potential reads in $D = 3$

$$\mu = \hbar\omega \, (15\langle N_0 \rangle a/\ell)^{2/5}, \tag{4.55}$$

where ℓ is the length scale (4.29) characterizing the trap. With this expression, Eq. (4.53) can be rewritten as

$$r_{\rm TF}/\ell = (15\langle N_0 \rangle a/\ell)^{1/5} \gg 1 \tag{4.56}$$

showing that, as expected, the repulsive interaction spreads out the condensate and thus decreases the condensate density at the center of the trap.

4.5 Bogoliubov Theory

The *Bogoliubov theory* goes beyond the mean-field approximation by taking into account the first corrections due to quantum fluctuations. To obtain these, we consider the shape of the potential energy density

$$\mathcal{V} = -\mu|\psi|^2 + \frac{1}{2}g|\psi|^4 \tag{4.57}$$

of the uniform theory (4.40) with $\mu > 0$ (see Fig. 1.2) so that the potential takes its minimum along a circle around the origin $\psi = 0$. To take this nontrivial minimum into account, we shift ψ by a (complex) constant v and introduce a new complex field χ by writing

$$\psi(x) = v + \chi(x). \tag{4.58}$$

The nonzero value

$$|v|^2 = \mu/g \tag{4.59}$$

minimizes the potential energy density and allows for the nontrivial ground state. The average of the complex field χ vanishes. In terms of the shifted field, the quadratic terms of the Lagrangian (4.40) may be cast in matrix form as

$$\mathcal{L}_0 = \frac{1}{2} X^\dagger K(\mathrm{i}\partial) X, \qquad X \equiv \begin{pmatrix} \chi \\ \chi^* \end{pmatrix}, \tag{4.60}$$

with

$$K(\mathrm{i}\partial) \equiv \begin{pmatrix} \mathrm{i}\hbar\partial_t + \hbar^2\nabla^2/2m + \mu - 2g|v|^2 & -gv^2 \\ -gv^{*2} & -\mathrm{i}\hbar\partial_t + \hbar^2\nabla^2/2m + \mu - 2g|v|^2 \end{pmatrix}. \tag{4.61}$$

The inverse of K defines the Feynman propagator $\Delta_\mathrm{F}(k)$,

$$\Delta_\mathrm{F}(k) = \frac{1}{\hbar^2\omega^2 - E^2(\mathbf{k}) + \mathrm{i}\eta} \begin{pmatrix} \hbar\omega + \epsilon(\mathbf{k}) - \mu + 2g|v|^2 & -gv^2 \\ -gv^{*2} & -\hbar\omega + \epsilon(\mathbf{k}) - \mu + 2g|v|^2 \end{pmatrix}, \tag{4.62}$$

with $E(\mathbf{k})$ the spectrum of the elementary excitations in a weakly interacting Bose gas at the absolute zero of temperature expressed in terms of v and μ

$$E(\mathbf{k}) = \sqrt{[\epsilon(\mathbf{k}) - \mu + 2g|v|^2]^2 - g^2|v|^4} \tag{4.63}$$

and $\epsilon(\mathbf{k}) = \hbar^2\mathbf{k}^2/2m$ the kinetic energy of noninteracting atoms. This spectrum reduces to the familiar *Bogoliubov spectrum* when the mean-field value $|v|^2 = \mu/g$ is inserted,

$$E(\mathbf{k}) = \sqrt{\epsilon^2(\mathbf{k}) + 2\mu\epsilon(\mathbf{k})}. \tag{4.64}$$

For long wave lengths, the Bogoliubov spectrum becomes linear

$$E(\mathbf{k}) \sim \hbar c|\mathbf{k}|, \tag{4.65}$$

with $c = \sqrt{\mu/m}$ a velocity, showing that the spectrum is *gapless*. This part of the spectrum describes a sound wave whose quanta are phonons with c the speed of sound. In the opposite limit of short wave lengths, the Bogoliubov spectrum becomes quadratic

$$E(\mathbf{k}) \sim \epsilon(\mathbf{k}) + \mu = \epsilon(\mathbf{k}) + g|v|^2. \tag{4.66}$$

This form is typical for particles of mass m without mutual interactions moving in a medium described by the constant potential μ. The condensate plays the role of the medium as follows from replacing μ with $g|v|^2$ at the right. The spectrum crosses over from linear to quadratic around the wave number

$$k_\times = 1/\xi, \tag{4.67}$$

with ξ the correlation length introduced in Eq. (1.74). For this wave number, $\epsilon(k_\times) \approx g|v|^2$. The phonon part of the dispersion curve shrinks to zero in the limit $g \to 0$ of a noninteracting Bose gas.

4.6 Renormalization

We next compute the particle number density and the speed of sound at the absolute zero of temperature perturbatively in terms of the coupling constant g. We do so by calculating the thermodynamic potential Ω for a static, uniform system. By Eqs. (2.19) and (2.202), with κ the (isothermal) compressibility which is related to the the speed of sound c through Eq. (2.203), both n and c can then be computed by taking derivatives with respect to the chemical potential.

In the approximation (4.60) of ignoring higher than second order in the fields, the integral over the X field becomes Gaussian. Carrying out this integral, which amounts to the saddle-point approximation first encountered in Sec. 3.4, we obtain for the partition function

$$Z = \exp\left(-\frac{\mathrm{i}}{\hbar}\int \mathrm{d}^d x\, \mathcal{V}_\mathrm{c}\right)\int \mathrm{D}X \exp\left(\frac{\mathrm{i}}{\hbar}\int \mathrm{d}^d x\, \mathcal{L}_0\right)$$
$$= \exp\left(-\frac{\mathrm{i}}{\hbar}\int \mathrm{d}^d x\, \mathcal{V}_\mathrm{c}\right)\mathrm{Det}^{-1/2}\left[K(p)\right], \tag{4.68}$$

where $\mathcal{V}_\mathrm{c}$ is the mean-field or classical value of the potential energy density (4.57),

$$\mathcal{V}_\mathrm{c} = -\frac{1}{2}\frac{\mu^2}{g}. \tag{4.69}$$

The use of a potential here rather than an action is to underscore that the system is assumed to be static and uniform. Setting

$$Z = \exp\left[-\frac{\mathrm{i}}{\hbar}\int \mathrm{d}^d x\,(\mathcal{V}_\mathrm{c} + \mathcal{V}_1)\right], \tag{4.70}$$

where use is made of the identity (2.55), $\det(M) = \exp[\mathrm{tr}\ln(M)]$, we find from Eq. (4.68) the one-loop potential energy density

$$\mathcal{V}_1 = -\frac{\mathrm{i}\hbar}{2}\mathrm{tr}\int \frac{\mathrm{d}^d k}{(2\pi)^d}\ln\left[K(k)\right]. \tag{4.71}$$

To evaluate the frequency integral, it is convenient to assume, without loss of generality, that v is real and to first take the derivative of this expression with respect to μ

$$\frac{\partial}{\partial\mu}\mathrm{tr}\int\frac{\mathrm{d}\omega}{2\pi}\ln\begin{pmatrix}\hbar\omega - \epsilon - \mu & -\mu \\ -\mu & -\hbar\omega - \epsilon - \mu\end{pmatrix} = -2\int\frac{\mathrm{d}\omega}{2\pi}\frac{\epsilon}{\hbar^2\omega^2 - E^2 + \mathrm{i}\eta}, \tag{4.72}$$

with $\eta = 0^+$ the usual small positive switching parameter that is to be set to zero after the frequency integral has been performed, and $E(\mathbf{k})$ the Bogoliubov spectrum (4.64). The resulting frequency integral can be evaluated with the help of contour integration, yielding

$$\int\frac{\mathrm{d}\omega}{2\pi}\frac{\epsilon}{\hbar^2\omega^2 - E^2 + \mathrm{i}\eta} = -\frac{\mathrm{i}}{2\hbar}\frac{\epsilon}{E}. \tag{4.73}$$

This in turn is readily integrated with respect to μ. Putting the pieces together, we obtain for $\mathcal{V}_1$:

$$\mathcal{V}_1 = \frac{1}{2}\int \frac{\mathrm{d}^D k}{(2\pi)^D} E(\mathbf{k}), \tag{4.74}$$

which is nothing but the zero-point energy of a weakly interacting Bose gas. Note that in contrast to fermions, see Eq. (3.118), bosons contribute to the zero-point energy with a positive sign.

The remaining integral over wave vectors $\mathbf{k}$ diverges in the ultraviolet and needs to be regularized. We do so by introducing a wave number cutoff Λ. The physical motivation for this regularization procedure is that the simple microscopic model (4.57) is only valid far below the cutoff. Above it, the model breaks down and new physics starts. The atoms, which far below the cutoff are described by the field $\psi(x)$, can then no longer be considered as elementary point particles. The cutoff separates the region where the model is no longer valid from the region where it is. As we will see, the unknown, excluded region is incorporated into the effective theory by redefining, or renormalizing a finite number of parameters of the theory. These are not calculable within the theory and must be determined by experiment.

In the large-Λ limit, the wave vector integral in Eq. (4.74) gives for $D = 3$

$$\mathcal{V}_1 = \frac{1}{12\pi^2}\mu\Lambda^3 - \frac{1}{4\pi^2}\frac{m}{\hbar^2}\mu^2\Lambda + \frac{8}{15\pi^2}\left(\frac{m}{\hbar^2}\right)^{3/2}\mu^{5/2} + O\left(\Lambda^{-1}\right), \tag{4.75}$$

where an irrelevant term ($\propto \Lambda^5$) independent of the chemical potential is ignored. Equation (4.75) contains two terms which diverge in the limit $\Lambda \to \infty$. Note that these divergences arise from a region where the model (4.40) is not applicable. They therefore cannot be of deep physical significance. As a consequence of the uncertainty principle, stating that large wave numbers correspond to small distances, terms arising from the ultraviolet region are always local and can be absorbed by redefining the parameters appearing in the original Lagrangian. Since we have $\mu = g|v|^2$, the terms in Eq. (4.75) diverging in the ultraviolet are indeed of a form already present in the original theory (4.57). The theory is therefore renormalizable (at least to this order). Specifically, the diverging terms

$$\mathcal{V}_\Lambda = \frac{1}{12\pi^2}g|v|^2\Lambda^3 - \frac{1}{4\pi^2}\frac{m}{\hbar^2}g^2|v|^4\Lambda \tag{4.76}$$

can be handled by defining the effective potential

$$\mathcal{V}_{\text{eff}} \equiv \mathcal{V}_c + \mathcal{V}_1 = -\mu_r|v|^2 + \frac{1}{2}g_r|v|^4 + \frac{8}{15\pi^2}\left(\frac{m}{\hbar^2}\right)^{3/2}\mu_r^{5/2}, \tag{4.77}$$

where to this order

$$\mu_r \equiv \mu - \frac{1}{12\pi^2}g\Lambda^3, \quad g_r \equiv g - \frac{1}{2\pi^2}\frac{m}{\hbar^2}g^2\Lambda. \tag{4.78}$$

In other words, the diverging terms can be interpreted as modifying, or dressing the bare parameters g, μ of the original theory. These modified, or renormalized parameters are to be identified with the parameters measured in experiment. For a renormalizable theory, all divergences can be absorbed into the parameters of the original theory.

In the last term of Eq. (4.77), the bare parameter μ is replaced by its (one-loop) renormalized counterpart μ_r, which is justified because the last term is already a one-loop result. This nonanalytic contribution $\propto m^{3/2}(g_r|v|^2)^{5/2}$ is typical for a gapless mode in five (*sic!*) spacetime dimensions. The effective number of spacetime dimensions is five here because in nonrelativistic quantum theories, where time derivatives appear in combination with two space derivatives, the time dimension counts double compared to the (three) space dimensions.

The result (4.77) could have been obtained directly without renormalization if we, instead of introducing a wave number cutoff to regularize the integrals, had employed analytic regularization. In such a regularization scheme, where, for example, integrals are analytically continued to arbitrary real values of the space dimension D, divergences proportional to powers of the cutoff never show up.

To illustrate the power of dimensional regularization, consider the simple integral

$$I \equiv \int \frac{\mathrm{d}^3 k}{(2\pi)^3} \frac{1}{\mathbf{k}^2 + M^2}, \tag{4.79}$$

with $M > 0$. Introducing a wave number cutoff, we find in the large-Λ limit

$$I = \frac{1}{2\pi^2}\Lambda - \frac{1}{4\pi}M + O\left(\Lambda^{-1}\right). \tag{4.80}$$

The first term on the right diverges in the ultraviolet, while the rest remains finite. When dimensional regularization is used instead by considering the integral in arbitrary D dimensions, it gives

$$I = \frac{\Gamma(1 - D/2)}{(4\pi)^{D/2}} \frac{1}{(M^2)^{1-D/2}}, \tag{4.81}$$

which for $D \to 3$ yields only the finite part of the previous result (4.80). Irrelevant ultraviolet divergences, i.e., terms which diverge with a strictly positive power of the wave number cutoff, are therefore indeed suppressed. As an aside, note that the dimensional-regularized result is somewhat surprising for a positive definite integral is assigned a negative value.

This is not to say that divergences do not appear at all in dimensional regularization. We came across such divergences already in the previous chapter. They derive from the fact that for $n = 0, 1, 2, \cdots$

$$\Gamma(-n + \varepsilon) = \frac{(-1)^n}{n!}\left[\frac{1}{\varepsilon} + \psi(n+1) + O(\varepsilon)\right], \tag{4.82}$$

where $\psi(x)$ is the digamma function

$$\psi(x) \equiv \frac{\mathrm{d}}{\mathrm{d}x} \ln \Gamma(x). \tag{4.83}$$

Comparison to the exemplary integral (4.81) shows that, typically, divergences in dimensional regularization appear as $1/\varepsilon$ poles, where ε is the deviation from the number of dimensions considered, provided it is even. This happens, for example, when the integral (4.81) is considered in $D = 2$ (see below). In a regularization with a wave number cutoff, such divergences are of the form $\ln(\Lambda/\kappa)$, with κ an inverse length scale. They are relevant also in the infrared because for fixed cutoff $\ln(\Lambda/\kappa) \to \infty$ when κ is taken to zero.

For the case at hand, the integral over the loop wave vectors in Eq. (4.74) yields in arbitrary space dimension D the regularized expression

$$\mathcal{V}_1 = -L_D \left(\frac{m}{\hbar^2}\right)^{D/2} \mu^{D/2+1}, \tag{4.84}$$

with the coefficient

$$L_D \equiv \frac{\Gamma(1-D/2)\Gamma(D/2+1/2)}{2\pi^{D/2+1/2}\Gamma(D/2+2)}. \tag{4.85}$$

In deriving this, we employed the integral representation (3.136) of the gamma function to write the square-root appearing in the Bogoliubov spectrum (4.64) as an integral over Schwinger proper time. By setting $D = 3$ here, we obtain only the finite part of Eq. (4.75), showing that indeed all terms diverging with a strictly positive power of the wave number cutoff are suppressed in this analytic regularization scheme. For this reason, we will omit the subscript "r" on μ and g when considering $D = 3$, or, more generally, $D \neq 2$.

For $D \to 2$ the regularized expression (4.84) diverges. To investigate this, we expand the potential (4.84) around $D = 2$ by setting $\varepsilon = 2 - D$ and letting $\varepsilon \to 0^+$. This gives

$$\mathcal{V}_1 = -\frac{1}{4\pi\varepsilon} \frac{m}{\hbar^2} \frac{\mu^2}{\kappa^\varepsilon} + O(\varepsilon^0), \tag{4.86}$$

with κ an arbitrary scale parameter of dimension inverse length which enters for dimensional reasons. If the Bogoliubov spectrum had not been gapless, but had an energy gap instead, this parameter would have appeared in Eq. (4.86) instead of κ. Comparing the one-loop contribution to the mean-field contribution (4.69), we conclude that Eq. (4.86) leads to a dressing of the bare coupling constant g, yielding the renormalized coupling g_r

$$\frac{1}{g_\mathrm{r}} = \frac{\kappa^\varepsilon}{g}\left(1 + \frac{1}{2\pi\varepsilon}\frac{m}{\hbar^2} g\kappa^{-\varepsilon}\right). \tag{4.87}$$

Here, the arbitrary inverse length scale κ introduced above is used to render the combination $(m/\hbar^2)g_r$ dimensionless. This result is to be compared to the renormalization of the electric charge in QED in $d = 4$, see Eq. (3.143). The chemical potential is not renormalized to this order. The low-frequency, long wave-length behavior of the theory is obtained by letting the scale parameter $\kappa \to 0^+$. For fixed bare coupling g, it then follows that the renormalized coupling flows to $g_r \to 2\pi(\hbar^2/m)\varepsilon$. For $D < 2$, or equivalently $\varepsilon > 0$, this fixed point is nontrivial. In the limit $D \to 2$, g_r tends to zero and the theory becomes Gaussian. The fixed point can be equally retrieved from the beta function,

$$\beta(g_r) \equiv \kappa \frac{\mathrm{d}g_r}{\mathrm{d}\kappa} = -\varepsilon g_r + \frac{1}{2\pi}\frac{m}{\hbar^2} g_r^2, \tag{4.88}$$

by setting it to zero. In the limit $D \to 2$, this flow equation yields

$$\frac{1}{g_r(\kappa')} - \frac{1}{g_r(\kappa)} = -\frac{1}{2\pi}\frac{m}{\hbar^2}\ln\left(\frac{\kappa'}{\kappa}\right). \tag{4.89}$$

As for the electric charge in QED, the physical coupling constant g_r becomes weaker and weaker at larger distances.

In so-called nonrenormalizable theories, the terms diverging in the ultraviolet are still local but not of a form present in the original Lagrangian. Whereas in former days such theories were rejected because their supposed lack of predictive power, the modern view is that there are no fundamental theories and that there is no basic difference between renormalizable and nonrenormalizable theories. Even a renormalizable theory like (4.40) should be extended to include all higher-order terms such as a $|\psi|^6$-term which are allowed by symmetry. These additional terms render the theory "nonrenormalizable". This does not, however, change the predictive power of the theory. The point is that when describing the physics at an inverse length scale κ far below the cutoff, the higher-order terms are suppressed by powers of κ/Λ, as follows from dimensional analysis. Far below the cutoff, the nonrenormalizable terms are therefore negligible.

4.7 One-Loop Corrections

By the expression (2.200) for the zero-temperature thermodynamic potential, with Z given in Eq. (4.70) and the combination $\mathcal{V}_{\text{eff}} = \mathcal{V}_c + \mathcal{V}_1$ given in Eq. (4.77), Ω reads in this approximation

$$\Omega = \int \mathrm{d}^D x \, \mathcal{V}_{\text{eff}}. \tag{4.90}$$

Using Eqs. (2.19), (2.202) and (2.203), we are now in a position to determine the particle number density and the speed of sound with the result for $D = 3$

$$n = \frac{\mu}{g}\left[1 - \frac{4}{3\pi^2}\left(\frac{m}{\hbar^2}\right)^{3/2} g\mu^{1/2}\right] \tag{4.91}$$

and

$$c^2 = \frac{\mu}{m}\left[1 + \frac{2}{3\pi^2}\left(\frac{m}{\hbar^2}\right)^{3/2} g\mu^{1/2}\right], \tag{4.92}$$

where the last equation is obtained by expanding in the coupling constant g, which is assumed to be small. These equations reveal that the expansion is more precisely one in terms of the dimensionless parameter $(m/\hbar^2)^{3/2} g\mu^{1/2}$. In lowest order, the speed of sound is given by $c = \sqrt{\mu/m}$, in agreement with the value extracted from the Bogoliubov spectrum (4.65). The slope obtained after calculating the first-order correction to the elementary excitation spectrum (4.65) turns out to be in agreement with Eq. (4.92). This connection has been proved to persist to all orders in perturbation theory, see also Sec. 4.9, and implies the equivalence of the elementary excitations and the collective density fluctuations in a weakly interacting Bose gas.

The expressions for arbitrary $D \neq 2$ can be obtained similarly, starting from Eq. (4.84). This gives

$$n = \frac{\mu}{g}\left[1 + \frac{1}{2}(D+2)L_D\left(\frac{m}{\hbar^2}\right)^{D/2} g\mu^{D/2-1}\right] \tag{4.93}$$

and

$$c^2 = \frac{\mu}{m}\left[1 - \frac{1}{4}(D+2)(D-2)L_D\left(\frac{m}{\hbar^2}\right)^{D/2} g\mu^{D/2-1}\right], \tag{4.94}$$

where the last expression is obtained by again expanding in the coupling constant g. Note that L_D changes sign at $D = 2$. The relative sign in Eq. (4.93) therefore changes there, while that in Eq. (4.94) does not because of an additional factor of $D - 2$.

Up to this point, we have taken the chemical potential to be the independent parameter, thereby assuming the presence of a reservoir with which the weakly interacting Bose gas can freely exchange atoms. The system can thus have any number of particles since only the average number is fixed by external conditions. From the experimental point of view it is, however, often more realistic to consider the particle number fixed. If this is the case, the particle number density n is to be considered as independent variable and the chemical potential should be expressed in terms of it. For $D = 3$ this can be achieved by inverting the relation (4.91):

$$\begin{aligned}\mu &= gn\left[1 + \frac{4}{3\pi^2} n^{1/2}\left(\frac{mg}{\hbar^2}\right)^{3/2}\right] \\ &= \frac{4\pi\hbar^2 na}{m}\left[1 + \frac{32}{3}\left(\frac{na^3}{\pi}\right)^{1/2}\right],\end{aligned} \tag{4.95}$$

where in the last step the (renormalized) coupling constant is expressed in terms of the s-wave scattering length, see Eq. (4.43).

For the speed of sound (4.92) expressed in terms of the particle number density, we find for $D = 3$

$$\begin{aligned} c^2 &= \frac{gn}{m}\left[1 + \frac{2}{\pi^2} n^{1/2}\left(\frac{mg}{\hbar^2}\right)^{3/2}\right] \\ &= \frac{4\pi\hbar^2 na}{m^2}\left[1 + \frac{16}{\sqrt{\pi}}\left(na^3\right)^{1/2}\right], \end{aligned} \tag{4.96}$$

showing that c^2 is linear in the coupling constant. Without the interparticle interaction characterized by g, the speed of sound would be zero. If the interaction is attractive, c^2 turns out negative according to Eq. (4.96), which implies an instability. In this case, the weakly interacting Bose gas gives way to a more condensed state of matter, such as a liquid, or collapses. Equation (4.96) also shows that the dimensionless expansion parameter has become $(na^3)^{1/2}$. Perturbation theory is applicable only when this parameter is small.

Equation (4.95), expressing the chemical potential in terms of the particle density, can be used to cast the effective potential (4.77) in terms of n. This gives by Eq. (4.90)

$$\frac{\Omega}{V} = -\frac{2\pi\hbar^2 n^2 a}{m}\left[1 + \frac{64}{5\sqrt{\pi}}\left(na^3\right)^{1/2}\right], \tag{4.97}$$

with V the volume of the system. Since the pressure P is related to the thermodynamic potential via $P = -\partial\Omega/\partial V$, Eq. (4.97) shows that the pressure is entirely due to the repulsive interaction, which makes physical sense.

For the average energy $\langle E \rangle$ at the absolute zero of temperature, the general expression (2.20) gives

$$\frac{\langle E \rangle}{V} = \frac{2\pi\hbar^2 n^2 a}{m}\left[1 + \frac{128}{15\sqrt{\pi}}\left(na^3\right)^{1/2}\right]. \tag{4.98}$$

For completeness, we also record the expressions for arbitrary $D \neq 2$,

$$\begin{aligned} \mu &= gn\left[1 - \frac{1}{2}(D-2)(D+2)L_D\, n^{D/2-1}\left(\frac{mg}{\hbar^2}\right)^{D/2}\right], \\ c^2 &= \frac{gn}{m}\left[1 - \frac{1}{4}D(D+2)L_D\, n^{D/2-1}\left(\frac{mg}{\hbar^2}\right)^{D/2}\right], \\ \frac{\Omega}{V} &= -\frac{1}{2}gn^2\left[1 - (D^2-6)L_D\, n^{D/2-1}\left(\frac{mg}{\hbar^2}\right)^{D/2}\right], \\ \frac{\langle E \rangle}{V} &= \frac{1}{2}gn^2\left[1 - 2L_D\, n^{D/2-1}\left(\frac{mg}{\hbar^2}\right)^{D/2}\right]. \end{aligned} \tag{4.99}$$

For $D = 1$, where the relevant parameter is $\gamma \equiv mg/\hbar^2 n$, the problem of a gas of Bose particles confined to move on a line and interacting via a repulsive delta-function potential has been solved exactly. Comparison to the exact solution shows that the Bogoliubov approximation is valid for small γ up to $\gamma \sim 2$. The Bogoliubov expression for the speed of sound is even indistinguishable from the true result up to $\gamma \sim 10$. This is remarkable as the Bogoliubov expression for the ground-state energy becomes negative for $\gamma > (3\pi/4)^2 \approx 5.55$.

Finally, we compute the fraction of particles residing in the condensate. To this end, v and μ need be kept as independent variables. We therefore use, instead of the Bogoliubov expression (4.64), where the mean-field equation $|v|^2 = \mu/g$ has been inserted, the more general expression (4.63) for the spectrum of the elementary excitations. With this more general expression, Eq. (4.74) for the one-loop potential is still valid, and so is Eq. (4.90). We thus obtain for the particle number density in $D = 3$

$$n = |v|^2 - \frac{1}{2}\frac{\partial}{\partial\mu}\int \frac{\mathrm{d}^3k}{(2\pi)^3} E(\mathbf{k})\bigg|_{\mu=g|v|^2}, \tag{4.100}$$

where the mean-field value $\mu = g|v|^2$ is to be substituted after taking the derivative with respect to the chemical potential. Ignoring an irrelevant diverging term, which amounts to a renormalization as before, we find to this order

$$n = |v|^2 + \frac{1}{3\pi^2}\left(\frac{mg}{\hbar^2}|v|^2\right)^{3/2} \tag{4.101}$$

or for the so-called *depletion* of the condensate

$$\frac{n}{n_0} - 1 \approx \frac{8}{3\sqrt{\pi}}\left(na^3\right)^{1/2}. \tag{4.102}$$

This equation shows that even at the absolute zero of temperature not all the particles reside in the condensate. Due to quantum fluctuations, particles are removed from the ground state and put in states of finite momentum. In strongly interacting superfluid ^{4}He, where quantum effects are more pronounced, less than 10% of the particles reside in the condensate, while for weakly interacting alkali Bose gases, this percentage is about 99%. Near a Feshbach resonance as in recent experiments on ^{85}Rb, the percentage can go down to about 90%. (Feshbach resonances are discussed in Sec. 10.2 below.)

As an aside, note that the last term in Eq. (4.101) can also be cast in terms of the Compton wave length $\lambda_C = 2\pi\hbar/mc$ of the Bose particles as

$$n = |v|^2 + \frac{8}{3\pi}\frac{1}{\lambda_C^3}, \tag{4.103}$$

where use is made of the the mean-field results in the second term, which is justified because that term is already a one-loop result.

4.8 Phonon Decay

Before closing the zero-temperature discussion of weakly interacting Bose gases, we show that a phonon can decay into a pair of phonons, see Fig. 4.1. To this end, we consider the Bogoliubov spectrum for small wave vectors,

$$E(\mathbf{k}) = \hbar c(|\mathbf{k}| + \gamma|\mathbf{k}|^3) \tag{4.104}$$

with $c = \sqrt{\mu/m}$ the speed of sound in lowest order and $\gamma \equiv \hbar^2/8\mu m > 0$. The conservation of energy-momentum in the decay process shown in Fig. 4.1,

$$E(\mathbf{k}) = E(\mathbf{k}') + E(\mathbf{k} - \mathbf{k}'), \tag{4.105}$$

then assumes the form

$$|\mathbf{k}| - |\mathbf{k}'| - |\mathbf{k} - \mathbf{k}'| = -\gamma\left(|\mathbf{k}|^3 - |\mathbf{k}'|^3 - |\mathbf{k} - \mathbf{k}'|^3\right). \tag{4.106}$$

Provided that the wave vector $\mathbf{k}$ of the original phonon is small so that the angle θ between $\mathbf{k}$ and the wave vector of one of the final phonons $\mathbf{k}'$ is also small, one can replace $|\mathbf{k} - \mathbf{k}'|$ at the left side of Eq. (4.106) with

$$\begin{aligned} |\mathbf{k} - \mathbf{k}'| &= \left[(|\mathbf{k}| - |\mathbf{k}'|)^2 + 2|\mathbf{k}||\mathbf{k}'|(1 - \cos\theta)\right]^{1/2} \\ &\approx |\mathbf{k}| - |\mathbf{k}'| + \frac{|\mathbf{k}||\mathbf{k}'|}{|\mathbf{k}| - |\mathbf{k}'|}(1 - \cos\theta), \end{aligned} \tag{4.107}$$

while on the right side of Eq. (4.106) one can set $|\mathbf{k} - \mathbf{k}'|^3 \approx (|\mathbf{k}| - |\mathbf{k}'|)^3$, giving

$$\gamma \approx \frac{1}{3}\frac{1 - \cos\theta}{(|\mathbf{k}| - |\mathbf{k}'|)^2}. \tag{4.108}$$

Since the right side is positive, as is γ, this equation allows for solutions, implying that phonons in a weakly interacting Bose gas can indeed decay into pairs of phonons.

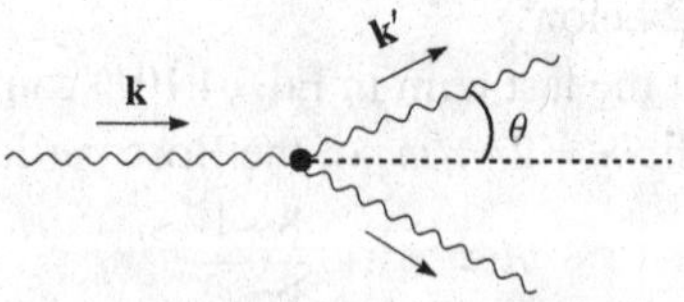

Fig. 4.1 Decay of a phonon with wave vector **k** in a pair of phonons.

4.9 Effective Theory

Since gapless modes in general require justification for their existence, we expect the gaplessness of the Bogoliubov spectrum to be a result of Goldstone's theorem. This is corroborated by the relativistic version of the theory. There, one finds two spectra, one corresponding to a massive Higgs particle which in the nonrelativistic limit becomes too heavy and decouples from the theory, and one corresponding to the Nambu-Goldstone mode of the spontaneously broken global U(1) symmetry. The latter reduces in the nonrelativistic limit to the Bogoliubov spectrum. As remarked before, the equivalence of elementary excitations and collective density fluctuations has been proved to all orders in perturbation.

To show this we derive the effective action governing the Nambu-Goldstone mode at low frequencies and long wave lengths. According to the program spelled out in Sec. 2.7, the effective action is obtained by considering the Legendre transform of the generator of connected diagrams $W[J, J^*]$, which is related to the partition function $Z[J, J^*]$ in the presence of external sources $J(x)$ and $J^*(x)$ through

$$Z[J, J^*] = Z[0, 0]\,\mathrm{e}^{(\mathrm{i}/\hbar)W[J,J^*]}. \tag{4.109}$$

With the sources included, the partition function (4.46) becomes

$$Z = \int \mathrm{D}\psi^*\mathrm{D}\psi \,\exp\left\{\frac{\mathrm{i}}{\hbar}\left[S + \int \mathrm{d}^d x\,(J^*\psi + J\psi^*)\right]\right\}. \tag{4.110}$$

Explicitly, the Legendre transform reads

$$\Gamma[\psi_\mathrm{c}, \psi_\mathrm{c}^*] = W[J, J^*] - \int \mathrm{d}^d x\,[J^*(x)\psi_\mathrm{c}(x) + J(x)\psi_\mathrm{c}^*(x)], \tag{4.111}$$

where $\psi_\mathrm{c}(x)$ is the ensemble average of the fluctuating field $\psi(x)$ in the presence of external sources $J(x)$ and $J^*(x)$,

$$\psi_\mathrm{c}(x) = \frac{\delta W[J, J^*]}{\delta J^*(x)}, \tag{4.112}$$

and with a similar definition for ψ_c^*.

The effective action we are seeking corresponds to setting

$$\psi_\mathrm{c}(x) = v\,\mathrm{e}^{\mathrm{i}\varphi(x)}, \tag{4.113}$$

with $\varphi(x)$ the Nambu-Goldstone field of the spontaneously broken U(1) invariance, and v minimizing the effective potential. The theory will, first of all, depend only on derivatives of the Nambu-Goldstone field so that the effective theory is invariant under global U(1) phase transformations under which $\varphi(x)$ is shifted, $\varphi(x) \to \varphi(x) + \alpha$. Moreover, the Nambu-Goldstone field will only appear in the combination

$$U(x) \equiv \hbar\partial_t\varphi + \frac{\hbar^2}{2m}(\nabla\varphi)^2 \tag{4.114}$$

so as to respect Galilei invariance under which $\varphi(x)$ transform as in Eq. (1.79). This means that the effective theory will be of the form (1.76) we obtained before in a purely classical setting plus higher-order powers of $U(x)$. Ignoring these higher-order contributions, we need to determine only two coefficients. This can be done using general arguments as follows.

From the Poisson bracket (1.83) it follows that φ is canonically conjugate to $-\hbar n$,

$$\{\varphi(x), -\hbar n(x')\} = \delta(x - x'), \tag{4.115}$$

with $n(x)$ denoting the average particle number density. The Hamilton equation (1.42) then becomes in this case

$$\frac{\delta H_{\rm eff}}{\delta n(x)} = -\hbar \partial_t \varphi(x). \tag{4.116}$$

The standard definition of the chemical potential μ, cf. Eq. (1.102), gives for the left side of this equation

$$\frac{\delta H_{\rm eff}}{\delta n(x)} = \mu(x) + \frac{1}{2} m \mathbf{v}_{\rm s}^2(x), \tag{4.117}$$

where $\mathbf{v}_{\rm s}$ is the superfluid velocity (1.81). Combining these results, we conclude that

$$\mu(x) = -U(x), \tag{4.118}$$

stating that the field $-U(x)$ plays the role of the (local) chemical potential in the effective theory. A similar relation we obtained before in Eq. (1.90) for an ideal classical fluid. From this relation it follows that the coefficient of the term linear in $U(x)$ must be $-\bar{n}$, with $\bar{n}$ the average particle number density of the system uniform in space and time so that

$$-\left.\frac{\partial \mathcal{L}_{\rm eff}}{\partial U}\right|_{U=0} = \bar{n}, \tag{4.119}$$

as required. This result in turn implies that the coefficient of the term quadratic in $U(x)$ must yield the compressibility κ introduced in Eq. (2.202):

$$\left.\frac{\partial^2 \mathcal{L}_{\rm eff}}{\partial U^2}\right|_{U=0} = -\left.\frac{\partial n}{\partial U}\right|_{U=0} = \bar{n}^2 \kappa. \tag{4.120}$$

In this way, the effective theory reads

$$\mathcal{L}_{\rm eff} = -\bar{n}\left[\hbar \partial_t \varphi + \frac{\hbar^2}{2m}(\nabla\varphi)^2\right] + \frac{\bar{n}}{2mc^2}\left[\hbar \partial_t \varphi + \frac{\hbar^2}{2m}(\nabla\varphi)^2\right]^2, \tag{4.121}$$

where, using Eq. (2.203), we expressed the compressibility in terms of the speed of sound c. Because the Nambu-Goldstone field appears only in combination with

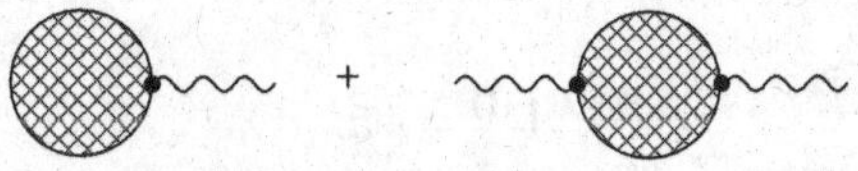

Fig. 4.2 Diagrammatic representation of the effective theory (4.121). The symbols are explained in the text.

derivatives, it tends to be uniform in space and time, and thereby lends rigidity to the system.

The effective theory (4.121) is represented diagrammatically in Fig. 4.2, where a wiggly line denotes the field U. Following the recipe given in Sec. 2.5, we arrive at the following Feynman rules.

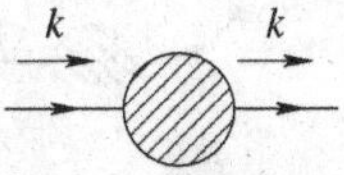

$$= \mathrm{i}\hbar\Delta_{\mathrm{F}}(k), \tag{4.122}$$

with $\Delta_{\mathrm{F}}(k)$ denoting the *full* Feynman propagator, which is related to the full connected two-point Green function $G_{\mathrm{c}}(k)$ through

$$G_{\mathrm{c}}(k) = \mathrm{i}\hbar\Delta_{\mathrm{F}}(k). \tag{4.123}$$

Because the field ψ is complex, the lines carry an additional arrow: a field ψ^* is represented by a line with an arrow pointing outwards, while for a field ψ the direction of the arrow is reversed. Real fields, such as U, are represented by arrowless lines. Since $-U$ couples to the theory in the same way as the chemical potential, the vertex of the theory is given by

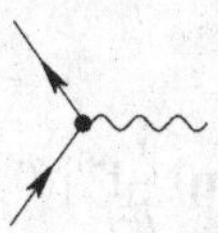

$$= -\frac{\mathrm{i}}{\hbar}. \tag{4.124}$$

The first diagram in Fig. 4.2 without the external line denotes $\mathrm{i}/\hbar$ times the proper vertex $\Gamma^{(1)}$. It can be expanded in terms of the the two-point connected Green function as

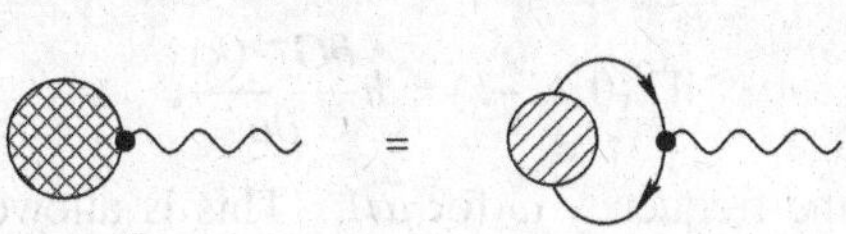

(4.125)

which translates into

$$\frac{\mathrm{i}}{\hbar}\Gamma^{(1)} = \int \frac{\mathrm{d}^d k}{(2\pi)^d}\left(-\frac{\mathrm{i}}{\hbar}\right)\mathrm{i}\hbar\Delta_{\mathrm{F}}(k). \tag{4.126}$$

The integral on the right gives, with the identification (4.123), the average particle number density $\bar{n}$,

$$\int \frac{\mathrm{d}^d k}{(2\pi)^d} G_{\mathrm{c}}(k) = \bar{n}, \tag{4.127}$$

so that

$$\Gamma^{(1)} = -\bar{n}, \tag{4.128}$$

where we recall that the bar over n is to indicate that the average particle number density is constant, representing the density of the system uniform in space and time, i.e., with $U(x)$ equal to zero.

The second diagram in Fig. 4.2 without the external lines denotes $\mathrm{i}/\hbar$ times the (0 0)-component of the *full* polarization tensor, Π^{00}, introduced in Sec. 2.8, at zero frequency transfer and small wave vectors $\mathbf{q}$. The diagram can be expanded as

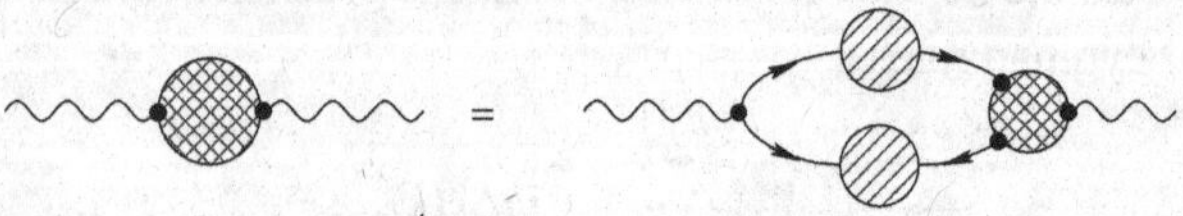

(4.129)

where

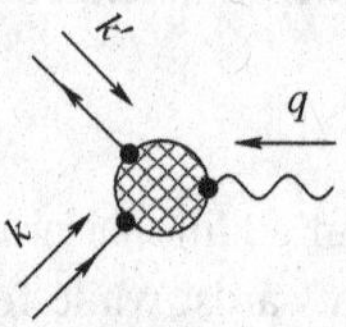

$$= \frac{\mathrm{i}}{\hbar}\Gamma^0(q,k,k')\,(2\pi)^d\delta(q+k+k') \tag{4.130}$$

denotes the three-point proper vertex $\Gamma^0(q,k,k')$. In algebraic form, the expansion translates into

$$\frac{\mathrm{i}}{\hbar}\lim_{\mathbf{q}\to 0}\Pi^{00}(0,\mathbf{q}) =$$
$$\lim_{\mathbf{q}\to 0}\int\frac{\mathrm{d}^d k}{(2\pi)^d}\left(-\frac{\mathrm{i}}{\hbar}\right)\mathrm{i}\hbar\Delta_{\mathrm{F}}(\omega,\mathbf{k}+\mathbf{q})\,\frac{\mathrm{i}}{\hbar}\Gamma^0[(0,-\mathbf{q}),(\omega,\mathbf{k}+\mathbf{q}),-k]\,\mathrm{i}\hbar\Delta_{\mathrm{F}}(k).$$

To make the connection with Eq. (4.121), we invoke the Ward identity (2.227) in the form

$$\mathrm{i}\Gamma^0(0,k,-k) = \hbar\frac{\partial G_{\mathrm{c}}^{-1}(k)}{\partial\mu}, \tag{4.131}$$

where we swapped the frequency ω for $\mu/\hbar$. This is allowed for the chemical potential enters the theory (4.40) in the same way as $\hbar\omega$. Then

$$\begin{aligned}\lim_{\mathbf{q}\to 0}\Pi^{00}(0,\mathbf{q}) &= -\int\frac{\mathrm{d}^d k}{(2\pi)^d}G_{\mathrm{c}}(k)\,\frac{\partial G_{\mathrm{c}}^{-1}(k)}{\partial\mu}\,G_{\mathrm{c}}(k)\\ &= \frac{\partial}{\partial\mu}\int\frac{\mathrm{d}^d k}{(2\pi)^d}G_{\mathrm{c}}(k)\\ &= \frac{\partial\bar{n}}{\partial\mu} = \bar{n}^2\kappa,\end{aligned} \tag{4.132}$$

which is nothing but the compressibility sum rule (2.204). In this way, it is shown that the diagrams 4.2 indeed represent the effective theory (4.121).

These diagrams can also be evaluated perturbatively in a loop expansion, to obtain, say, expressions for the particle number density and the speed of sound as in Eq. (4.91) and (4.92), respectively. In doing so, one encounters in addition to ultraviolet divergences, which can be handled by renormalization, also infrared divergences. These arise because the Bogoliubov spectrum is gapless. When however all contributions are added together (at least to one-loop order), these divergences are seen to cancel. The explicit one-loop computation also shows that $\hbar\partial_t\varphi$ always appears in the combination (4.114) dictated by Galilei invariance. Terms that would spoil this feature drop out.

The expressions for the average particle number and current density that follow from the effective theory (4.121) are of the same form as in ideal fluid hydrodynamics, see Eq. (1.117),

$$n(x) = \bar{n} - \frac{\bar{n}}{c^2}\left[\frac{\hbar}{m}\partial_t\varphi(x) + \frac{1}{2}\mathbf{v}_s^2(x)\right] \tag{4.133}$$

$$\mathbf{j}(x) = n(x)\mathbf{v}_s(x), \tag{4.134}$$

with $\mathbf{v}_s$ the superfluid velocity (1.81). On taking the gradient of Eq. (4.118), we obtain the Euler equation governing the superfluid flow

$$m\partial_t\mathbf{v}_s + \frac{1}{2}m\nabla\mathbf{v}_s^2 = -\nabla\mu, \tag{4.135}$$

or, using that in the absence of vortices $\nabla\times\mathbf{v}_s = 0$,

$$m\frac{\mathrm{d}\mathbf{v}_s}{\mathrm{d}t} = -\nabla\mu, \tag{4.136}$$

where the total derivative $\mathrm{d}/\mathrm{d}t = \partial_t + \mathbf{v}_s\cdot\nabla$ is the derivative (1.94) following the flow. These equations are the exact analogs of Eqs. (1.92) and (1.93) found in ideal fluid hydrodynamics.

As a consistency check, we compute the static structure factor (2.212) from the effective theory. Starting from the relation (2.213), we obtain

$$\begin{aligned}\bar{n}S(\mathbf{k}) &= \int\frac{\mathrm{d}\omega}{2\pi}\langle\delta n(k)\delta n(-k)\rangle \\ &= \left(\frac{\hbar\bar{n}}{mc^2}\right)^2\int\frac{\mathrm{d}\omega}{2\pi}\omega^2\langle\varphi(k)\varphi(-k)\rangle\rangle,\end{aligned} \tag{4.137}$$

where in the last step Eq. (4.133) is used with the nonlinear term omitted. The correlation function on the right can be expressed as

$$\langle\varphi(k)\varphi(-k)\rangle\rangle = \mathrm{i}\hbar D_{\mathrm{F}}(k), \tag{4.138}$$

with D_F denoting the Feynman propagator of the phase field,

$$D_F(k) = \frac{mc^2}{\hbar^2 \bar{n}} \frac{1}{\omega^2 - c^2 k^2 + i\eta}, \tag{4.139}$$

as can be read off from the quadratic terms in the Lagrangian density (4.121). It then follows that apart from an irrelevant diverging contribution

$$S(\mathbf{k}) = \frac{\hbar|\mathbf{k}|}{2mc}, \tag{4.140}$$

in agreement with the previous result (2.212).

Since the Nambu-Goldstone field in the effective theory (4.121) always comes with a derivative, the nonlinear terms carry additional factors of $\hbar|\mathbf{k}|/mc$, with $|\mathbf{k}|$ the wave number. These terms can therefore be ignored provided the wave number is smaller than the inverse coherence length $|\mathbf{k}| < 1/\xi$ with

$$\xi \equiv \frac{\hbar}{\sqrt{2m\mu}} = \frac{1}{\sqrt{2}} \frac{\hbar}{mc}, \tag{4.141}$$

first introduced in Eq. (1.74). In the case of ^{4}He, the coherence length, or Compton wave length $\lambda_C = 2\pi\hbar/mc$, is of order 1 nm. In this system, the bound $|\mathbf{k}| < 1/\xi$, below which the nonlinear terms can be neglected, coincides with the region where the elementary excitation spectrum is linear and the description solely in terms of a sound mode is applicable.

The astute reader may worry about an apparent mismatch in the number of degrees of freedom in the normal and the Bose-Einstein condensed, or superfluid state. Whereas the normal state is described by a complex ψ field that has two real components, the superfluid state is described by only a single real scalar field φ. The resolution of this paradox lies in the form of the spectrum of the modes. In the normal state, the spectrum (ignoring interactions) $E(\mathbf{k}) = \epsilon(\mathbf{k}) - \mu$, with $\mu < 0$, is quadratic in $\mathbf{k}$ so that only positive energies appear in the Fourier decomposition, and a complex field is needed to describe such a mode (see Sec. 2.4). In the superfluid state, where the spectrum $E(\mathbf{k}) = \pm c|\mathbf{k}|$ is linear in $\mathbf{k}$ for long wave lengths, the counting goes differently. The Fourier decomposition now contains positive as well as negative energies and, as we saw in Sec. 2.4, a single real field suffice to describe this mode. That is to say, although the number of fields differs, the number of degrees of freedom is the same in both states.

4.10 Finite Temperature

We next consider a uniform weakly interacting Bose gas just below the temperature where Bose-Einstein condensation sets in. We wish to obtain an effective description of the system in equilibrium in this temperature regime. The field featuring in this effective theory is the static, time-averaged field (2.44), corresponding to the zero-frequency mode in the Fourier series (2.41). The effective theory is obtained by perturbatively integrating out the irrelevant degrees of freedom in the microscopic theory, which in this case are the dynamic $n \neq 0$ frequency modes. Although by themselves irrelevant as far as the long-distance behavior of the system in equilibrium is concerned, these modes do determine the coefficients of the effective theory.

To appreciate the difference between the zero-frequency and the dynamic modes, we return to the (unregularized) expression (2.69) cast in the equivalent form

$$n(x) = \frac{1}{\beta x} - \frac{1}{2} + \frac{2}{\beta} \sum_{n=1}^{\infty} \frac{x}{\hbar^2 \omega_n^2 + x^2}, \tag{4.142}$$

where the sum is now over the positive integers only. The first term on the right represents the zero-frequency ($n = 0$) contribution, while the second, which is irrelevant for our purposes, cancels the $T = 0$ contribution contained in the sum. These two terms, corresponding to the first terms in a Laurent expansion of the Bose-Einstein distribution function,

$$n(x) = \frac{1}{e^{\beta x} - 1} = \frac{1}{\beta x} - \frac{1}{2} + O(\beta x), \tag{4.143}$$

are special as they are the only terms surviving the high-temperature limit $\beta \to 0$. The sum in Eq. (4.142) represents the contributions of the dynamic modes ($n > 0$). The splitting in zero-frequency and dynamic modes is in particular useful when calculating the thermodynamic potential of a weakly interacting Bose gas.

As an illustration, we first consider an ideal Bose gas above the condensation temperature specified by the thermodynamic potential (2.23). After an integration by parts, Ω takes the form in D space dimensions

$$\frac{\Omega}{V} = -\frac{1}{\beta \lambda_T^D} F_{D/2+1}(\alpha), \tag{4.144}$$

where λ_T is the thermal wave length (2.33), and

$$F_{D/2+1}(\alpha) \equiv \frac{2}{\Gamma(D/2+1)} \int_0^\infty \mathrm{d}q \frac{q^{D+1}}{e^{q^2+\alpha} - 1} \tag{4.145}$$

with q the dimensionless loop variable defined by $q^2 = \beta \hbar^2 \mathbf{k}^2 / 2m$.

On comparing this result to the alternative evaluation (4.3) of the thermodynamic potential, where the integration over the wave vectors has already been carried out at the expense of a remaining sum typical for a fugacity series, it follows that

$$F_{D/2+1}(\alpha) = g_{D/2+1}(z), \tag{4.146}$$

where $g_{D/2+1}(z)$, with $z = \exp(-\alpha)$, is the Bose function (4.4). The equivalence of both results is readily established by expanding the integrand at the left of Eq. (4.146) in a geometric series and using the integral representation of the gamma function.

The zero-frequency mode, corresponding to the first term in the Laurent expansion of the integrand in Eq. (4.145), gives the contribution

$$F^{(0)}_{D/2+1}(\alpha) = \frac{2}{\Gamma(D/2+1)} \int_0^\infty \mathrm{d}q \frac{q^{D+1}}{q^2+\alpha} = \alpha^{D/2}\Gamma(-D/2), \tag{4.147}$$

in agreement with Eq. (4.12). For $D = 3$, it leads to the nonanalytic dependence $\sim \alpha^{3/2}$ on α. The superscript "0" on $F_{D/2+1}(\alpha)$ indicates the zero-frequency contribution.

The rest of the terms in the Laurent expansion of the integrand in Eq. (4.145), corresponding to the dynamic modes, are analytic. It is a generic property of dynamic modes that they produce only analytic contributions. Each term thus generated in Eq. (4.145) contains an integral of the form

$$\int_0^\infty \frac{\mathrm{d}q}{q} \frac{q^{2t}}{\mathrm{e}^{pq^2}-1} \tag{4.148}$$

and derivatives thereof with respect to the parameter p (which is set to unity at the end). For $t \leq 1$, these loop integrals diverge in the infrared. We handle the divergences using zeta function regularization by analytically continuing the following equation

$$2\int_0^\infty \frac{\mathrm{d}q}{q} \frac{q^{2t}}{\mathrm{e}^{pq^2}-1} = \Gamma(t)\zeta(t)p^{-t}, \tag{4.149}$$

with t initially chosen large enough so that the integral converges, to arbitrary values of t. With this regularization scheme, the expansion of the function (4.145) assumes the form

$$F_{D/2+1}(\alpha) = F^{(0)}_{D/2+1}(\alpha) + \sum_{l=0}^{\infty} \frac{\zeta(D/2)}{\Gamma(l+1)}(-\alpha)^l, \tag{4.150}$$

with $F^{(0)}_{D/2+1}(\alpha)$ the zero-frequency contribution (4.147). Since $\alpha = -\beta\mu$, this expansion can be viewed as a high-temperature expansion.

We next turn to a weakly interacting Bose gas. As in Sec. 4.6, where we studied the thermodynamic potential at the absolute zero of temperature, we assume the condensate to be a constant. The field v introduced in (4.58) then depends only on temperature. The zero-temperature result (4.71) must be adjusted in two ways. First, the rules outlined in Sec. 2.2 must be applied to transform the zero-temperature theory into one at finite temperature. Second, the zero-temperature value (4.59) for v cannot be used at finite temperature. We therefore leave for the moment $v(T)$ undetermined. With these adjustments, the one-loop potential becomes at finite temperature

$$\mathcal{V}_1(T) = \frac{1}{2}\frac{1}{\beta}\sum_n \int \frac{\mathrm{d}^3k}{(2\pi)^3} \mathrm{tr}\, \ln[K(\mathrm{i}\omega_n, \mathbf{k})], \tag{4.151}$$

where K is the matrix introduced in Eq. (4.61). The mean-field value of the potential energy density (4.57) is given by

$$\mathcal{V}_\mathrm{c}(T) = -\mu|v(T)|^2 + \frac{1}{2}g|v(T)|^4. \tag{4.152}$$

As at zero temperature, $|v(T)|^2$ denotes the number density of particles in the condensate.

We first carry out the sum over the Matsubara frequencies $\hbar\omega_n = 2\pi n\beta^{-1}$ in Eq. (4.151). We proceed in the same way as in the zero-temperature limit and first take the derivative with respect to the chemical potential:

$$\frac{\partial}{\partial\mu}\mathrm{tr}\, \ln(K) = -2\frac{\epsilon(\mathbf{k}) - \mu + 2g|v(T)|^2}{\hbar^2\omega_n^2 + E^2(\mathbf{k})}, \tag{4.153}$$

where $E(\mathbf{k})$ now denotes the elementary excitation spectrum (4.63) with v replaced with $v(T)$. Using the identity

$$\begin{aligned} \frac{1}{\beta}\sum_n \frac{1}{\hbar^2\omega_n^2 + E^2} &= \frac{1}{2E}\coth\left(\tfrac{1}{2}\beta E\right) \\ &= \frac{1}{E}\left(\frac{1}{2} + \frac{1}{\mathrm{e}^{\beta E} - 1}\right), \end{aligned} \tag{4.154}$$

which follows immediately from the identity (2.69), we find

$$\frac{1}{\beta}\frac{\partial}{\partial\mu}\sum_n \mathrm{tr}\, \ln(K) = -\frac{1}{E}[\epsilon - \mu + 2g|v(T)|^2]\coth\left(\tfrac{1}{2}\beta E\right). \tag{4.155}$$

Because the combination in front of the hyperbolic cotangent is precisely the derivative of E,

$$\frac{\partial}{\partial\mu}E = -\frac{1}{E}[\epsilon - \mu + 2g|v(T)|^2] \tag{4.156}$$

and also

$$\frac{1}{2}\coth\left(\tfrac{1}{2}x\right) = \frac{\mathrm{d}}{\mathrm{d}x}\left[\tfrac{1}{2}x + \ln\left(1 - \mathrm{e}^{-x}\right)\right], \tag{4.157}$$

up to an irrelevant constant, Eq. (4.155) is readily integrated with respect to μ. In this way, we obtain by Eqs. (4.151) and (4.90) for the one-loop thermodynamic potential

$$\frac{\Omega_1(v)}{V} = \int \frac{\mathrm{d}^Dk}{(2\pi)^D}\left[\frac{1}{2}E(\mathbf{k}) + \frac{1}{\beta}\ln\left(1 - \mathrm{e}^{-\beta E(\mathbf{k})}\right)\right]. \tag{4.158}$$

The first term in Eq. (4.158) is the zero-temperature result, which has been analyzed in Sec. 4.6. Here, we study the second term just below the condensation temperature $T_{\rm BEC}$ where Bose-Einstein condensation sets in by expanding it in a high-temperature series. The expansion is justified only if $T_{\rm BEC}$ is in the high-temperature regime. This, as we shall see shortly, is indeed the case for a weakly interacting Bose gas.

To proceed, we write, specializing to $D = 3$ for the moment

$$\begin{aligned} J &\equiv \int \frac{d^3k}{(2\pi)^3} \ln\left(1 - e^{-\beta E(\mathbf{k})}\right) \\ &= \frac{4}{\pi^{1/2}} \frac{1}{\lambda_T^3} \int_0^\infty dq q^2 \ln\left(1 - e^{-\beta E(q)}\right), \end{aligned} \tag{4.159}$$

with q the dimensionless loop variable introduced below Eq. (4.145), and

$$\beta E(q) = \sqrt{(q^2 + \alpha + 2\upsilon)^2 - \upsilon^2}, \tag{4.160}$$

where $\upsilon \equiv \beta g |v|^2$ and $\alpha = -\beta\mu$. To obtain the contributions of the dynamic modes, we make use of the observation earlier in this section that such modes generate only analytic contributions. This means that, as far as these contributions are concerned, the integrand in Eq. (4.159) can be expanded in a Taylor series as

$$\begin{aligned} q^2 \ln\left(1 - e^{-\beta E(q)}\right) &= q^2 \ln\left(1 - e^{-q^2}\right) + \frac{q^2}{e^{q^2} - 1}(\alpha + 2\upsilon) \\ &- \frac{1}{2} \frac{q^2 e^{q^2}}{(e^{q^2} - 1)^2}(\alpha + 2\upsilon)^2 - \frac{1}{2} \frac{1}{e^{q^2} - 1}\upsilon^2 + O\left(\beta^3\right). \end{aligned} \tag{4.161}$$

The leading contribution in this high-temperature expansion corresponds to replacing the finite-temperature Bogoliubov spectrum (4.63) with the free single-particle spectrum $\epsilon(\mathbf{k})$. This means that the energy of an elementary excitation at high temperatures is approximately given by the free-particle energy. The expansion yields by Eq. (4.149)

$$\begin{aligned} J = J^{(0)} - \frac{1}{2}\Gamma(\tfrac{3}{2})[\zeta(\tfrac{5}{2}) - \zeta(\tfrac{3}{2})(\alpha + 2\upsilon) + \tfrac{1}{2}\zeta(\tfrac{1}{2})(\alpha + 2\upsilon)^2 \\ + \zeta(\tfrac{1}{2})\upsilon^2 + O\left(\beta^3\right)], \end{aligned} \tag{4.162}$$

where $J^{(0)}$ stands for the contribution to the integral of the zero-frequency mode. In addition to the expansion parameter α of a noninteracting Bose gas, a second dimensionless expansion parameters appears here, *viz.* υ. Both parameters are proportional to β.

At two loop, we consider the bubble diagram shown in Fig. 4.3. The symmetry factor of the diagram is 2 for interchanging the two bubbles does not change its

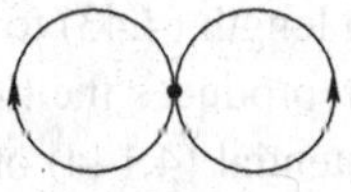

Fig. 4.3 Two-loop contribution to the thermodynamic potential Ω of a weakly interacting Bose gas.

topology. The Feynman rules can be obtained by adapting the recipe given in Sec. 2.5 to finite temperature. Specifically, the vertex follows from noting that the partition function is of the form (2.48) with the Euclidean action $S_{\rm E}$ containing the interaction term $\frac{1}{2}g\int d^4x|\chi|^4$. By taking two derivatives with respect to χ and two with respect to χ^* so that no field is left, we obtain as vertex

$$\equiv -2\frac{g}{\hbar}. \tag{4.163}$$

The finite-temperature propagator $\Delta_{\rm E}(k)$ follows from the zero-temperature counterpart (4.62) by applying the rule (2.43) and replacing ω with $i\omega_n$, and also by including an overall minus sign implied by the rule (2.39). An internal line represents $\hbar\Delta_{\rm E}(k)$ by Eq. (2.81). With these rules, the two-loop diagram gives as contribution to the thermodynamic potential

$$\frac{\Omega_2(v)}{V} = -gI^2, \tag{4.164}$$

where I denotes the one-loop integral

$$I = \int \frac{d^3k}{(2\pi)^3}\frac{\epsilon(\mathbf{k}) - \mu + 2g|v|^2}{E(\mathbf{k})}\frac{1}{e^{\beta E(\mathbf{k})} - 1}. \tag{4.165}$$

Here, use is made of the identity (4.154) to carry out the sum over the Matsubara frequencies and the zero-temperature part has been omitted. Expanding the integrands in a high-temperature series so as to obtain the contributions from the dynamic modes, we arrive at the expression

$$\frac{\Omega_2(v)}{V} = \frac{g}{\lambda_T^6}\left[\zeta^2(\tfrac{3}{2}) - 2\zeta(\tfrac{1}{2})\zeta(\tfrac{3}{2})(\alpha + 2v)\right]. \tag{4.166}$$

Added together, the zero-, one-, and two-loop contributions give

$$\frac{\Omega_0(v) + \Omega_1(v) + \Omega_2(v)}{V} = c_0 - r|v|^2 + \frac{u}{2}|v|^4, \tag{4.167}$$

with

$$c_0 = -\frac{1}{\beta\lambda_T^3}\left\{\zeta(\tfrac{5}{2}) - 2\zeta^2(\tfrac{3}{2})\delta - \zeta(\tfrac{3}{2})\alpha\left[1 - 4\zeta(\tfrac{1}{2})\delta\right] + \tfrac{1}{2}\zeta(\tfrac{1}{2})\alpha^2\right\}, \tag{4.168}$$

a v-independent term and

$$\delta \equiv \frac{g}{2}\frac{\beta}{\lambda_T^3} = \frac{a}{\lambda_T}, \tag{4.169}$$

the ratio of the s-wave scattering length (4.43) to the thermal wave length (2.33). In the limit $\delta \to 0$, Eq. (4.168) reproduces the first contributions of the dynamic modes to the thermodynamic potential (4.144) of an ideal Bose gas with $F_{D/2+1}$ ($D = 3$) specified in Eq. (4.150), albeit with $\alpha > 0$ now. Furthermore,

$$-\beta r = \left[\alpha + 4\zeta(\tfrac{3}{2})\delta\right]\left[1 - 4\zeta(\tfrac{1}{2})\delta\right], \tag{4.170a}$$

$$u = g\left[1 - 12\zeta(\tfrac{1}{2})\delta\right]. \tag{4.170b}$$

It is important to note that the coefficients given in Eqs. (4.170) express the dressing of the zero-temperature parameters μ and g by fluctuations in the dynamic modes.

The condensation temperature $T_{\rm BEC}$ is determined by the vanishing of the coefficient r of the quadratic term, yielding

$$\alpha_{\rm c} = -4\zeta(\tfrac{3}{2})\delta_{\rm c}, \tag{4.171}$$

which gives $T_{\rm BEC}$ in terms of the chemical potential. Note that $\zeta(\frac{3}{2}) > 0$, while $\zeta(\frac{1}{2}) < 0$. When the particle number is fixed, μ must be replaced with n as independent variable. It is most easily obtained from the thermodynamic potential (4.167) at $T = T_{\rm BEC}$ where v vanishes:

$$n = -\frac{1}{V}\frac{\partial\Omega}{\partial\mu}\bigg|_{T=T_{\rm BEC}} = \frac{1}{\lambda^3_{T_{\rm BEC}}}\left\{\zeta(\tfrac{3}{2})\left[1 - 4\zeta(\tfrac{1}{2})\delta_{\rm c}\right] - \zeta(\tfrac{1}{2})\alpha_{\rm c}\right\}. \tag{4.172}$$

By Eq. (4.171), the last two correction terms precisely cancel, and the expression reduces to that of an ideal Bose gas:

$$n = \zeta(\tfrac{3}{2})/\lambda^3_{T_{\rm BEC}}. \tag{4.173}$$

In other words, to this order, fluctuations in the dynamic modes of the weakly interacting theory do not lead to a shift in the condensation temperature of an ideal Bose gas. The shift obtained at one loop is precisely canceled by the two-loop contribution. In the next section, it is shown that the zero-frequency modes, which will be treated nonperturbatively, do lead to a shift. For completeness, we mention that for arbitrary $2 < D < 4$, Eq. (4.171) and (4.173) generalize to

$$\alpha_{\rm c} = -4\zeta(D/2)\delta_{\rm c}, \quad n = \zeta(D/2)/\lambda^D_{T_{\rm BEC}}, \tag{4.174}$$

where now $\delta = g\beta/2\lambda^D_T$.

With Eq. (4.173), the chemical potential at $T_{\rm BEC}$ given in Eq. (4.171) can be rewritten in a form similar to the zero-temperature relation (4.51) in the absence of an external potential ($U_{\rm ex} = 0$) as

$$\mu(T_{\rm BEC}) = 2gn = \frac{8\pi\hbar^2 a}{m}n. \tag{4.175}$$

Both forms can be summarized by

$$\mu(T) = g(2n - n_0), \tag{4.176}$$

where $n_0 = n$ in the zero-temperature limit, while n_0 tends to zero at the condensation temperature. This relation, known as the *Hugenholtz-Pines relation*, is an exact result at both temperatures. It follows that the chemical potential remains positive at the condensation temperature of a weakly interacting Bose gas. This is in contrast to an ideal Bose gas (obtained by letting $g \to 0$), where the chemical potential tends to zero when the condensation temperature is approached from above. Whereas the chemical potential of an ideal Bose gas remains zero all the way down to zero temperature, that of a weakly interacting Bose gas decreases from $8\pi\hbar^2 an/m$ at $T_{\rm BEC}$ to $4\pi\hbar^2 an/m$ at zero temperature.

To justify the high-temperature expansion, we use the Hugenholtz-Pines relation to replace the particle number density in Eq. (4.173) with the chemical potential, giving

$$k_{\rm B}T_{\rm BEC} = \frac{2\pi\hbar^2}{m}\left(\frac{\mu}{2\zeta(3/2)g}\right)^{2/3}. \tag{4.177}$$

The right side is written solely in terms of parameters contained in the Lagrangian (4.40). Since this temperature is large for g small, the high-temperature expansion is consistent with the weak-coupling assumption of perturbation theory.

The value of $v(T)$, which so far has been left unspecified, follows from extremizing Eq. (4.167),

$$|v(T)|^2 = r/u \approx \mu/g - 2\zeta(\tfrac{3}{2})/\lambda_T^3 \tag{4.178}$$

to leading order. It physically denotes the particle number density n_0 in the condensate. Since to this order the single-particle spectrum can be replaced with that of an ideal Bose gas, see below Eq. (4.161), the number density n_1 of particles not in the condensate is given by

$$n_1 = \zeta(\tfrac{3}{2})/\lambda_T^3, \tag{4.179}$$

and the total number density $n = n_0 + n_1$ by

$$n = \mu/g - \zeta(\tfrac{3}{2})/\lambda_T^3. \tag{4.180}$$

Just below the condensation temperature, Eq. (4.178) can be expanded as

$$\begin{aligned}|v(T)|^2 &= 3\zeta(\tfrac{3}{2})\lambda_{T_{\rm BEC}}^{-3}\,(1 - T/T_{\rm BEC}) + O\left[(1 - T/T_{\rm BEC})^2\right] \\ &= 3n\,(1 - T/T_{\rm BEC}) + O\left[(1 - T/T_{\rm BEC})^2\right],\end{aligned} \tag{4.181}$$

showing that when the condensation temperature is approached from below, the condensate smoothly vanishes.

We end this section by investigating the spectrum of the elementary excitations at finite temperature. The most important characteristic of the zero-temperature Bogoliubov spectrum (4.64) is that it vanishes linearly in the long wave-length limit. This property is a direct consequence of the spontaneously broken U(1) symmetry. We therefore expect it to persist at any temperature below $T_{\rm BEC}$. Now, simply substituting the value (4.178) for $|v(T)|^2$ in the finite-temperature Bogoliubov spectrum (4.63), one finds that it has an energy gap. To resolve this paradox, note that whereas Eq. (4.178) includes fluctuations in the dynamic modes, the spectrum (4.63) does not, which is inconsistent. To fix it, we return to Eq. (4.170a) and note that it physically reflects a change in the chemical potential $\mu \to r$ due to thermal fluctuations in the dynamic modes. This change must be included in the finite-temperature spectrum (4.63). To leading order in $1/T$, Eq. (4.170a) gives

$$r = \mu - 2\zeta(\tfrac{3}{2})g/\lambda_T^{3/2} \tag{4.182}$$

and $g|v|^2 = r$. In this way, the finite-temperature spectrum (4.63) becomes

$$E(\mathbf{k}) = \sqrt{\epsilon^2(\mathbf{k}) + 2g|v(T)|^2\epsilon(\mathbf{k})}, \tag{4.183}$$

which is indeed gapless. When written in terms of r, this expression becomes completely analogous to the zero-temperature Bogoliubov expression (4.64). The speed of sound that follows from this spectrum at finite temperature reads

$$c^2(T) = \frac{g}{m}|v(T)|^2 \approx 3\frac{g}{m}n(1 - T/T_{\rm BEC}), \tag{4.184}$$

just below the condensation temperature. It vanishes when the temperature approaches $T_{\rm BEC}$ from below, in accord with the observation that the gapless Nambu-Goldstone mode vanishes at the condensation temperature. Note that also at finite temperature, the speed of sound squared is linear in the coupling constant and vanishes when the interparticle repulsion is turned off. If the interaction becomes attractive, c^2 turns negative, which hints at an instability. The system either converts to a more condensed state of matter, such as a liquid, or collapses.

4.11 Large-N Expansion

In the previous section, the dynamic modes of a weakly interacting Bose gas were found to two-loop order not to shift the condensation temperature of a noninteracting gas. That perturbative analysis was shown to be equivalent to a high-temperature expansion. In this section, the effect of the zero-frequency modes is studied in the infinite-temperature limit, where the dynamic modes decouple and the theory reduces to a static one. In this limit, the finite-temperature action (divided by $\hbar$),

$$\mathcal{S} \equiv \frac{1}{\hbar} S_{\rm E} = \frac{1}{\hbar} \int_0^{\hbar\beta} {\rm d}\tau \int {\rm d}^D x \mathcal{L}_{\rm E}, \tag{4.185}$$

with $\mathcal{L}_{\rm E}$ the Euclidean Lagrangian density obtained from its zero-temperature counterpart (4.40) by using the rule (2.39), reduces to

$$\mathcal{S} = \int {\rm d}^D x \left[\frac{1}{2} (\partial_i \phi_a)^2 + \frac{1}{2} r \phi^2 + \frac{u}{4!} \left(\phi^2\right)^2 \right]. \tag{4.186}$$

Here, the integral over Euclidean time simply produces a factor $\hbar\beta$. The fields ϕ_a ($a = 1, 2$), which only depend on space coordinates, are defined through

$$\psi \equiv \sqrt{(m/\hbar^2) k_{\rm B} T} \, (\phi_1 + {\rm i}\phi_2), \tag{4.187}$$

while $r \equiv -2m\mu/\hbar^2$, and

$$u \equiv 12 g (m^2/\hbar^4) k_{\rm B} T = 96\pi^2 a / \lambda_T^2 \tag{4.188}$$

by Eq. (4.43), while $\phi^2 = \sum_a \phi_a \phi_a$. The shift in the condensation temperature is computed indirectly by using the expression (4.8) for the condensation temperature of an ideal Bose gas. It follows from this expression that the shift in $T_{\rm BEC}$ at fixed particle number density is related to a change in the particle number density at fixed temperature by

$$\left. \frac{\Delta T_{\rm BEC}}{T_{\rm BEC}} \right|_n = - \frac{2}{D} \left. \frac{\Delta n}{n} \right|_{T_{\rm BEC}}, \tag{4.189}$$

with $n = \langle |\psi|^2 \rangle = (m/\hbar^2) k_{\rm B} T \langle \phi^2 \rangle$.

To circumvent infrared divergences, the effect of the zero-frequency modes on the condensation temperature will be calculated nonperturbatively in a so-called $1/N$ expansion, where the two-component scalar field ϕ is extended to N components ($a = 1, \cdots, N$) with N large. The inverse of N then provides the theory with a small parameter so that calculations can be ordered according to the powers of $1/N$ with Nu held fixed, instead of the number of loops, as in ordinary perturbation theory. For an ideal Bose gas, n is proportional to N so that the right side of Eq. (4.173) must be replaced with

$$n = \frac{N}{2} \zeta(\tfrac{3}{2}) / \lambda_{T_{\rm BEC}}^3 \tag{4.190}$$

for general N.

The large-N expansion is facilitated by introducing an auxiliary field α via a Hubbard-Stratonovich transformation to decouple the quartic interaction term. The partition function Z then assumes the form

$$\begin{aligned} Z &= \int {\rm D}\phi \, {\rm e}^{-\mathcal{S}} \\ &= \int {\rm D}\phi \int {\rm D}\alpha \exp\left\{ - \int {\rm d}^D x \left[\frac{1}{2} (\partial_i \phi_a)^2 + \frac{1}{2} \alpha \phi^2 - \frac{3}{2u} (\alpha - r)^2 \right] \right\}. \end{aligned} \tag{4.191}$$

The integral over ϕ has become a simple Gaussian which gives for the partition function $Z = \int \mathrm{D}\alpha \exp(-\mathcal{S}_\alpha)$ with

$$\mathcal{S}_\alpha = -\frac{3}{2u}\int \mathrm{d}^D x\,(\alpha - r)^2 + \frac{N}{2}\,\mathrm{Tr}\ln\left(-\nabla^2 + \alpha\right) \tag{4.192}$$

the effective action governing the auxiliary field. This expression yields as propagator for the α field

$$D'_\mathrm{F}(\mathbf{k}) = -\frac{1}{3/u + (N/2)\Pi(\mathbf{k})}, \tag{4.193}$$

where the last term in the denominator,

$$\Pi(\mathbf{k}) \equiv \int \frac{\mathrm{d}^D q}{(2\pi)^D}\frac{1}{\mathbf{q}^2(\mathbf{q}+\mathbf{k})^2}, \tag{4.194}$$

follows from writing $\ln[-\nabla^2 + \alpha(\mathbf{x})] = \ln(-\nabla^2) + \ln[1 - \alpha(\mathbf{x})/\nabla^2]$, and expanding the second term in derivatives, see Sec. 3.1. The minus sign in Eq. (4.193) arises because the auxiliary field is purely imaginary. The propagator (4.193) summarizes an infinite series of Feynman diagrams (see Fig. 4.4), forming a geometric series of bubble insertions,

$$D'_\mathrm{F}(\mathbf{k}) = -\frac{u}{3} + \left(-\frac{u}{3}\right)^2 \frac{N}{2}\Pi(\mathbf{k}) + \left(-\frac{u}{3}\right)^3 \left[\frac{N}{2}\Pi(\mathbf{k})\right]^2 + \cdots, \tag{4.195}$$

where $D_\mathrm{F}(\mathbf{k}) \equiv -\frac{1}{3}u$ is the bare propagator of the α field that can be read off from the Lagrangian in Eq. (4.191). The rest of the Feynman rules we need can also be read off from the Lagrangian in Eq. (4.191). In particular, the propagator of each of the components of the ϕ field reads

$$\Delta_\mathrm{F}(\mathbf{p}) = \frac{1}{\mathbf{p}^2 + r + \Sigma(\mathbf{p}) - \Sigma(0)}, \tag{4.196}$$

with the self-energy given by the Feynman diagram 4.5 to leading order in the $1/N$ expansion,

$$\Sigma(\mathbf{p}) - \Sigma(0) = -\int \frac{\mathrm{d}^D k}{(2\pi)^D} D'_\mathrm{F}(\mathbf{k})\left(\frac{1}{(\mathbf{k}+\mathbf{p})^2 + r} - \frac{1}{\mathbf{k}^2 + r}\right). \tag{4.197}$$

Fig. 4.4 The infinite series of Feynman diagrams leading to the dressed propagator (4.193) of the α field. The heavy dashed line denotes the dressed propagator $D'_\mathrm{F}(\mathbf{k})$ and the light dashed line denotes the bare propagator $D_\mathrm{F}(\mathbf{k})$. The bubble stands for $(N/2)\Pi(\mathbf{k})$, where the factor N arises from the sum over all field components, while 2 is the symmetry factor of the bubble.

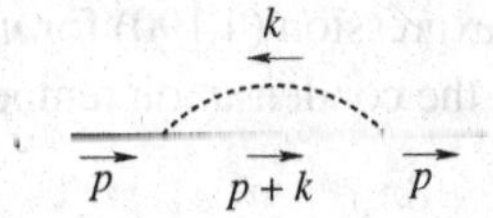

Fig. 4.5 The leading-order contribution in $1/N$ to the self-energy of the ϕ field.

By subtracting $\Sigma(0)$, we carried out a mass renormalization such that $\Delta_F^{-1}(0) = r$. At the critical point, $r = 0$ to this order.

The shift $\Delta n = (m/\hbar^2)k_B T_{BEC}\Delta\langle\phi^2\rangle$ in the particle number density at T_{BEC} resulting from turning on the interaction follows in leading order in the $1/N$ expansion from Eq. (4.196) as

$$\Delta\langle\phi^2\rangle = -N\int\frac{d^D p}{(2\pi)^D}\frac{1}{\mathbf{p}^4}\left[\Sigma(\mathbf{p}) - \Sigma(0)\right]. \tag{4.198}$$

The integrals are best evaluated using the basic formulas (3.43) and (3.44) of dimensional regularization, giving

$$\Pi(\mathbf{k}) = \frac{\Gamma(2-D/2)\,\Gamma^2(D/2-1)}{(4\pi)^{D/2}\quad\Gamma(D-2)}|\mathbf{k}|^{D-4} \tag{4.199}$$

and

$$\int\frac{d^D p}{(2\pi)^D}\frac{1}{\mathbf{p}^4}\left(\frac{1}{(\mathbf{p}+\mathbf{k})^2} - \frac{1}{\mathbf{k}^2}\right) = \frac{\Gamma(3-D/2)}{(4\pi)^{D/2}}\frac{\Gamma(D/2-1)\Gamma(D/2-2)}{\Gamma(D-3)}|\mathbf{k}|^{D-6} \tag{4.200}$$

where the last term in the integrand yields zero in dimensional regularization. Finally,

$$\int\frac{d^D k}{(2\pi)^D}\frac{|\mathbf{k}|^{D-6}}{(6/Nu) + b|\mathbf{k}|^{D-4}} = -\frac{1}{2}\frac{\Omega_D}{(2\pi)^D}\Gamma\left[(D-2)/(D-4)\right]\Gamma\left[(D-6)/(D-4)\right] \times (6/Nu)^{(D-2)/(D-4)}\, b^{2(3-D)/(D-4)}, \tag{4.201}$$

where $\Omega_D = 2\pi^{D/2}/\Gamma(D/2)$ and b is the coefficient of $|\mathbf{k}|^{D-4}$ in Eq. (4.199). The last integral diverges in the physical limit $D \to 3$ for in $D = 3-\varepsilon$, $\Gamma[(D-2)/(D-4)]/\Gamma[(D-6)/(D-4)] = -1/\varepsilon + O(\varepsilon)$. Combined with the $1/\Gamma(D-3)$ factor appearing in Eq. (4.200), this produces a finite result in $D = 3$. Pasting the pieces together, one obtains for $D \to 3$

$$\Delta\langle\phi^2\rangle = -\frac{1}{96\pi^2}Nu. \tag{4.202}$$

From the dependence $(Nu)^{(D-2)/(4-D)}$ in arbitrary D it follows that only in $D = 3$ this result is linear in Nu. Also note that the coefficient b does not enter the final result as it is raised to the power $2(3-D)/(D-4)$ which is zero in $D = 3$. By

Eq. (4.189) with the ideal-gas expression (4.190) for n, the result (4.202) leads to the interaction-induced shift in the condensation temperature

$$\begin{aligned}\frac{\Delta T_{\rm BEC}}{T_{\rm BEC}} &= \frac{8\pi}{3\zeta(3/2)} \frac{a}{\lambda_{T_{\rm BEC}}} \\ &= \frac{8\pi}{3\zeta^{4/3}(3/2)} a n^{1/3} \approx 2.33\, a n^{1/3}.\end{aligned} \tag{4.203}$$

Although this result is valid only for large N, it is independent of this parameter. The next-to-leading order in $1/N$ was found to give only a moderate correction to this leading result of order 25% for $N = 2$. Monte Carlo studies typically agree within a factor of 2 with the estimate (4.203).

4.12 Two-Fluid Model

The Bogoliubov theory, describing a weakly interacting Bose gas, does not apply to strongly interacting superfluid ^{4}He. Both systems share, however, many features. In this section, a two-fluid description of a weakly interacting Bose gas is given to underscore these similarities. In the context of superfluid ^{4}He, the two-fluid model is a phenomenological model that successfully describes many of its startling properties. It is based on the assumption that the system can be separated in a condensate and a normal liquid consisting of elementary excitations. Two types of elementary excitations can be identified. The first consists of phonons, the quanta of sound waves, which we already met in the theory of weakly interacting Bose gases. Excitations of the second type were dubbed *rotons* by Landau. They are sometimes pictured as almost free particle excitations surrounded by a cloud of phonons. Strictly speaking, it is impossible to divide the elementary excitations into two types as both are part of a single-branch spectrum, consisting of phonons at long wave lengths and rotons at shorter wave lengths.

Figure 4.6 shows an experimental curve of the elementary excitation spectrum in superfluid ^{4}He obtained in recent high-precision inelastic neutron scattering measurements. In the long wave-length limit the spectrum is linear,

$$\lim_{|\mathbf{k}|\to 0} E(\mathbf{k}) = \hbar c |\mathbf{k}|, \tag{4.204}$$

with a slope determined by the speed of sound. The local minimum of the spectrum indicates the presence of rotons. More specifically, it denotes the minimum of energy Δ_0 needed to excite them. The two parts of the curve are separated by a maximum which renders rotons stable against decay into phonons. The plateau of the spectrum at larger wave numbers is explained by the decay of single rotons

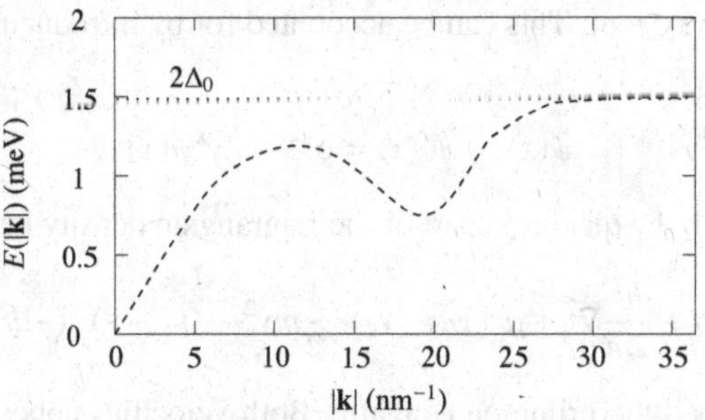

Fig. 4.6 Elementary excitation spectrum $E(|\mathbf{k}|)$ in superfluid ^{4}He as a function of the wave number $|\mathbf{k}|$. Reprinted figure with permission from A. Sakhel and H. R. Glyde, Phys. Rev. **B70**, 144511 (2004). Copyright 2004 by the American Physical Society.

into pairs of rotons. The energy of the plateau, indicated by the horizontal line, is $2\Delta_0$ and corresponds to the energy needed to produce such a pair.

To facilitate converting to other units, we record the values 1eV $\approx 1.602 \times 10^{-19}$J and $k_B \approx 1.380 \times 10^{-23}$J/K so that 1eV corresponds to approximately 1.160×10^4K

As shown in Eq. (4.102), the Bogoliubov theory predicts a depletion of the condensate. No matter how weak, the interparticle repulsion always removes particles from the condensate, even at zero temperature. In superfluid ^{4}He, where the interaction is strong, an even stronger depletion of the condensate is expected. It has been estimated that only about 8% of the particles condense in the zero wave-vector state. Despite this, it turns out that at zero temperature all the particles participate in the superfluid motion. Apparently, the condensate drags the elementary excitations along with it.

To show this for a weakly interacting Bose gas, assume that the entire system moves with a velocity $\mathbf{v}$ relative to the laboratory system. As in standard hydrodynamics, the time derivative in the frame following the motion of the fluid is $\partial_t + \mathbf{v} \cdot \nabla$, see Eq. (1.94). If inserted in the Lagrangian (4.40) of a weakly interacting Bose gas, it assumes the form

$$\mathcal{L} = \mathrm{i}\hbar\psi^*(\partial_t + \mathbf{v} \cdot \nabla)\psi - \frac{\hbar^2}{2m}\nabla\psi^* \cdot \nabla\psi + \mu\psi^*\psi - \frac{1}{2}g|\psi|^4. \tag{4.205}$$

The extra term involves the total momentum of the system $\int \mathrm{d}^D x\, \psi^*(-\mathrm{i}\hbar\nabla)\psi$. The velocity $\mathbf{v}$ multiplying this term is on the same footing as the chemical potential μ which multiplies the particle number $\int \mathrm{d}^D x\, \psi^*\psi$. Whereas μ is associated with particle number conservation, $\mathbf{v}$ is related to the conservation of momentum, see Sec. 1.3.

In the two-fluid picture, the condensate can move with a different velocity $\mathbf{v}_s$

as the rest of the system. This can be accounted for by introducing new fields, cf. Eq. (1.62)

$$\psi(x) \to \psi'(x) = \mathrm{e}^{-\mathrm{i}(m/\hbar)\mathbf{v}_\mathrm{s}\cdot\mathbf{x}}\psi(x) \tag{4.206}$$

in terms of which the quadratic part of the Lagrangian density becomes

$$\mathcal{L}_0 = \psi^*\left[\mathrm{i}\hbar\partial_t + \frac{\hbar^2}{2m}\nabla^2 + \mu + m\mathbf{v}\cdot\mathbf{v}_\mathrm{s} - \frac{1}{2}m\mathbf{v}_\mathrm{s}^2 + (\mathbf{v}_\mathrm{s} - \mathbf{v})\cdot(-\mathrm{i}\hbar\nabla)\right]\psi, \tag{4.207}$$

where we dropped the prime on ψ again. Both velocities appear in this expression. Apart from the change $\mathbf{v} \to \mathbf{v} - \mathbf{v}_\mathrm{s}$ in the last term on the right, the field transformation results in a change of the chemical potential

$$\mu \to \mu_{\mathbf{v}_\mathrm{s}} \equiv \mu + m\mathbf{v}\cdot\mathbf{v}_\mathrm{s} - \frac{1}{2}m\mathbf{v}_\mathrm{s}^2. \tag{4.208}$$

The equations for the Bogoliubov spectrum and the thermodynamic potential are readily written down for the present case provided these two changes are kept in mind. In particular, the one-loop potential (4.158) now reads

$$\mathcal{V}_1(T) = \frac{1}{2}\int\frac{\mathrm{d}^D k}{(2\pi)^D}E_{\mathbf{v}_\mathrm{s}}(\mathbf{k}) + \frac{1}{\beta}\int\frac{\mathrm{d}^D k}{(2\pi)^D}\ln\left(1 - \mathrm{e}^{-\beta[E_{\mathbf{v}_\mathrm{s}}(\mathbf{k})-(\mathbf{v}_\mathrm{s}-\mathbf{v})\cdot\hbar\mathbf{k}]}\right), \tag{4.209}$$

where $E_{\mathbf{v}_\mathrm{s}}(\mathbf{k})$ is the Bogoliubov spectrum (4.63) with the replacement (4.208). The mean-field value of the potential energy density $\mathcal{V}_\mathrm{c}(T)$ is given by Eq. (4.152) with the same replacement. The average momentum density T^{0i}, or equivalently, mass current density $g^i = T^{0i}$ [see Eq. (1.70)], is obtained in this approximation by taking the derivative of the potential energy density $\mathcal{V}_\mathrm{c}(T) + \mathcal{V}_1(T)$ with respect to $-\mathbf{v}$. This yields

$$\mathbf{g} = mn\mathbf{v}_\mathrm{s} - \int\frac{\mathrm{d}^D k}{(2\pi)^D}\frac{\hbar\mathbf{k}}{\exp\{\beta[E_{\mathbf{v}_\mathrm{s}}(\mathbf{k}) - (\mathbf{v}_\mathrm{s} - \mathbf{v})\cdot\hbar\mathbf{k}]\} - 1}, \tag{4.210}$$

where use is made of the equation

$$\frac{\partial\mu_{\mathbf{v}_\mathrm{s}}}{\partial\mathbf{v}} = m\mathbf{v}_\mathrm{s}. \tag{4.211}$$

The last term at the right side of Eq (4.210) is the contribution of the elementary excitations. In the zero-temperature limit, this term vanishes, and $\mathbf{g} = mn\mathbf{v}_\mathrm{s}$. This equation, involving the total particle number density n, shows that, at zero temperature, indeed all the particles are involved in the superflow, although only a fraction of them resides in the condensate. When the condensate moves with the same velocity as the rest of the system ($\mathbf{v}_\mathrm{s} = \mathbf{v}$), the last term in Eq. (4.210) vanishes again, now by symmetry.

Assuming the difference between the normal and superfluid velocities small, we can expand the last term in Eq. (4.210) to linear order in this difference, giving

$$\mathbf{g} = \rho \mathbf{v}_\mathrm{s} + \frac{\beta}{D} \int \frac{\mathrm{d}^D k}{(2\pi)^D} \hbar^2 \mathbf{k}^2 \frac{\mathrm{e}^{\beta E(\mathbf{k})}}{(\mathrm{e}^{\beta E(\mathbf{k})} - 1)^2} (\mathbf{v} - \mathbf{v}_\mathrm{s}), \tag{4.212}$$

where we introduced the average total mass density of the fluid $\rho \equiv mn$, and set higher-order terms in $\mathbf{v}$ or $\mathbf{v}_\mathrm{s}$ to zero in $E_{\mathbf{v}_\mathrm{s}}$. The result (4.212) therefore corresponds to linear response. In deriving it, use is made of the relation

$$\int \frac{\mathrm{d}^D k}{(2\pi)^D} k^i k^j f(k) = \frac{1}{D} \delta^{ij} \int \frac{\mathrm{d}^D k}{(2\pi)^D} \mathbf{k}^2 f(|\mathbf{k}|), \tag{4.213}$$

with $f(|\mathbf{k}|)$ an arbitrary function depending only on the length of $\mathbf{k}$. The last term at the right side of Eq. (4.212), denoting the contribution of the elementary excitations, defines the normal mass density ρ_n,

$$\rho_\mathrm{n} \equiv \frac{\beta}{D} \int \frac{\mathrm{d}^D k}{(2\pi)^D} \hbar^2 \mathbf{k}^2 \frac{\mathrm{e}^{\beta E(\mathbf{k})}}{(\mathrm{e}^{\beta E(\mathbf{k})} - 1)^2}. \tag{4.214}$$

Writing

$$\rho = \rho_\mathrm{s} + \rho_\mathrm{n} \tag{4.215}$$

for the total mass density, we can cast Eq. (4.212) in the form

$$\mathbf{g} = \rho_\mathrm{s} \mathbf{v}_\mathrm{s} + \rho_\mathrm{n} \mathbf{v}, \tag{4.216}$$

with the last term on the right denoting the mass current, or equivalently, the momentum carried by the excitations. These are the basic equations of the two-fluid model. The model was introduced by Tisza using ideas of London to give a phenomenological description of superfluid ^{4}He, and was extended and refined by Landau. It not only successfully explained various startling experimental properties of the strongly interacting system, but also predicted new phenomena that were later confirmed by experiment.

For temperatures just below T_BEC, the integral in Eq. (4.214) can be treated in a high-temperature expansion as before. The leading contribution, which corresponds to replacing the finite-temperature Bogoliubov spectrum (4.63) with the free single-particle spectrum $\epsilon(\mathbf{k})$, reads

$$\rho_\mathrm{n} = \rho \, (T/T_\mathrm{BEC})^{D/2}, \tag{4.217}$$

where T_BEC denotes the condensation temperature (4.173).

In the opposite limit of low temperatures, phonons are the dominant excitations. The finite-temperature Bogoliubov spectrum can then be replaced with the phonon spectrum, and

$$\rho_\mathrm{n} = \frac{2}{\pi^{(D+1)/2}} \Gamma\left[(D+3)/2\right] \zeta(D+1) \frac{(k_\mathrm{B} T)^{D+1}}{\hbar^D c^{D+2}} \tag{4.218}$$

valid for $k_\mathrm{B} T << mc^2$.

4.13 Impurities

We next consider the effect of impurities. In particular, we investigate how random disorder affects the condensate and superfluidity. We restrict ourselves to the absolute zero of temperature.

Impurities are accounted for by including a random field $\phi(\mathbf{x})$ in the theory,

$$\mathcal{L}_{\square} = \phi(\mathbf{x})\, |\psi(x)|^2. \tag{4.219}$$

The field ϕ, being static, depends only on space coordinates and can also be viewed as an external static potential. It physically represents randomly located scattering centers of random strength, which can be attractive or repulsive. We assume that the disorder is sufficiently dilute and that it is Gaussian distributed with zero average,

$$P(\phi) = \exp\left[-\frac{1}{2\zeta}\int \mathrm{d}^D x\, \phi^2(\mathbf{x})\right]. \tag{4.220}$$

The distribution is characterized by the parameter ζ of dimension $\mathrm{kg}^2\mathrm{m}^{4+D}s^{-4}$. We will deal only with *quenched* or frozen impurities. In this case, the average $\langle O(\psi^*, \psi)\rangle$ of some observable $O(\psi^*, \psi)$, built from the fields ψ and ψ^*, is calculated in two steps. In the first, the ensemble average $\langle O(\psi^*, \psi)\rangle_\phi$ is computed for a given or frozen configuration $\phi(\mathbf{x})$. That is, the average is taken with respect to the partition function

$$Z[\phi] = \int \mathrm{D}\psi^* \mathrm{D}\psi \, \exp\left(\frac{\mathrm{i}}{\hbar}\int \mathrm{d}^d x\, \mathcal{L}\right), \tag{4.221}$$

where $\mathcal{L}$ now stands for the Lagrangian density (4.40) with the term (4.219) included. In the second step, the averaging over the random field is carried out

$$\langle O(\psi^*, \psi)\rangle = \int \mathrm{D}\phi\, P(\phi) \langle O(\psi^*, \psi)\rangle_\phi. \tag{4.222}$$

In terms of the shifted field (4.58), the random term (4.219) reads

$$\mathcal{L}_{\square} = \phi(\mathbf{x})\left[|v|^2 + |\chi(x)|^2 + v\chi^*(x) + v^*\chi(x)\right]. \tag{4.223}$$

The first two terms lead to an irrelevant change in the chemical potential so that only the last two terms, which can be cast in the form

$$\mathcal{L}_{\square} = \phi(\mathbf{x})\Upsilon^\dagger X, \qquad \Upsilon \equiv \begin{pmatrix} v \\ v^* \end{pmatrix}, \tag{4.224}$$

need be considered. In the Bogoliubov approximation, where higher than second order in the fields are ignored, the partition function (4.221), written as a functional integral over the shifted field X, becomes

$$Z[\phi] = \exp\left(-\frac{\mathrm{i}}{\hbar}\int \mathrm{d}^d x\, \mathcal{V}_\mathrm{c}\right)\int \mathrm{D}X \exp\left[\frac{\mathrm{i}}{\hbar}\int \mathrm{d}^d x\,(\mathcal{L}_0 + \mathcal{L}_{\square})\right], \tag{4.225}$$

with $\mathcal{L}_0$ given by Eq. (4.60). The integral over X, which is Gaussian in the approximation, yields an additional term to the effective potential in Eq. (4.70)

$$\int \mathrm{d}^d x \mathcal{V}_{\boxdot} = \frac{1}{2} \int \mathrm{d}^d x \mathrm{d}^d x' \, \phi(\mathbf{x}) \Upsilon^\dagger \Delta_\mathrm{F}(x - x') \Upsilon \phi(\mathbf{x}'), \tag{4.226}$$

where Δ_F is the the Feynman propagator (4.62). Fourier transforming this contribution gives

$$\int \mathrm{d}^D x \mathcal{V}_{\boxdot} = \frac{1}{2} \int \frac{\mathrm{d}^D k}{(2\pi)^D} |\phi(\mathbf{k})|^2 \, \Upsilon^\dagger \Delta_\mathrm{F}(0, \mathbf{k}) \Upsilon, \tag{4.227}$$

where it is used that the Fourier transform of a real field $\phi(\mathbf{x})$ satisfies $\phi(-\mathbf{k}) = \phi^*(\mathbf{k})$. Since the random field is Gaussian distributed, see Eq. (4.220), we have

$$\langle \phi(\mathbf{x}) \phi(\mathbf{x}') \rangle = \zeta \delta(\mathbf{x} - \mathbf{x}'), \tag{4.228}$$

or

$$\left\langle |\phi(\mathbf{k})|^2 \right\rangle = V\zeta. \tag{4.229}$$

The remaining integration over the loop wave vectors in Eq. (4.227) is readily carried out to yield in $2 < D < 4$ space dimensions

$$\langle \mathcal{V}_{\boxdot} \rangle = -\frac{\Gamma(1 - D/2)}{(2\pi)^{D/2}} \left(\frac{m}{\hbar^2} \right)^{D/2} |v|^2 \left(3g|v|^2 - \mu \right)^{D/2-1} \zeta. \tag{4.230}$$

Because the one-loop potential $\mathcal{V}_1$ in Eq. (4.70) led to a depletion of the condensate, the contribution (4.230) is expected to lead to an additional depletion. Taking the derivative with respect to (minus) the chemical potential, we find

$$n_\zeta = -\frac{\partial \langle \mathcal{V}_{\boxdot} \rangle}{\partial \mu} = \frac{\Gamma(2 - D/2)}{4\pi^{D/2}} \left(\frac{m}{\hbar^2} \right)^{D/2} g^{D/2-2} n_0^{D/2-1} \zeta, \tag{4.231}$$

where $n_0 = |v|^2$ denotes the number density of particles residing in the condensate. The mean-field value $|v|^2 = \mu/g$ is used only after the derivative has been taken. This is justified to this order because the expression (4.230) is a one-loop result. The contribution (4.231) is to be added to the right side of Eq. (4.101),

$$n = n_0 + n_1 + n_\zeta, \tag{4.232}$$

where n_1 stands for the last term in that equation. For a given total particle number density n, less particles than in the pure system are contained in the condensate. The divergence of n_ζ in the limit $g \to 0$ for $D < 4$ signals the collapse of the disordered system when the interparticle repulsion is removed.

We next calculate the mass current $\mathbf{g}$ to determine the superfluid mass density ρ_s, i.e., the mass density flowing with the superfluid velocity $\mathbf{v}_\mathrm{s}$. In the preceding section, we found that in the absence of impurities and at the absolute zero of temperature all the particles participate in the superflow and move on the average with the velocity $\mathbf{v}_s$. This is expected to break down in the presence of impurities. To determine the change in the superfluid mass density due to impurities, we replace $\mathcal{L}_0$ in the partition function (4.225) with the Lagrangian density (4.207) to obtain a modified contribution (4.227) to the effective potential. Taking the derivative with respect to the externally imposed velocity, $-\mathbf{v}$, we obtain to linear order in the difference $\mathbf{v} - \mathbf{v}_s$

$$\mathbf{g} = \rho_s \mathbf{v}_s + \rho_n \mathbf{v}, \tag{4.233}$$

with the zero-temperature superfluid and normal mass density

$$\rho_s = m\left(n - \frac{4}{D} n_\zeta\right), \quad \rho_n = \frac{4}{D} m n_\zeta, \tag{4.234}$$

respectively. These results show, first of all, that impurities generate normal fluid, even at the absolute zero of temperature. Moreover, as the normal density is for $D < 4$ a factor $4/D$ larger than the mass density mn_ζ removed from the condensate by the impurities, they generate more normal fluid than they took from the condensate. This implies that part of the zero wave-vector state belongs for $2 < D < 4$ not to the condensate, but to the normal fluid. This fraction of the zero wave-vector state is trapped by the impurities, i.e., is localized. If the depletion of the condensate by the impurities is roughly $D/4$, then ρ_s vanishes and superfluidity is completely destroyed.

4.14 Bose-Hubbard Model

In recent years, experimental groups have succeeded in loading ultracold atomic gases in optical lattices—periodic intensity patterns formed by the interference of two or more laser beams. A typical potential landscape felt by the atoms can be of the form of a 3D generalization of a 2D egg carton,

$$V(\mathbf{x}) = V_0 \sum_{i=1}^{3} \sin^2(|\mathbf{k}| x^i), \tag{4.235}$$

with $|\mathbf{k}|$ the wave number of the laser light. When the potential depth V_0 is steep enough, each of the potential wells confine a number of atoms. The centers of the wells may be pictured as lattice sites of the optical lattice. By adjusting the intensity of the laser beams, experimentalists have unique control over the underlying periodic structure. By lowering the intensity, they are able to dial the system from being in the so-called Mott insulating state, where all the atoms are pinned to the lattice sites, to the superfluid state. The atoms thus go from one extreme of being completely localized in the Mott insulating state, to the other extreme of being completely delocalized in the superfluid state. The two ground states are separated by a phase transition. Taking place very close to the absolute zero of temperature, the transition provides an example of a *quantum phase transition*. Such transitions do not involve thermal fluctuations, as do ordinary thermal phase transitions, but involve quantum fluctuations.

The insulating state is described by the *Bose-Hubbard model*. In the framework of canonical quantization, this lattice model is defined by the Hamiltonian

$$H_{\rm BHM} = -\frac{1}{2}\sum_{\mathbf{x}\mathbf{x}'} t_{\mathbf{x}\mathbf{x}'}(a^\dagger_{\mathbf{x}} a_{\mathbf{x}'} + a^\dagger_{\mathbf{x}'} a_{\mathbf{x}}) + \sum_{\mathbf{x}} \left[-\mu\hat{n}_{\mathbf{x}} + \tfrac{1}{2}U\hat{n}_{\mathbf{x}}(\hat{n}_{\mathbf{x}} - 1)\right], \tag{4.236}$$

where the sum $\sum_{\mathbf{x}}$ is over all lattice sites and $t_{\mathbf{x}\mathbf{x}'}$ is the so-called hopping matrix, which we assume to be symmetric in its indices so that

$$\sum_{\mathbf{x}\mathbf{x}'} t_{\mathbf{x}\mathbf{x}'} a^\dagger_{\mathbf{x}} a_{\mathbf{x}'} = \sum_{\mathbf{x}\mathbf{x}'} t_{\mathbf{x}\mathbf{x}'} a^\dagger_{\mathbf{x}'} a_{\mathbf{x}}. \tag{4.237}$$

For simplicity, we consider a hypercubic lattice of volume L^D with $(L/a)^D$ lattice sites specified by their position vector $\mathbf{x}$ and a is the lattice spacing. The operator $a^\dagger_{\mathbf{x}}$ creates a boson at site $\mathbf{x}$, while $a_{\mathbf{x}}$ annihilates a boson at that site, and $\hat{n}_{\mathbf{x}} \equiv a^\dagger_{\mathbf{x}} a_{\mathbf{x}}$ is the particle number operator which counts the number of bosons at site $\mathbf{x}$. This operator is given a "hat" to avoid confusion. The creation and annihilation operators satisfy the commutation relations

$$\left[a_{\mathbf{x}}, a^\dagger_{\mathbf{x}'}\right] = \delta_{\mathbf{x},\mathbf{x}'}, \tag{4.238}$$

$$[a_{\mathbf{x}}, a_{\mathbf{x}'}] = \left[a^\dagger_{\mathbf{x}}, a^\dagger_{\mathbf{x}'}\right] = 0. \tag{4.239}$$

They are straightforward generalizations of those appearing in the algebraic description of a harmonic oscillator, see Sec. 2.4. There, $a^\dagger$ creates a quantum of vibration, while a annihilates such a quantum. An eigenvalue of the operator $\hat{n} = a^\dagger a$ denotes the number of quanta present in a state. Since a quantum-mechanical problem is equivalent to a field theory in zero space dimensions, the creation and annihilation operators of a harmonic oscillator do not carry a space index. The first term in the Hamiltonian (4.236) describes the hopping of an atom from site $\mathbf{x}'$ to site $\mathbf{x}$. Finally, μ is the chemical potential, and $U > 0$ the on-site interparticle repulsion. The zero-temperature phase diagram is as follows.

In the limit $t_{\mathbf{x}\mathbf{x}'}/U \to 0$, hopping is completely suppressed and the Hamiltonian decouples into $(L/a)^D$ identical copies, one for each lattice site. Each of these copies is a simple quantum-mechanical problem with Hamiltonian

$$H_0 = -\mu\hat{n} + \frac{1}{2}U\hat{n}(\hat{n} - 1). \tag{4.240}$$

The on-site energy

$$\epsilon_n = -\mu n + \frac{1}{2}Un(n - 1) \tag{4.241}$$

is minimized when each site is occupied by an integer number of bosons. More specifically, within the interval $n - 1 < \mu/U < n$, exactly n integer number of bosons occupy each site. The occupation number jumps discontinuously whenever μ/U goes through a positive integer. Precisely at the integer values $\mu/U = n$, the ground states with n and $n+1$ particles per site are degenerate. If the chemical potential is negative, $n = 0$. We denote these ground states by $|n\rangle$ with, see Eq. (2.101),

$$|n\rangle = \frac{1}{\sqrt{n!}}(a^\dagger)^n|0\rangle \tag{4.242}$$

and $|0\rangle$ the vacuum. As for a harmonic oscillator, $a^\dagger|n\rangle = \sqrt{n+1}|n+1\rangle$, $a|n\rangle = \sqrt{n}|n-1\rangle$ for $n \geq 1$, and $a|0\rangle = 0$ so that $\hat{n}|n\rangle = n|n\rangle$ and $H_0|n\rangle = \epsilon_n|n\rangle$. Within an interval in which the particles are pinned to the lattice sites, the single-particle excitation spectrum is gapped, and the system is a Mott insulator with zero compressibility, $\kappa = 0$. Indeed, with the chemical potential fixed to the value $\mu/U = n - \frac{1}{2} + \delta$, with $-\frac{1}{2} < \delta < \frac{1}{2}$, the energy required to add a particle to a site is

$$\begin{aligned} E_\text{p} &= \epsilon_{n+1} - \epsilon_n = -\mu + Un \\ &= U(\tfrac{1}{2} - \delta) > 0, \end{aligned} \tag{4.243}$$

while the energy required to remove a particle from a site, which is equivalent to adding a hole to a site, is

$$\begin{aligned} E_\text{h} &= \epsilon_{n-1} - \epsilon_n = \mu + U(n-1) \\ &= U(\tfrac{1}{2} + \delta) > 0. \end{aligned} \tag{4.244}$$

In the following, the Feynman propagator $\Delta_\text{F}(t)$ in the state $|n\rangle$ of this quantum-mechanical problem is needed. In terms of creation and annihilation operators it is defined by the time-ordered product

$$\Delta_\text{F}(t) \equiv -\frac{\text{i}}{\hbar}\langle n|\,\text{T}\,a(t)a^\dagger(0)|n\rangle \tag{4.245}$$

as for a harmonic oscillator, where T is the time-ordering operator

$$\text{T}\,a(t)a^\dagger(0) \equiv \begin{cases} a(t)a^\dagger(0) & \text{for } t > 0 \\ a^\dagger(0)a(t) & \text{for } 0 > t, \end{cases} \tag{4.246}$$

and

$$a(t) = \text{e}^{(\text{i}/\hbar)Ht}\,a(0)\,\text{e}^{-(\text{i}/\hbar)Ht} \tag{4.247}$$

is the annihilation operator in the Heisenberg picture where operators are time dependent. In the ground state $|n\rangle$ with exactly n particles per site, the Feynman propagator reads, by the rules given below Eq. (4.242), explicitly

$$\Delta_\text{F}(t) = -\frac{\text{i}}{\hbar}\left[\theta(t)(n+1)\text{e}^{(\text{i}/\hbar)(\epsilon_n - \epsilon_{n+1})t} + \theta(-t)n\text{e}^{(\text{i}/\hbar)(\epsilon_{n-1} - \epsilon_n)t}\right], \tag{4.248}$$

with $\theta(t)$ denoting the Heaviside step function which is used for time ordering. For later use, note that a derivative with respect to the chemical potential brings down a factor of time,

$$\frac{\partial}{\partial\mu}\Delta_{\rm F}(t) = {\rm i}\frac{t}{\hbar}\Delta_{\rm F}(t). \tag{4.249}$$

After Fourier transforming, the propagator assumes the form

$$\Delta_{\rm F}(\omega) = \frac{n+1}{\hbar\omega - Un + \mu + {\rm i}\,{\rm sgn}(\omega)\eta} - \frac{n}{\hbar\omega - U(n-1) + \mu + {\rm i}\,{\rm sgn}(\omega)\eta}, \tag{4.250}$$

with $\eta = 0^+$ as usual. The relative minus sign between the two terms on the right has the same origin as in the propagator (2.94) of a harmonic oscillator. Note that in the interval $n - 1 < \mu/U < n$ under consideration, the first term has a pole at the positive energy $E = E_{\rm p} \equiv -\mu + Un > 0$, while the second term has a pole at the negative energy $E = -E_{\rm h} \equiv -\mu + U(n-1) < 0$.

Outside the limit $t_{\mathbf{xx}'}/U \to 0$, the particles can either be pinned or hop through the lattice and thus delocalize, depending on the values of the parameters. Being at zero temperature, delocalized bosons condense in a superfluid state. The single-particle spectrum is gapless here and the system compressible ($\kappa \neq 0$). With increasing $t_{\mathbf{xx}'}/U$, the intervals in which the particles are pinned to the lattice sites become smaller and eventually vanish at some critical value. The phase boundary is a result of the competition between the repulsive interaction, which tends to localize the particles, and the hopping term, which tends to delocalize them. To determine its precise location, we assume that hopping takes place only between nearest neighbors $\mathbf{x}$ and $\mathbf{x}'$, and set $t_{\mathbf{xx}'} = t$, with t the hopping parameter. For $\mathbf{x}$ and $\mathbf{x}'$ not nearest neighbors, $t_{\mathbf{xx}'} = 0$.

Returning to the functional integral approach, we write the partition function Z as

$$Z = \int {\rm D}\phi^* {\rm D}\phi \exp\left\{\frac{\rm i}{\hbar}\int {\rm d}t \left[L_0 + \frac{t}{2}a^D \sum_{\langle\mathbf{xx}'\rangle} (\phi^*_{\mathbf{x}}\phi_{\mathbf{x}'} + \phi^*_{\mathbf{x}'}\phi_{\mathbf{x}})\right]\right\}, \tag{4.251}$$

where the angle brackets in $\sum_{\langle\mathbf{xx}'\rangle}$ indicate that the sum runs only over nearest-neighbor sites. In the functional integral approach, $\phi_{\mathbf{x}}(t)$ is an ordinary, time-dependent lattice field which replaces the operator $a_{\mathbf{x}}$ in canonical quantization. More specifically,

$$a_{\mathbf{x}} \mathrel{\hat=} a^{D/2}\phi_{\mathbf{x}}, \quad a^\dagger_{\mathbf{x}} \mathrel{\hat=} a^{D/2}\phi^*_{\mathbf{x}}, \tag{4.252}$$

where the powers of the lattice spacing a are included to facilitate taking the continuum limit. In Eq. (4.251),

$$L_0 = a^D \sum_{\mathbf{x}} {\rm i}\hbar\phi^*_{\mathbf{x}}\partial_t\phi_{\mathbf{x}} - H_0 \tag{4.253}$$

is the Lagrangian corresponding to the Hamiltonian H_0 in the absence of hopping given by the last terms in Eq. (4.236) with $\hat{n}_{\mathbf{x}}$ replaced with $a^D\phi^*_{\mathbf{x}}\phi_{\mathbf{x}}$.

The generalization $D_F(k)$ of the Feynman propagator (4.250) in the presence of hopping is given by

$$D_F^{-1}(k) = \Delta_F^{-1}(\omega) - \epsilon_{\mathbf{k}}, \tag{4.254}$$

with $\epsilon_{\mathbf{k}} = -2t\sum_i \cos(k^i a)$, as can be read off from the quadratic terms in Eq. (4.251). The single-particle excitation spectra follow as

$$E^{\pm}_{\mathbf{k}} = -\mu + \frac{1}{2}\epsilon_{\mathbf{k}} + \frac{U}{2}(2n-1) \pm \frac{1}{2}\sqrt{\epsilon^2_{\mathbf{k}} + 2U(2n+1)\epsilon_{\mathbf{k}} + U^2}, \tag{4.255}$$

where $E_{\mathrm{p}} \equiv E^+$ is the energy required to add a particle to the insulating state, while $E_{\mathrm{h}} \equiv -E^-$ is the energy required to remove a particle from that state, or, what is equivalent, to add a hole. The kinetic energy $\epsilon_{\mathbf{k}} \propto t$ reflects the fact that the particles and holes can hop through the lattice. For small $\epsilon_{\mathbf{k}}$, the spectra become

$$E^{\pm}_{\mathbf{k}} \approx \begin{cases} -\mu + Un + (n+1)\epsilon_{\mathbf{k}} \\ -\mu + U(n-1) - n\epsilon_{\mathbf{k}}. \end{cases} \tag{4.256}$$

To facilitate an expansion in $t_{\mathbf{x}\mathbf{x}'}/U$, we decouple the hopping term by means of a Hubbard-Stratonovich transformation,

$$Z = \int \mathrm{D}\phi^*\mathrm{D}\phi \int \mathrm{D}\psi^*\mathrm{D}\psi \times \exp\left\{\frac{\mathrm{i}}{\hbar}\int \mathrm{d}t\left[L_0 + a^D\sum_{\mathbf{x}}(\psi_{\mathbf{x}}\phi^*_{\mathbf{x}} + \psi^*_{\mathbf{x}}\phi_{\mathbf{x}}) - a^D\sum_{\mathbf{x}\mathbf{x}'} t^{-1}_{\mathbf{x}\mathbf{x}'}\psi^*_{\mathbf{x}}\psi_{\mathbf{x}'}\right]\right\}. \tag{4.257}$$

On substituting the Euler-Lagrange equation for the auxiliary field,

$$\psi_{\mathbf{x}} = \sum_{\mathbf{x}'} t_{\mathbf{x}\mathbf{x}'}\phi_{\mathbf{x}'}, \tag{4.258}$$

back into the Lagrangian, the original form is recovered. The partition function can now be written as a functional integral over the auxiliary fields as

$$Z = \int \mathrm{D}\psi^*\mathrm{D}\psi\, \mathrm{e}^{(\mathrm{i}/\hbar)S[\psi^*,\psi]} \tag{4.259}$$

with the action

$$S[\psi^*,\psi] = -\int \mathrm{d}t a^D \sum_{\mathbf{x}\mathbf{x}'} t^{-1}_{\mathbf{x}\mathbf{x}'}\psi^*_{\mathbf{x}}\psi_{\mathbf{x}'} - \mathrm{i}\hbar\ln\langle \mathrm{e}^{(\mathrm{i}/\hbar)\int \mathrm{d}t a^D \sum_{\mathbf{x}}(\psi_{\mathbf{x}}\phi^*_{\mathbf{x}} + \psi^*_{\mathbf{x}}\phi_{\mathbf{x}})}\rangle_0 \tag{4.260}$$

and where the average is taken with respect to the Lagrangian (4.253). In lowest order, cf. Eq. (2.107),

$$\langle \mathrm{e}^{(\mathrm{i}/\hbar)\int \mathrm{d}t a^D \sum_{\mathbf{x}}(\psi_{\mathbf{x}}\phi^*_{\mathbf{x}} + \psi^*_{\mathbf{x}}\phi_{\mathbf{x}})}\rangle_0 = \exp\left[\frac{-\mathrm{i}}{\hbar}a^D\sum_{\mathbf{x}}\int \mathrm{d}t\mathrm{d}t'\,\psi^*_{\mathbf{x}}(t)\Delta_F(t-t')\psi_{\mathbf{x}}(t')\right] \tag{4.261}$$

with $\Delta_{\rm F}(t)$ the propagator introduced in Eq. (4.245). For fields slowly varying in time, we can write

$$\psi_{\mathbf{x}}(t') = \psi_{\mathbf{x}}(t) + (t'-t)\partial_t\psi_{\mathbf{x}}(t) + \frac{1}{2}(t'-t)^2\partial_t^2\psi_{\mathbf{x}}(t) + \cdots, \tag{4.262}$$

and the action (4.260) leads to the Lagrangian

$$\begin{aligned} L(\psi^*,\psi) = {} & \mathrm{i}\hbar\frac{\partial}{\partial\mu}\Delta_{\rm F}(\omega=0)a^D\sum_{\mathbf{x}}\psi^*_{\mathbf{x}}\partial_t\psi_{\mathbf{x}} - a^D\sum_{\mathbf{x}\mathbf{x}'}t^{-1}_{\mathbf{x}\mathbf{x}'}\psi^*_{\mathbf{x}}\psi_{\mathbf{x}'} \\ & - \Delta_{\rm F}(\omega=0)a^D\sum_{\mathbf{x}}\psi^*_{\mathbf{x}}\psi_{\mathbf{x}} - \frac{u}{2}a^{2D}\sum_{\mathbf{x}}(\psi^*_{\mathbf{x}}\psi_{\mathbf{x}})^2 + \cdots \end{aligned} \tag{4.263}$$

by the identities

$$\int \mathrm{d}t\, \Delta_{\rm F}(t) = \Delta_{\rm F}(\omega=0) \tag{4.264}$$

$$\int \mathrm{d}t\, t\Delta_{\rm F}(t) = -\mathrm{i}\hbar\frac{\partial}{\partial\mu}\Delta_{\rm F}(\omega=0), \tag{4.265}$$

where in the last line use is made of Eq. (4.249). The quartic $\sum_{\mathbf{x}}(\psi^*_{\mathbf{x}}\psi_{\mathbf{x}})^2$ and higher-order terms in $\psi_{\mathbf{x}}$ can be calculated by expanding the left side of Eq. (4.261) to higher order. The coefficient u of the quartic term is expected to be positive for stability. The quadratic term in the Taylor expansion (4.262) leads to a term quadratic in time derivatives, $a^D\sum_{\mathbf{x}}\partial_t\psi^*_{\mathbf{x}}\partial_t\psi_{\mathbf{x}}$.

The phase boundary between the insulating and superfluid phases corresponds to the condition

$$\frac{1}{2Dt} + \Delta_{\rm F}(\omega=0) = 0, \tag{4.266}$$

where the quadratic term $\psi^*_{\mathbf{x}}\psi_{\mathbf{x}}$ changes sign. The factor $2D$ is the coordination number of a hypercubic lattice in D dimensions. When the coefficient of the quadratic term becomes positive, $\psi_{\mathbf{x}}$ develops a finite average. Since for a uniform system

$$\langle\psi\rangle = 2Dt\langle\phi\rangle \tag{4.267}$$

by Eq. (4.258), ψ serves as a superfluid order parameter. Explicitly, by Eq. (4.250), the phase boundary is specified in lowest order by

$$\frac{1/2D}{t/U} = \frac{n}{\mu/U-(n-1)} - \frac{n+1}{\mu/U-n}, \tag{4.268}$$

see Fig. 4.7. The phase diagram shows lobelike Mott insulating states in the $t-\mu$ plane. Within a Mott lobe, each lattice site is occupied by exactly the same *integer* number of particles, and the compressibility vanishes. The locations of the tips of the lobes follow from extremizing this condition with respect to μ/U, giving

$$(\mu/U)_{\rm c} = \sqrt{n(n+1)} - 1 \tag{4.269}$$

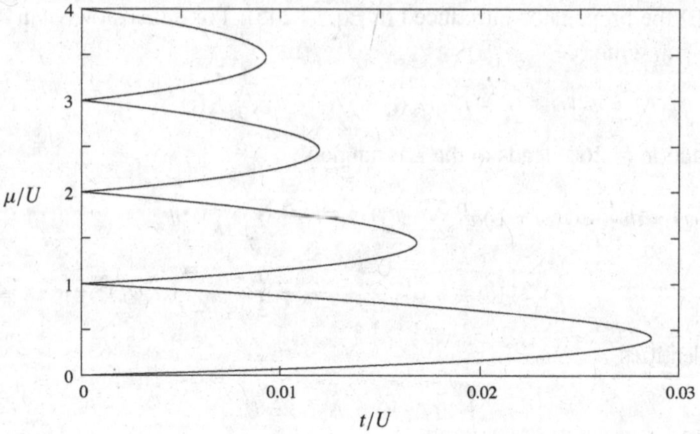

Fig. 4.7 Phase diagram of the Bose-Hubbard model (4.236) in lowest order.

and

$$(U/2Dt)_c = 2n + 1 + 2\sqrt{n(n+1)}. \qquad (4.270)$$

Two types of phase transitions, belonging to different universality classes, can be distinguished in the phase diagram. The generic transition results from adding or subtracting a small number of particles to or from the incompressible Mott insulating state. Boosting a term linear in time derivatives, the theory describing this quantum phase transition is nonrelativistic in nature. The quadratic term in time derivatives is irrelevant with respect to the linear term and can be dropped. A second type occurs at fixed integer particle number density and takes place at the tip of the lobes. Such a transition is triggered by decreasing U/t so as to enable the bosons to overcome the on-site repulsion. Taking place at the tip of the lobes, $(\partial/\partial\mu)\Delta_F(\omega = 0) = 0$ and the term linear in time derivatives vanishes. As a result, the term quadratic in time derivatives must now be included in the theory. Since the spatial derivatives also appear quadratic in lowest order, the theory describing these quantum phase transitions is relativisticlike. These transitions belong to the O(2) universality class in $d = D + 1$ dimensions.

Notes

(i) First-hand accounts of the creation of the first atomic Bose-Einstein condensates can be found in [Cornell and Wieman (2002)] and [Ketterle (2002)].

(ii) In addition to various monographs, numerous introductions to and reviews of BEC appeared in the past decade, see, for example, [Dalfovo *et al.* (1999)] and [Pethick and Smith (2002)]. For a guide to the literature, see the resource letter [Hall (2003)].

(iii) The critical properties of an ideal Bose gas were established by Gunton and Buckingham (1968). This topic is covered in the textbook [Pathria (1996)].

(iv) The classical field theory that came to be called Gross-Pitaevskii theory was put forward in [Gross (1958)] and [Pitaevskii (1958)].

(v) In [Edwards and Burnett (1995)], the Thomas-Fermi approximation is compared to the numerical solution of the Gross-Pitaevskii theory.

(vi) Bogoliubov's ground-braking paper on superfluidity appeared in [Bogoliubov (1947)].

(vii) For a discussion on the connection between the coupling constant in the Bogoliubov theory and the the s-wave scattering length, see [Hugenholtz (1965)].

(viii) A general analysis of nonrelativistic effective Lagrangians is given by Leutwyler (1994).

(ix) The one-dimensional Bose gas with a repulsive two-particle delta-function potential was solved exactly in [Lieb and Liniger (1963); Lieb (1963)].

(x) The Bogoliubov theory, both at zero and at finite temperature, are covered in the textbooks [Abrikosov *et al.* (1963)] and [Fetter and Walecka (1971)]. Much of our presentation of the Bogoliubov theory can be traced back to Popov, who pioneered the functional integral approach to statistical mechanics and condensed matter in addition to gauge theories, see [Popov (1983, 1987)].

(xi) Sections 4.9 and 4.10 closely follow [Schakel (1994, 2003)]. The relation (4.118), connecting the chemical potential to the time rate of change of the phase of the order parameter, was first established in the context of superconductivity by Josephson (1962). Its importance in the context of superfluidity was pointed out by Anderson (1966).

(xii) The relativistic weakly interacting Bose gas has been studied in [Kapusta (1981); Haber and Weldon (1981, 1982); Bernstein and Dodelson (1991)], and in [Benson *et al.* (1991)]. The last two references also discuss in detail the nonrelativistic limit.

(xiii) The renormalization aspects of the Bogoliubov theory at the absolute zero of temperature were studied in [Uzunov (1981)]. For a survey of these results and their applications to quantum phase transitions, see [Shopova and Uzunov (2003)] that also includes a clear discussion of the critical properties of Bose-Einstein condensation.

(xiv) The modern view of renormalization theory, which we adopt, is based on concepts developed in the context of statistical mechanics. For a lucid discussion of this view, see [Cao and Schweber (1993); Cao (1993)], and [Schweber (1993)]. In these references, the modern perspective is also put in historical context. For an application of this approach to quantum gravity—a notoriously nonrenormalizable theory, see [Donoghue (1994)].

(xv) The zero-temperature Bogoliubov spectrum was shown by Beliaev (1958) to remain gapless when one-loop quantum corrections are included. This was subsequently proved to hold to all orders in perturbation theory by Hugenholtz and Pines (1959).

(xvi) The equivalence, in the long wave-length limit, of the single particle excitation and the collective density fluctuation in a weakly interacting Bose gas was proved by Gavoret and Nozières (1964). In that reference, also the argument leading to the result (4.132) was first given. Our presentation also borrows from [Fradkin (1967)].

(xvii) The interaction-induced shift in the condensation temperature of a dilute Bose gas was calculated for large N by Baym *et al.* (2000). The next-to-leading order in $1/N$ was computed by Arnold and Tomášik (2000). Various numerical investigations have been carried out, see, for example, the Monte Carlo study [Kashurnikov *et al.* (2001)]. For introductions to the $1/N$ expansion, see, for example, [Coleman (1982)], reprinted in [Coleman (1988)], and [Polyakov (1987)]. Much of the technique used to calculate the shift for large N can be found in [Ma (1973, 1976)].

(xviii) The two-fluid model was introduced by Tisza (1938). It was based in part on London's proposal London (1938) that the superfluid, or λ-transition in liquid ^{4}He was analogous to Bose-Einstein condensation in dilute gases. Landau's landmark paper on the subject appeared in [Landau (1941)]. Feynman put forward his ideas in [Feynman (1954)].

(xix) The history of superfluidity is described in [Hoddeson *et al.* (1992); Griffin (1999)] and [Balibar (2007)], while a nontechnical introduction to the topic can be found in [Leggett (1989, 1999)]. Classic books on superfluidity include [London (1954)] and [Nozières and Pines (1990)].

(xx) The saturation of the elementary excitation spectrum in superfluid ^{4}He at short wave lengths, $E(\mathbf{k}) \rightarrow 2\Delta_0$, was predicted by Pitaevskii (1959) as signaling the decay of a roton into two rotons, each requiring a minimum of energy Δ_0. The experimental curve in Fig. 4.6 is reprinted with permission from [Sakhel and Glyde (2004)].

(xxi) The argument leading to Eqs. (4.207) and (4.210) is based on [Brown (1992)].

(xxii) The effect of impurities on a weakly interacting Bose gas was established by Huang and Meng (1992). For an introduction to disordered systems, see [Ma (1976)].

(xxiii) Following a suggestion by Jaksch *et al.* (1998), Greiner *et al.* (2002) first observed the Mott insulator-to-superfluid transition in ultracold atomic gases loaded into an optional lattice. The transition was predicted and first described by Fisher *et al.* (1989). Our presentation is entirely based on this seminal paper. For an introduction to quantum phase transitions, see [Sondhi *et al.* (1997)].

Chapter 5

Vortices in 2D

This chapter covers vortices in two space dimensions, where, at scales large compared to their core size, they may be pictured as pointlike objects. They are studied both at the absolute zero of temperature as well as at finite temperature where they trigger the so-called Berezinskii-Kosterlitz-Thouless (BKT), or vortex-unbinding phase transition. The mappings onto the classical Coulomb gas and the sine-Gordon model are given, the renormalization group equations governing the BKT transition derived, and the physical consequences for a superfluid ^{4}He film discussed.

5.1 Vortex Dynamics

As starting point for the description of vortices in a superfluid film at the absolute zero of temperature, we take the effective theory (4.121), which is also valid in two space dimensions. Vortices are incorporated as in Sec. 1.11 through minimal coupling to a Villain potential $A^{\rm V}_\mu$, see Eq. (1.158) with ϕ replaced with $(\hbar/m)\varphi$. Consider N vortices of winding number w_α ($\alpha = 1, \cdots, N$) centered at $\mathbf{x}_1(t), \cdots, \mathbf{x}_N(t)$. The Villain potential describing these vortices satisfies the relation

$$\nabla \times \mathbf{A}^{\rm V}(x) = -\sum_\alpha \kappa_\alpha \delta[\mathbf{x} - \mathbf{x}_\alpha(t)], \tag{5.1}$$

where $\nabla \times \mathbf{A} \equiv \epsilon^{ij}\partial_i A^j$ and ϵ^{ij} is the antisymmetric Levi-Civita symbol in two space dimensions with $\epsilon^{12} = 1$. Moreover, $\kappa_\alpha = (2\pi\hbar/m)w_\alpha$ is the circulation (1.125) of the αth vortex which is quantized in units of $2\pi\hbar/m$. The curl of the superfluid velocity field (1.81), which after minimal coupling reads

$$\mathbf{v}_{\rm s} = \frac{\hbar}{m}\nabla\varphi - \mathbf{A}^{\rm V}, \tag{5.2}$$

then gives

$$\nabla \times \mathbf{v}_s = \sum_\alpha \kappa_\alpha \delta[\mathbf{x} - \mathbf{x}_\alpha(t)]. \tag{5.3}$$

The sum over the indices labeling the vortices will always be made explicit. Invariance under the gauge transformations (1.161) can be used to pick a certain gauge as in electrodynamics. In the temporal gauge $\Phi^V = 0$, Eq. (5.1) can be solved to yield

$$A^{V,i}(x) = \epsilon^{ij} \sum_\alpha \kappa_\alpha \delta^j[x, L_\alpha(t)] \tag{5.4}$$

where $\delta[x, L_\alpha(t)]$ is a delta function on the line $L_\alpha(t)$ starting at the center $\mathbf{x}_\alpha(t)$ of the αth vortex and running to spatial infinity along an arbitrary path:

$$\delta^i[x, L_\alpha(t)] \equiv \int_{L_\alpha(t)} \mathrm{d}x'^i \, \delta(\mathbf{x} - \mathbf{x}'). \tag{5.5}$$

With the Villain potential included, the effective theory produces, as we now show, the same result (1.156) for the particle number density profile in the presence of a static vortex as the Gross-Pitaevskii theory.

The linearized field equation in the presence of static vortices, obtained from the effective theory (4.121) with $\nabla\varphi$ replaced with $\nabla\varphi - (m/\hbar)\mathbf{A}^V$ and $\partial_t\varphi$ set to zero, is of the form (1.170), with the solution

$$\varphi(\mathbf{x}) = -\frac{m}{\hbar} \int \mathrm{d}^2x' \, G(\mathbf{x} - \mathbf{x}') \nabla' \cdot \mathbf{A}^V(\mathbf{x}'). \tag{5.6}$$

Here, $G(\mathbf{x})$ now denotes the Green function of the *two-dimensional* Laplace operator

$$G(\mathbf{x}) = \int \frac{\mathrm{d}^2k}{(2\pi)^2} \frac{\mathrm{e}^{\mathrm{i}\mathbf{k}\cdot\mathbf{x}}}{\mathbf{k}^2}, \tag{5.7}$$

satisfying $-\nabla^2 G(\mathbf{x}) = \delta(\mathbf{x})$. Because this Green function plays an important role in the following, we evaluate the integral explicitly. Note that it diverges both in the infrared as well as in the ultraviolet. The infrared divergence is handled by subtracting the value at the origin:

$$G(\mathbf{x}) - G(0) = \int \frac{\mathrm{d}^2k}{(2\pi)^2} \frac{\mathrm{e}^{\mathrm{i}\mathbf{k}\cdot\mathbf{x}} - 1}{\mathbf{k}^2}, \tag{5.8}$$

while the ultraviolet divergence is regularized by introducing a wave number cutoff $1/a$. Then

$$\begin{aligned} G(\mathbf{x}) - G(0) &= \frac{1}{(2\pi)^2} \int_0^{1/a} \frac{\mathrm{d}|\mathbf{k}|}{|\mathbf{k}|} \int_0^{2\pi} \mathrm{d}\theta \left(\mathrm{e}^{\mathrm{i}|\mathbf{k}|r\cos\theta} - 1\right) \\ &= \frac{1}{2\pi} \int_0^{1/a} \frac{\mathrm{d}|\mathbf{k}|}{|\mathbf{k}|} \left[J_0(|\mathbf{k}|r) - 1\right], \end{aligned} \tag{5.9}$$

where $r \equiv |\mathbf{x}|$ and J_0 is the zeroth-order Bessel function of the first kind. The remaining integral can be split into two parts as

$$G(\mathbf{x}) - G(0) = \frac{1}{2\pi}\left\{\int_0^1 \frac{\mathrm{d}q}{q}\,[J_0(q) - 1] + \int_1^{r/a} \mathrm{d}q \frac{J_0(q)}{q} - \int_1^{r/a} \frac{\mathrm{d}q}{q}\right\}, \tag{5.10}$$

with $q \equiv |\mathbf{k}|r$ the new integration variable. Asymptotically, for $r >> a$, the upper limit in the second integral can be taken to infinity, and

$$G(\mathbf{x}) - G(0) = -\frac{1}{2\pi}\ln(r/a) + C, \tag{5.11}$$

where the constant C takes the value

$$\begin{aligned} 2\pi C &= \int_0^1 \frac{\mathrm{d}q}{q}\,[J_0(q) - 1] + \int_1^{\infty} \mathrm{d}q \frac{J_0(q)}{q} \\ &= -\gamma + \ln 2 > 0. \end{aligned} \tag{5.12}$$

The explicit form (5.11) of the Green function gives for the superfluid velocity field (5.2), cf. Eq. (1.81),

$$v_{\mathrm{s}}^i(\mathbf{x}) = -\frac{1}{2\pi}\epsilon^{ij}\sum_\alpha \kappa_\alpha \frac{x^j - x_\alpha^j}{|\mathbf{x} - \mathbf{x}_\alpha|^2}, \tag{5.13}$$

which is valid for $\mathbf{x}$ sufficiently far away from the vortex cores. Specializing to the case of a single static vortex centered at the origin, we arrive on substituting this solution in Eq. (4.133) at the density profile

$$n(\mathbf{x}) = \bar{n}\left(1 - w^2\frac{\xi^2}{r^2}\right). \tag{5.14}$$

This expression is identical to the solution (1.156) obtained from the Gross-Pitaevskii theory, thereby illustrating that, through the use of Villain potentials, vortices are correctly accounted for in the effective theory.

We proceed by investigating the dynamics of vortices in a superfluid film. We consider only the first part of the effective theory (4.121). By ignoring the higher-order terms, we approximate the superfluid by an incompressible fluid for which the particle number density is constant, $n(x) = \bar{n}$, see Eq. (4.133). As before, the temporal gauge $\Phi^{\mathrm{V}} = 0$ is picked, and $\nabla\varphi$ is replaced with $\nabla\varphi - (m/\hbar)\mathbf{A}^{\mathrm{V}}$, with the Villain potential specified in Eq. (5.1). The solution of the resulting field equation for φ is again of the form (5.6), but it now depends on time through the Villain potential for $\mathbf{x}_\alpha = \mathbf{x}_\alpha(t)$ so that, for example,

$$\begin{aligned} \partial_t A^{\mathrm{V},i}(x) &= \epsilon^{ij}\sum_\alpha \kappa_\alpha \partial_t \delta^j[x, L_\alpha(t)] \\ &= -\epsilon^{ij}\sum_\alpha \kappa_\alpha \dot{x}_\alpha^j(t)\,\delta[\mathbf{x} - \mathbf{x}_\alpha(t)], \end{aligned} \tag{5.15}$$

where use is made of the rule

$$\partial_t \int_{x_\alpha(t)}^{\infty} \mathrm{d}x\, f(x) = -f[x_\alpha(t)]\, \dot{x}_\alpha(t). \tag{5.16}$$

If substituted back into the action $S \equiv \int \mathrm{d}^3x\, \mathcal{L}_{\rm eff}$, the solution yields the action governing the vortices. For the first term in the effective theory (4.121), we obtain in this way

$$\begin{aligned} S_{\rm kin} &= -\hbar\bar{n} \int \mathrm{d}^3x\, \partial_t\varphi \\ &= m\bar{n} \int \mathrm{d}^3x\, \mathrm{d}^2x'\, G(\mathbf{x}-\mathbf{x}')\nabla' \cdot \partial_t \mathbf{A}^{\rm V}(t,\mathbf{x}') \\ &= -m\bar{n}\epsilon^{ij} \sum_\alpha \kappa_\alpha \int \mathrm{d}^3x\, \partial_i G[\mathbf{x}-\mathbf{x}_\alpha(t)]\dot{x}^j_\alpha(t) \end{aligned} \tag{5.17}$$

after partial integration. The remaining integral over space gives

$$\int \mathrm{d}^2x\, \partial_i G(\mathbf{x}-\mathbf{x}_\alpha) = \frac{1}{2} x^i_\alpha \tag{5.18}$$

apart from an additive constant as can be checked by taking the derivative $\partial/\partial x^i_\alpha$ of this equation. The other terms in the effective theory can be treated similarly, with the result

$$S = m\bar{n} \int \mathrm{d}t \left[-\frac{1}{2}\sum_\alpha \kappa_\alpha \mathbf{x}_\alpha \times \dot{\mathbf{x}}_\alpha - \frac{1}{2}\sum_{\alpha,\beta} \kappa_\alpha\kappa_\beta G(\mathbf{x}_\alpha - \mathbf{x}_\beta) \right], \tag{5.19}$$

with $\mathbf{x} \times \mathbf{y} \equiv \epsilon^{ij}x^iy^j$ in two space dimensions. The dynamic term in this action is unusual, and leads, as will be shown shortly, to a twisted canonical structure reminiscent of that found in the Landau problem of a charged particle confined to move in a plane perpendicular to an applied magnetic field, see Sec. 3.8. The potential energy term in the action (5.19) shows that vortices interact through a 2D Coulomb interaction. The "electric" charge of the vortices is given by their circulation. Since the potential energy depends quadratically on κ, vortices of higher winding number carry a much larger energy than singly quantized vortices. It is therefore justified to consider in the following only vortices of unit winding number, $w_\alpha = \pm 1$.

We isolate the self-interaction by writing the last term of the action as

$$\sum_{\alpha,\beta} \kappa_\alpha\kappa_\beta G(\mathbf{x}_\alpha - \mathbf{x}_\beta) = \sum_{\alpha,\beta} \kappa_\alpha\kappa_\beta \left[G(\mathbf{x}_\alpha - \mathbf{x}_\beta) - G(0)\right] + \left(\sum_\alpha \kappa_\alpha\right)^2 G(0). \tag{5.20}$$

Because $G(0)$ diverges in the infrared as $G(0) \propto \ln(R/a)$, with R the linear size of the system, the charges must add up to zero so that the action is finite. From now on we will therefore assume overall charge neutrality, $\sum_\alpha w_\alpha = 0$. That is, half of the vortices have winding number $w = +1$, while the other vortices have winding number $w = -1$, and N must be an even integer. By Eq. (5.11) the action thus becomes

$$S = m\bar{n} \int \mathrm{d}t \left[-\frac{1}{2} \sum_\alpha \kappa_\alpha \mathbf{x}_\alpha \times \dot{\mathbf{x}}_\alpha + \frac{1}{4\pi} \sum_{\alpha \neq \beta} \kappa_\alpha \kappa_\beta \ln\left(|\mathbf{x}_\alpha - \mathbf{x}_\beta|/a\right) \right], \tag{5.21}$$

up to an irrelevant additive constant proportional to the constant C introduced in Eq. (5.12). The parameter a, whose inverse serves as an ultraviolet wave-number cutoff, physically represents the vortex core size.

To display the canonical structure, we rewrite the first term of the Lagrangian corresponding to the action (5.21) as

$$L_{\rm kin} = -m\bar{n} \sum_\alpha \kappa_\alpha x_\alpha^1 \dot{x}_\alpha^2, \tag{5.22}$$

ignoring a total derivative. It follows that the canonical conjugate to the second component x_α^2 of the center coordinate $\mathbf{x}_\alpha$ is essentially its first component

$$\frac{\partial L_{\rm kin}}{\partial \dot{x}_\alpha^2} = -m\bar{n}\kappa_\alpha x_\alpha^1. \tag{5.23}$$

This implies that phase space coincides with real space, and leads, upon quantizing, to the commutation relation

$$[x_\alpha^1, x_\beta^2] = \frac{\mathrm{i}}{w_\alpha} \ell^2 \delta_{\alpha\beta}, \tag{5.24}$$

where

$$\ell \equiv 1/\sqrt{2\pi\bar{n}} \tag{5.25}$$

is a characteristic length whose definition is such that $2\pi\ell^2$ denotes the average area occupied by a particle of the superfluid film. Remarkably, the right side of the commutation relation is independent of Planck's constant. The commutation relation leads to a Heisenberg uncertainty in the location of the vortex centers given by

$$\Delta x_\alpha^1 \Delta x_\alpha^2 \geq \frac{\ell^2}{2|w_\alpha|}, \tag{5.26}$$

which is inverse proportional to the particle number density.

Using that the number of quantum states per unit phase space is $1/2\pi\hbar$, we conclude that for the case at hand, the available number of states in an area S_α of *real* space is

$$\nu_{S_\alpha} = |w_\alpha|\, \bar{n} S_\alpha. \tag{5.27}$$

In other words, the number of states per unit area available to a vortex is proportional to the particle number density.

This phenomenon that phase space coincides with real space also arises in the Landau problem. There, it leads to the degeneracy $|eB|/2\pi\hbar c$ of each Landau level, where $e(> 0)$ is the unit of charge. In terms of the magnetic flux quantum

$$\Phi_0 = 2\pi\hbar c/e, \tag{5.28}$$

the Landau degeneracy can be rewritten as $|B|/\Phi_0 = n_\otimes$, with $n_\otimes$ denoting the flux number density. That is, whereas the degeneracy in the case of vortices in a superfluid film is determined by the particle number density, for the Landau problem it is determined by the flux number density. Using this analogy, we infer that the characteristic length (5.25) translates into

$$\ell_B = 1/\sqrt{2\pi\bar{n}_\otimes} \tag{5.29}$$

which is precisely the magnetic length of the Landau problem.

The first term in the action (5.21) not only gives rise to a twisted canonical structure, it in addition is responsible for a geometric phase $\gamma(\Gamma)$ that the wave function of a vortex acquires when it traverses a closed path Γ. To obtain this geometric phase, we first consider the case of a point particle of charge $q = -e(< 0)$ that moves adiabatically around a close path Γ in the presence of an applied magnetic field. The wave function of the particle picks up an extra factor, known as the *Aharonov-Bohm* phase,

$$\mathrm{e}^{\mathrm{i}\gamma(\Gamma)} = \exp\left(-\frac{\mathrm{i}e}{\hbar c}\oint_\Gamma \mathrm{d}\mathbf{x}\cdot\mathbf{A}\right) = \mathrm{e}^{-2\pi\mathrm{i}\Phi(\Gamma)/\Phi_0}, \tag{5.30}$$

where $\mathbf{A}$ is the vector potential describing the external magnetic field and $\Phi(\Gamma) \equiv BS(\Gamma)$ is the magnetic flux through the area $S(\Gamma)$ spanned by the loop Γ. The geometric phase $\gamma(\Gamma)$ is seen to count (2π times) the number of flux quanta enclosed by the path Γ. By virtue of the above analogy, it follows that the geometric phase picked up by the wave function of the αth vortex when it adiabatically moves around a closed path in the superfluid film counts ($2\pi w_\alpha$ times) the number of superfluid particles enclosed by the path.

The action (5.21) yields the well-known equations of motion for point vortices in an incompressible two-dimensional superfluid:

$$\dot{x}^i_\beta(t) = -\frac{1}{2\pi}\epsilon^{ij}\sum_{\alpha\neq\beta}\kappa_\alpha\frac{x^j_\beta(t) - x^j_\alpha(t)}{|\mathbf{x}_\beta(t) - \mathbf{x}_\alpha(t)|^2}. \tag{5.31}$$

From this we conclude that $\dot{\mathbf{x}}_\beta(t) = \mathbf{v}_\mathrm{s}\left[\mathbf{x}_\beta(t)\right]$, where $\mathbf{v}_\mathrm{s}(x)$ is the superfluid velocity (5.13) with the time-dependence of the vortex centers included. This nicely illustrates a result due to Helmholtz for ideal fluids, stating that a vortex moves with the fluid, i.e., at the local velocity produced by the other vortices in the system.

5.2 2D Coulomb Gas

The action (5.21) reveals that vortices in a superfluid film interact through a 2D Coulomb potential. To further investigate such vortices it seems therefore fitting to first consider a classical 2D Coulomb gas consisting of $N = N_+ + N_-$ identical particles with N_+ particles having electric charge q and N_- particles having electric charge $-q$. The canonical partition function Z_N is given by

$$Z_N = \sum_{N_+,N_-}{}' \frac{z^N}{N_+!N_-!} \prod_\alpha \int \frac{\mathrm{d}^2 x_\alpha}{a^2} \mathrm{e}^{-\beta H}, \tag{5.32}$$

where the factors of a in the integration measure are included for dimensional reasons, and H is the Hamiltonian

$$H = \frac{1}{2} \sum_{\alpha,\beta} q_\alpha q_\beta U(\mathbf{x}_\alpha - \mathbf{x}_\beta). \tag{5.33}$$

Here, $U(\mathbf{x}) \equiv 2\pi G(\mathbf{x})$ with $G(\mathbf{x})$ the Green function (5.7). The factor 2π is the two-dimensional analog of the factor 4π appearing in standard three-dimensional electrostatics using Gaussian units. In 2D, the electric charge squared q^2 is of dimension energy. Moreover, $z \equiv \exp(-\beta\epsilon_c)$ in Eq. (5.32) is the fugacity, with ϵ_c a positive constant denoting the core energy of a charge. The energy required to create a single charge receives in addition to an electrostatic also a nonelectrostatic contribution. In the core energy ϵ_c only the latter is included. The electrostatic contribution will be dealt with shortly. The factors $N_+!$ and $N_-!$ take into account that the positive and negative charges are indistinguishable. Finally, the sum in Eq. (5.33) is over configurations with all possible values of N_+ and N_-, satisfying the constraint $N = N_+ + N_-$. The prime on the sum is to remind the reader of this constraint.

By repeating the steps leading to Eq. (5.21), we conclude that the charges must add up to zero so as to obtain a nonzero partition function. From now on we will therefore assume overall charge neutrality, $\sum_\alpha q_\alpha = 0$, so that $N_+ = N_- = N/2$, where N must be an even integer. With Eq. (5.11), the Hamiltonian then becomes

$$H = -\frac{1}{2} \sum_{\alpha \neq \beta} q_\alpha q_\beta \ln\left(|\mathbf{x}_\alpha - \mathbf{x}_\beta|/a\right), \tag{5.34}$$

where the contribution $-\pi C q^2$, with C the constant introduced in Eq. (5.12), is now absorbed in the core energy. In obtaining this contribution, use is made of the identity

$$\sum_{\alpha \neq \beta} q_\alpha q_\beta = -\sum_\alpha q_\alpha^2 = -Nq^2 \tag{5.35}$$

which follows from charge neutrality. Having dealt with the self-interaction, we limit the integrations in Eq. (5.32) over the positions $\mathbf{x}_\alpha$ of the charges ($\alpha = 1, \cdots, N$) to those regions where the charges are more than a distance a apart, $|\mathbf{x}_\alpha - \mathbf{x}_\beta| > a$.

It proves convenient to introduce dimensionless integration variables by letting

$$\mathbf{x}_\alpha \to \check{\mathbf{x}}_\alpha \equiv \mathbf{x}_\alpha / R, \tag{5.36}$$

with R the linear size of the system. The partition function then assumes the form

$$Z_N = hR^{2N} \exp\left[\frac{\beta}{2} \sum_{\alpha \neq \beta} q_\alpha q_\beta \ln(R/a)\right], \tag{5.37}$$

with h an R-independent factor

$$h = \frac{1}{[(N/2)!]^2} \frac{z^N}{a^{2N}} \prod_\alpha \int \mathrm{d}^2 \check{x}_\alpha \exp\left[\frac{\beta}{2} \sum_{\alpha \neq \beta} q_\alpha q_\beta \ln(|\check{\mathbf{x}}_\alpha - \check{\mathbf{x}}_\beta|)\right]. \tag{5.38}$$

Introducing the two-dimensional area $A = \pi R^2$ of the system, we can rewrite the partition function as

$$Z_N = h a^{\beta N q^2/2} \left(\frac{A}{\pi}\right)^{N - \beta N q^2/4}. \tag{5.39}$$

By virtue of the thermodynamic relation (2.25) with $\Omega = -\beta^{-1} \ln Z$,

$$P = k_\mathrm{B} T \frac{\partial}{\partial A} \ln(Z_N), \tag{5.40}$$

the pressure P of a two-dimensional Coulomb gas follows as

$$P = \frac{N}{A}\left(k_\mathrm{B} T - \tfrac{1}{4} q^2\right). \tag{5.41}$$

This equation of state is similar to that of an ideal gas given in Eq. (2.34), save for a shift in the temperature scale by

$$k_\mathrm{B} T_\mathrm{BKT} \equiv \frac{1}{4} q^2. \tag{5.42}$$

To appreciate the physical relevance of the temperature T_BKT, consider a single pair of oppositely charged particles. Let $\mathbf{x}_\alpha$ denote the location of the charge q and $\mathbf{y}_\alpha$ denote that of the charge $-q$. By Eq. (5.32), this dipole contributes a factor to the partition function of the form

$$\begin{aligned} \int \mathrm{d}^2 x_\alpha \mathrm{d}^2 y_\alpha \exp\left[-\beta q^2 \ln\left(|\mathbf{x}_\alpha - \mathbf{y}_\alpha|/a\right)\right] &\sim R^4 \mathrm{e}^{-\beta q^2 \ln(R/a)} \\ &\sim R^{4-\beta q^2}, \end{aligned} \tag{5.43}$$

as follows on dimensional grounds. Below $T_{\rm BKT}$, where $4 - \beta q^2 < 0$, the contribution is negligible in the limit $R \to \infty$. This can be understood by picturing the dipole as a tightly bound pair so that the charges neutralize each other. That is to say, the low-temperature state appears to be devoid of free charges, implying that the Coulomb gas becomes insulating at sufficiently low temperatures. For temperatures above $T_{\rm BKT}$, on the other hand, the contribution of a single pair diverges in the limit $R \to \infty$, meaning that the creation of well-separated dipoles is favored. The high-temperature state would therefore consist of a plasma of unbound charges. The correctness of this phase diagram of a 2D Coulomb gas, with an insulating dipole phase at low temperatures and a conducting plasma phase at high temperatures, will be verified in detail in the following section.

5.3 Berezinskii-Kosterlitz-Thouless Transition

Since dipoles are seen to play an important role, it is expedient to express the partition function in terms of dipole variables. Ignoring for the moment possible interactions between the dipoles, we may write the relevant part of the partition function as

$$Z = \sum_{M=0}^{\infty} \frac{z^{2M}}{M!} \prod_{\alpha=1}^{M} \frac{1}{a^4} \int \mathrm{d}^2 x_\alpha \mathrm{d}^2 y_\alpha \exp\left[-\beta \sum_{\alpha=1}^{M} q^2 \ln\left(|\mathbf{x}_\alpha - \mathbf{y}_\alpha|/a\right)\right], \tag{5.44}$$

where $\mathbf{x}_\alpha$ and $\mathbf{y}_\alpha$ label the positions of the two charges forming the αth dipole, and, as before, $z = \exp(-\beta\epsilon_c)$ is the fugacity of a single charge. In Eq. (5.44), the sum $\sum_M$ is taken over the number of dipoles M, showing that a *grand canonical* ensemble is considered, and the factor $1/M!$ is included to prevent overcounting. The integrals over the positions of the charges are restricted to the regions $|\mathbf{x}_\alpha - \mathbf{y}_\alpha| > a$. Finally, for dimensional reasons, factors of $1/a$ are include in the integration measures. Since interactions between the dipoles are ignored, the partition function (5.44) is readily evaluated, with the result

$$Z = \exp\left(2\pi A \frac{z^2}{a^4} \int \mathrm{d}r\, r e^{-\beta V(r)}\right), \tag{5.45}$$

with A the area of the system and $V(r)$ the potential energy of a pair of equal and opposite charges with separation r,

$$V(r) = q^2 \ln(r/a). \tag{5.46}$$

For the average number $n_{\circ\!-\!\circ}$ of dipoles per unit area, we obtain in this approximation

$$n_{\circ\!-\!\circ} = -\frac{1}{2\beta A} \frac{\partial \ln(Z)}{\partial \epsilon_c} = 2\pi \frac{z^2}{a^4} \int \mathrm{d}r\, r e^{-\beta V(r)}, \tag{5.47}$$

using that $-2\epsilon_c$ represents the chemical potential of a dipole.

Consider now a pair of equal and opposite charges with large separation. Since this pair has a large potential energy, it is expected that dipoles with smaller separation are created that screen the charges of the original pair, thereby reducing the total potential energy. In other words, the space between the two original charges becomes occupied with polarized dipoles. Such polarization effects are conveniently described by introducing a scale-dependent *dielectric function* $\epsilon(r)$. With polarization effects included, the potential energy of a pair with separation r is no longer given by Eq. (5.46), but is replaced with the expression

$$V_{\text{eff}}(r) = \int_a^r \frac{\mathrm{d}r'}{r'} \frac{q^2}{\epsilon(r')} \tag{5.48}$$

which reduces to the previous one in the limit $\epsilon(r) \to 1$. The average number density $n_{\circ\!-\!\circ}(r)$ of dipoles with separation r is given by

$$n_{\circ\!-\!\circ}(r) = \frac{z^2}{a^2} \mathrm{e}^{-\beta V_{\text{eff}}(r)}, \tag{5.49}$$

cf. Eq. (5.47).

We proceed to obtain an explicit expression for the dielectric function $\epsilon(r)$. As in standard electrostatics, it is related to the electric susceptibility χ through

$$\epsilon(r) = 1 + 2\pi\chi(r). \tag{5.50}$$

The factor 2π is the the two-dimensional counterpart of the factor 4π in three dimensions. The electric susceptibility is in turn determined by the polarizability $\alpha(r)$ of a dipole in the following way:

$$\chi(r) = 2\pi \frac{1}{a^2} \int_a^r \mathrm{d}r' r' n_{\circ\!-\!\circ}(r')\alpha(r'). \tag{5.51}$$

In 2D, the polarizability has the dimension of length square. To calculate $\alpha(r)$ we use its defining equation

$$\mathbf{p} = \alpha \mathbf{E}, \tag{5.52}$$

where $\mathbf{p}$ is the dipole moment and $\mathbf{E}$ the applied electric field. In such a field, a dipole has an extra potential energy given by

$$U(\mathbf{E}) = -pE\cos\theta, \tag{5.53}$$

with $E = |\mathbf{E}|$, $p = |\mathbf{p}| = qr$, and θ the angle between the applied field and the dipole. Whence, the polarizability of an isolated dipole is given by

$$\alpha(r) = \frac{\partial}{\partial E} \langle p\cos\theta \rangle \Big|_{E=0}, \tag{5.54}$$

where the angle brackets denote the angular average weighed by the Boltzmann factor $\exp[-\beta U(\mathbf{E})]$. Explicitly,

$$\alpha(r) = \frac{\beta p^2}{2\pi}\int_0^{2\pi} \mathrm{d}\theta \cos^2\theta = \frac{1}{2}\beta q^2 r^2. \tag{5.55}$$

Using this approximate expression in Eq. (5.51), we obtain for the scale-dependent dielectric function

$$\epsilon(r) = 1 + 2\pi^2 q^2 \beta \int_a^r \frac{\mathrm{d}r'}{r'} z^2(r'), \tag{5.56}$$

where we introduced the scale-dependent fugacity $z(r)$ through

$$z^2(r) \equiv z^2(a)\frac{r^4}{a^4}\mathrm{e}^{-\beta V_{\mathrm{eff}}(r)}, \tag{5.57}$$

which tends to $z(a) \equiv z$ in the limit $r \to a$. More explicitly,

$$z^2(r) = z^2(a)\exp\left[4\ln\left(\frac{r}{a}\right) - \beta\int_a^r \frac{\mathrm{d}r'}{r'}\frac{q^2}{\epsilon(r')}\right]. \tag{5.58}$$

It is customary to also introduce a scale-dependent temperature through $\beta(r) \equiv \beta(a)/\epsilon(r)$, with $\beta(a) \equiv \beta$, in terms of which Eq. (5.56) assumes the form

$$\beta^{-1}(r) = \beta^{-1}(a) + 2\pi^2 q^2 \int_a^r \frac{\mathrm{d}r'}{r'} z^2(r'). \tag{5.59}$$

Note that in this approach the charge q is not renormalized.

The integral equations (5.59) and (5.58) may be converted into differential equations by differentiating them with respect to the parameter λ defined as

$$\lambda \equiv \ln(r/a). \tag{5.60}$$

This gives the famous Kosterlitz renormalization group equations

$$\frac{\mathrm{d}\beta^{-1}}{\mathrm{d}\lambda} = 2\pi^2 q^2 z^2 + O(z^4) \tag{5.61}$$

$$\frac{\mathrm{d}z}{\mathrm{d}\lambda} = \left(2 - \tfrac{1}{2}q^2\beta\right)z + O(z^3). \tag{5.62}$$

The correction terms are included to show that these renormalization group equations are valid only when $z(r)$ is small, i.e., when the dipoles form a dilute gas. They yield the approximate solution

$$z(r) = z(a)\,(r/a)^{2-q^2\beta(a)/2} \tag{5.63}$$

$$\beta^{-1}(r) = \beta^{-1}(a) + \pi^2 q^2 \frac{z^2(r)}{2 - q^2\beta(a)/2}, \tag{5.64}$$

where the two bare parameters $z(a)$ and $\beta(a)$ label the renormalization group trajectories. We are interested in the behavior of the system at large length scales.

In the limit $r >> a$, the renormalized fugacity $z(r)$ tends to zero in the low-temperature state, where $2 - q^2\beta(a)/2 < 0$, implying that the creation of a single charge costs an infinite amount of energy. This corroborates the statement that the low-temperature state is devoid of free charges. The fugacity becomes scale independent right at the BKT temperature (5.42).

In the high-temperature state, $z(r)$ increases with r, and eventually exceeds the limit of validity of the approximate solutions (5.64). To show that this state is a plasma state of Coulomb charges, we calculate the density $\varrho(r)$ of charges of either sign within an annulus $a < r' < r$. Using the lowest-order approximation (5.46) for the potential energy V_{eff}, we find, cf. Eq. (5.47)

$$\begin{aligned}\varrho(r) &= 2\pi\frac{z(a)}{a^4}\int_a^r \mathrm{d}r'\, r' \mathrm{e}^{-\beta(a)V(r')/2} \\ &= 4\pi\frac{z(a)}{a^2}\frac{1}{4-q^2\beta(a)}\left[\left(\frac{r}{a}\right)^{2-q^2\beta(a)/2} - 1\right].\end{aligned} \tag{5.65}$$

To lowest order in z we can, by Eq. (5.64), replace $\beta(a)$ with $\beta(r)$. In the limit $a \to 0$, we then obtain for the density of charges

$$\varrho(r) = 4\pi\frac{z(r)}{a^2}\frac{1}{4-q^2\beta(r)}, \tag{5.66}$$

with the fugacity $z(r)$ given by Eq. (5.63). Since in the high-temperature state $z(r) \neq 0$, the charge density is nonzero, showing that this state is indeed a plasma state.

This renormalization group analysis of the approximate solution shows that in lowest approximation, the BKT temperature is given by Eq. (5.42). Returning to the Kosterlitz renormalization group equations (5.62), we conclude that the true BKT temperature, which incorporates polarization effects, is given by $\beta(r) = 4/q^2$, i.e., by Eq. (5.42) with $T = T(a)$ replaced with $T(r)$. At this critical temperature, the right side of Eq. (5.62) changes sign, leading to qualitatively different behavior of the dielectric function $\epsilon(r)$.

The differential equations (5.61) and (5.62) simplify when written in terms of the variables

$$x \equiv 2 - \frac{1}{2}q^2\beta, \quad y \equiv 2\pi^2 z, \tag{5.67}$$

where they become

$$\frac{\mathrm{d}y}{\mathrm{d}\lambda} = xy, \quad \frac{\mathrm{d}x}{\mathrm{d}\lambda} = y^2\left(1 - \tfrac{1}{2}x\right)^2 \approx y^2. \tag{5.68}$$

This set is easily solved by dividing the two equations:

$$\frac{\mathrm{d}y}{\mathrm{d}x} = \frac{x}{y}. \tag{5.69}$$

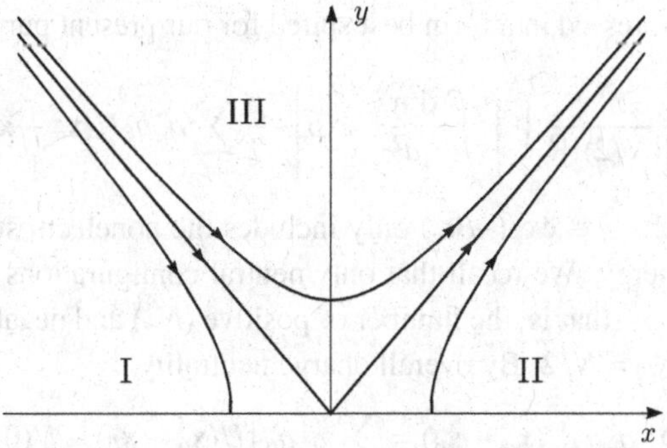

Fig. 5.1 Kosterlitz renormalization group flow.

The solutions of this equation, which is valid only near the point $x = y = 0$, are hyperbolas:

$$x^2 - y^2 = \text{const.} \tag{5.70}$$

The renormalization flow one obtains in this way is depicted in Fig. 5.1. The arrows indicate the direction of the flow with increasing length scale r. The two straight lines $y = \pm x$, separate the x-y half-plane in three different regions. The critical temperature corresponds to $x = 0$. In region I, to the left of the separatrix $y = -x$, the fugacity y is renormalized to zero. This low-temperature state is therefore characterized by a line of fixed points. In the vicinity of the critical temperature and close to the separatrix, the relation $y \approx -x$ implies that a finite but small fugacity reduces the temperature to

$$\beta \approx \beta_{\mathrm{BKT}} + 4\pi^2 z(a)/q^2, \tag{5.71}$$

a trend which Fig. 5.1 clearly reflects.

In the regions II and III, y increases with r and quickly exceeds the limit of validity of the renormalization group equations. Whereas region I is characterized by the absence of free charges, the regions II and III correspond to a plasma state of Coulomb charges.

5.4 Sine-Gordon Model

We so far in this chapter have tacitly adopted the corpuscular point of view, picturing the Coulomb charges as particles. We now turn to the equivalent field-theoretic description.

As starting point we take the *grand canonical* partition function of the classical Coulomb gas expressed in a form best suited for our present purposes,

$$Z = \sum_{N=0}^{\infty} \frac{z^N}{[(N/2)!]^2} \prod_{\alpha=1}^{N} \int \frac{d^2 x_\alpha}{a^2} \exp\left[-\frac{\beta}{2} \sum_{\alpha,\beta} q_\alpha q_\beta U(\mathbf{x}_\alpha - \mathbf{x}_\beta)\right], \tag{5.72}$$

where the fugacity $z \equiv \exp(-\beta\epsilon_c)$ only includes the nonelectrostatic contribution ϵ_c to the core energy. We recall that only neutral configurations contribute to the partition function—that is, the number of positive (N_+) and negative (N_-) charges is equal, $N_+ = N_- = N/2$. By overall charge neutrality,

$$\sum_{\alpha,\beta} q_\alpha q_\beta U(\mathbf{x}_\alpha - \mathbf{x}_\beta) = \sum_{\alpha \neq \beta} q_\alpha q_\beta \left[U(\mathbf{x}_\alpha - \mathbf{x}_\beta) - U(0)\right]. \tag{5.73}$$

The integrations in Eq. (5.72) are restricted to regions where the charges are more than a distance a apart, $|\mathbf{x}_\alpha - \mathbf{x}_\beta| > a$. The expression (5.72) is more general than the one in Eq. (5.44) where possible interactions between the dipoles were initially ignored. In Eq. (5.72), there is no similar restriction. By considering a grand canonical ensemble, we assume that charges in the classical Coulomb gas can be created by thermal fluctuations.

The key to deriving the field-theoretic description is the observation below Eq. (5.33) that the Coulomb potential is the inverse of the Laplace operator ∇^2, apart from a multiplicative factor. This makes it possible to represent the exponential function in Eq. (5.72) as a functional integral over an auxiliary field ϕ as

$$\exp\left[-\frac{\beta}{2} \sum_{\alpha,\beta} q_\alpha q_\beta U(\mathbf{x}_\alpha - \mathbf{x}_\beta)\right] = \int \mathrm{D}\phi \exp\left\{-\int d^2 x \left[\frac{1}{4\pi\beta}(\nabla\phi)^2 + \mathrm{i}\rho\phi\right]\right\} \tag{5.74}$$

with $\rho(\mathbf{x}) = \sum_\alpha q_\alpha \delta(\mathbf{x} - \mathbf{x}_\alpha)$ the charge density. In this way, the partition function becomes

$$Z = \sum_{N=0}^{\infty} \frac{z^N}{[(N/2)!]^2} \prod_{\alpha=1}^{N} \int \frac{d^2 x_\alpha}{a^2} \int \mathrm{D}\phi \exp\left\{-\int d^2 x \left[\frac{1}{4\pi\beta}(\nabla\phi)^2 + \mathrm{i}\rho\phi\right]\right\}. \tag{5.75}$$

In a mean-field treatment, the functional integral over the auxiliary field ϕ is approximated by the saddle point, defined by the solution of the field equation

$$\frac{\mathrm{i}}{\beta}\nabla^2\phi = -2\pi\rho. \tag{5.76}$$

With the definition $\Phi \equiv \mathrm{i}\phi/\beta$, this equation assumes the form of Gauss's law in two dimensions, with Φ the electrostatic scalar potential. The auxiliary field can therefore be thought of as representing the fluctuating scalar potential.

By charge neutrality, we have the identity

$$\left[\int d^2x \left(e^{iq\phi(\mathbf{x})} + e^{-iq\phi(\mathbf{x})}\right)\right]^N = \frac{N!}{[(N/2)!]^2} \prod_{\alpha=1}^{N} \int d^2x_\alpha e^{-i\sum_\alpha q_\alpha \phi(\mathbf{x}_\alpha)}, \tag{5.77}$$

where it is recalled that N denotes the total number of charges which is an even number. The factor $N!/[(N/2)!]^2$ is the number of charge-neutral terms contained in the binominal expansion of the left side. Using Eq. (5.77) backwards, we can rewrite the partition function (5.75) as

$$\begin{aligned} Z &= \sum_{N=0}^{\infty} \frac{z^N}{N!} \int D\phi \exp\left[-\frac{1}{4\pi\beta} \int d^2x (\nabla\phi)^2\right] \left[2 \int \frac{d^2x}{a^2} \cos(q\phi)\right]^N \\ &= \int D\phi \exp\left\{-\int d^2x \left[\frac{1}{4\pi\beta}(\nabla\phi)^2 - \frac{2z}{a^2}\cos(q\phi)\right]\right\}. \end{aligned} \tag{5.78}$$

The field theory appearing on the right, which is known as the *sine-Gordon model*, provides a field-theoretic description of a classical Coulomb gas. The Euler-Lagrange equation for the auxiliary field resulting from this expression reads

$$\frac{1}{\beta}\nabla^2\phi = 2\pi q \frac{z}{a^2}\left(e^{iq\phi} - e^{-iq\phi}\right). \tag{5.79}$$

On comparison with the previous field equation (5.76), it follows that the right side represents (-2π times) the charge density. In terms of the scalar potential Φ, Eq. (5.79) takes the form of a Poisson-Boltzmann equation

$$\nabla^2\Phi = -2\pi q\left(\frac{z}{a^2}e^{-\beta q\Phi} - \frac{z}{a^2}e^{\beta q\Phi}\right), \tag{5.80}$$

describing, at least for temperatures above the BKT temperature, a charged plasma of positive and negative charges with local density

$$n_\pm(\mathbf{x}) = \frac{z}{a^2}\, e^{\mp\beta q\Phi(\mathbf{x})}, \tag{5.81}$$

respectively. In the limit $\Phi \to 0$, the density of both positive and negative charges is given by z/a^2. Equation (5.80) provides a self-consistent equation for the scalar potential Φ. Its solution gives, through Eq. (5.81), the spatial distribution of the charges.

Since Eq. (5.78) is an equivalent representation of the Coulomb gas, the conclusions of the previous section should also apply to the sine-Gordon model. In particular, the BKT temperature $k_B T_{BKT} = q^2/4$ should reappear as a critical temperature of the model. To show this, we consider the ratio $Z(z)/Z(0)$ for a finite system of linear size R. For z small, the ratio can be expanded in a power series in z as

$$\frac{Z(z)}{Z(0)} \approx 1 + \frac{2z^2}{a^4}\int^R d^2x d^2y \langle\cos[q\phi(\mathbf{x})]\cos[q\phi(\mathbf{y})]\rangle_0, \tag{5.82}$$

where the average is taken with respect to the free theory corresponding to setting $z = 0$ in Eq. (5.78). By Eq. (5.74), the Gaussian integral yields

$$\begin{aligned}\langle\cos[q\phi(\mathbf{x})]\cos[q\phi(\mathbf{y})]\rangle_0 &= \frac{1}{2}\langle e^{iq\phi(\mathbf{x})}e^{-iq\phi(\mathbf{y})}\rangle_0 \\ &= \frac{1}{2}\,e^{q^2\beta[U(\mathbf{x}-\mathbf{y})-U(0)]} = \frac{1}{2}\,(a/|\mathbf{x}-\mathbf{y}|)^{q^2\beta},\end{aligned} \tag{5.83}$$

where the first line follows because only charge-neutral configurations contribute. The constant C introduced in Eq. (5.11) is irrelevant for our present purposes, and has been ignored in the last step. In this way, the ratio becomes

$$\frac{Z(z)}{Z(0)} \approx 1 + \frac{z^2}{a^{4-q^2\beta}}\int^R d^2x d^2y\,\frac{1}{|\mathbf{x}-\mathbf{y}|^{q^2\beta}}. \tag{5.84}$$

Consider now the change of scale, $R \to bR$ with $b > 1$. Then

$$\frac{Z(z)}{Z(0)} \approx 1 + \frac{z^2}{a^{4-q^2\beta}}b^{4-q^2\beta}\int^R d^2x' d^2y' \frac{1}{|\mathbf{x}'-\mathbf{y}'|^{q^2\beta}}, \tag{5.85}$$

where new integration variables $\mathbf{x} \to \mathbf{x}' = \mathbf{x}/b$ and $\mathbf{y} \to \mathbf{y}' = \mathbf{y}/b$ are introduced so that the integrals are again up to R. It follows that this transformation leads to a renormalization of the fugacity

$$z \to z' = b^{\chi} z, \qquad \chi \equiv D - \frac{1}{2}q^2\beta, \tag{5.86}$$

with $D = 2$ and χ the so-called *scaling dimension* of the fugacity z. Since z appears in the combination $z\int d^D x \cos[q\phi(\mathbf{x})]$ in the Hamiltonian, it follows that $\cos[q\phi(\mathbf{x})]$, or equivalently, $\exp[\pm iq\phi(\mathbf{x})]$ has scaling dimension $y \equiv D - \chi$ which is $y = \frac{1}{2}q^2\beta$ here, in agreement with Eq. (5.83).

If $\chi > 0$, the fugacity becomes larger as we scale to larger distances. The cosine interaction is said to be a *relevant perturbation*. If, on the other hand, $\chi < 0$, i.e., $T < T_{\rm BKT}$, the fugacity scales to zero. This is precisely what was found in Eq. (5.64) of the previous section. In the low-temperature state, the interaction term of the sine-Gordon model, being proportional to z, gets suppressed in the large-scale limit $r >> a$, i.e., it becomes *irrelevant*. The long-distance behavior of the model in this state is therefore well described by a free gapless theory given by the first term of the model, $(\nabla\phi)^2/4\pi\beta$. An isolated charge cannot exist in this gapless state because its energy would diverge in the infrared.

We next turn to the high-temperature state. An important physical characteristic of a charged plasma is that it screens external charges. This so-called Debye screening can be illustrated by considering an external point charge at the origin. Assuming $q\Phi << k_BT$ so that $\sinh(\beta q\Phi) \approx \beta q\Phi$, we can approximate the Poisson-Boltzmann equation (5.80) by the linear equation

$$(\nabla^2 - q_D^2)\Phi(\mathbf{x}) = -2\pi q_0\delta(\mathbf{x}), \quad q_D^2 \equiv 4\pi q^2\beta\frac{z}{a^2}, \tag{5.87}$$

with q_0 the external charge. The wave number $q_{\rm D}$ appearing in Eq. (5.87) provides the high-temperature state with an infrared cutoff so that a single charge now has a finite energy and thus a finite probability to be created. The solution of the linearized Poisson-Boltzmann equation is given by $\Phi(\mathbf{x}) = q_0 K_0(q_{\rm D} r)$ with K_0 the zeroth-order modified Bessel function of the second kind and with $r = |\mathbf{x}|$. The mass term in Eq. (5.87) denotes (2π times) the charge density $\rho_{\rm ind}$ induced by the external charge, i.e.,

$$\rho_{\rm ind}(\mathbf{x}) = -\frac{1}{2\pi} q_0 q_{\rm D}^2 K_0(q_{\rm D} r). \tag{5.88}$$

Since asymptotically

$$K_0(q_{\rm D} r) \sim \sqrt{\frac{\pi}{2 q_{\rm D} r}}\, {\rm e}^{-q_{\rm D} r}. \tag{5.89}$$

the inverse wave number, $1/q_{\rm D}$, plays the role of a screening length—the so-called *Debye screening length*. Integrating $\rho_{\rm ind}(\mathbf{x})$ over the entire system, we see that the total induced charge,

$$\int {\rm d}^2 x \, \rho_{\rm ind} = -q_0 \tag{5.90}$$

completely screens the external charge, no matter its value.

5.5 Superfluid ^{4}He Films

In this section, it will be shown that a liquid ^{4}He film exhibits a similar BKT phase transition as a classical Coulomb gas in 2D. The role of charges is played by vortices which, as already remarked, are pointlike in 2D at scales large compared to their core size. The superfluid low-temperature state is characterized by tightly bound vortex-antivortex pairs. At the BKT temperature, these pairs unbind and thereby disorder the superfluid state. The disordered state consists of a plasma of unbound vortices.

The phase transition is an equilibrium transition and so the time dependence can be ignored. As starting point of our analysis we take the Hamiltonian

$$H = \frac{1}{2}\rho_{\rm s} \int {\rm d}^2 x \, \mathbf{v}_{\rm s}^2, \tag{5.91}$$

where $\rho_{\rm s}$ is the superfluid mass density which we assume to be a function of temperature only, and $\mathbf{v}_{\rm s}$ is the superfluid velocity (5.2) with the Villain potential $\mathbf{A}^{\rm V}$ accounting for vortices through the static counterpart of Eq. (5.1). The Hamiltonian follows from the effective theory (4.121) valid at $T = 0$ by setting the time-dependence to zero, ignoring the higher-order terms, and by replacing the zero-temperature expression for the superfluid mass density $m\bar{n}$ with $\rho_s(T)$. The latter is a response function, which indicates what fraction of the particles participates in the superflow. In general, this fraction differs from the fraction of particles condensed in the ground state. For a pure system at the absolute zero of temperature, although not all particles are condensed, $\rho_s(0) = m\bar{n}$ and all particles participate in the superflow, see Sec. 4.12.

The canonical partition function describing a superfluid ^{4}He film at finite temperature with N vortices of which N_+ have winding number $w = +1$ and N_- have winding number $w = -1$ is given as a functional integral over the superfluid velocity potential φ by

$$Z_N = \sum_{N_+,N_-}{}' \frac{z^N}{N_+!N_-!} \prod_\alpha \int \frac{d^2 x_\alpha}{a^2} \int D\varphi \, e^{-\beta H}, \tag{5.92}$$

with H the Hamiltonian (5.91). The factor $N_+!$ ($N_-!$) accounts for the fact that the (anti)vortices are indistinguishable, and $\int dx_\alpha$ denotes the integration over the positions of the vortices. The sums over the number N_+ of vortices and the number N_- of antivortices are restricted by the constraint that the total number of vortices adds up to N, $N_+ + N_- = N$. This is indicated by the prime on the sum. The factors of a, which is of order the vortex core size, are included in the integration measures for dimensional reasons. To account for the core energy ϵ_c, we included the fugacity $z \equiv \exp(-\beta\epsilon_c)$. The functional integral over φ is Gaussian and can therefore be evaluated by simply substituting the Euler-Lagrange equation for φ back into the Hamiltonian as was done in deriving the action (5.19). We in this way arrive at the expression

$$Z_N = \sum_{N_+,N_-}{}' \frac{z^N}{N_+!N_-!} \prod_\alpha \int \frac{d^2 x_\alpha}{a^2} \exp\left[-\pi\beta\frac{\hbar^2}{m^2}\rho_s \sum_{\alpha,\beta} w_\alpha w_\beta U(\mathbf{x}_\alpha - \mathbf{x}_\beta)\right], \tag{5.93}$$

where $U(\mathbf{x})$ was introduced in Eq. (5.33). This is precisely the canonical partition function describing a classical Coulomb gas in 2D with charges $q_\alpha = q w_\alpha = \pm q$, where

$$q \equiv \sqrt{2\pi(\hbar^2/m^2)\rho_s}. \tag{5.94}$$

That is to say, the theory discussed in the preceding sections can be taken over *verbatim* to describe a liquid ^{4}He film. In particular, the film is predicted to undergo a BKT phase transition at the critical temperature

$$k_B T_{BKT} = \frac{1}{4} q^2 = \frac{\pi}{2} \frac{\hbar^2}{m^2} \rho_s \tag{5.95}$$

triggered by the unbinding of vortex-antivortex pairs. The BKT temperature is well below the bulk transition temperature. At T_{BKT}, the two-dimensional superfluid mass density ρ_s drops discontinuously to zero. The jump $\Delta\rho_s$ measured in units of $T_{\rm BKT}$ is predicted by this theory to be universal,

$$\frac{\Delta\rho_{\rm s}}{k_{\rm B}T_{\rm BKT}} = \frac{8\pi}{\kappa_0^2}, \tag{5.96}$$

depending only on the unit of circulation $\kappa_0 \equiv 2\pi\hbar/m$. Stated differently, although the superfluid mass density may vary from sample to sample, the curves $\rho_{\rm s}(T)$ all terminate on a line with universal slope $8\pi k_{\rm B}/\kappa_0^2$ as the temperature approaches $T_{\rm BKT}$ from below.

Above the vortex-unbinding transition temperature, but still below the bulk transition temperature, a liquid ^{4}He film is in a plasma state of unbound vortices described by the sine-Gordon model (5.78). In the low-temperature state, vortices play no role as they are tightly bound in vortex-antivortex pairs. In the large-scale limit, the interaction term of the model becomes irrelevant and can be ignored so that the superfluid state in 2D is described by the gapless scalar field ϕ.

5.6 Duality

To see the connection between the sine-Gordon field ϕ and the superfluid velocity potential φ, we recast the field equation (5.76) in the form

$$\frac{\rm i}{\beta}\nabla^2\phi = -\frac{m}{\hbar}q\nabla\times\mathbf{v}_{\rm s}, \tag{5.97}$$

with $\rho(\mathbf{x}) \equiv \sum_\alpha q_\alpha\delta(\mathbf{x}-\mathbf{x}_\alpha)$ and where the right side follows from Eq. (5.3). Upon integrating this equation, we arrive at the relation

$$\partial_i\varphi - \frac{m}{\hbar}A^{{\rm V},i} = \frac{\rm i}{q\beta}\epsilon^{ij}\partial_j\phi, \tag{5.98}$$

up to an irrelevant integration constant. Such a relation, involving the antisymmetric Levi-Civita symbol, is typical for *dual* variables. It nicely illustrates that although the dual variable ϕ is a regular field, it nevertheless contains information about the vortices which in the original formulation are described through the singular Villain potential $\mathbf{A}^{\rm V}$. The sine-Gordon model is said to be dual to the original model (5.91). In general, a dual mapping transforms a formulation expressed in one set of variables into an other formulation in terms of other variables. Being dual to each other, either the original or the dual theory can be used for computation, whichever is more convenient for the problem at hand.

As an illustration, we compute the superfluid current-current correlation function $\langle g_{\rm s}^i(\mathbf{k})g_{\rm s}^j(-\mathbf{k})\rangle$ in both formulations. Here,

$$\mathbf{g}_{\rm s} = \rho_{\rm s}\mathbf{v}_{\rm s} \tag{5.99}$$

is the superfluid momentum density, or equivalently, the superfluid mass current density. In the superfluid state, where vortices can be ignored, the Villain potential $\mathbf{A}^{\mathrm{V}}$ can be set to zero in the original theory, which thereby reduces to a free gapless scalar theory. With a source term $\mathbf{J}$ added to the Hamiltonian (5.91), the generating functional $Z[\mathbf{J}]$ reads,

$$Z[\mathbf{J}] = \int \mathrm{D}\varphi \exp\left[-\int \mathrm{d}^2x \left(\tfrac{1}{2}\beta\rho_{\mathrm{s}}\mathbf{v}_{\mathrm{s}}^2 + \mathbf{g}_{\mathrm{s}} \cdot \mathbf{J}\right)\right], \tag{5.100}$$

with $\mathbf{v}_{\mathrm{s}} = (\hbar/m)\nabla\varphi$ now. The current-current correlation function is readily calculated by differentiating $Z[\mathbf{J}]$ twice with respect to $\mathbf{J}$, and setting the source to zero at the end. This gives, after passing to momentum space,

$$\langle g_{\mathrm{s}}^i(\mathbf{k})g_{\mathrm{s}}^j(-\mathbf{k})\rangle = \frac{\rho_{\mathrm{s}}}{\beta}\frac{k^ik^j}{\mathbf{k}^2}. \tag{5.101}$$

The superfluid response has only a longitudinal component proportional to $k^ik^j/\mathbf{k}^2$. This is different from the normal fluid's response, which in addition has a transverse component proportional to $\delta^{ij} - k^ik^j/\mathbf{k}^2$ with the same coefficient as the longitudinal component. Specifically,

$$\begin{aligned}\langle g^i(\mathbf{k})g^j(-\mathbf{k})\rangle &= \frac{\rho}{\beta}\frac{k^ik^j}{\mathbf{k}^2} + \frac{\rho_{\mathrm{n}}}{\beta}\left(\delta^{ij} - \frac{k^ik^j}{\mathbf{k}^2}\right) \\ &= \frac{\rho_{\mathrm{s}}}{\beta}\frac{k^ik^j}{\mathbf{k}^2} + \frac{\rho_{\mathrm{n}}}{\beta}\delta^{ij},\end{aligned} \tag{5.102}$$

where $\rho = \rho_{\mathrm{s}} + \rho_{\mathrm{n}}$ is the total mass density (4.215).

To evaluate this correlation function in the dual formulation, the original theory must be dualized with a source term included. In the superfluid state, where vortices can be ignored, this amounts to linearizing the theory (5.100) with the help of a Hubbard-Stratonovich transformation,

$$Z[\mathbf{J}] = \int \mathrm{D}\mathbf{e}\,\mathrm{D}\varphi \exp\left[-\int \mathrm{d}^2x \left(\frac{1}{2\beta}\mathbf{e}^2 + \mathrm{i}\sqrt{\rho_s}\,\mathbf{v}_{\mathrm{s}} \cdot \mathbf{e} + \mathbf{g}_{\mathrm{s}} \cdot \mathbf{J}\right)\right]. \tag{5.103}$$

The original form (5.100) is recovered by integrating out the auxiliary field $\mathbf{e}$ again, which is tantamount to substituting the Euler-Lagrange equation

$$\mathbf{e} = -\mathrm{i}\beta\sqrt{\rho_s}\,\mathbf{v}_{\mathrm{s}} \tag{5.104}$$

back into the form (5.103). The integral over the velocity potential φ yields the constraint $\nabla \cdot \mathbf{e} - \mathrm{i}\sqrt{\rho_{\mathrm{s}}}\,\nabla \cdot \mathbf{J} = 0$, which is solved by

$$e^i - \mathrm{i}\sqrt{\rho_{\mathrm{s}}}J^i = \frac{1}{\sqrt{2\pi}}\epsilon^{ij}\partial_j\phi, \tag{5.105}$$

where the coefficient on the right is chosen for consistency with the normalization in Eq. (5.78). Note that on substituting this solution with $\mathbf{J} = 0$ into the

Euler-Lagrange equation (5.104), the relation (5.98) with $\mathbf{A}^{\mathrm{V}} = 0$ is recovered. Replacing the functional integral over the auxiliary field $\mathbf{e}$ with one over ϕ, we finally arrive at the dual theory in the presence of a source term

$$Z[\mathbf{J}] = \int \mathrm{D}\phi \exp\left[-\frac{1}{2\beta}\int \mathrm{d}^2x \left(\frac{1}{\sqrt{2\pi}}\epsilon^{ij}\partial_j\phi + \mathrm{i}\sqrt{\rho_\mathrm{s}}J^i\right)^2\right]. \tag{5.106}$$

Note the appearance of $1/\beta$ rather than β in the exponential, meaning that the temperature gets inverted when going from the original to the dual theory. This is typical for a dual mapping. Proceeding as in the original theory, we find

$$\langle g_\mathrm{s}^i(\mathbf{k})g_\mathrm{s}^j(-\mathbf{k})\rangle = \frac{\rho_\mathrm{s}}{\beta}\delta^{ij} - \frac{1}{2\pi}\frac{\rho_\mathrm{s}}{\beta^2}\epsilon^{ik}\epsilon^{jl}k^k k^l\langle\phi(\mathbf{k})\phi(-\mathbf{k})\rangle, \tag{5.107}$$

where the first term is the diamagnetic contribution that arises because a term proportional to $\mathbf{J}^2$ appears in the dual theory. The average in the second term is to be taken with respect to the partition function (5.106) with the source set to zero. The expression (5.107) then reduces to Eq. (5.101) obtained in the original theory by use of the 2D identity

$$\epsilon^{ik}\epsilon^{jl} = \delta^{ij}\delta^{kl} - \delta^{il}\delta^{kj}. \tag{5.108}$$

This demonstrates that calculations performed in the original or the dual theory produce the same results, as was to be expected for both formulations describe the same physical system.

Notes

(i) The dynamics of vortices in superfluid films is described in, for example, [Yourgrau and Mandelstam (1979); Popov (1987)], and in [Lund (1991)]. See also [Lamb (1932)].

(ii) The phase change of the wave function of a superfluid film as a vortex moves around a closed path was established by Haldane and Wu (1985).

(iii) The exact equation of state (5.41) of a classical 2D Coulomb gas was derived by Salzberg and Prager (1963) and independently by May (1967).

(iv) For the mapping of the Coulomb gas onto the sine-Gordon model, see [Polyakov (1987)] where also applications of this mapping in higher than two dimensions are discussed.

(v) The original papers on the BKT transition are [Berezinskii (1970, 1971)], and [Kosterlitz and Thouless (1973)].

(vi) The renormalization group equations governing the BKT transition were derived by Kosterlitz (1974). Our presentation is based also on [Huang and Polonyi (1991)]. For the electrostatic concepts used, see [Jackson (1975)].

(vii) The universal jump in the superfluid mass density of 2D superfluids was predicted by Nelson and Kosterlitz (1977) and was experimentally established by Bishop and Reppy (1980).

Chapter 6

Fermi Gases

This chapter is devoted to degenerate Fermi systems in the normal state. Because of the Pauli principle, the behavior of such systems is completely different from their bosonic counterparts. The presence of a Fermi surface manifests itself also at a technical level, and calls for different techniques to evaluate integrals.

6.1 Interacting Fermi Gas

We start this chapter by considering a dilute Fermi gas with a short-range repulsive interaction at the absolute zero of temperature. We assume that the interaction is instantaneous and can be described by a spin-independent two-particle potential $V(\mathbf{x}, \mathbf{x}')$. The Lagrangian then reads

$$L = L_0 - \frac{1}{2} \int \mathrm{d}^D x \mathrm{d}^D x' \psi^*_\sigma(t, \mathbf{x}) \psi_\sigma(t, \mathbf{x}) V(\mathbf{x} - \mathbf{x}') \psi^*_{\sigma'}(t, \mathbf{x}') \psi_{\sigma'}(t, \mathbf{x}'), \tag{6.1}$$

with the Grassmann fields ψ^*_σ and ψ_σ describing fermions of spin $\sigma = \uparrow, \downarrow$, and L_0 the Lagrangian of an ideal Fermi gas,

$$L_0 = \int \mathrm{d}^D x \, \psi^*_\sigma \left(\mathrm{i}\hbar \frac{\partial}{\partial t} + \frac{\hbar^2}{2m} \nabla^2 + \mu \right) \psi_\sigma. \tag{6.2}$$

We wish to compute the ground-state energy $\langle E \rangle$ in perturbation theory. We do so by evaluating the thermodynamic potential Ω, which at finite temperature is related to the partition function Z through Eq. (2.8), i.e., through $\Omega = -\beta^{-1} \ln Z$. At zero temperature, where imaginary time $0 \leq \tau \leq \hbar\beta$ is replaced with real time $0 \leq t \leq \mathcal{T}$, this relation translates into

$$\Omega = \mathrm{i} \frac{\hbar}{\mathcal{T}} \ln Z, \tag{6.3}$$

with Z now the zero-temperature partition function which can be represented by the function integral

$$Z = \int \mathrm{D}\psi^* \mathrm{D}\psi \, \mathrm{e}^{(\mathrm{i}/\hbar) S}, \tag{6.4}$$

and $\langle E\rangle = \Omega + \mu\langle N\rangle$ by Eq. (2.20) with $\langle N\rangle$ the average particle number formally given by Eq. (2.19). To arrive at a diagrammatic representation of Ω, recall that the generator $(\mathrm{i}/\hbar)W$ of connected vacuum diagrams is related to Z through Eq. (2.149), i.e., through $(\mathrm{i}/\hbar)W = \ln(Z)$ apart from an irrelevant constant term. With Eq. (6.3) it then follows that $-(\mathrm{i}/\hbar)\Omega/V$ is given by the sum of all connected vacuum diagrams. In reaching this conclusion, we divided by the spacetime volume factor $\mathcal{T}V$ which reflects our convention of overall momentum conservation, see Sec. 2.5.

In addition to the Feynman rule (2.130) for the propagator with $K(k) = \hbar\omega - \xi(\mathbf{k})$, or

$$S_{\mathrm{F}}(k) = \frac{1}{\hbar\omega - \xi(\mathbf{k}) + \mathrm{i}\,\mathrm{sgn}(\omega)\eta}, \tag{6.5}$$

where $\xi(\mathbf{k}) \equiv \epsilon(\mathbf{k}) - \mu$ is the single-particle excitation energy $\epsilon(\mathbf{k}) = \hbar^2\mathbf{k}^2/2m$ of the free theory measured relative to the chemical potential μ, we have, instead of the rule (2.133) for a local interaction, now

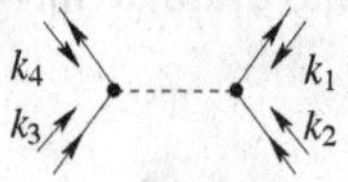

$$\equiv -\frac{\mathrm{i}}{\hbar}(2\pi)^d V(\mathbf{k})\delta(k_1 + k_2 + k_3 + k_4), \tag{6.6}$$

where $\mathbf{k} = \mathbf{k}_3 + \mathbf{k}_4$.

In zeroth order, the thermodynamic potential is that of an ideal gas,

$$\Omega_1 = \mathrm{i}\frac{\hbar}{\mathcal{T}}\,\mathrm{Tr}\ln\left(p_0 - \frac{1}{2m}\mathbf{p}^2 + \mu\right), \tag{6.7}$$

or

$$\frac{\Omega_1}{V} = \mathrm{i}\hbar g\int\frac{\mathrm{d}^d k}{(2\pi)^d}\left[\hbar\omega - \xi(\mathbf{k})\right], \tag{6.8}$$

where the prefactor $g = 2$ arises from the sum over two spin directions in the closed loop. As for the bosonic effective potential (4.71), the frequency integral here is best evaluated by first taking the derivative with respect to μ,

$$\begin{aligned}\frac{\partial}{\partial\mu}\hbar\int\frac{\mathrm{d}\omega}{2\pi}\ln\left[\hbar\omega - \xi(\mathbf{k})\right] &= \hbar\int\frac{\mathrm{d}\omega}{2\pi}\frac{\mathrm{e}^{\mathrm{i}\omega 0^+}}{\hbar\omega - \xi(\mathbf{k}) + \mathrm{i}\,\mathrm{sgn}(\omega)\eta} \\ &= \mathrm{i}\theta[-\xi(\mathbf{k})]\end{aligned} \tag{6.9}$$

where we included the usual convergence factor $\exp(\mathrm{i}\omega 0^+)$, see below Eq. (2.69). The argument of the step function changes sign at $|\mathbf{k}| = k_{\mathrm{F}}$, which in zeroth order is related to μ simply through $\mu = \hbar^2 k_{\mathrm{F}}^2/2m$. The right side of Eq. (6.9) is readily integrated with respect to μ to yield

$$\begin{aligned}\frac{\Omega_1}{V} &= g\int\frac{\mathrm{d}^D k}{(2\pi)^D}\xi(\mathbf{k})\theta[-\xi(\mathbf{k})] \\ &= \frac{g}{10\pi^2}\frac{\hbar^2 k_{\mathrm{F}}^5}{2m} - \mu n = -\frac{2}{5}\mu n,\end{aligned} \tag{6.10}$$

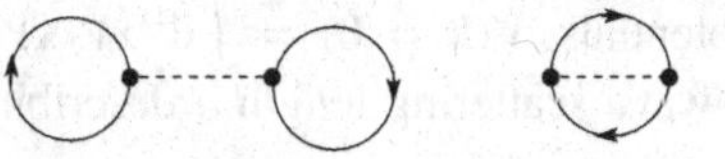

Fig. 6.1 The direct or Hartree (left) and exchange (right) vacuum diagrams.

with

$$n = g \int \frac{d^D k}{(2\pi)^D} \theta[-\xi(\mathbf{k})] = \frac{g}{6\pi^2} k_F^3 \tag{6.11}$$

the total average particle number density.

To first order in the interaction potential V, the two diagrams in Fig. 6.1, collectively known as the Hartree-Fock vacuum diagrams, contribute. The first diagram, known as the direct or *Hartree term*, gives

$$-\frac{i}{\hbar}\frac{\Omega_2^d}{V} = (-1)^2 g^2 \frac{1}{2} \int \frac{d^d k}{(2\pi)^d} \frac{d^d k'}{(2\pi)^d} i\hbar S_F(k) \left(-\frac{i}{\hbar}\right) V(0) i\hbar S_F(k'), \tag{6.12}$$

where 2 is the symmetry factor of the diagram. The factor $(-1)^2$ arises because each fermion loop gives a minus sign, and the prefactor $g^2 = 4$ results from the sum over two spin directions in each of the two loops. With the convergence factors $\exp(i\omega 0^+)$ included in the propagators, the frequency integrals produce step functions by Eq. (3.14) so that

$$\frac{\Omega_2^d}{V} = \frac{1}{2} g^2 \int \frac{d^D k}{(2\pi)^D} \frac{d^D k'}{(2\pi)^D} \theta[-\xi(\mathbf{k})]\theta[-\xi(\mathbf{k}')] V(0). \tag{6.13}$$

The second diagram in Fig. 6.1 , known as the *exchange term*, gives the two-loop contribution

$$-\frac{i}{\hbar}\frac{\Omega_2^x}{V} = -g\frac{1}{2} \int \frac{d^d k}{(2\pi)^d} \frac{d^d k'}{(2\pi)^d} i\hbar S_F(k) \left(-\frac{i}{\hbar}\right) V(\mathbf{k}-\mathbf{k}') i\hbar S_F(k'), \tag{6.14}$$

where the factor of 2 is again the symmetry factor of the diagram. The frequency integrals can be evaluated as before to give

$$\frac{\Omega_2^x}{V} = -\frac{1}{2} g \int \frac{d^D k}{(2\pi)^D} \frac{d^D k'}{(2\pi)^D} \theta[-\xi(\mathbf{k})]\theta[-\xi(\mathbf{k}')] V(\mathbf{k}-\mathbf{k}'). \tag{6.15}$$

Combined, these contributions give as energy density of the ground state to this order

$$\frac{\langle E \rangle}{V} = \frac{3}{5}\mu n + \frac{1}{2} g \int \frac{d^D k}{(2\pi)^D} \frac{d^D k'}{(2\pi)^D} \theta[-\xi(\mathbf{k})]\theta[-\xi(\mathbf{k}')] \left[g V(0) - V(\mathbf{k}-\mathbf{k}')\right]. \tag{6.16}$$

For nonsingular potentials, $V(\mathbf{k}=0)=\int d^D x V(\mathbf{x})$ can be characterized as in Eq. (4.43) by the s-wave scattering length a describing the scattering of two particles in vacuum as

$$V(0)=4\pi\hbar^2 a/m, \tag{6.17}$$

specializing to $D=3$. For a dilute Fermi gas, the Fermi wave number k_{F} is small, and $V(\mathbf{k}-\mathbf{k}')$ can be approximated by $V(0)$. Equation. (6.16) can then be expressed as

$$\frac{\langle E\rangle}{\langle N\rangle}=\mu\left(\frac{3}{5}+\frac{2}{3\pi}k_{\mathrm{F}}a\right) \tag{6.18}$$

with $\langle N\rangle = nV$ the average total particle number. From this, the chemical potential μ follows as

$$\mu=\frac{\partial\langle E\rangle}{\partial\langle N\rangle}=\mu\left(1+\frac{4}{3\pi}k_{\mathrm{F}}a\right) \tag{6.19}$$

to this order.

6.2 Impurities

In Sec. 4.13, it was shown that impurities lead to a collapse of an ideal Bose gas, and that a repulsive interaction between atoms is needed to stabilize the disordered system. Due to the Pauli exclusion principle, no such interparticle repulsion is needed for Fermi systems. We therefore consider an ideal gas of fermions with spin $\frac{1}{2}$ described by the Lagrangian (6.2) augmented by a (nonmagnetic) interaction term of the form

$$\mathcal{L}_{\boxdot}=\sum_{a=1}^{N_{\boxdot}}\psi^*_\sigma(x)u(\mathbf{x}-\mathbf{x}_a)\psi_\sigma(x). \tag{6.20}$$

Here, the sum is over the $N_{\boxdot}$ impurities, and the static potential $u(\mathbf{x}-\mathbf{x}_a)$ represents the interaction due to a frozen impurity located at $\mathbf{x}_a$. It is assumed that the impurities scatter the fermions elastically so that no energy transfer takes place. It is further assumed that the separate impurities act independently and are randomly distributed throughout the sample. The free fermion propagator $S_{\mathrm{F}}(k)$, defined by the Lagrangian density (6.2), is given by Eq. (6.5). An impurity vertex will be denoted by a cross,

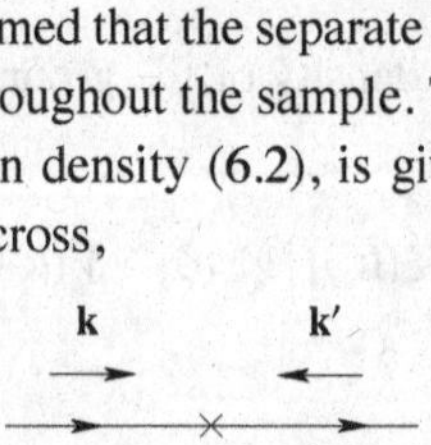

$$=\frac{\mathrm{i}}{\hbar}\sum_a u(-\mathbf{k}'-\mathbf{k})\mathrm{e}^{-\mathrm{i}(\mathbf{k}'+\mathbf{k})\cdot\mathbf{x}_a}, \tag{6.21}$$

where $u(\mathbf{q})$, with $\mathbf{q} \equiv -\mathbf{k}' - \mathbf{k}$ denoting the wave vector transfer, is the Fourier transform of $u(\mathbf{x})$. Note that according to the rules spelled out in Sec. 2.5, all the frequencies and wave vectors are taken to flow into the vertex. Since the sample is assumed to be uniform on a macroscopic scale, we will average over the position of each impurity:

$$\left\langle e^{i\mathbf{q}\cdot\mathbf{x}_a} \right\rangle \equiv \frac{1}{V} \int d^3 x_a \, e^{i\mathbf{q}\cdot\mathbf{x}_a}, \tag{6.22}$$

with V the volume of the system. It is only after the impurity averaging is carried out that the theory becomes translational invariant.

The self-energy $\Sigma(k)$ is defined through

$$S'_F(k) = \frac{1}{\hbar\omega - \xi(\mathbf{k}) - \Sigma(k)}, \tag{6.23}$$

where $S'_F(k)$ denotes the *full* propagator. Diagrammatically,

$$-\frac{i}{\hbar}\Sigma(\omega, \mathbf{k}, \mathbf{k}') = \tag{6.24}$$

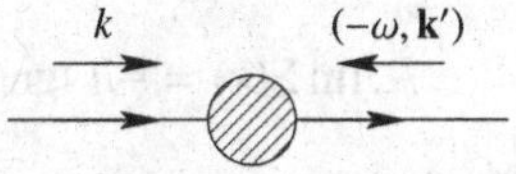

where in case of translational invariance $\Sigma(\omega, \mathbf{k}, \mathbf{k}') = \Sigma(k)(2\pi)^3\delta(\mathbf{k} + \mathbf{k}')$. The simplest self-energy diagram is given by Eq. (6.21). Taking the impurity average, we obtain as contribution

$$\begin{aligned} \Sigma^{(1)}(k)(2\pi)^3\delta(\mathbf{k} + \mathbf{k}') &= -\sum_a u(-\mathbf{k}' - \mathbf{k}) \left\langle e^{-i(\mathbf{k}'+\mathbf{k})\cdot\mathbf{x}_a} \right\rangle \\ &= -n_\square u(0)(2\pi)^3\delta(\mathbf{k} + \mathbf{k}'), \end{aligned} \tag{6.25}$$

with $n_\square = N_\square / V$ denoting the impurity density. The constant $-n_\square u(0)$ can be absorbed by renormalizing the chemical potential and will henceforth be ignored by setting it to zero. As a result, the impurity average of the next simple diagram

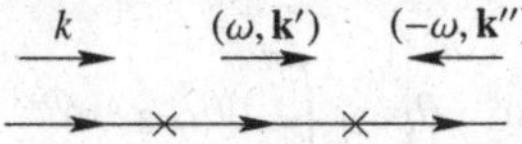

which describes the scattering by two different impurities, located at $\mathbf{x}_a$ and $\mathbf{x}_b$ say, will be zero. However, if scattering takes place by the same impurity atom, i.e., if $\mathbf{x}_a = \mathbf{x}_b$,

$$-\frac{i}{\hbar}\Sigma^{(2)}(\omega, \mathbf{k}, \mathbf{k}'') \equiv \tag{6.26}$$

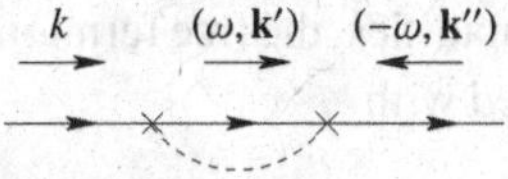

the impurity average of the diagram is nonzero,

$$\begin{aligned} &-\frac{i}{\hbar}\Sigma^{(2)}(\omega, \mathbf{k}, \mathbf{k}'') \\ &= \sum_a \int \frac{d^3 k'}{(2\pi)^3} \left\langle \frac{i}{\hbar} u(\mathbf{k}' - \mathbf{k}) e^{i(\mathbf{k}'-\mathbf{k})\cdot\mathbf{x}_a} i\hbar S_F(\omega, \mathbf{k}') \frac{i}{\hbar} u(-\mathbf{k}'' - \mathbf{k}') e^{-i(\mathbf{k}''+\mathbf{k}')\cdot\mathbf{x}_a} \right\rangle \\ &= n_\square \int \frac{d^3 k'}{(2\pi)^3} \left(\frac{i}{\hbar}\right)^2 |u(\mathbf{k}' - \mathbf{k})|^2 i\hbar S_F(\omega, \mathbf{k}')(2\pi)^3\delta(\mathbf{k} + \mathbf{k}''). \end{aligned} \tag{6.27}$$

The two vertices in the diagram (6.26) are connected by a dashed line to indicate that both refer to the same impurity atom. In the last step, it is used that for real interactions $u(\mathbf{x}), u(-\mathbf{q}) = u^*(\mathbf{q})$. The impurity averaging is seen to produce again translational invariance. Up to this order, the self-energy reads

$$\Sigma^{(2)}(k) = n_\Box \int \frac{\mathrm{d}^3 k'}{(2\pi)^3} |u(\mathbf{k}' - \mathbf{k})|^2 \left\{ \frac{\mathrm{P}}{\hbar\omega - \xi(\mathbf{k}')} - \mathrm{i}\pi \,\mathrm{sgn}(\omega) \delta[\hbar\omega - \xi(\mathbf{k}')] \right\}, \tag{6.28}$$

where use is made of Eq. (2.207). In the limit $\omega \to 0$, the principal part yields a real constant which can be absorbed by renormalizing the chemical potential and need therefore not be considered. In this limit, the main contribution arises from momenta near the Fermi surface so that the imaginary part of the self-energy, given by the second term in the integrand of Eq. (6.28), can be written as

$$\mathrm{Im}\,\Sigma(k) = -\hbar \,\mathrm{sgn}(\omega)/2\tau_\mathrm{F}, \tag{6.29}$$

where τ_F is defined by

$$\hbar/\tau_\mathrm{F} \equiv 2\pi n_\Box \nu(0) \int \frac{\mathrm{d}\Omega_\mathbf{q}}{4\pi} |u(\hat{\mathbf{q}})|^2, \tag{6.30}$$

with $\nu(0) = mk_\mathrm{F}/2\pi^2\hbar^2$ the density of states per spin degree of freedom at the Fermi surface, and $\int \mathrm{d}\Omega_\mathbf{q}$ the angle integrals

$$\int \mathrm{d}\Omega_\mathbf{q} \equiv \int_0^\pi \mathrm{d}\theta \sin\theta \int_0^{2\pi} \mathrm{d}\varphi, \tag{6.31}$$

or in D space dimensions

$$\int \mathrm{d}\Omega_\mathbf{q} \equiv \int_0^\pi \mathrm{d}\theta_{D-2} \sin^{D-2}\theta_{D-2} \int_0^\pi \mathrm{d}\theta_{D-3} \sin^{D-3}\theta_{D-3} \cdots \int_0^{2\pi} \mathrm{d}\varphi, \tag{6.32}$$

while $\hat{\mathbf{q}} \equiv \mathbf{q}/|\mathbf{q}|$ stands for the unit vector in the direction of $\mathbf{q}$. Physically, τ_F denotes the average lifetime of fermions near the Fermi surface. It follows that in the presence of impurities, the free fermion propagator (6.5) acquires an imaginary part and is replaced with

$$S_\mathrm{F}^\Box(k) = \frac{1}{\hbar\omega - \xi(\mathbf{k}) + \mathrm{i}\hbar \,\mathrm{sgn}(\omega)/2\tau_\mathrm{F}}. \tag{6.33}$$

For completeness, we also record the finite-temperature counterpart of this propagator in Euclidean (E) spacetime

$$S_\mathrm{E}^\Box(k) = \frac{1}{-\mathrm{i}\hbar\omega_n + \xi(\mathbf{k}) - \mathrm{i}\hbar \,\mathrm{sgn}(\omega_n)/2\tau_\mathrm{F}}, \tag{6.34}$$

with $\omega_n = \pi(2n+1)/\hbar\beta$ denoting the fermionic Matsubara frequencies.

In the spectral representation, the propagator (6.33) reads

$$S_F^{\Box}(k) = \int d\omega' \frac{\rho(\omega', \mathbf{k})}{\omega - \omega'}, \tag{6.35}$$

with $\rho(k)$ the spectral density

$$\begin{aligned}\rho(k) &\equiv \frac{i}{2\pi\hbar}\left(\frac{1}{\omega - \xi(\mathbf{k})/\hbar + i/2\tau_F} - \frac{1}{\omega - \xi(\mathbf{k})/\hbar - i/2\tau_F}\right) \\ &= \frac{1}{2\pi\hbar\tau_F}\frac{1}{[\omega - \xi(\mathbf{k})]^2/\hbar^2 + 1/4\tau_F^2}\end{aligned} \tag{6.36}$$

which assumes a Lorentzian shape because of the impurities. The free propagator of the pure system is recovered in the limit of infinite lifetime $\tau_F \to \infty$, where the spectral density becomes a delta function $\delta[\hbar\omega - \xi(\mathbf{k})]$.

To see the physical implication of a finite fermion lifetime, consider the spatial Fourier transform of $S_F^{\Box}(k)$ for $\omega > 0$,

$$S_F^{\Box}(\omega, \mathbf{x}) = \int \frac{d^3k}{(2\pi)^3} \frac{1}{\hbar\omega - \xi(\mathbf{k}) + i\hbar/2\tau_F} e^{i\mathbf{k}\cdot\mathbf{x}}. \tag{6.37}$$

For $\hbar\omega, \hbar/\tau_F << \mu$, the propagator can be approximated by

$$S_F^{\Box}(\omega, \mathbf{x}) \approx \frac{\nu(0)}{k_F r} \int d\xi \frac{\sin(|\mathbf{k}|r)}{\hbar\omega - \xi + i\hbar/2\tau_F} \tag{6.38}$$

with $|\mathbf{k}| \approx k_F + (m/\hbar^2 k_F)\xi$ and $r \equiv |\mathbf{x}|$, and where the integration range has been extended from $(-\mu, \infty)$ to $(-\infty, \infty)$ with only negligible error. The remaining integral can be evaluated by contour integration, with the result

$$S_F^{\Box}(\omega, \mathbf{x}) = -\pi\nu(0)\frac{e^{ik_F r}}{k_F r} e^{-r/2\ell}, \tag{6.39}$$

where $\ell \equiv v_F \tau_F$ is the average free path of fermions near the Fermi surface, and where the ω dependence on the right has been neglected by virtue of the assumption $\hbar\omega << \mu$. This result shows that a finite fermion lifetime produces an exponential damping factor for distances larger that ℓ.

In the next section, we calculate the conductivity in this model by means of the derivative expansion method presented in Sec. 3.1.

6.3 Conductivity

For an isotropic system, the conductivity σ relates the electric current density $-e\mathbf{j}_e$ in a metal, with $q = -e(< 0)$ the negative charge of the electrons, to the electric field $\mathbf{E}$ via Ohm's law

$$-e\mathbf{j}_e = \sigma\mathbf{E}. \tag{6.40}$$

For simplicity we assume the system to be uniform in space so that

$$\mathbf{E} = -\frac{1}{c}\frac{\partial}{\partial t}\mathbf{A}(t). \tag{6.41}$$

To leading order, the effect of impurities, which eventually leads to a finite conductivity, is taken into account by replacing the free electron propagator S_{F} with $S_{\mathrm{F}}^{\boxminus}$ given in Eq. (6.33). The system is coupled to an electric background field by minimal substitution

$$\nabla \to \nabla + \mathrm{i}\frac{e}{\hbar c}\mathbf{A}. \tag{6.42}$$

Since the Lagrangian density is quadratic in the electron fields, the functional integral

$$Z[\mathbf{A}] = \int \mathrm{D}\psi^* \mathrm{D}\psi \, \exp\left(\frac{\mathrm{i}}{\hbar}\int \mathrm{d}^4 x \, \mathcal{L}\right) \tag{6.43}$$

is readily evaluated. The result can be written in the form

$$Z[\mathbf{A}] = \mathrm{e}^{(\mathrm{i}/\hbar)\Gamma_1[\mathbf{A}]} \tag{6.44}$$

with the one-loop effective action

$$\Gamma_1[\mathbf{A}] = -\mathrm{i}\hbar \, \mathrm{Tr} \ln \left[K(p) - \Lambda(\mathbf{A})\right], \tag{6.45}$$

where the differential operator $K(p)$, with $p_\mu \equiv \mathrm{i}\hbar\partial_\mu$, is the inverse of the propagator (6.33), and $\Lambda(\mathbf{A})$ was introduced in Eq. (3.11):

$$\Lambda(\mathbf{A}) \equiv \frac{e}{2mc}(\mathbf{p}\cdot\mathbf{A} + \mathbf{A}\cdot\mathbf{p}) + \frac{e^2}{2mc^2}\mathbf{A}^2. \tag{6.46}$$

Recall from Sec. 3.1 that we adopt the convention that the momentum operator p_μ acts on all the fields to its right, while the ordinary derivative ∂_μ acts only on the first field to its right. The minus sign in Eq. (6.45) derives from the fermion loop. Proceeding as in Sec. 3.2, we cast the effective action in the form (3.10) with the propagator (6.33) replacing the propagator $S_{\mathrm{F}}(k)$ of the pure system.

The terms of interest to us are quadratic in $\mathbf{A}$. By virtue of Eq. (2.183),

$$\frac{e}{c}\langle \mathbf{j}_{\mathrm{e}}(x)\rangle = -\frac{\delta\Gamma_1[\mathbf{A}]}{\delta\mathbf{A}(x)}, \tag{6.47}$$

these terms produce the part in the current linear in $\mathbf{A}$. In the language of linear response, we must calculate the polarization tensor introduced in Eq. (2.186). By Eq. (2.192)

$$\langle j_{\mathrm{e}}^i(k)\rangle = -\frac{e}{c}\Pi^{ij}(k)A^j(k) \tag{6.48}$$

after Fourier transforming. Comparison of this expression with the Fourier transform of Eq. (6.40) gives the relation

$$\sigma(k)\delta^{ij} = -\mathrm{i}e^2 \frac{1}{\omega} \Pi^{ij}(k), \tag{6.49}$$

or, using that the polarization tensor is transverse with respect to both its indices, see Eq. (2.188),

$$\sigma(k) = -\mathrm{i}e^2 \frac{\omega}{\mathbf{k}^2} \Pi^{00}(k). \tag{6.50}$$

Given the sum rules (2.198) and (2.204) for the polarization tensor, this relation implies the following sum rules for σ:

$$\lim_{\omega \to \infty} \sigma(k) = \mathrm{i} \frac{e^2 n}{m} \frac{1}{\omega}, \tag{6.51}$$

$$\lim_{\mathbf{k} \to 0} \lim_{\omega \to 0} \sigma(k) = -\mathrm{i} \frac{e^2 n}{m} \frac{\omega}{c_\mathrm{s}^2 \mathbf{k}^2}. \tag{6.52}$$

The first term quadratic in $\mathbf{A}$ originates from the $\ell = 1$ term in Eq. (3.10),

$$\Gamma_1^{(\ell=1)} = \mathrm{i}\hbar \frac{e^2}{2mc^2} g \int \mathrm{d}^d x \frac{\mathrm{d}^d k}{(2\pi)^d} S_\mathrm{F}^{\square}(k) \mathbf{A}^2, \tag{6.53}$$

where $g = 2$ is the spin degeneracy factor. The integral over k yields by Eq. (3.13) the average electron number n, which is a constant because the system is assumed to be uniform, and

$$\Gamma_1^{(\ell=1)} = -\frac{e^2 n}{2mc^2} \int \mathrm{d}^d x \, \mathbf{A}^2. \tag{6.54}$$

As in the pure case, we expect this diamagnetic term, which contains no derivatives, to cancel against another contribution.

The other relevant term quadratic in $\mathbf{A}$ is contained in $\Gamma_1^{(\ell=2)}$. Applying the shift rule (3.7), we obtain, cf. Eq. (3.17)

$$\Gamma_1^{(\ell=2)} = \mathrm{i}\hbar \frac{e^2}{2m^2c^2} g \int \mathrm{d}^d x \frac{\mathrm{d}^d k}{(2\pi)^d} S_\mathrm{F}^{\square}(k) S_\mathrm{F}^{\square}(\omega - q_0, \mathbf{k}) \hbar k^i \hbar k^j A^i A^j, \tag{6.55}$$

where we introduced the notation $q_0 \equiv \mathrm{i}\partial_t$ for the ordinary time derivative. Because the system is assumed to be uniform, spatial derivatives yield zero. The second propagator in this expression is given by Eq. (6.33) with ω replaced with $\omega - q_0$ in both the first *and* the last term in the numerator.

We first consider the time-independent part of this contribution to see if it cancels the diamagnetic contribution and write

$$\frac{1}{[\hbar\omega - \xi + \mathrm{i}\hbar\, \mathrm{sgn}(\omega)/2\tau_\mathrm{F}]^2} = -\frac{\partial}{\partial\mu} \frac{1}{\hbar\omega - \xi + \mathrm{i}\hbar\, \mathrm{sgn}(\omega)/2\tau_\mathrm{F}}. \tag{6.56}$$

With the mandatory convergence factor $\exp(i\omega 0^+)$ included [see below Eq. (2.69)], the frequency integral yields

$$\frac{\partial}{\partial \mu}\hbar \int \frac{d\omega}{2\pi} \frac{e^{i\omega 0^+}}{\hbar\omega - \xi + i\hbar\,\text{sgn}(\omega)/2\tau_F} = i\frac{\partial}{\partial \mu}\theta(-\xi) = i\delta(-\xi). \tag{6.57}$$

The delta function forces the momenta to be on the Fermi surface, and

$$\int \frac{d^D k}{(2\pi)^D}\delta[-\xi(\mathbf{k})]k^i k^j = \frac{\delta^{ij}}{D}\nu(0)k_F^2. \tag{6.58}$$

Since in arbitrary space dimension D

$$n = g\int \frac{d^D k}{(2\pi)^D}\theta[-\xi(\mathbf{k})] = g\frac{\Omega_D}{(2\pi)^D}\int_0^{k_F} d|\mathbf{k}||\mathbf{k}|^{D-1} = g\frac{\Omega_D}{(2\pi)^D}\frac{1}{D}k_F^D \tag{6.59}$$

$$\nu(0) = \int \frac{d^D k}{(2\pi)^D}\delta[-\xi(\mathbf{k})] = \frac{\Omega_D}{(2\pi)^D}\frac{m}{\hbar^2}k_F^{D-2}, \tag{6.60}$$

where the particle number density includes the spin degeneracy factor $g = 2$, while the density of states does not, we have

$$n = \frac{1}{D}\frac{\hbar^2 k_F^2}{m}g\nu(0). \tag{6.61}$$

The contribution (6.55) with q_0 set to zero therefore produces the term

$$\Gamma_1^{(\ell=2)}(q_0 = 0) = \frac{e^2 n}{2mc^2}\int d^d x\,\mathbf{A}^2, \tag{6.62}$$

which exactly cancels the diamagnetic contribution (6.54).

For $q_0 \neq 0$, it turns out that the easiest way to compute the integrals is to evaluate the wave vector integral before the frequency integral. Replacing the integral over wave vectors by one over the energy ξ, and noting that the main contribution comes from momenta close to the Fermi surface, we find

$$I^{ij} \equiv \int \frac{d^3 k}{(2\pi)^D} S_F^{\square}(k)S_F^{\square}(\omega - q_0, \mathbf{k})k^i k^j \approx \frac{k_F^2}{D}\delta^{ij}\nu(0)I \tag{6.63}$$

with

$$I \equiv \int_{-\infty}^{\infty} d\xi \frac{1}{\hbar\omega - \xi + i\hbar\,\text{sgn}(\omega)/2\tau_F}\frac{1}{\hbar\omega - \hbar q_0 - \xi + i\hbar\,\text{sgn}(\omega - q_0)/2\tau_F}. \tag{6.64}$$

Here, the integration range is extended from $(-\mu, \infty)$ to $(-\infty, \infty)$. Because of the rapid convergence of the integral, this produces only a negligible error. The integral can be evaluated with the help of contour integration. When the two poles are on the same side of the real ξ axis, the integral is zero. Only when the two poles are on opposite sides, a finite result obtains, *viz.*

$$I = \frac{2\pi i}{\hbar}\frac{1}{|q_0| + i/\tau_F}\left[\theta(-\omega)\theta(\omega - q_0) + \theta(\omega)\theta(-\omega + q_0)\right], \tag{6.65}$$

where the products of two step functions derive from the constraints on the locations of the two poles for $\omega < 0$ and $\omega > 0$. Because of the step functions, the remaining frequency integral is trivial and gives for small q_0

$$\Gamma_1^{(\ell=2)}(q_0) = -\frac{e^2 n}{2mc^2} \int \mathrm{d}^d x \, \mathbf{A} \cdot \frac{|q_0|}{|q_0| + \mathrm{i}/\tau_\mathrm{F}} \mathbf{A}. \tag{6.66}$$

By virtue of the defining equation (6.40) of the conductivity, Eq. (6.66) leads to

$$\sigma(q_0) = \mathrm{i}\frac{e^2 n}{m} \frac{\mathrm{sgn}(q_0)}{|q_0| + \mathrm{i}/\tau_\mathrm{F}}. \tag{6.67}$$

In the DC limit $q_0 \to 0^+$, this reduces to the Drude formula $\sigma = e^2 n \tau_\mathrm{F}/m$, while in the opposite limit $q_0 \to \infty$, it reproduces the sum rule (6.51).

The expression for a nonuniform system reads for $q_0 > 0$

$$\sigma(q) = \frac{\mathrm{i}e^2 n}{m} \frac{1}{q_0 + \mathrm{i}/\tau_\mathrm{F} - c_\mathrm{s}^2 \mathbf{q}^2/q_0} \tag{6.68}$$

as can be inferred from the following heuristic argument. From the steps leading to Eq. (2.210) it was concluded that for a compressible system, the sum rules are exhausted by the spectrum $\omega^2(\mathbf{q}) = c_\mathrm{s}^2 \mathbf{q}^2$ of the associated sound wave, with c_s the speed of sound. The additional term in the denominator at the right side of Eq. (6.68) is obtained by replacing q_0 in Eq. (6.67) with $q_0 - c_\mathrm{s}^2 \mathbf{q}^2/q_0$ as prescribed by the sound wave spectrum. The conductivity in this way automatically satisfies the compressibility sum rule (6.52).

6.4 Kondo Effect

We next discuss how a magnetic impurity atom in a metal affects the conductivity. Kondo showed that when the lifetime of the conduction electrons is calculated beyond the Born approximation, logarithmic terms come in. This leads, in the case of antiferromagnetic coupling, to an increase of the resistance as the temperature is lowered, thus giving an explanation for the experimentally observed minimum in the resistance of dilute magnetic alloys, which until then was not understood.

The Kondo model for a single impurity fixed in space is specified by the Lagrangian density

$$\begin{aligned}\mathcal{L} = {} & \psi_\sigma^*(x)\left(\mathrm{i}\hbar\partial_t + \frac{\hbar^2}{2m}\nabla^2 + \mu\right)\psi_\sigma(x) + \chi_\tau^*(t)\mathrm{i}\hbar\partial_t\chi_\tau(t) \\ & - J\psi_\sigma^*(x)\frac{\sigma^i_{\sigma\sigma'}}{2}\psi_{\sigma'}(x)\chi_\tau^*(t)\frac{\tau^i_{\tau\tau'}}{2}\chi_{\tau'}(t),\end{aligned} \tag{6.69}$$

with the Grassmann fields $\psi_\sigma, \psi_\sigma^*, \sigma = \uparrow, \downarrow$ describing the conduction electrons, and $\chi_\tau, \chi_\tau^*, \tau = \uparrow, \downarrow$ describing the magnetic impurity atom, while $J\,(> 0)$ is the (antiferromagnetic) coupling constant. The fields χ_τ and χ_τ^* have no space dependence as the impurity is assumed to be fixed in space. The combination

$$\mathbf{s}(x) \equiv \psi_\sigma^*(x)\frac{\boldsymbol{\sigma}_{\sigma\sigma'}}{2}\psi_{\sigma'}(x), \tag{6.70}$$

with $\boldsymbol{\sigma}$ the Pauli matrices

$$\sigma^1 = \begin{pmatrix} 0 & 1 \\ 1 & 0 \end{pmatrix}, \quad \sigma^2 = \begin{pmatrix} 0 & -\mathrm{i} \\ \mathrm{i} & 0 \end{pmatrix}, \quad \sigma^3 = \begin{pmatrix} 1 & 0 \\ 0 & -1 \end{pmatrix}, \tag{6.71}$$

denotes the spin density of the conduction electrons (in units of $\hbar$), while a similar definition with ψ and $\boldsymbol{\sigma}$ replaced with χ and $\boldsymbol{\tau}$, which are Pauli matrices too, describes the spin of the impurity. The restriction to a single impurity is justified because, as in the preceding section, the important contributions derive form repeated scattering by the same impurity. We study the model in perturbation theory at finite temperatures.

The finite-temperature Feynman rules are given by

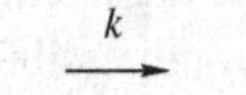

$$= \frac{\hbar}{-\mathrm{i}\hbar\omega_n + \xi(\mathbf{k})}, \tag{6.72}$$

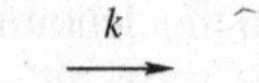

$$= \frac{\hbar}{-\mathrm{i}\hbar\omega_n}, \tag{6.73}$$

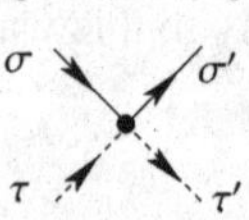

$$= -\frac{J}{\hbar}\frac{\sigma^i_{\sigma\sigma'}}{2}\frac{\tau^i_{\tau\tau'}}{2}, \tag{6.74}$$

where the conservation of frequency and wave vector at the vertex is understood. A fermion loop carries a minus sign and each internal line carries a sum over Matsubara frequencies and an integral over wave vectors:

$$\frac{1}{\hbar\beta}\sum_{n=-\infty}^{\infty}\int\frac{\mathrm{d}^3k}{(2\pi)^3}. \tag{6.75}$$

The Feynman diagram of interest to us is the one-loop correction to the vertex:

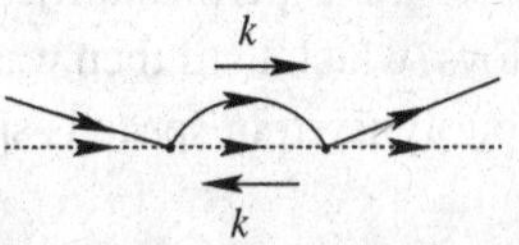

$$= -\frac{1}{\hbar}\Gamma_{\sigma\sigma',\tau\tau'}(0) \tag{6.76}$$

at zero frequency and wave-vector transfer. The diagram physically describes how an incoming conduction electron is scattered twice by the fixed magnetic impurity. In each scattering event, the spins can be flipped. The diagram stands for

$$-\frac{1}{\hbar}\Gamma_{\sigma\sigma',\tau\tau'}(0) = \frac{1}{16}\frac{J^2}{\hbar^2}(\sigma^i\sigma^j)_{\sigma\sigma'}(\tau^i\tau^j)_{\tau\tau'}\frac{1}{\hbar\beta}\sum_n\int\frac{\mathrm{d}^3k}{(2\pi)^3}\frac{\hbar}{-\mathrm{i}\hbar\omega_n + \xi(\mathbf{k})}\frac{\hbar}{\mathrm{i}\hbar\omega_n}. \tag{6.77}$$

Using the identity $\sigma^i\sigma^j = \delta^{ij} + i\epsilon^{ijk}\sigma^k$, we obtain for the spin part of the diagram

$$(\sigma^i\sigma^j)_{\sigma\sigma'}(\tau^i\tau^j)_{\tau\tau'} = 3\delta_{\sigma\sigma'}\delta_{\tau\tau'} - 2\sigma^i_{\sigma\sigma'}\tau^i_{\tau\tau'}. \tag{6.78}$$

To carry out the wave vector integral, we write

$$\int \frac{d^3k}{(2\pi)^3} \frac{1}{-i\hbar\omega_n + \xi(k)} = \int d\xi\, \nu(\xi) \frac{1}{-i\hbar\omega_n + \xi}, \tag{6.79}$$

with $\nu(\xi)$ the density of states, and take a simple Lorentzian of width $2D$ for ν:

$$\nu(\xi) = \frac{1}{\pi}\frac{1}{V}\frac{D}{\xi^2 + D^2}. \tag{6.80}$$

Here, the energy scale D is of order the Fermi energy. The resulting expression is

$$\Gamma_{\sigma\sigma',\tau\tau'}(0) = \frac{J^2}{8}\left(3\delta_{\sigma\sigma'}\delta_{\tau\tau'} - 2\sigma^i_{\sigma\sigma'}\tau^i_{\tau\tau'}\right)\frac{1}{\beta}\sum_{n=0}^{\infty}\frac{1}{\hbar\omega_n + D}\frac{1}{\hbar\omega_n}. \tag{6.81}$$

The sum over the Matsubara frequencies (2.47), $\omega_n = \pi(2n+1)/\hbar\beta$, can be evaluated in closed form in terms of the digamma function $\psi(z)$,

$$\psi(z) = -\gamma - \frac{1}{z} + \sum_{m=1}^{\infty}\left(\frac{1}{m} - \frac{1}{m+z}\right), \tag{6.82}$$

with the special value $\psi(\frac{1}{2}) = -\gamma - 2\ln(2)$, as

$$\Gamma_{\sigma\sigma',\tau\tau'}(0) = -\frac{\lambda^2}{16\nu(0)}\left(3\delta_{\sigma\sigma'}\delta_{\tau\tau'} - 2\sigma^i_{\sigma\sigma'}\tau^i_{\tau\tau'}\right)\left[\psi\left(\frac{1}{2}\right) - \psi\left(\frac{1}{2} + \frac{D}{2\pi k_B T}\right)\right] \tag{6.83}$$

where $\lambda \equiv \nu(0)J$, with $\nu(0) \equiv 1/\pi V D$ the density of states at the Fermi surface for which $\xi = 0$, denotes the dimensionless Kondo coupling constant. By virtue of the asymptotic form of the digamma function,

$$\lim_{x\to\infty}\psi(x) = \ln(x) - \frac{1}{2x} + O\left(\frac{1}{x^2}\right), \tag{6.84}$$

the right side of Eq. (6.83) diverges as $\ln(D/k_B T)$ in the low-temperature limit.

Since $\Gamma_{\sigma\sigma',\tau\tau'}(0)$ contains a contribution proportional to $\sigma^i\tau^i$, we conclude that it leads to a dressing, or renormalization of the bare coupling constant λ

$$\begin{aligned}\lambda_r &= \lambda\left\{1 + \frac{\lambda}{2}\left[\psi\left(\frac{1}{2}\right) - \psi\left(\frac{1}{2} + \frac{D}{2\pi k_B T}\right)\right]\right\} \\ &\approx \lambda\left[1 - \frac{\lambda}{2}\ln\left(\frac{2e^\gamma D}{\pi k_B T}\right)\right],\end{aligned} \tag{6.85}$$

where the approximation is valid for $D/k_B T >> 1$. The correction term becomes of the same order as the leading term when T reaches the Kondo temperature T_K,

$$k_B T_K \equiv D e^{-2/\lambda}. \tag{6.86}$$

At this temperature, perturbation theory breaks down.

The inverse lifetime τ_F^{-1} of electrons at the Fermi surface is given by the imaginary part of the self-energy diagram, cf. Eq. (6.29):

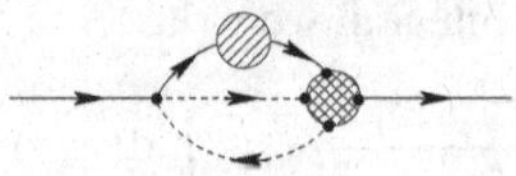

which is proportional to $\lambda\lambda_r$. By Eq. (6.67) it follows that, to this order, the resistance behaves as:

$$\sigma^{-1} \sim \lambda_r^2 \left[1 + \frac{\lambda_r}{2} \ln\left(\frac{D}{k_B T} \right) \right], \tag{6.87}$$

where irrelevant constants have been suppressed. This is Kondo's result explaining the experimentally observed $\ln(T)$ behavior of the resistance, known as the *Kondo effect*.

To study this further we apply renormalization group theory by considering the effect of changing the bandwidth D introduced in Eq. (6.80). Taking the derivative of the renormalized coupling strength (6.85) with respect to D at fixed bare coupling λ, we find

$$D\frac{\partial \lambda_r}{\partial D} = -\frac{\lambda_r^2}{2}, \tag{6.88}$$

showing how the coupling $\lambda_r(D)$ must scale with D so as to leave the bare coupling constant λ unchanged. Equation (6.88) is readily integrated, with the result, for small coupling,

$$\frac{1}{\lambda_r(D')} - \frac{1}{\lambda_r(D)} = \frac{1}{2} \ln\left(\frac{D'}{D} \right). \tag{6.89}$$

The sign on the right is opposite to the sign found in QED, see Eq. (3.146). Because of this, the coupling strength $\lambda_r(D')$ *decreases* with increasing bandwidth D'. In fact, since $\lambda_r(D')$ tends to zero as $D' \to \infty$, the theory is *asymptotically free*—a property it shares with quantum chromodynamics (QCD), the theory of the strong nuclear interactions. At high energies, the perturbative approach is therefore perfectly valid.

The situation at low energies is completely different. As D' decreases, the coupling strength $\lambda_r(D')$ increases. At the energy scale $D' = k_B T_K$, with T_K the Kondo temperature (6.86), the perturbative result (6.89) diverges for antiferromagnetic coupling ($J > 0$). In this strong-coupling regime, the theory shows *confinement* in the sense that the impurity atom forms a spin singlet with the spins of the surrounding conduction electrons. The spin of the impurity atom is thus completely screened and scattering is as for a nonmagnetic impurity. For $T < T_K$ the resistance is indeed experimentally found to saturate. Confinement at low energies, which is sometimes referred to as *infrared slavery*, is the flip side of asymptotic freedom at high energies. At strong coupling, perturbation theory breaks down, of course, and qualitatively different methods are needed in this regime.

6.5 Degenerate Electron Gas

We next consider an electron gas at the absolute zero of temperature in the presence of a uniform, positively charged background so that the system is overall charge neutral. The electrons are described by the Lagrangian (6.1) with

$$V(\mathbf{x}-\mathbf{x}') = \frac{e^2}{|\mathbf{x}-\mathbf{x}'|} \tag{6.90}$$

representing the instantaneous Coulomb repulsion. With the help of a Hubbard-Stratonovich transformation, and the observation that

$$G(\mathbf{x}) \equiv \frac{1}{4\pi}\frac{1}{r} = \int \frac{\mathrm{d}^3k}{(2\pi)^3}\frac{\mathrm{e}^{\mathrm{i}\mathbf{k}\cdot\mathbf{x}}}{\mathbf{k}^2}, \tag{6.91}$$

with $r \equiv |\mathbf{x}|$, is the Green function of the three-dimensional Laplace operator $-\nabla^2 G(\mathbf{x}) = \delta(\mathbf{x})$, the theory can be rewritten as a gauge theory

$$\mathcal{L} = \psi_\sigma^*(x)\left[\mathrm{i}\hbar\partial_t + \frac{\hbar^2}{2m}\nabla^2 + \mu + e\Phi(x)\right]\psi_\sigma(x) + \frac{1}{8\pi}[\nabla\Phi(x)]^2, \tag{6.92}$$

involving only the scalar potential $\Phi(x)$. Because the Coulomb potential is instantaneous only spatial derivatives of Φ arise. Its field equation,

$$\nabla^2\Phi(x) = 4\pi e\, j_\mathrm{e}^0(x), \tag{6.93}$$

with $j_\mathrm{e}^0 = \psi_\sigma^*\psi_\sigma$ the electron number density, is Gauss' law.

Formulated as a gauge theory, the partition function,

$$Z = \int \mathrm{D}\Phi\,\mathrm{D}\psi^*\mathrm{D}\psi\,\mathrm{e}^{(\mathrm{i}/\hbar)S}, \tag{6.94}$$

involves in addition to the functional integral over the Grassmann fields also a functional integral over the scalar potential. When carrying out the Gaussian integral over Φ, thereby undoing the Hubbard-Stratonovich transformation, we recover the original Lagrangian Eq. (6.1) with the Coulomb potential (6.90).

The advantage of the gauge formulation is that the functional integral over the electron fields becomes Gaussian and can be formally evaluated to give the effective action

$$\begin{aligned}\Gamma[\Phi_\mathrm{c}] &= -\mathrm{i}\hbar\,\mathrm{Tr}\ln[1 + eS_\mathrm{F}\Phi_\mathrm{c}(x)] + \frac{1}{8\pi}\int \mathrm{d}^4x[\nabla\Phi_\mathrm{c}(x)]^2 \\ &= \mathrm{i}\hbar\,\mathrm{Tr}\sum_{\ell=1}^{\infty}\frac{1}{\ell}[-eS_\mathrm{F}\Phi_\mathrm{c}(x)]^\ell + \frac{1}{8\pi}\int \mathrm{d}^4x[\nabla\Phi_\mathrm{c}(x)]^2\end{aligned} \tag{6.95}$$

as follows from Eqs. (2.137) and (2.138) with A replaced with $e\Phi_c$. Here, Φ_c denotes the average of the fluctuating scalar potential. The first term in the expansion gives by Eq. (3.13)

$$\Gamma^{(\ell=1)} = en \int d^4 x \Phi_c(x). \tag{6.96}$$

This term is canceled by the positively charged background. As a shortcut, the same result can be obtained by simply setting the spatial average of the Coulomb potential to zero. In terms of Fourier components

$$\Phi_c(\mathbf{k}) = \int d^3 x\, \Phi_c(\mathbf{x}) e^{-i\mathbf{k}\cdot\mathbf{x}} \tag{6.97}$$

this is equivalent to taking

$$\Phi_c(\mathbf{k} = 0) = 0. \tag{6.98}$$

We will in the following adopt this shortcut.

The terms quadratic in the scalar potential read, cf. Eq. (6.55),

$$\Gamma^{(\ell=2)} = \frac{i}{2}\hbar e^2 g \int d^4 x \frac{d^4 k}{(2\pi)^d} \Phi_c(x) S_F(k) S_F(k - i\partial)\Phi_c(x) + \frac{1}{8\pi}\int d^4 x [\nabla\Phi_c(x)]^2, \tag{6.99}$$

where $g = 2$ counts the number of internal degrees of freedom. It derives from the sum over two spin directions in the closed loop. This contribution can formally be rewritten in terms of the polarization tensor π^{00} introduced in Eq. (2.186) as

$$\Gamma^{(\ell=2)} = \int \frac{d^4 q}{(2\pi)^4} \Phi_c(-q) \left[\frac{\mathbf{q}^2}{8\pi} + \frac{1}{2} e^2 \pi^{00}(q)\right] \Phi_c(q) \tag{6.100}$$

after Fourier transforming, where $q = (q_0, \mathbf{q})$. The change $q \to -q$ of the integration variable shows that the polarization tensor is even in q, $\pi^{00}(-q) = \pi^{00}(q)$. By Eq. (6.99), $\pi^{00}(q)$ reads explicitly in lowest order

$$\pi^{00}(q) = ig\hbar \int \frac{d^4 k}{(2\pi)^4} S_F(k) S_F(k+q). \tag{6.101}$$

According to the general definition of the Feynman propagator in Sec. 2.4, the expression in square brackets in the effective action (6.99) defines the Feynman propagator $D'_F(q)$ of the scalar potential through

$$D'^{-1}_F(q) = \frac{\mathbf{q}^2}{4\pi} + e^2 \pi^{00}(q). \tag{6.102}$$

A relation of this type, involving the propagators of the interacting and noninteracting theory, is known as a *Dyson equation*. It can be equivalently stated as

$$D'_F(q) = \frac{4\pi}{\mathbf{q}^2} - \frac{4\pi e^2}{\mathbf{q}^2} \pi^{00}(q) D'_F(q), \tag{6.103}$$

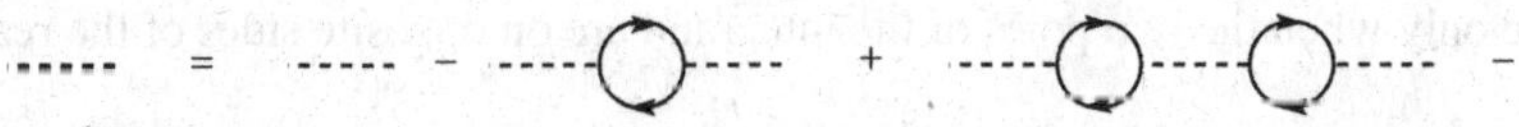

Fig. 6.2 The infinite series of Feynman diagrams leading to the dressed propagator (6.104) of the scalar potential. The heavy dashed line denotes the dressed propagator $D'_F(q)$ and the light dashed line denotes the bare propagator $4\pi/\mathbf{q}^2$. The bubble stands for $\Pi^{00}(q)$. The minus signs arise from the fermion loops.

with the solution

$$D'_F(q) = \frac{4\pi/\mathbf{q}^2}{1 + (4\pi e^2/\mathbf{q}^2)\pi^{00}(q)}. \tag{6.104}$$

This dressed propagator is the sum of an infinite series of Feynman diagrams (see Fig. 6.2), which form, as in the large-N expansion discussed in Sec. 4.11, a geometric series,

$$D'_F(q) = \frac{4\pi}{\mathbf{q}^2} - \left(\frac{4\pi e^2}{\mathbf{q}^2}\right)^2 \pi^{00}(q) + \left(\frac{4\pi e^2}{\mathbf{q}^2}\right)^3 \left[\pi^{00}(q)\right]^2 + \cdots. \tag{6.105}$$

The approximation where only bubble or ring diagrams are considered is referred to in condensed matter theory as the *random phase approximation*.

The solution (6.104) can be conveniently written in terms of the dielectric function $\epsilon(q)$ as

$$D'_F(q) = \frac{1}{\epsilon(q)} \frac{4\pi}{\mathbf{q}^2}, \tag{6.106}$$

with

$$\epsilon(q) \equiv 1 + \frac{4\pi e^2}{\mathbf{q}^2}\pi^{00}(q) \tag{6.107}$$

the expression for the dielectric function in the random phase approximation. The spectrum $\omega(\mathbf{q})$ of the excitation described by the propagator $D'_F(q)$ follows from setting the denominator to zero:

$$\epsilon[\omega(\mathbf{q}), \mathbf{q}] = 0. \tag{6.108}$$

By Eq. (6.50), the dielectric function can be equivalently expressed in terms of the conductivity as

$$\epsilon(q) = 1 + 4\pi \mathrm{i}\frac{\sigma(q)}{q_0}. \tag{6.109}$$

We next compute $\pi^{00}(q)$ explicitly in lowest order, starting with the frequency integral. Given the form (6.5) of the propagator, four cases need to be considered, depending on whether $\xi(\mathbf{k})$ and $\xi(\mathbf{k} + \mathbf{q})$ are positive or negative. The integral is finite only when the two poles of the integrand are on opposite sides of the real ω axis:

$$\pi^{00}(q) = -g \int \frac{d^3k}{(2\pi)^3} \left[\frac{\theta(k_F - |\mathbf{k}|)\theta(|\mathbf{k}+\mathbf{q}| - k_F)}{\hbar q_0 + \xi(\mathbf{k}) - \xi(\mathbf{k}+\mathbf{q}) + i\eta} - \frac{\theta(|\mathbf{k}| - k_F)\theta(k_F - |\mathbf{k}+\mathbf{q}|)}{\hbar q_0 + \xi(\mathbf{k}) - \xi(\mathbf{k}+\mathbf{q}) - i\eta} \right], \quad (6.110)$$

where the step functions derive from the constraints on the locations of the two poles. By letting $\mathbf{k} \to -\mathbf{k} - \mathbf{q}$ in the second term, the result can be rewritten as

$$\pi^{00}(q) = -g \int \frac{d^3k}{(2\pi)^3} \theta(k_F - |\mathbf{k}|)\theta(|\mathbf{k}+\mathbf{q}| - k_F) \times \left(\frac{1}{\hbar q_0 + \xi(\mathbf{k}) - \xi(\mathbf{k}+\mathbf{q}) + i\eta} - \frac{1}{\hbar q_0 + \xi(\mathbf{k}+\mathbf{q}) - \xi(\mathbf{k}) - i\eta} \right), \quad (6.111)$$

showing that $\pi^{00}(q)$ is even in q_0. Without loss of generality, we therefore assume $q_0 > 0$.

The wave vector integrals can be best evaluated for the real and imaginary part of $\pi^{00}(q)$ separately. Using the identity (2.207) and the property $\theta(-x) = 1 - \theta(x)$, we find for the real part

$$\mathrm{Re}\,\pi^{00}(q) = -\mathrm{P}\,g \int \frac{d^3k}{(2\pi)^3} \theta(k_F - |\mathbf{k}|)\,[1 - \theta(k_F - |\mathbf{k}+\mathbf{q}|)] \times \frac{2[\xi(\mathbf{k}+\mathbf{q}) - \xi(\mathbf{k})]}{(\hbar q_0)^2 - [\xi(\mathbf{k}+\mathbf{q}) - \xi(\mathbf{k})]^2}. \quad (6.112)$$

Since the product of the two step functions is symmetric under the interchange of $\mathbf{k}$ and $\mathbf{k} + \mathbf{q}$, while the last factor in the integrand is antisymmetric under this interchange, that contribution vanishes, and

$$\mathrm{Re}\,\pi^{00}(q) = -\mathrm{P}\,g \int \frac{d^3k}{(2\pi)^3} \theta(k_F - |\mathbf{k}|) \times \left(\frac{1}{\hbar q_0 + \xi(\mathbf{k}) - \xi(\mathbf{k}+\mathbf{q})} - \frac{1}{\hbar q_0 + \xi(\mathbf{k}+\mathbf{q}) - \xi(\mathbf{k})} \right), \quad (6.113)$$

where the second term follows from the first by letting $q_0 \to -q_0$. The remaining integrals are elementary and yield the *Lindhard function*

$$\mathrm{Re}\,\pi^{00}(q) = \frac{1}{4} g\nu(0) \left[1 + \frac{k_F}{|\mathbf{q}|} \left(1 - \frac{(q_0 + \hbar \mathbf{q}^2/2m)^2}{v_F^2 \mathbf{q}^2} \right) \times \ln\left(\left| \frac{q_0 + v_F|\mathbf{q}| + \hbar\mathbf{q}^2/2m}{q_0 - v_F|\mathbf{q}| + \hbar\mathbf{q}^2/2m} \right| \right) + (q_0 \to -q_0) \right], \quad (6.114)$$

with $v_{\rm F} \equiv \hbar k_{\rm F}/m$ the Fermi velocity and $\nu(0)$ the density of states (6.60) per spin degree of freedom in $D = 3$.

The imaginary part of the polarization tensor can be inferred from the real part by recalling that for a complex number $z = |z|e^{i\theta}$,

$$\ln(z) = \ln(|z|) + i\theta. \tag{6.115}$$

Each of the four logarithms in Eq. (6.114) is thus accompanied by a term $i\pi$ whenever the argument turns negative. This happens when

$$q_0 = r v_{\rm F}|\mathbf{q}| + s\hbar\mathbf{q}^2/2m \tag{6.116}$$

with $r, s = \pm 1$. By inspection, we in this way find for the imaginary part

$$\mathrm{Im}\,\pi^{00}(q) = \frac{\pi}{4} g\nu(0)\frac{k_{\rm F}}{|\mathbf{q}|}\begin{cases} 2q_0/k_{\rm F}v_{\rm F}, & q_0 \leq +v_{\rm F}|\mathbf{q}| - \hbar\mathbf{q}^2/2m \\ 1 - (q_0 - \hbar\mathbf{q}^2/2m)^2/v_{\rm F}^2\mathbf{q}^2, & \\ \qquad |v_{\rm F}|\mathbf{q}| - \hbar\mathbf{q}^2/2m| \leq q_0 \leq +v_{\rm F}|\mathbf{q}| + \hbar\mathbf{q}^2/2m & \\ 0, & q_0 \geq +v_{\rm F}|\mathbf{q}| + \hbar\mathbf{q}^2/2m \\ 0, & q_0 \leq -v_{\rm F}|\mathbf{q}| + \hbar\mathbf{q}^2/2m \end{cases} \tag{6.117}$$

where the absolute value in the second condition is necessary for distinguishing the cases $|\mathbf{q}| < 2k_{\rm F}$ and $|\mathbf{q}| > 2k_{\rm F}$. Remember that, without loss of generality, we assume that $q_0 > 0$. The first condition, therefore, implies $|\mathbf{q}| < 2k_{\rm F}$, while the last one implies $|\mathbf{q}| > 2k_{\rm F}$.

In the static limit, $q_0 \to 0$, the Lindhard function is real and Eq. (6.114) reduces to

$$\begin{aligned} \pi^{00}(\mathbf{q}) &= \frac{1}{2} g\nu(0)\left[1 + \frac{k_{\rm F}}{|\mathbf{q}|}\left(1 - \frac{\mathbf{q}^2}{4k_{\rm F}^2}\right)\ln\left(\frac{|\mathbf{q}| + 2k_{\rm F}}{||\mathbf{q}| - 2k_{\rm F}|}\right)\right] \\ &= g\nu(0)\left(1 - \frac{1}{12}\frac{\mathbf{q}^2}{k_{\rm F}^2}\right) + \mathcal{O}\left(\mathbf{q}^4\right). \end{aligned} \tag{6.118}$$

The result in the limit $\mathbf{q} \to 0$ is in agreement with the compressibility sum rule (2.204) for an ideal Fermi gas, with $c_{\rm s}^2 = \frac{1}{3}v_{\rm F}^2$ the speed of sound squared. With only the first term of the series included, the dielectric function (6.107) becomes

$$\lim_{\mathbf{q}\to 0} \epsilon(\mathbf{q}) = 1 + q_{\rm FT}^2/\mathbf{q}^2, \tag{6.119}$$

with $q_{\rm FT}$ the *Thomas-Fermi* wave number given by

$$q_{\rm FT}^2 \equiv 4\pi e^2 g\nu(0), \tag{6.120}$$

and the propagator (6.106) takes the simple form

$$D_{\rm F}'(\mathbf{q}) = \frac{4\pi}{\mathbf{q}^2 + q_{\rm FT}^2}. \tag{6.121}$$

This implies that in a degenerate electron gas, the Coulomb potential is short range

$$D'_{\rm F}(\mathbf{x}) = \frac{e^{-q_{\rm TF} r}}{r}, \tag{6.122}$$

falling off exponentially fast at distances $r \equiv |\mathbf{x}|$ larger than $1/q_{\rm TF}$. Physically, when a negatively charged test particle is inserted into the system, it will push the surrounding electrons away. Because of the presence of the positively charged uniform background, this effectively results in a positively charged screening cloud around the test particle. From distances larger than the screening length $\lambda \sim 1/q_{\rm TF}$, the test particle appears neutral as its Coulomb potential is exponentially suppressed. Since the dielectric function diverges in the long wave-length limit $\mathbf{q} \to 0$,

$$\lim_{\mathbf{q}\to 0} \frac{1}{\epsilon(\mathbf{q})} = 0, \tag{6.123}$$

which implies that the screening is perfect.

Keeping q_0 fixed and letting $\mathbf{q} \to 0$, we find in accord with the f sum rule (2.198)

$$\lim_{\mathbf{q}\to 0} \pi^{00}(q) = -\frac{n}{m}\frac{\mathbf{q}^2}{q_0^2}\left(1 + \frac{3}{5}\frac{v_{\rm F}^2\mathbf{q}^2}{q_0^2}\right), \tag{6.124}$$

where $n = k_{\rm F}^3/3\pi^2$ denotes the average particle number density. With this expression, the dielectric function (6.107) becomes

$$\lim_{\mathbf{q}\to 0} \epsilon(q) = 1 - \frac{\omega_{\rm p}^2}{q_0^2}\left(1 + \frac{3}{5}\frac{v_{\rm F}^2\mathbf{q}^2}{q_0^2}\right), \tag{6.125}$$

with $\omega_{\rm p}$ the plasma frequency

$$\omega_{\rm p}^2 \equiv \frac{4\pi e^2 n}{m}. \tag{6.126}$$

By Eq. (6.108), this result implies the spectrum

$$\omega^2(\mathbf{q}) = \omega_{\rm p}^2 + \frac{3}{5} v_{\rm F}^2 \mathbf{q}^2, \tag{6.127}$$

valid for small wave number. This plasma mode physically represents undamped oscillations in the charge density which are intimately connected to charge screening. When electrons are pushed away from the negatively charged test particle, they will, in general, overshoot. As a result, they will be pulled back towards the region they came from, overshoot again, etc. In this way, the charge disturbance puts its surrounding electrons in simple harmonic motion with frequency $\omega(\mathbf{q})$. The plasma frequency denotes the lowest attainable frequency, $\lim_{\mathbf{q}\to 0} \omega(\mathbf{q}) = \omega_{\rm p}$. The estimate (6.127) of the plasma dispersion coefficient multiplying $\mathbf{q}^2$, which is based on the Lindhard function, i.e., the lowest-order expression for π^{00} and the random phase approximation, compares unexpectedly well with experimental values for simple metals.

To bring out the close connection between screening and plasma oscillations, we express the inverse screening length $q_{\rm FT}$ in terms of the plasma frequency as

$$q_{\rm FT} = \omega_{\rm p}/c_{\rm s}, \tag{6.128}$$

where $c_{\rm s} = v_{\rm F}/\sqrt{3}$ is the speed of sound of an ideal Fermi gas obtained from the degenerate electron gas by setting $e \to 0$. The result (6.128) is quite general and remains valid even for mutually interacting electrons, provided $c_{\rm s}$ denotes the true speed of sound of the corresponding neutral system obtained by letting $e \to 0$. This follows from the compressibility sum rule (2.204) and the definition (6.107) of the dielectric function, which imply

$$\lim_{\mathbf{q}\to 0} \epsilon(\mathbf{q}) = 1 + \frac{4\pi e^2}{\mathbf{q}^2} n^2 \kappa = 1 + \frac{\omega_{\rm p}^2}{c_{\rm s}^2 \mathbf{q}^2}, \tag{6.129}$$

with κ the compressibility.

The relation (6.128) can, incidentally, also be applied to a classical Coulomb plasma. With the speed of sound (2.205) of a perfect neutral classical gas, it gives the inverse of the Debye screening length

$$q_{\rm D}^2 = \frac{4\pi e^2 n}{k_{\rm B} T}, \tag{6.130}$$

and Eq. (6.129) reduces to the Debye-Hückel form

$$\lim_{\mathbf{q}\to 0} \epsilon(\mathbf{q}) = 1 + \frac{4\pi e^2 n}{k_{\rm B} T} \frac{1}{\mathbf{q}^2}. \tag{6.131}$$

To further underscore the close connection between screening and plasma oscillations, we write the dielectric function in terms of the spectral representation (2.206) of the polarization tensor as

$$\epsilon(q) = 1 - \frac{4\pi e^2}{\mathbf{q}^2} \frac{1}{\hbar} \int_0^\infty {\rm d}\omega' \left(\frac{1}{q_0 - \omega' + i\eta} - \frac{1}{q_0 + \omega' + i\eta} \right) S(\omega', \mathbf{q}). \tag{6.132}$$

Then, with the approximation

$$\frac{1}{\epsilon(\mathbf{q})} \approx 1 - \frac{4\pi e^2}{\mathbf{q}^2} \pi^{00}(\mathbf{q}), \tag{6.133}$$

the condition (6.123) for perfect screening leads to the sum rule

$$\lim_{\mathbf{q}\to 0} \int_0^\infty {\rm d}q_0 \frac{S(q)}{\hbar q_0} = \frac{\mathbf{q}^2}{8\pi e^2}, \tag{6.134}$$

which is to be compared to the corresponding sum rule (2.209) for a neutral system. The second sum rule satisfied by the dynamic structure factor $S(q)$ of the neutral system, the f sum rule, also applies here:

$$\int_0^\infty \mathrm{d}q_0 \hbar q_0 S(q) = \frac{n\hbar^2 \mathbf{q}^2}{2m}. \tag{6.135}$$

Both sum rules are exhausted by the plasma mode in the long wave-length limit,

$$\lim_{\mathbf{q}\to 0} S(q) = \frac{n\hbar \mathbf{q}^2}{2m\omega_\mathrm{p}} \delta(q_0 - \omega_\mathrm{p}), \tag{6.136}$$

in the same way that the corresponding sum rules for the neutral system are exhausted by the gapless sound wave. For the static structure factor (2.212) this implies that

$$\lim_{\mathbf{q}\to 0} S(\mathbf{q}) = \frac{1}{n}\int_0^\infty \mathrm{d}q_0 S(q) = \frac{\hbar \mathbf{q}^2}{2m\omega_\mathrm{p}}. \tag{6.137}$$

In the charged system, the sound mode of the neutral Fermi system is replaced by the plasma mode. The main difference between these two types of collective modes is that whereas sound waves are gapless, plasma oscillations have a minimal frequency. The loss of the gapless mode is accompanied by the change of the bare long-range Coulomb potential into a short-range interaction.

6.6 Thermodynamic Potential

We next perturbatively compute the thermodynamic potential of a degenerate electron gas. The Lagrangian (6.92) yields the following zero-temperature Feynman rules:

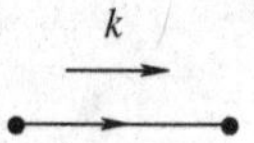

$$= \mathrm{i}\hbar S_\mathrm{F}(k), \tag{6.138}$$

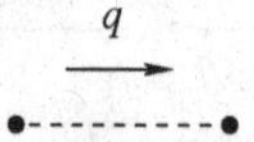

$$= \mathrm{i}\hbar D_\mathrm{F}(q), \tag{6.139}$$

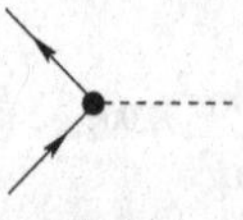

$$= \mathrm{i}\frac{e}{\hbar}, \tag{6.140}$$

with $D_\mathrm{F}(q) = 4\pi/\mathbf{q}^2$. It is understood that four-momentum is conserved at the vertex. Each fermion loop carries a minus sign and with each internal line is associated an integral over frequency and wave vectors.

The first contribution to the thermodynamic potential is identical to that of an ideal Fermi gas given in Eq. (2.24) for finite T. Introducing the density of states (2.28), we can write this one-loop contribution as

$$\begin{aligned}\Omega_1 &= -\frac{V}{\beta}g\int \mathrm{d}\epsilon\, \nu(\epsilon)\ln\left(1+\mathrm{e}^{-\beta(\epsilon-\mu)}\right)\\ &= -\frac{2}{3}\frac{V}{\beta}g\int \mathrm{d}\epsilon\, \nu(\epsilon)\epsilon\frac{1}{\mathrm{e}^{\beta(\epsilon-\mu)}+1},\end{aligned} \tag{6.141}$$

where the prefactor $g = 2$ arises from the sum over two spin directions in the closed loop, while the last line follows by integrating by parts. It shows that this contribution represents the kinetic energy of the Fermi sea. In the limit $T \to 0$, this contribution reduces to

$$\begin{aligned}\Omega_1 &= -\frac{2}{3}Vg\int\frac{\mathrm{d}^3k}{(2\pi)^3}\frac{\hbar^2\mathbf{k}^2}{2m}\theta(|\mathbf{k}|-k_\mathrm{F})\\ &= -\frac{2}{5}\mu\langle N\rangle,\end{aligned} \tag{6.142}$$

where in the first line the wave vector integrals are reinstalled. The result (6.142) is in agreement with Eq. (6.10), which was obtained directly at $T = 0$. For the ground-state energy $\langle E\rangle$, which is related to the thermodynamic potential through Eq. (2.20), this result gives at the absolute zero of temperature

$$\frac{\langle E_1\rangle}{\langle N\rangle} = \frac{3}{5}\mu \tag{6.143}$$

for the average energy per particle, with N denoting the total particle number $N = N_\uparrow + N_\downarrow$.

The two-loop contributions are given by the Feynman diagrams in Fig. 6.1. The direct term is zero by Eq. (6.98). The exchange term gives the two-loop contribution

$$-\frac{\mathrm{i}}{\hbar}\frac{\Omega_2^\mathrm{x}}{V} = -\frac{g}{2}\left(\frac{\mathrm{i}e}{\hbar}\right)^2\int\frac{\mathrm{d}^4k}{(2\pi)^4}\frac{\mathrm{d}^4q}{(2\pi)^4}\mathrm{i}\hbar\frac{4\pi}{|\mathbf{k}-\mathbf{q}|^2}\mathrm{i}\hbar S_\mathrm{F}(k)\mathrm{i}\hbar S_\mathrm{F}(q), \tag{6.144}$$

where the factor of 2 is the symmetry factor of the diagram. With the convergence factors $\exp(\mathrm{i}\omega 0^+)$ included in the propagators, the frequency integrals produce by Eq. (3.14) step functions,

$$\frac{\Omega_2^\mathrm{x}}{V} = -\frac{g}{2}4\pi e^2\int\frac{\mathrm{d}^3k}{(2\pi)^3}\frac{\mathrm{d}^3q}{(2\pi)^3}\frac{1}{|\mathbf{k}-\mathbf{q}|^2}\theta(k_\mathrm{F}-|\mathbf{k}|)\theta(k_\mathrm{F}-|\mathbf{q}|). \tag{6.145}$$

Rescaling the wave vectors by k_F and denoting the angle between the vectors $\mathbf{k}$

and $\mathbf{q}$ by θ, we obtain

$$\begin{aligned}\frac{\Omega_2^{\mathrm{x}}}{V} &= -\frac{g}{2}4\pi e^2 k_{\mathrm{F}}^4 \frac{4\pi}{(2\pi)^3}\int_0^1 \mathrm{d}|\mathbf{k}|\,\mathbf{k}^2 \frac{2\pi}{(2\pi)^3}\int_0^1 \mathrm{d}|\mathbf{q}|\,\mathbf{q}^2 \\ &\quad\times \int_0^{\pi} \mathrm{d}\theta \sin\theta \frac{1}{\mathbf{k}^2+\mathbf{q}^2-2|\mathbf{k}||\mathbf{q}|\cos\theta} \\ &= -\frac{g}{2}4\pi e^2 k_{\mathrm{F}}^4 \frac{1}{8\pi^4}\int_0^1 \mathrm{d}|\mathbf{k}| \int_0^1 \mathrm{d}|\mathbf{q}|\,|\mathbf{k}||\mathbf{q}| \ln\left(\frac{|\mathbf{k}|+|\mathbf{q}|}{||\mathbf{k}|-|\mathbf{q}||}\right) \\ &= -\frac{g}{2}\frac{1}{4\pi^3}e^2 k_{\mathrm{F}}^4,\end{aligned} \tag{6.146}$$

where in the last step it is used that

$$\int_0^1 \mathrm{d}k \int_0^1 \mathrm{d}q\, kq \ln\left(\frac{k+q}{|k-q|}\right) = \frac{1}{2}. \tag{6.147}$$

As next contribution to the thermodynamic potential we consider the ring or bubble diagram shown in Fig. 6.3. which forms the second diagram in an infinite

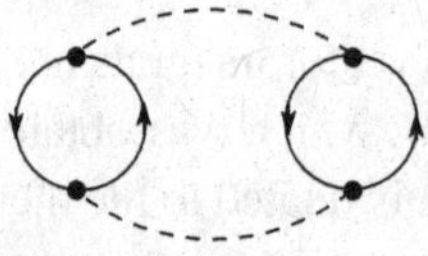

Fig. 6.3 The three-loop ring or bubble diagram.

series started by the exchange diagram. Consecutive terms in the series have one additional bubble insertion. The three-loop diagram (6.3) gives

$$-\frac{\mathrm{i}}{\hbar}\frac{\Omega_3^{\mathrm{r}}}{V} = \frac{1}{4}\int \frac{\mathrm{d}^4 q}{(2\pi)^4}\left[\mathrm{i}\hbar \frac{4\pi\,\mathrm{i}e^2}{\mathbf{q}^2}\frac{}{\hbar}\Pi^{00}(q)\right]^2, \tag{6.148}$$

where the factor of 4 is the symmetry factor of the diagram and π^{00} is the polarization tensor (6.101) which was formally introduced in Eq. (2.186). The integral over q_0 is readily evaluated with the help of contour integration and the representation (6.111) of the polarization tensor, with the result

$$\begin{aligned}\frac{\Omega_3^{\mathrm{r}}}{V} &= -\frac{3}{16\pi^5}\frac{e^2}{a_0}n \int \mathrm{d}^3 q \int \mathrm{d}^3 k \int \mathrm{d}^3 k' \\ &\quad\times \frac{1}{\mathbf{q}^4}\frac{\theta(|\mathbf{k}+\mathbf{q}|-1)\theta(1-|\mathbf{k}|)\theta(|\mathbf{k}'+\mathbf{q}|-1)\theta(1-|\mathbf{k}'|)}{\mathbf{q}^2+\mathbf{q}\cdot(\mathbf{k}'+\mathbf{k})}\end{aligned} \tag{6.149}$$

where $a_0 \equiv \hbar^2/me^2$ is the Bohr radius and where all the wave vectors are rescaled by k_{F}. As we will now demonstrate, the integral over $\mathbf{q}$ diverges logarithmically for small wave vectors. Concentrating on this divergence, we approximate the integral

$$I \equiv \int \mathrm{d}^3k \int \mathrm{d}^3k' \frac{\theta(|\mathbf{k}+\mathbf{q}|-1)\theta(1-|\mathbf{k}|)\theta(|\mathbf{k}'+\mathbf{q}|-1)\theta(1-|\mathbf{k}'|)}{\mathbf{q}^2+\mathbf{q}\cdot(\mathbf{k}'+\mathbf{k})} \tag{6.150}$$

for small $\mathbf{q}$ by

$$I \approx \frac{(2\pi)^2}{|\mathbf{q}|}\int_0^1 \mathrm{d}(\cos\theta_k)\int_0^1 \mathrm{d}(\cos\theta_{k'})\int_{1-|\mathbf{q}|\cos\theta_k}^1 \mathrm{d}|\mathbf{k}|\,\mathbf{k}^2 \int_{1-|\mathbf{q}|\cos\theta_{k'}}^1 \mathrm{d}|\mathbf{k}'|\,\mathbf{k}'^2 \times \frac{1}{|\mathbf{k}|\cos\theta_k+|\mathbf{k}'|\cos\theta_{k'}}, \tag{6.151}$$

where it is used that $|\mathbf{k}+\mathbf{q}| \approx |\mathbf{k}|+|\mathbf{q}|\cos\theta_k$, with θ_k denoting the angle between the vectors $\mathbf{k}$ and $\mathbf{q}$, and similarly $|\mathbf{k}'+\mathbf{q}| \approx |\mathbf{k}'|+|\mathbf{q}|\cos\theta_{k'}$. Since

$$\int_{1-|\mathbf{q}|\cos\theta_{k'}}^1 \mathrm{d}|\mathbf{k}'|\,\mathbf{k}'^2 \frac{1}{|\mathbf{k}|\cos\theta_k+|\mathbf{k}'|\cos\theta_{k'}} \approx \frac{|\mathbf{q}|\cos\theta_{k'}}{|\mathbf{k}|\cos\theta_k+\cos\theta_{k'}}, \tag{6.152}$$

with a similar approximation for the integral over $|\mathbf{k}|$, I takes the form

$$\begin{aligned} I &\approx (2\pi)^2|\mathbf{q}|\int_0^1 \mathrm{d}(\cos\theta_k)\int_0^1 \mathrm{d}(\cos\theta_{k'})\frac{\cos\theta_k\cos\theta_{k'}}{\cos\theta_k+\cos\theta_{k'}} \\ &\approx \frac{2}{3}(1-\ln 2)(2\pi)^2|\mathbf{q}|. \end{aligned} \tag{6.153}$$

When substituted in Eq. (6.149), this gives the expression

$$\frac{\Omega_3^{\mathrm{r}}}{V} \approx -\frac{2}{\pi^2}(1-\ln 2)\frac{e^2}{a_0}n\int \mathrm{d}|\mathbf{q}|\frac{1}{|\mathbf{q}|}, \tag{6.154}$$

which indeed diverges logarithmically in the limit $|\mathbf{q}| \to 0$. However, when screening effects are taking into account, which amounts to summing up all ring diagrams, the wave number q_{FT} introduced in Eq. (6.120) provides an infrared cutoff, and the expression becomes finite

$$\begin{aligned} \frac{\Omega_3^{\mathrm{r}}}{V} &\approx -\frac{2}{\pi^2}(1-\ln 2)\frac{e^2}{a_0}n\int_{q_{\mathrm{FT}}/k_{\mathrm{F}}}^1 \mathrm{d}|\mathbf{q}|\frac{1}{|\mathbf{q}|} \\ &\approx \frac{2}{\pi^2}(1-\ln 2)\frac{e^2}{a_0}n\ln\left(\frac{q_{\mathrm{FT}}}{k_{\mathrm{F}}}\right), \end{aligned} \tag{6.155}$$

to logarithmic accuracy. Adding the contributions (6.142), (6.146), and (6.155) together, we obtain the first terms of a series

$$\frac{\Omega}{\langle N\rangle} = -\frac{2}{5}\frac{\hbar^2k_{\mathrm{F}}^2}{2m} - \frac{3}{4\pi}e^2k_{\mathrm{F}} + \frac{2}{\pi^2}(1-\ln 2)\frac{me^4}{\hbar^2}\ln\left(\frac{q_{\mathrm{FT}}}{k_{\mathrm{F}}}\right) + \cdots, \tag{6.156}$$

with expansion parameter $me^2/\hbar^2k_{\mathrm{F}}$. The series is usually written in terms of the ratio of the two length scales present,

$$r_{\mathrm{s}} \equiv r_0/a_0, \tag{6.157}$$

with r_0 the typical interparticle spacing defined through

$$n = \frac{1}{(4\pi/3)}\frac{1}{r_0^3}. \tag{6.158}$$

The ground-state energy per particle then reads to this order

$$\begin{aligned}\frac{\langle E\rangle}{\langle N\rangle} &= \frac{e^2}{2a_0}\left[\frac{3}{5}\left(\frac{9\pi}{4}\right)^{2/3}\frac{1}{r_s^2} - \frac{3}{2}\left(\frac{3}{2\pi}\right)^{2/3}\frac{1}{r_s} + \frac{2}{\pi^2}(1-\ln 2)\ln r_s + O\left(r_s^0\right)\right] \\ &= \frac{e^2}{2a_0}\left[\frac{2.2099}{r_s^2} - \frac{0.9163}{r_s} + 0.0621\ln r_s + O\left(r_s^0\right)\right],\end{aligned} \tag{6.159}$$

where it is used that

$$k_F = \left(3\pi^2 n\right)^{1/3} = \left(\frac{9\pi}{4}\right)^{1/3}\frac{1}{r_0}. \tag{6.160}$$

Note that because of the minus sign in front of the exchange term, that contribution lowers the ground-state energy per electron compared to an ideal Fermi gas.

The expansion (6.159) is valid for small r_s, which corresponds to high electron number densities n. At very high densities, the first term, representing the kinetic energy of the Fermi sea, dominates, and the Coulomb interaction can be ignored. Typical densities for valence electrons in 3D metals correspond to values $r_s = 2$–6 so that the expression (6.159) cannot be expected to provide reliable estimates of the ground-state energy of real metals. The compressibility calculated from the expression (6.159) turns negative around $r_s = 6$ and, thus, signals the onset of an instability in the electron gas.

In the limit $r_s \to \infty$ of low electron number densities, the Coulomb interaction energy dominates the kinetic energy. In this regime, the electrons are expected to crystallize in the ground state, and to form what is known as a *Wigner lattice* to minimize the repulsive Coulomb interaction. The ground-state energy is then given by the zero-point energy of the lattice vibrations in addition to the electrostatic energy.

6.7 Self-Energy

The electron self-energy $\Sigma(k)$ is in lowest order given by the diagram

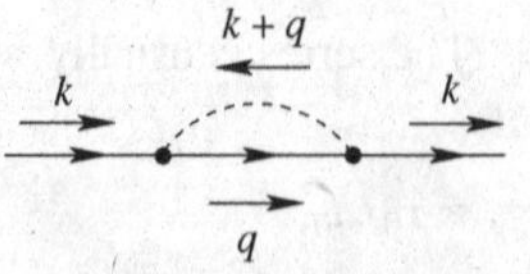

$$= -\frac{i}{\hbar}\Sigma_1(k) \tag{6.161}$$

which translates into

$$-\frac{\mathrm{i}}{\hbar}\Sigma_1(k) = \left(\frac{\mathrm{i}e}{\hbar}\right)^2 \int \frac{\mathrm{d}^4q}{(2\pi)^4}\mathrm{i}\hbar\frac{4\pi}{|\mathbf{k}-\mathbf{q}|^2}\mathrm{i}\hbar S_{\mathrm{F}}(q). \tag{6.162}$$

With the mandatory convergence factor included in the electron propagator, the frequency integral is again readily evaluated with the result

$$-\frac{\mathrm{i}}{\hbar}\Sigma_1(k) = 4\pi e^2\frac{\mathrm{i}}{\hbar}\int \frac{\mathrm{d}^3q}{(2\pi)^3}\frac{\theta(k_{\mathrm{F}}-|\mathbf{q}|)}{|\mathbf{k}-\mathbf{q}|^2}. \tag{6.163}$$

The remaining integrals are elementary, and give

$$\Sigma_1(k) = -\frac{e^2}{\pi}k_{\mathrm{F}}\left(1+\frac{k_{\mathrm{F}}^2-\mathbf{k}^2}{2k_{\mathrm{F}}|\mathbf{k}|}\ln\left|\frac{k_{\mathrm{F}}+|\mathbf{k}|}{k_{\mathrm{F}}-|\mathbf{k}|}\right|\right). \tag{6.164}$$

At the Fermi surface, this lowest-order expression is seen to diverge logarithmically. When substituted in Eq. (6.23), it leads to the revised spectrum

$$E(\mathbf{k}) = \xi(\mathbf{k}) + \Sigma_1(\mathbf{k}). \tag{6.165}$$

The divergence obtained in lowest order would imply that the effective electron mass m^* at the Fermi surface, defined through

$$\left.\frac{\partial E(\mathbf{k})}{\partial k}\right|_{k_{\mathrm{F}}} = \frac{\hbar^2 k_{\mathrm{F}}}{m^*}, \tag{6.166}$$

vanishes, in disagreement with observation, showing that m^* is of order electron mass m. The divergence obtained in lowest order perturbation theory disappears when screening effects are taken into account. Specifically, with the screened interaction, Eq. (6.163) is replaced with

$$\begin{aligned}\Sigma(k) &= -4\pi e^2\int\frac{\mathrm{d}^3q}{(2\pi)^3}\frac{\theta(k_{\mathrm{F}}-|\mathbf{q}|)}{|\mathbf{k}-\mathbf{q}|^2+q_{\mathrm{FT}}^2}\\ &= -\frac{e^2}{2\pi}\frac{1}{|\mathbf{k}|}\int_0^{k_{\mathrm{F}}}\mathrm{d}|\mathbf{q}|\,|\mathbf{q}|\ln\left(\frac{(|\mathbf{k}|+|\mathbf{q}|)^2+q_{\mathrm{FT}}^2}{(|\mathbf{k}|-|\mathbf{q}|)^2+q_{\mathrm{FT}}^2}\right)\\ &= -\frac{2}{\pi}e^2k_{\mathrm{F}}\left[1-\frac{q_{\mathrm{FT}}}{k_{\mathrm{F}}}\arctan\left(\frac{k_{\mathrm{F}}}{q_{\mathrm{FT}}}\right)-\frac{1}{3}\frac{k_{\mathrm{F}}^2\mathbf{k}^2}{\left(k_{\mathrm{F}}^2+q_{\mathrm{FT}}^2\right)^2}+O\left(\mathbf{k}^4\right)\right],\end{aligned} \tag{6.167}$$

and the effective electron mass becomes indeed finite,

$$m^* = m\left(1+\frac{4}{3\pi}\frac{1}{a_0}\frac{k_{\mathrm{F}}^3}{(k_{\mathrm{F}}^2+q_{\mathrm{FT}}^2)^2}\right)^{-1}. \tag{6.168}$$

This illustrates that although diagrams by themselves may diverge in the infrared, summing an infinite set of them may produce a finite result.

Notes

(i) Most of the topics discussed in this chapter are covered in the textbooks [Abrikosov *et al.* (1963)] and [Fetter and Walecka (1971)].

(ii) Kondo's seminal paper on the subject now bearing his name appeared in [Kondo (1964)]. For another early paper, see [Abrikosov (1965)].

(iii) Our derivation of the scaling (6.88) of the coupling constant in the Kondo effect is equivalent to Anderson's "poor man's scaling approach" [Anderson (1970)]. For a further discussion of the scaling properties, the reader is referred to the book [Anderson (1984)]. The strong-coupling regime and in particular the cross-over between the strong- and weak-coupling limit was numerically solved by Wilson (1975) by renormalization group methods. For reviews of the Kondo effect, see [Nozières (1974)] and [Affleck (2005)]. The topic is also covered in the textbook [Mattuck (1976)].

(iv) A classic book on degenerate electron gases is [Pines and Nozières (1966)]. For a short introduction, highlighting the role of the Lindhard function, see [Smith (1983)]. For a presentation using quantum field theory, see [Chang (1990)].

(v) The Thomas-Fermi approximation was introduced by Thomas (1927) and Fermi (1928).

(vi) The logarithmic term in the expression (6.159) for the ground-state energy of a degenerate electron gas was first obtained by Macke (1950), while the next term in the series, which is a constant, was first determined by Gell-Mann and Brueckner (1957).

(vii) The random phase approximation was introduced in [Bohm and Pines (1953); Pines (1953)].

(viii) The possibility that electrons at low densities may form a lattice was proposed by Wigner (1938).

Chapter 7

Magnetic Order in Fermi Systems

In this chapter, we study magnetism at the absolute zero of temperature, starting from a microscopic fermionic lattice model, the so-called Hubbard model. Besides displaying ferromagnetic order, the Hubbard model also shows antiferromagnetic order for a certain range of values of the coupling constant and filling factor. In both states, the SO(3) spin rotation symmetry is spontaneously broken down to the residual SO(2) symmetry group of rotations about the direction of the (staggered) magnetization. The derivative expansion, presented in Sec. 3.1, will be used in both the ferromagnetic and the antiferromagnetic state to derive the effective theories describing the corresponding spin waves. Parametrizing the coset space SO(3)/SO(2), which forms a two-sphere S^2, the spin waves are the Nambu-Goldstone modes emerging from the spontaneous breakdown of the spin rotation symmetry. Although both effective theories are built from a unit vector $\mathbf{n}$ with three components tracing out the surface of the two-sphere, they differ in that ferromagnetic spin waves have a quadratic spectrum $\omega \propto \mathbf{k}^2$, whereas antiferromagnetic spin waves have a linear spectrum $\omega^2 \propto \mathbf{k}^2$.

7.1 Hubbard Model

The fermionic Hubbard model is a lattice model describing valance band electrons hopping from one lattice site, representing an atom, to another. For definiteness, we consider hypercubic lattices. The electrons interact with each other via a repulsive Coulomb potential. It is assumed that this Coulomb interaction is highly screened with a screening length of order the lattice spacing a so that it can be approximated by an on-site repulsion. In the framework of canonical quantization, the Hamiltonian is given by the fermionic counterpart of the Bose-Hubbard model

(4.236)

$$H = -\frac{1}{2}\sum_{\mathbf{x}\mathbf{x}'} t_{\mathbf{x}\mathbf{x}'}(a^\dagger_{\sigma,\mathbf{x}} a_{\sigma,\mathbf{x}'} + a^\dagger_{\sigma,\mathbf{x}'} a_{\sigma,\mathbf{x}}) + \sum_{\mathbf{x}} (-\mu \hat{n}_{\mathbf{x}} + U \hat{n}_{\uparrow,\mathbf{x}} \hat{n}_{\downarrow,\mathbf{x}}). \tag{7.1}$$

The operator $a^\dagger_{\sigma,\mathbf{x}}$ creates an electron of spin $\sigma = \uparrow, \downarrow$ at the lattice site specified by the position vector $\mathbf{x}$, while $a_{\sigma,\mathbf{x}}$ annihilates such an electron at that site, and $\hat{n}_{\mathbf{x}} \equiv \sum_\sigma \hat{n}_{\sigma,\mathbf{x}}$, where $\hat{n}_{\sigma,\mathbf{x}} = a^\dagger_{\sigma,\mathbf{x}} a_{\sigma,\mathbf{x}}$ counts the number of electrons of spin σ at site $\mathbf{x}$. In contrast to the bosonic model, here, the creation and annihilation operators anticommute to reflect the fermionic character of the electrons. Specifically,

$$\left\{a_{\sigma,\mathbf{x}}, a^\dagger_{\sigma',\mathbf{x}'}\right\} = \delta_{\sigma,\sigma'}\delta_{\mathbf{x},\mathbf{x}'}, \tag{7.2}$$

$$\{a_{\sigma,\mathbf{x}}, a_{\sigma',\mathbf{x}'}\} = \left\{a^\dagger_{\sigma,\mathbf{x}}, a^\dagger_{\sigma',\mathbf{x}'}\right\} = 0. \tag{7.3}$$

The first term in the Hamiltonian (7.1), with the symmetric real matrix $t_{\mathbf{x}\mathbf{x}'}$, describes the hopping of an electron from lattice site $\mathbf{x}'$ to $\mathbf{x}$. In the following, we assume that hopping takes place only between nearest neighbors $\mathbf{x}$ and $\mathbf{x}'$, and set $t_{\mathbf{x}\mathbf{x}'} = t$, with t the hopping parameter. When $\mathbf{x}$ and $\mathbf{x}'$ are not nearest neighbors $t_{\mathbf{x}\mathbf{x}'} = 0$. The last term in the Hubbard Hamiltonian, with coupling constant $U > 0$, represents the on-site Coulomb repulsion. Finally, the parameter μ denotes the chemical potential. Because the electrons are tightly bound to the lattice sites, the Hubbard model is an example of a *tight-binding model*. It is popular because in the limit of strong on-site Coulomb repulsion $U \to \infty$ it requires a large energy to put two electrons (with opposite spin, of course) on the same site. The ground state has consequently no doubly occupied sites. At half filling, i.e., when the number of electrons equals the number of lattice sites, no electrons are transported. An insulator of this type, which does not involve completely filled bands, as is usually the case, is called a *Mott insulator*.

In the opposite limit of weak Coulomb repulsion $U \to 0$, the theory describes free fermions hopping on the lattice. In this limit, we expand the annihilation and creation operators as

$$\begin{aligned} a_{\sigma,\mathbf{x}} &= \frac{1}{\sqrt{N}}\sum_{\mathbf{k}} a_{\sigma,\mathbf{k}} e^{+i\mathbf{k}\cdot\mathbf{x}}, \\ a^\dagger_{\sigma,\mathbf{x}} &= \frac{1}{\sqrt{N}}\sum_{\mathbf{k}} a^\dagger_{\sigma,\mathbf{k}} e^{-i\mathbf{k}\cdot\mathbf{x}}, \end{aligned} \tag{7.4}$$

with the inverse transformations

$$\begin{aligned} a_{\sigma,\mathbf{k}} &= \frac{1}{\sqrt{N}}\sum_{\mathbf{x}} a_{\sigma,\mathbf{x}} e^{-i\mathbf{k}\cdot\mathbf{x}}, \\ a^\dagger_{\sigma,\mathbf{k}} &= \frac{1}{\sqrt{N}}\sum_{\mathbf{x}} a^\dagger_{\sigma,\mathbf{x}} e^{+i\mathbf{k}\cdot\mathbf{x}}. \end{aligned} \tag{7.5}$$

The sum over the wave vectors is restricted to the first Brillouin zone defined by $-\pi/a \leq k^i \leq \pi/a$, and N denotes the number of values the wave vector $\mathbf{k}$ can take. It is related to the volume V of the hypercubic lattice through $V = Na^D$, with a the lattice spacing, see Notation and Conventions. The definitions are such that the anticommutation relation (7.2) translates into

$$\left\{a_{\sigma,\mathbf{k}}, a^{\dagger}_{\sigma',\mathbf{k}'}\right\} = \delta_{\sigma,\sigma'}\delta_{\mathbf{k},\mathbf{k}'}, \tag{7.6}$$

where use is made of the orthogonality condition

$$\sum_{\mathbf{x}} \mathrm{e}^{-\mathrm{i}(\mathbf{k}-\mathbf{k}')\cdot\mathbf{x}} = N\delta_{\mathbf{k},\mathbf{k}'}. \tag{7.7}$$

The quadratic terms of the Hubbard Hamiltonian then assume the form

$$H_0 = \sum_{\mathbf{k}} a^{\dagger}_{\sigma,\mathbf{k}} \left(\epsilon_{\mathbf{k}} - \mu\right) a_{\sigma,\mathbf{k}}, \tag{7.8}$$

with the kinetic energy

$$\epsilon_{\mathbf{k}} = -2t \sum_i \cos(k^i a). \tag{7.9}$$

In the limit $N \to \infty$, the sum over $\mathbf{k}$ can be replaced with the integral

$$\sum_{\mathbf{k}} \to V \int \frac{\mathrm{d}^D k}{(2\pi)^D}. \tag{7.10}$$

At half filling, there are as many negative energy states as there are positive energy states so that the chemical potential is zero. The Fermi surface of the noninteracting system is then determined by the equation $\epsilon_{\mathbf{k}} = 0$. Figure 7.1 shows, as an example, the Fermi sea and its "surface" on a square lattice.

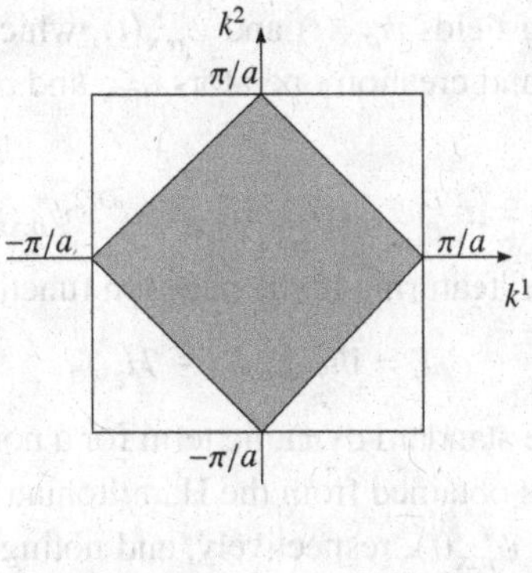

Fig. 7.1 Fermi sea for the tight-binding model on a square lattice at half filling.

Since we are interested in the magnetic properties of the Hubbard model, it is expedient to rewrite it in a way that explicitly involves the spin operator (in units of $\hbar$),

$$\mathbf{S}_\mathbf{x} \equiv a^\dagger_{\sigma,\mathbf{x}} \frac{\boldsymbol{\sigma}_{\sigma\sigma'}}{2} a_{\sigma',\mathbf{x}}, \tag{7.11}$$

where $\boldsymbol{\sigma}$ are the Pauli matrices (6.71). This rewriting is based on the identity

$$\sigma^\alpha_{\sigma\sigma'}\sigma^\alpha_{\tau\tau'} = 2\delta_{\sigma\tau'}\delta_{\sigma'\tau} - \delta_{\sigma\sigma'}\delta_{\tau\tau'}, \tag{7.12}$$

which can be verified by direct computation. On the left, a sum over $\alpha = 1, 2, 3$ is implied. By Eq. (7.2),

$$S^\alpha_\mathbf{x} S^\alpha_\mathbf{x} = -\frac{3}{2}\hat{n}_{\uparrow,\mathbf{x}}\hat{n}_{\downarrow,\mathbf{x}} + \frac{3}{4}\hat{n}_\mathbf{x} \tag{7.13}$$

so that the interaction term can be cast in the equivalent form

$$H_\mathrm{I} \equiv U\hat{n}_{\uparrow,\mathbf{x}}\hat{n}_{\downarrow,\mathbf{x}} = -\frac{2}{3}U\mathbf{S}^2_\mathbf{x} + \frac{1}{2}U\hat{n}_\mathbf{x}. \tag{7.14}$$

The last term may be absorbed by redefining the chemical potential so that the full Hubbard Hamiltonian may be equivalently written as

$$H = -\frac{t}{2}\sum_{\langle\mathbf{x}\mathbf{x}'\rangle}(a^\dagger_{\sigma,\mathbf{x}}a_{\sigma,\mathbf{x}'} + a^\dagger_{\sigma,\mathbf{x}'}a_{\sigma,\mathbf{x}}) - \sum_\mathbf{x}\left(\mu\hat{n}_\mathbf{x} + \tfrac{2}{3}U\mathbf{S}^2_\mathbf{x}\right), \tag{7.15}$$

where the angle brackets in $\sum_{\langle\mathbf{x}\mathbf{x}'\rangle}$ indicate that the sum runs only over nearest-neighbor sites. The minus sign in front of the last term with $U > 0$ implies that the interaction energy is minimized if the spin is maximized at each lattice site.

The partition function Z of the model can be represented as a functional integral over Grassmann fields $\psi_\sigma(x)$ and $\psi^*_\sigma(x)$ describing the electrons as

$$Z = \int \mathrm{D}\psi^*\mathrm{D}\psi \exp\left(\frac{\mathrm{i}}{\hbar}\int \mathrm{d}t\, a^D \sum_\mathbf{x} \mathcal{L}\right). \tag{7.16}$$

Here, the anticommuting fields $\psi_{\sigma,\mathbf{x}}(t)$ and $\psi^*_{\sigma,\mathbf{x}}(t)$, which also depend on time, replace the annihilation and creation operators $a_{\sigma,\mathbf{x}}$ and $a^\dagger_{\sigma,\mathbf{x}}$ of the canonical approach. Specifically,

$$a_{\sigma,\mathbf{x}} \mathrel{\hat{=}} a^{D/2}\psi_{\sigma,\mathbf{x}}(t), \quad a^\dagger_{\sigma,\mathbf{x}} \mathrel{\hat{=}} a^{D/2}\psi^*_{\sigma,\mathbf{x}}(t). \tag{7.17}$$

The Lagrangian density $\mathcal{L}$ featuring in the partition function is given by

$$\mathcal{L} = \mathrm{i}\hbar\psi^*_\sigma\partial_t\psi_\sigma - \mathcal{H}, \tag{7.18}$$

where the first term is the standard dynamic term for a nonrelativistic theory. The Hamiltonian density $\mathcal{H}$ is obtained from the Hamiltonian (7.15) by replacing $a_{\sigma,\mathbf{x}}$ and $a^\dagger_{\sigma,\mathbf{x}}$ with $\psi_{\sigma,\mathbf{x}}(t)$ and $\psi^*_{\sigma,\mathbf{x}}(t)$, respectively, and noting that on the lattice $H = a^D\sum_\mathbf{x}\mathcal{H}$.

We proceed by introducing an auxiliary vector field $\boldsymbol{\phi}$ via a Hubbard-Stratonovich transformation to linearize the interaction term in the Lagrangian,

$$\exp\left(\frac{\mathrm{i}}{\hbar}\frac{2}{3}Ua^D \int \mathrm{d}^d x\, \mathbf{s}^2\right) = \int \mathrm{D}\boldsymbol{\phi} \exp\left[\frac{\mathrm{i}}{\hbar}\int \mathrm{d}^d x\left(\boldsymbol{\phi}\cdot\mathbf{s} - \frac{3}{8}\frac{1}{Ua^D}\boldsymbol{\phi}^2\right)\right], \tag{7.19}$$

where

$$\mathbf{s}(x) \equiv \psi^*_\sigma(x)\frac{\boldsymbol{\sigma}_{\sigma\sigma'}}{2}\psi_{\sigma'}(x) \tag{7.20}$$

denotes the spin density in units of $\hbar$, cf. Eq. (7.11). It is assumed that the lattice spacing a is sufficiently small so that the sum $a^D\sum_{\mathbf{x}}$ can be replaced with the integral $\int \mathrm{d}^D x$. This transformation leads to a Lagrangian bilinear in the Grassmann fields

$$\mathcal{L} = \mathcal{L}_0 + \boldsymbol{\phi}\cdot\mathbf{s} - \frac{3}{8}\frac{1}{Ua^D}\boldsymbol{\phi}^2, \tag{7.21}$$

where $\mathcal{L}_0$ denotes the noninteracting part. Formally, the auxiliary field, which has the dimension of energy and satisfies the Euler-Lagrange equation

$$\boldsymbol{\phi} = \frac{4}{3}Ua^D\mathbf{s}, \tag{7.22}$$

acts as an applied magnetic field that couples to the electron spin. The partition function now becomes

$$Z = \int \mathrm{D}\boldsymbol{\phi}\, \mathrm{Det}\left(p_0 - \epsilon(\mathbf{p}) + \mu + \tfrac{1}{2}\boldsymbol{\phi}\cdot\boldsymbol{\sigma}\right)\exp\left(-\frac{\mathrm{i}}{\hbar}\frac{3}{8}\frac{1}{Ua^D}\int \mathrm{d}^d x\, \boldsymbol{\phi}^2\right) \tag{7.23}$$

after the Gaussian integral over the Grassmann fields has been evaluated. The kinetic energy operator $\epsilon(\mathbf{p})$ is defined by Eq. (7.9) with $\hbar\mathbf{k}$ replaced with $\mathbf{p} = -\mathrm{i}\hbar\nabla$ in the continuum, while $p^0 = p_0 = \mathrm{i}\hbar\partial_t$. The argument of the determinant is a 2×2 matrix in spin space as is evident from the last term. The other terms are to be thought of as being multiplied by the 2×2 unit matrix.

7.2 Ferromagnetic State

The representation (7.23) of the partition function leads to the one-fermion-loop effective action

$$\Gamma_1[\boldsymbol{\phi}_\mathrm{c}] = -\mathrm{i}\hbar\,\mathrm{Tr}\ln\left[p_0 - \epsilon(\mathbf{p}) + \mu + \tfrac{1}{2}\boldsymbol{\phi}_\mathrm{c}\cdot\boldsymbol{\sigma}\right] \tag{7.24}$$

with $\boldsymbol{\phi}_\mathrm{c}$ the average field, cf. Eq. (7.22)

$$\boldsymbol{\phi}_\mathrm{c}(x) = \frac{4}{3}Ua^D\langle\mathbf{s}(x)\rangle, \tag{7.25}$$

which is proportional to the magnetization. As always, the trace Tr in Eq. (7.24) includes the trace over spin indices.

To explore the phase diagram of the Hubbard model, it suffice to consider the case where the average spin is uniform in space and time, $\langle \mathbf{s}(x) \rangle = \bar{\mathbf{s}}$. By Eq. (7.25), the auxiliary field is then constant too, $\boldsymbol{\phi}_c(x) = \mathbf{v}$, and its direction defines the spin quantization axis globally. The effective action (7.24) now splits into two separated parts

$$\Gamma_1 = -\mathrm{i}h \sum_\sigma \mathrm{Tr} \ln \left[p_0 - \epsilon(\mathbf{p}) + \mu_\sigma \right], \tag{7.26}$$

where

$$\mu_{\uparrow,\downarrow} \equiv \mu \pm \frac{1}{2} v \tag{7.27}$$

with $v = |\mathbf{v}| > 0$, is the effective chemical potential for the spin-up and the spin-down electrons, respectively. Note that $\mu = \frac{1}{2}(\mu_\uparrow + \mu_\downarrow)$ and $v = \mu_\uparrow - \mu_\downarrow$, and also $\mu_\sigma = \hbar^2 k_{\mathrm{F},\sigma}^2 / 2m$, with $\hbar k_{\mathrm{F},\sigma}$ the Fermi momentum for electrons with spin $\sigma = \uparrow, \downarrow$.

The average number density $\bar{n}_\sigma$ of electrons with spin $\sigma = \uparrow, \downarrow$ is readily obtained from Eq. (7.26) by taking the derivative with respect to μ_σ:

$$\bar{n}_\sigma = -\mathrm{i}\hbar \int \frac{\mathrm{d}^d k}{(2\pi)^d} \frac{\mathrm{e}^{\mathrm{i}\omega 0^+}}{\hbar\omega - \xi_\sigma(\mathbf{k}) + \mathrm{i}\,\mathrm{sgn}(\omega)\eta}, \tag{7.28}$$

where $\xi_\sigma(\mathbf{k}) \equiv \epsilon(\mathbf{k}) - \mu_\sigma$, and where we included the usual convergence factor $\exp(\mathrm{i}\omega 0^+)$ and applied the iη prescription, see Secs. 2.3 and 2.4. This integral gives, cf. Eq. (3.15)

$$\bar{n}_\sigma = \int \frac{\mathrm{d}^D k}{(2\pi)^D} \theta\left[-\xi_\sigma(\mathbf{k}) \right] = \int_0^{\mu_\sigma} \mathrm{d}\epsilon \, \nu(\epsilon), \tag{7.29}$$

with $\nu(\epsilon)$ the density of states per spin degree of freedom. If the average total electron number density $\bar{n}$ is fixed, $\bar{n}_{\uparrow,\downarrow}$ fulfill the constraint $\bar{n}_\uparrow + \bar{n}_\downarrow = \bar{n}$. In terms of $\bar{n}_\sigma$, the average spin density (in units of $\hbar$) reads

$$\bar{s}^3 = \frac{1}{2}(\bar{n}_\uparrow - \bar{n}_\downarrow) = \frac{1}{2} \int_{\mu_\downarrow}^{\mu_\uparrow} \mathrm{d}\epsilon \, \nu(\epsilon), \tag{7.30}$$

where we have taken the spin quantization axis along the third direction. Through Eqs. (7.25) and (7.27), both $\bar{s}^3$ and $\mu_{\uparrow,\downarrow}$ depend on v. Equation (7.30) therefore provides a self-consistent equation for the magnetization. One solution is given by the trivial one with zero magnetization, $v = 0$, which describes the paramagnetic state. At a critical value U_c of the Hubbard coupling, determined by the so-called *Stoner criterion*

$$1 = \frac{2}{3} \nu(0) \, U_c a^D, \tag{7.31}$$

where $\nu(0)$ denotes the density of states per spin degree of freedom at the Fermi surface, a second solution develops with a nonzero magnetization. At this critical point, the ground-state energy of the paramagnetic state with $\mathbf{v} = 0$ becomes larger than that of the ferromagnetic state with $\mathbf{v} \neq 0$ so that the system becomes unstable towards developing a finite magnetization. As a result, the SO(3) spin rotation symmetry gets spontaneously broken down to rotations about the preferred direction defined by the magnetization. As it is zero in the symmetric, paramagnetic state and nonzero in the ordered, ferromagnetic state, $\mathbf{v}$, which is proportional to the magnetization, serves as *order parameter* for distinguishing the two states in the precise sense formulated by Landau in 1937.

7.3 Ferromagnetic Spin Waves

We next derive the effective theory governing ferromagnetic spin waves. We thereto set

$$\boldsymbol{\phi}_{\rm c}(x) = v\mathbf{n}(x) \tag{7.32}$$

in the effective action (7.24), where the unit vector $\mathbf{n}$ denotes the direction of the local magnetization, and $v = |\mathbf{v}| > 0$ is a constant. By keeping the magnitude of the order parameter fixed and allowing only the direction of the order parameter to change, we focus on the Nambu-Goldstone modes. In the context of superconductivity, this limit is known as the London limit.

To facilitate the calculation, we rotate the unit vector $\mathbf{n}$ in the fixed third direction of spin space via a unitarity transformation,

$$\mathbf{n}(x) \cdot \boldsymbol{\sigma} = s(x)\sigma^3 s^\dagger(x). \tag{7.33}$$

Since this vector is spacetime dependent, the rotation matrix s achieving this will also vary in space and time. We proceed as in Sec. 3.6, and decompose the Grassmann field ψ_σ in the Lagrangian density (7.21) as

$$\psi(x) = s(x)\chi(x), \tag{7.34}$$

cf. (3.89). The one-fermion-loop effective action (7.24) is then replaced with

$$\Gamma_1 = -i\hbar\,\mathrm{Tr}\ln\left[p_0 - V_0 - \epsilon(\mathbf{p} + \mathbf{V}) + \mu + \tfrac{1}{2}v\sigma^3\right], \tag{7.35}$$

with V_μ a (pure) SU(2) "gauge field"

$$V_\mu \equiv -i\hbar s^\dagger \partial_\mu s = V_\mu^\alpha \sigma^\alpha \tag{7.36}$$

and $V_\mu^\alpha = \frac{1}{2}\,\mathrm{tr}\, V_\mu \sigma^\alpha$ the expansion coefficients of V_μ in terms of the generators σ^α of the SU(2) Lie algebra. Equation (7.35) is the higher-dimensional counterpart of the $(0 + 1)$-dimensional effective action (3.94) discussed in the context of the Berry phase.

We are interested in terms in the expansion of the one-fermion-loop effective action (7.35) which are at most quadratic in the gauge fields and contain no additional derivatives for these produce the lowest-order terms in the effective description of the spin waves. Note that

$$\begin{aligned}(\partial_\mu \mathbf{n})^2 &= \frac{1}{2}\operatorname{tr}(\partial_\mu n_\alpha \sigma^\alpha)^2 = \frac{1}{2}\operatorname{tr}\left[\partial_\mu (s\sigma^3 s^\dagger)\right]^2 \\ &= \frac{1}{\hbar^2}\operatorname{tr}\left[V_\mu{}^2 - (\sigma^3 V_\mu)^2\right],\end{aligned} \tag{7.37}$$

where tr denotes the trace over the spin indices. In deriving this it is used that the rotation matrix s introduced in Eq. (7.33) is unitary, i.e., $ss^\dagger = s^\dagger s = 1$.

We first concentrate on the V_0 terms and set $\mathbf{V}$ equal to zero for the moment. The contribution linear in $V^0 = V_0$ is given by

$$\Gamma_1^{(\ell=1)} = \mathrm{i}\hbar \operatorname{Tr} S_\mathrm{F} V_0, \tag{7.38}$$

where S_F stands for the propagator, cf. (3.93)

$$S_\mathrm{F}(k) = \frac{\hbar\omega - \xi(\mathbf{k}) - \frac{1}{2}v\sigma^3}{[\hbar\omega - \xi(\mathbf{k})]^2 - \frac{1}{4}v^2}, \tag{7.39}$$

with $\xi(\mathbf{k}) = \epsilon(\mathbf{k}) - \mu$. This gives for the corresponding Lagrangian density

$$\mathcal{L}_1^{(\ell=1)} = -\mathrm{i}\hbar V_0^3 \int \frac{\mathrm{d}^d k}{(2\pi)^d} \frac{v}{[\hbar\omega - \xi(\mathbf{k})]^2 - \frac{1}{4}v^2}. \tag{7.40}$$

In terms of the effective chemical potentials (7.27), the denominator in the integrand can be equivalently written as $(\hbar\omega - \xi)^2 - \frac{1}{4}v^2 = (\hbar\omega - \xi_\uparrow)(\hbar\omega - \xi_\downarrow)$. With the usual i$\eta$ prescription of Sec. 2.4 applied, the frequency integral is readily evaluated using contour integration, with the result

$$\mathrm{i}\hbar \int \frac{\mathrm{d}\omega}{2\pi} \frac{v}{(\hbar\omega - \xi)^2 - \frac{1}{4}v^2 + \mathrm{i}\,\mathrm{sgn}(\omega)\eta} = \theta(-\xi_\uparrow) - \theta(-\xi_\downarrow), \tag{7.41}$$

where $\theta(-\xi_\uparrow) - \theta(-\xi_\downarrow) = \theta(-|\xi| + \frac{1}{2}v)$, and

$$\begin{aligned}\mathcal{L}_1^{(\ell=1)} &= -V_0^3 \int_{\mu_\downarrow}^{\mu_\uparrow} \mathrm{d}\epsilon\, v(\epsilon) = -(\bar{n}_\uparrow - \bar{n}_\downarrow)V_0^3 \\ &= \mathrm{i}\hbar\bar{s}^3 \operatorname{tr}\sigma^3 s^\dagger \partial_t s,\end{aligned} \tag{7.42}$$

with $\hbar\bar{s}^3$ the average spin, see Eq. (7.30). Comparison with Eq. (3.95) shows that this term linear in time derivatives is nothing but a Berry phase, or Wess-Zumino term. By Eq. (3.99), the associated action can be written in terms of $\mathbf{n}$ as

$$S_\mathrm{WZ} = -\hbar\bar{s}^3 \int_0^1 \mathrm{d}u \int \mathrm{d}^d x\, \mathbf{n}\cdot(\partial_u \mathbf{n} \times \partial_t \mathbf{n}), \tag{7.43}$$

where it is recalled that u is an extra variable $0 \leq u \leq 1$ introduced to put this contribution in a regular form. The unit vector $\mathbf{n}$ also depends on this variable in a way such that

$$\mathbf{n}(x,1) = \mathbf{n}(x), \quad \mathbf{n}(x,0) = (0,0,1). \tag{7.44}$$

We proceed with the terms quadratic in $\mathbf{V}$. To this end, we expand the kinetic energy $\epsilon(\mathbf{k}+\mathbf{V}/\hbar)$ in a Taylor series as

$$\epsilon(\mathbf{k}+\mathbf{V}/\hbar) = \epsilon(\mathbf{k}) + \frac{1}{\hbar}\nabla_{\mathbf{k}}\epsilon\cdot\mathbf{V} + \frac{1}{2}\frac{1}{\hbar^2}\frac{\partial^2\epsilon}{\partial k^i\partial k^j}V^iV^j + \cdots, \tag{7.45}$$

with $\nabla_{\mathbf{k}} \equiv (\partial/\partial k^1, \partial/\partial k^2 \cdots, \partial/\partial k^D)$. The first term in the expansion of the one-fermion-loop effective action (7.35) then leads to

$$\mathcal{L}_1^{(\ell=1)} = \frac{i}{2}\frac{1}{\hbar}\,\mathrm{tr}\int\frac{\mathrm{d}^dk}{(2\pi)^d}S_{\mathrm{F}}(k)\frac{\partial^2\epsilon}{\partial k^i\partial k^j}V^iV^j \tag{7.46}$$

or, after the frequency integral has been evaluated,

$$\begin{aligned}\mathcal{L}_1^{(\ell=1)} &= -\frac{1}{4}\frac{1}{\hbar^2}\,\mathrm{tr}\int\frac{\mathrm{d}^Dk}{(2\pi)^D}\sum_\sigma\theta[-\xi_\sigma(\mathbf{k})]\frac{\partial^2\epsilon}{\partial k^i\partial k^j}V^iV^j \\ &= -\frac{1}{4}\frac{1}{\hbar^2}\,\mathrm{tr}\int\frac{\mathrm{d}^Dk}{(2\pi)^D}\sum_\sigma\delta[-\xi_\sigma(\mathbf{k})]\,(\nabla_{\mathbf{k}}\epsilon\cdot\mathbf{V})^2,\end{aligned} \tag{7.47}$$

where the last line follows by integrating by parts.

The second term in the expansion of the effective action (7.35) gives as contribution to lowest order in derivatives

$$\begin{aligned}\mathcal{L}_1^{(\ell=2)} = \frac{i}{2}\frac{1}{\hbar}\int\frac{\mathrm{d}^dk}{(2\pi)^d}\Bigg[&\frac{(\hbar\omega-\xi)^2}{\left[(\hbar\omega-\xi)^2-\frac{1}{4}v^2\right]^2}\,\mathrm{tr}\,(\nabla_{\mathbf{k}}\epsilon\cdot\mathbf{V})^2 \\ &+\frac{\frac{1}{4}v^2}{\left[(\hbar\omega-\xi)^2-\frac{1}{4}v^2\right]^2}\,\mathrm{tr}\,(\nabla_{\mathbf{k}}\epsilon\cdot\sigma_3\mathbf{V})^2\Bigg].\end{aligned} \tag{7.48}$$

The frequency integral in the first term,

$$I \equiv \hbar\int\frac{\mathrm{d}\omega}{2\pi}\frac{(\hbar\omega-\xi)^2}{\left[(\hbar\omega-\xi)^2-\frac{1}{4}v^2\right]^2}, \tag{7.49}$$

can be evaluated by splitting it in two

$$\begin{aligned}I &= \hbar\int\frac{\mathrm{d}\omega}{2\pi}\frac{1}{(\hbar\omega-\xi)^2-\frac{1}{4}v^2} + \frac{1}{4}v^2\hbar\int\frac{\mathrm{d}\omega}{2\pi}\frac{1}{\left[(\hbar\omega-\xi)^2-\frac{1}{4}v^2\right]^2} \\ &= -i\frac{1}{v}\theta(-|\xi|+\tfrac{1}{2}v) + \frac{1}{2}v\frac{\partial}{\partial v}\hbar\int\frac{\mathrm{d}\omega}{2\pi}\frac{1}{(\hbar\omega-\xi)^2-\frac{1}{4}v^2} \\ &= -\frac{i}{2}\frac{1}{v}\theta(-|\xi|+\tfrac{1}{2}v) - \frac{i}{4}\delta(-|\xi|+\tfrac{1}{2}v),\end{aligned} \tag{7.50}$$

where in the second step use is made of the integral (7.41). In doing so, the frequency integral in the second term of Eq. (7.48) has been evaluated at the same time. Adding the contributions (7.47) and (7.48), we obtain

$$\mathcal{L}_1^{(\ell=1,2)} = -\frac{1}{4}\frac{1}{\hbar^2}\,\mathrm{tr}\int\frac{\mathrm{d}^D k}{(2\pi)^D}\left\{\frac{1}{2}\sum_\sigma \delta(-\xi_\sigma) - \frac{1}{v}[\theta(-\xi_\uparrow) - \theta(-\xi_\downarrow)]\right\} \times\left[(\nabla_{\mathbf{k}}\epsilon\cdot\mathbf{V})^2 - (\nabla_{\mathbf{k}}\epsilon\cdot\sigma_3\mathbf{V})^2\right]. \tag{7.51}$$

For an isotropic lattice, this simplifies by Eq. (7.37) to

$$\mathcal{L}_1^{(\ell=1,2)} = -\frac{1}{2}\rho c^2(\partial_i\mathbf{n})^2, \tag{7.52}$$

with the product of the so-called *spin-wave stiffness* ρ and velocity squared c^2 given by

$$\rho c^2 = \frac{1}{2D}\int\frac{\mathrm{d}^D k}{(2\pi)^D}\left\{\frac{1}{2}\sum_\sigma \delta(-\xi_\sigma) - \frac{1}{v}[\theta(-\xi_\uparrow) - \theta(-\xi_\downarrow)]\right\}(\nabla_{\mathbf{k}}\epsilon)^2. \tag{7.53}$$

In the limit $v \to 0$, where

$$\theta(-\xi_{\uparrow,\downarrow}) = \theta(-\xi) \pm \frac{1}{2}v\delta(-\xi) + \frac{1}{8}v^2\delta'(-\xi) \pm \frac{1}{48}v^3\delta''(-\xi), \tag{7.54}$$

the right side of Eq. (7.53) vanishes as it must for the spin-wave stiffness is zero in the paramagnetic state. An expansion in v gives

$$\rho c^2 = \frac{1}{24D}v^2\int\frac{\mathrm{d}^D k}{(2\pi)^D}\delta''(-\xi)(\nabla_{\mathbf{k}}\epsilon)^2 + O(v^4). \tag{7.55}$$

The integral is readily evaluated in the continuum limit $a \to 0$, where the kinetic energy (7.9) becomes parabolic

$$\epsilon(\mathbf{k}) = \hbar^2\mathbf{k}^2/2m^*, \tag{7.56}$$

apart from an additive constant which can be absorbed by redefining the chemical potential. Here, m^* denotes the mass of the single-fermion excitations which need not, and in general will not, coincide with the free electron mass. In this limit,

$$\rho c^2 = \frac{D(D-2)}{192}\frac{\hbar^2}{m^*}\frac{\bar{n}}{\mu^2}v^2 + O(v^4), \tag{7.57}$$

where $\bar{n}$ and μ denote the particle number density (6.59) and chemical potential of a uniform ideal Fermi gas ($v = 0$).

Combined, Eqs. (7.43) and (7.52) give as one-fermion-loop effective action emerging in a pure metal with a parabolic conduction band

$$\Gamma_1 = \int\mathrm{d}^d x\left[-\hbar\bar{s}^3\int_0^1 \mathrm{d}u\,\mathbf{n}\cdot(\partial_u\mathbf{n}\times\partial_t\mathbf{n}) - \frac{1}{2}\rho c^2(\partial_i\mathbf{n})^2\right], \tag{7.58}$$

governing the ferromagnetic spin waves, or the Nambu-Goldstone mode at low frequencies and long wave lengths. The Euler-Lagrange equation obtained from this action is the celebrated Landau-Lifshitz equation

$$\hbar \bar{s}^3 \frac{\partial}{\partial t}\mathbf{n} = \rho c^2 \mathbf{n} \times \nabla^2 \mathbf{n} \tag{7.59}$$

which leads to a quadratic spectrum $\omega \propto \mathbf{k}^2$ for the ferromagnetic spin waves, whose quanta are called *magnons*. The term on the left originates from the Wess-Zumino term. In deriving that term, it is used that for a unit vector in $D = 3$, $\delta\mathbf{n} \cdot (\partial_u \mathbf{n} \times \partial_t \mathbf{n}) = 0$ so that

$$\begin{aligned} \delta \int_0^1 \mathrm{d}u \int \mathrm{d}^4 x\, \mathbf{n} \cdot (\partial_u \mathbf{n} \times \partial_t \mathbf{n}) &= \int_0^1 \mathrm{d}u \int \mathrm{d}^4 x [\mathbf{n} \cdot (\partial_u \delta\mathbf{n} \times \partial_t \mathbf{n}) + \mathbf{n} \cdot (\partial_u \mathbf{n} \times \partial_t \delta\mathbf{n})] \\ &= \int_0^1 \mathrm{d}u \int \mathrm{d}^4 x \{\partial_u [\mathbf{n} \cdot (\delta\mathbf{n} \times \partial_t \mathbf{n})] + \partial_t [\mathbf{n} \cdot (\partial_u \mathbf{n} \times \delta\mathbf{n})]\} \\ &= -\int \mathrm{d}^4 x\, [\delta\mathbf{n} \cdot (\mathbf{n} \times \partial_t \mathbf{n})], \end{aligned} \tag{7.60}$$

where in the last step we dropped a total time derivative and used Eq. (7.44) as well as the identity $\mathbf{a} \cdot (\mathbf{b} \times \mathbf{c}) = -\mathbf{b} \cdot (\mathbf{a} \times \mathbf{c})$. Also the identity $\mathbf{n} \times (\mathbf{n} \times \partial_t \mathbf{n}) = -\partial_t \mathbf{n}$ is used in deriving the term on the left of the Landau-Lifshitz equation.

7.4 Antiferromagnetic State

In Sec. 7.2, it was shown that for sufficiently large U, the paramagnetic ground state of the Hubbard model is unstable towards the formation of the ferromagnetic state characterized by a finite uniform magnetization. In this section, the stability of the paramagnetic state towards the formation of the *antiferromagnetic Néel* state is considered, where the magnetization flips sign when going from one lattice site to a neighboring one. It turns out that for the half-filled Hubbard model on a square lattice, the paramagnetic state is unstable towards the formation of the Néel state however weak the coupling may be.

In terms of the auxiliary field $\boldsymbol{\phi}_c$ satisfying Eq. (7.25), the Néel state on a square lattice is specified as follows:

$$\boldsymbol{\phi}_c(x) = (-1)^{(x^1+x^2)/a} v\mathbf{n}(x) = v\cos(\mathbf{Q} \cdot \mathbf{x})\mathbf{n}(x), \tag{7.61}$$

where $v = |\mathbf{v}| > 0$ is a constant as before, $\mathbf{n}$ a three-component unit vector in the direction of the staggered magnetization, and $\mathbf{Q} \equiv (\pi/a, \pi/a)$. This wave vector, mapping the lower left part of the Fermi "surface" in Fig. 7.1 onto the upper right part, is known as a *nesting vector*. Recall that lattice sites on the square lattice are specified by their position vector $\mathbf{x} = (x^1, x^2)$ so that x^1/a and x^2/a, with a the lattice spacing, are integers. The Lagrangian density (7.21) with $\boldsymbol{\phi}$ replaced with $\boldsymbol{\phi}_{\rm c}$ becomes

$$\mathcal{L} = \psi^*_\sigma(x)\left[\mathrm{i}\hbar\partial_t - \epsilon(-\mathrm{i}\nabla) + \mu + \tfrac{1}{2}v\cos(\mathbf{Q}\cdot\mathbf{x})\,\mathbf{n}\cdot\boldsymbol{\sigma}\right]\psi_\sigma(x) - \frac{3}{8}\frac{1}{Ua^2}\boldsymbol{\phi}_{\rm c}^2(x), \quad (7.62)$$

or in Fourier-transformed form

$$\mathcal{L} = \frac{1}{2}\Psi^\dagger(k)\begin{pmatrix} \hbar\omega - \xi(\mathbf{k}) & \tfrac{1}{2}v\mathbf{n}\cdot\boldsymbol{\sigma} \\ \tfrac{1}{2}v\mathbf{n}\cdot\boldsymbol{\sigma} & \hbar\omega + \xi(\mathbf{k}) \end{pmatrix}\Psi(k) - \frac{3}{8}\frac{1}{Ua^2}\boldsymbol{\phi}_{\rm c}^2(k), \quad (7.63)$$

where $\xi(\mathbf{k}) \equiv \epsilon(\mathbf{k}) - \mu$ is the kinetic energy relative to the chemical potential and where for computational convenience we introduced the four-component spinor

$$\Psi(k) \equiv \begin{pmatrix} \psi(\omega, \mathbf{k}) \\ \psi(\omega, \mathbf{k} - \mathbf{Q}) \end{pmatrix}, \quad \psi(k) \equiv \begin{pmatrix} \psi_\uparrow(k) \\ \psi_\downarrow(k) \end{pmatrix}. \quad (7.64)$$

For hopping between nearest neighbors only, $\epsilon(\mathbf{k})$ is given by Eq. (7.9) and satisfies $\epsilon(\mathbf{k} + \mathbf{Q}) = -\epsilon(\mathbf{k})$. The Feynman propagator $S_{\rm F}(k)$ for the Ψ field, defined by the quadratic part of the Lagrangian density (7.63), reads

$$\begin{aligned} S_{\rm F}(k) &= \begin{pmatrix} \hbar\omega - \xi(\mathbf{k}) & \tfrac{1}{2}v\mathbf{n}\cdot\boldsymbol{\sigma} \\ \tfrac{1}{2}v\mathbf{n}\cdot\boldsymbol{\sigma} & \hbar\omega + \xi(\mathbf{k}) \end{pmatrix}^{-1} \\ &= \frac{1}{\hbar^2\omega^2 - E^2(\mathbf{k}) + \mathrm{i}\eta}\begin{pmatrix} \hbar\omega + \xi(\mathbf{k}) & -\tfrac{1}{2}v\mathbf{n}\cdot\boldsymbol{\sigma} \\ -\tfrac{1}{2}v\mathbf{n}\cdot\boldsymbol{\sigma} & \hbar\omega - \xi(\mathbf{k}) \end{pmatrix}, \end{aligned} \quad (7.65)$$

with $E(\mathbf{k})$ the spectrum of the elementary excitations

$$E(\mathbf{k}) = \sqrt{\xi^2(\mathbf{k}) + \tfrac{1}{4}v^2}. \quad (7.66)$$

In the Néel state, the fermionic excitations are seen to acquire an energy gap $\Delta \equiv v/2$ at the Fermi "surface" of the paramagnetic state, which at half filling ($\mu = 0$) is defined by the equation $\epsilon(\mathbf{k}) = 0$, leading to $\xi(\mathbf{k}) = 0$. In other words, the states in the vicinity of the Fermi surface disappear, and the metallic state gives way to an insulating state.

We proceed to extract a self-consistent equation for the energy gap, called the *gap equation*. To this end, we cast Eq. (7.25) in terms of v as

$$\frac{\partial}{\partial v}\Gamma = 0, \quad (7.67)$$

where $\Gamma = S_{\rm c} + \Gamma_1$ is the effective action in a background $\mathbf{n}$ field with

$$\Gamma_1 = -\frac{\mathrm{i}}{2}\hbar\,\mathrm{Tr}\ln\begin{pmatrix} p_0 - \xi(\mathbf{p}) & \tfrac{1}{2}v\mathbf{n}\cdot\boldsymbol{\sigma} \\ \tfrac{1}{2}v\mathbf{n}\cdot\boldsymbol{\sigma} & p_0 + \xi(\mathbf{p}) \end{pmatrix}, \quad (7.68)$$

and

$$S_c = -\frac{3}{8}\frac{1}{U}\int dt \sum_{\mathbf{x}} \phi_c^2 = -\frac{3}{8}\frac{1}{U}\int dt \sum_{\mathbf{x}} v^2 \cos^2(\mathbf{Q}\cdot\mathbf{x})$$
$$= -\frac{3}{16}\frac{1}{U}\int dt \sum_{\mathbf{x}} v^2. \tag{7.69}$$

More explicitly, the gap equation reads

$$\frac{3}{8}\frac{1}{Ua^2}v = -\frac{i}{4}\hbar\,\mathrm{tr}\int \frac{d^3k}{(2\pi)^3} S_F(k)\begin{pmatrix} 0 & \mathbf{n}\cdot\sigma \\ \mathbf{n}\cdot\sigma & 0 \end{pmatrix}, \tag{7.70}$$

with S_F the Feynman propagator (7.65), and where use is made of the relation

$$\delta\,\mathrm{Tr}\ln A = \mathrm{Tr}\,A^{-1}\delta A. \tag{7.71}$$

After the trace over the discrete indices is taken, Eq. (7.70) assumes the form

$$\frac{3}{8}\frac{1}{Ua^2} = \frac{i}{2}\hbar\int \frac{d^3k}{(2\pi)^3}\frac{1}{\hbar^2\omega^2 - E^2(\mathbf{k}) + i\eta}. \tag{7.72}$$

The frequency integral is readily evaluated, with the result

$$\frac{3}{8}\frac{1}{Ua^2} = \int' \frac{d^2k}{(2\pi)^2}\frac{1}{E(\mathbf{k})}, \tag{7.73}$$

where the prime on the integral is to indicate that only wave vectors in the upper right quadrant of the first Brillouin zone are included,

$$\int' \frac{d^2k}{(2\pi)^2} \equiv \frac{1}{(2\pi)^2}\int_0^{\pi/a} dk^1 \int_0^{\pi/a-k^1} dk^2. \tag{7.74}$$

By symmetry, the other quadrants give the same contribution. This has been accounted for in Eq. (7.73) by including a factor of 4 at the right side. The integral obtains its main contributions from wave vectors around the Fermi "surface" of the paramagnetic state $k^1a + k^2a = \pi$ for which $\epsilon(\mathbf{k}) = 0$. Using the trigonometric formulas

$$\cos(k^1a) + \cos(k^2a) = 2\cos\left(\frac{k^1a + k^2a}{2}\right)\cos\left(\frac{k^1a - k^2a}{2}\right), \tag{7.75}$$

$$\cos\left(\frac{k^1a + k^2a}{2}\right) = \sin\left(\frac{\pi - (k^1a + k^2a)}{2}\right) \approx \frac{1}{2}\left[\pi - (k^1a + k^2a)\right], \tag{7.76}$$

together with the approximation of setting the last cosine in Eq. (7.75) to unity, we arrive at the linear approximation for the kinetic energy $\epsilon(\mathbf{k}) \approx 2t\left[(k^1 + k^2)a - \pi\right]$. The wave vector integration can then be carried out in closed form to yield in the weak-coupling limit $U << t$:

$$\frac{1}{U} = \frac{1}{3\pi t}\left[\ln\left(\frac{8\pi t}{v}\right) - 1\right], \tag{7.77}$$

or

$$v \propto t\,\mathrm{e}^{-3\pi t/U}, \tag{7.78}$$

showing that a nontrivial solution to the gap equation exists for arbitrary small on-site Coulomb repulsion. This is markedly different from what was found in Sec. 7.2 for the ferromagnetic state, where U must be larger than a critical value to support the ordered state.

In the strong-coupling limit $U >> t$, $E(\mathbf{k}) \approx \frac{1}{2}v$, and the gap equation (7.73) gives

$$\Delta = U/3, \tag{7.79}$$

showing that the energy gap of the fermionic excitations is proportional to the on-site Coulomb repulsion in this limit.

7.5 Antiferromagnetic Spin Waves

We next derive the effective theory describing antiferromagnetic spin waves. As in the ferromagnetic case, we rotate the unit vector in the fixed third direction of spin space by using Eq. (7.33). The propagator (7.65) then simplifies to

$$S_{\mathrm{F}}(k) = \frac{1}{\hbar^2\omega^2 - E^2(\mathbf{k}) + i\eta}\begin{pmatrix} \hbar\omega + \xi(\mathbf{k}) & \frac{1}{2}v\sigma^3 \\ \frac{1}{2}v\sigma^3 & \hbar\omega - \xi(\mathbf{k}) \end{pmatrix}. \tag{7.80}$$

Introducing the decomposition (7.34), which amounts to inserting unity $1 = ss^\dagger$ into the argument of the logarithm in Eq. (7.68) and using the cyclic property of the trace, we obtain as one-fermion-loop effective action

$$\Gamma_1 = -\frac{\mathrm{i}}{2}\hbar\,\mathrm{Tr}\ln\begin{pmatrix} p_0 - V_0 - \xi(\mathbf{p}+\mathbf{V}) & -\frac{1}{2}v\sigma^3 \\ -\frac{1}{2}v\sigma^3 & p_0 - V_0 + \xi(\mathbf{p}+\mathbf{V}) \end{pmatrix}, \tag{7.81}$$

with V^μ as in Eq. (7.36).

We first concentrate on the V_0 terms and set $\mathbf{V}$ equal to zero for the moment. It is easily checked that the contribution linear in V_0 generated by the first term in the expansion of the logarithm in Eq. (7.81) vanishes. This is different from the ferromagnetic case, where such a term is present, see Eq. (7.42). The second term in the expansion,

$$\Gamma_1^{(\ell=2)}[V_0] = \frac{\mathrm{i}}{2}\hbar\,\mathrm{Tr}\left[S_{\mathrm{F}}\begin{pmatrix} V_0 & 0 \\ 0 & V_0 \end{pmatrix}\right]^2, \tag{7.82}$$

yields the contribution

$$\mathcal{L}_1^{(\ell=2)}(V_0) = \mathrm{i}\hbar\,\mathrm{tr}\int\frac{\mathrm{d}^3k}{(2\pi)^3}\frac{1}{(\hbar^2\omega^2 - E^2 + i\eta)^2}\left[(\hbar^2\omega^2 + \xi^2)V_0^2 + \tfrac{1}{4}v^2(\sigma^3 V_0)^2\right],$$

which upon performing the frequency integral gives for the corresponding Lagrangian density

$$\mathcal{L}_1^{(\ell=2)}(V_0) = \frac{1}{16} v^2 \,\mathrm{tr} \int \frac{\mathrm{d}^2 k}{(2\pi)^2} \frac{1}{E^3} [{V_0}^2 - (\sigma^3 V_0)^2], \tag{7.83}$$

showing that the various contributions conspire together so as to generate the combination ${V_0}^2 - (\sigma^3 V_0)^2$.

We proceed with the terms quadratic in $\mathbf{V}$ obtained by expanding the kinetic energy $\epsilon(\mathbf{k} + \mathbf{V}/\hbar)$ in a Taylor series as in Eq.(7.45). Here, the quadratic term leads to a non-zero contribution already in the first term in the expansion of the logarithm:

$$\mathcal{L}_1^{(\ell=1)}(\mathbf{V}) = \frac{\mathrm{i}}{2} \frac{1}{\hbar} \,\mathrm{tr} \int \frac{\mathrm{d}^3 k}{(2\pi)^3} S_{\mathrm{F}}(k) V^i V^j \frac{\partial^2 \epsilon}{\partial k^i \partial k^j} \begin{pmatrix} 1 & 0 \\ 0 & -1 \end{pmatrix}. \tag{7.84}$$

Carrying out the frequency integral, we obtain after integrating by parts

$$\mathcal{L}_1^{(\ell=1)}(\mathbf{V}) = \frac{1}{2} \frac{1}{\hbar^2} \,\mathrm{tr} \int \frac{\mathrm{d}^2 k}{(2\pi)^2} \frac{\epsilon}{E} V^i V^j \frac{\partial^2 \epsilon}{\partial k^i \partial k^j} = -\frac{1}{8} \frac{v^2}{\hbar^2} \,\mathrm{tr} \int \frac{\mathrm{d}^2 k}{(2\pi)^2} \frac{1}{E^3} (\mathbf{V} \cdot \nabla_{\mathbf{k}} \epsilon)^2, \tag{7.85}$$

where $\nabla_{\mathbf{k}} \equiv (\partial/\partial k^1, \partial/\partial k^2)$.

The final contribution of interest to us originates from the linear term in the Taylor series (7.45) and arises in the second term of the expansion of the logarithm. It reads

$$\begin{aligned} \mathcal{L}_1^{(\ell=2)}(\mathbf{V}) &= \frac{\mathrm{i}}{2} \frac{1}{\hbar} \,\mathrm{tr} \int \frac{\mathrm{d}^3 k}{(2\pi)^3} \left[S_{\mathrm{F}}(k) \mathbf{V} \cdot \nabla_{\mathbf{k}} \epsilon \begin{pmatrix} 1 & 0 \\ 0 & -1 \end{pmatrix} \right]^2 \\ &= \frac{1}{16} \frac{v^2}{\hbar^2} \,\mathrm{tr} \int \frac{\mathrm{d}^2 k}{(2\pi)^2} \frac{1}{E^3} \left[(\mathbf{V} \cdot \nabla_{\mathbf{k}} \epsilon)^2 + \left(\sigma^3 \mathbf{V} \cdot \nabla_{\mathbf{k}} \epsilon \right)^2 \right]. \end{aligned} \tag{7.86}$$

For an isotropic lattice,

$$\int \frac{\mathrm{d}^D k}{(2\pi)^D} \frac{\partial}{\partial k^i} \frac{\partial}{\partial k^j} = \frac{1}{D} \int \frac{\mathrm{d}^2 k}{(2\pi)^2} \left(\frac{\partial}{\partial k^l} \right)^2 \delta^{ij}, \tag{7.87}$$

with D the number of space dimensions, so that adding the last two contributions, we obtain the combination $\mathbf{V}^2 - (\sigma^3 \mathbf{V})^2$.

Pasting the contributions together and using the identity (7.37), we finally arrive at the *nonlinear sigma model* describing the antiferromagnetic spin waves or the Nambu-Goldstone mode at low frequencies and long wave lengths

$$\mathcal{L}_1 = \frac{1}{2} \rho \left[(\partial_t \mathbf{n})^2 - c^2 (\partial_i \mathbf{n})^2 \right], \tag{7.88}$$

with the spin-wave stiffness ρ and velocity c formally given by the integrals

$$\rho = \frac{1}{8}\frac{v^2}{\hbar^2}\int_{\mathbf{k}}\frac{d^2k}{(2\pi)^2}\frac{1}{E^3} \tag{7.89}$$

$$\rho c^2 = \frac{1}{16}v^2\int_{\mathbf{k}}\frac{d^2k}{(2\pi)^2}\frac{(\nabla_{\mathbf{k}}\,\epsilon)^2}{E^3}, \tag{7.90}$$

and where it is recalled that $\mathbf{n}$ is a unit vector, $\mathbf{n}^2 = 1$. The nonlinear sigma model gives rise to a linear spectrum $\omega^2 \propto \mathbf{k}^2$ for the antiferromagnetic spin waves, which is to be compared with the quadratic spectrum $\omega \propto \mathbf{k}^2$ found for the ferromagnetic spin waves.

Notes

(i) The tight-binding model (7.1), now bearing his name, was introduced by Hubbard (1963, 1964a,b).

(ii) The ferromagnetic state of the Hubbard model is briefly covered in the textbook [Mattuck (1976)]. A more extensive treatment, which includes the antiferromagnetic spin waves, is given in the textbook [Fradkin (1991)] from which we have freely borrowed.

(iii) Spin waves in ferromagnets were predicted by Bloch (1930).

(iv) The Landau-Lifshitz equation was introduced in [Landau and Lifshitz (1935)].

(v) The concept of an order parameter is due to Bloch (1932).

(vi) The general notion of an order parameter in connection with changes of symmetry in phase transitions was put forward by Landau (1937).

(vii) For a discussion of the difference between ferro- and antiferromagnetic spin waves, see [Anderson (1984)]. The importance of a Berry phase for obtaining a quadratic spectrum in the ferromagnetic case was pointed out by Wen and Zee (1988).

Chapter 8

Superconductivity

This chapter presents the theory of superconductivity, starting from the renowned microscopic theory of Bardeen, Cooper, and Schrieffer (BCS). According to BCS theory, superconductivity emerges from an instability of the conduction electrons in a metal towards the formation of pairs. At sufficiently low temperatures, electrons on opposite sides of the Fermi surface pair up to form so-called Cooper pairs which subsequently condense. To obtain an effective description of the superconducting state, the fermionic degrees of freedom are integrated out in favor of a bosonic field describing the Cooper pair condensate. The resulting effective theory is the celebrated Ginzburg-Landau theory which predates the BCS theory by several years. The derivative expansion method is applied to obtain the effective theory not only near the transition temperature but also in the zero-temperature regime. Most striking physical properties of superconductors are extracted from the effective theory. Subjects which are often considered to be somewhat involved, such as the time-dependent Ginzburg-Landau theory and gapless superconductivity, are easily accounted for in this approach.

8.1 BCS Theory

The basis of the BCS theory of superconductivity is the observation that at sufficiently low temperatures the metallic state is unstable. Whenever an attractive interaction is present between them, no matter how weak, electrons on opposite sides of the Fermi surface eventually form Cooper pairs. In conventional superconductors, the attractive interaction between the conduction electrons is mediated by the phonons associated with the underlying crystal lattice formed by the atoms.

A local Lagrangian density describing such a system reads

$$\mathcal{L} = \sum_{\sigma=\uparrow,\downarrow} \psi_\sigma^*(x)\left(\mathrm{i}\hbar\frac{\partial}{\partial t} + \frac{\hbar^2}{2m}\nabla^2 + \mu\right)\psi_\sigma(x) - \lambda\psi_\uparrow^*(x)\psi_\downarrow^*(x)\psi_\downarrow(x)\psi_\uparrow(x), \qquad (8.1)$$

where the terms quadratic in the fields, which we denote by $\mathcal{L}_0$, describe an ideal Fermi gas, and the last term with coupling constant $\lambda < 0$ is a local interaction term. It represents the effective, phonon mediated, attraction between electrons. The field $\psi_\sigma(x)$ is an anticommuting Grassmann field that describes electrons of mass m and spin up ($\sigma = \uparrow$) and spin down ($\sigma = \downarrow$), and μ is the chemical potential. The Lagrangian density (8.1) is invariant under global U(1) transformations, under which

$$\psi_\sigma(x) \to \psi'_\sigma(x) = \mathrm{e}^{\mathrm{i}\alpha}\psi_\sigma(x), \qquad (8.2)$$

with α a constant transformation parameter. Notwithstanding its simple form, the microscopic model (8.1) is a good starting point to describe conventional superconductors. This is because the interaction term allows for the formation of Cooper pairs in a relative s-wave orbital state which below a transition temperature condense and thereby spontaneously break the global U(1) symmetry. This in turn gives rise to a gapless Nambu-Goldstone mode which, after incorporating the electromagnetic fields, lies at the root of most characteristic and startling properties of superconductors.

The field describing the Cooper pairs naturally satisfies Landau's definition of an order parameter. First, its average is zero in the symmetric, disordered state and nonzero in the state with broken symmetry. Second, it directly signals whether the global U(1) symmetry is spontaneously broken. Since an order parameter represents the essential degrees of freedom in the ordered state, it is expedient to express the partition function as a functional integral over these fields. In the next section, this program is carried out for the BCS theory.

8.2 Gap Equation

For computational convenience, we introduce Nambu's notation and rewrite the Lagrangian density (8.1) in terms of a two-component spinor

$$\Psi = \begin{pmatrix} \psi_\uparrow \\ \psi_\downarrow^* \end{pmatrix}, \quad \Psi^\dagger = (\psi_\uparrow^*, \psi_\downarrow). \qquad (8.3)$$

The free part $\mathcal{L}_0$ becomes in this notation

$$\mathcal{L}_0 = \Psi^\dagger \begin{pmatrix} \mathrm{i}\hbar\partial_t + \hbar^2\nabla^2/2m + \mu & 0 \\ 0 & \mathrm{i}\hbar\partial_t - \hbar^2\nabla^2/2m - \mu \end{pmatrix} \Psi, \qquad (8.4)$$

where it is used that the fields anticommute, and total derivatives have been neglected.

To be able to integrate out the fermionic degrees of freedom, the zero-temperature partition function represented as a functional integral,

$$Z = \int \mathrm{D}\Psi^\dagger \mathrm{D}\Psi \exp\left(\frac{\mathrm{i}}{\hbar}\int \mathrm{d}^d x\, \mathcal{L}\right), \tag{8.5}$$

must be written in a form bilinear in the electron fields. This is achieved by applying a Hubbard-Stratonovich transformation to cast the quartic interaction term as a functional integral over auxiliary fields Δ and Δ^*

$$\exp\left(-\frac{\mathrm{i}}{\hbar}\lambda \int \mathrm{d}^d x\, \psi_\uparrow^* \psi_\downarrow^* \psi_\downarrow \psi_\uparrow\right) = \int \mathrm{D}\Delta^* \mathrm{D}\Delta \exp\left[-\frac{\mathrm{i}}{\hbar}\int \mathrm{d}^d x \left(\Delta^* \psi_\downarrow \psi_\uparrow + \psi_\uparrow^* \psi_\downarrow^* \Delta - \frac{1}{\lambda}|\Delta|^2\right)\right], \tag{8.6}$$

where, as always, an overall normalization factor is omitted. The field equation for Δ,

$$\Delta = \lambda \psi_\downarrow \psi_\uparrow, \tag{8.7}$$

shows that the auxiliary field physically describes electron pairs forming a spin-singlet in a relative s-wave orbital state. It would therefore be more precise to include the spin labels and denote the auxiliary field by $\Delta_{\downarrow\uparrow}$. Because $\psi_\uparrow$ and $\psi_\downarrow$ anticommute, the auxiliary field, which is a spin singlet, is antisymmetric in these indices. The partition function now assumes the form

$$Z = \int \mathrm{D}\Psi^\dagger \mathrm{D}\Psi \int \mathrm{D}\Delta^* \mathrm{D}\Delta \, \exp\left(\frac{\mathrm{i}}{\hbar}\frac{1}{\lambda}\int \mathrm{d}^d x\, |\Delta|^2\right) \times \exp\left[\frac{\mathrm{i}}{\hbar}\int \mathrm{d}^d x\, \Psi^\dagger \begin{pmatrix} \mathrm{i}\hbar\partial_t + \hbar^2\nabla^2/2m + \mu & -\Delta \\ -\Delta^* & \mathrm{i}\hbar\partial_t - \hbar^2\nabla^2/2m - \mu \end{pmatrix} \Psi\right]. \tag{8.8}$$

Changing the order of integration and performing the Gaussian integration over the Grassmann fields, we obtain

$$Z = \int \mathrm{D}\Delta^* \mathrm{D}\Delta\, \mathrm{e}^{(\mathrm{i}/\hbar)S[\Delta^*,\Delta]}, \tag{8.9}$$

where $S[\Delta^*, \Delta]$ stands for the action

$$S[\Delta^*, \Delta] \equiv \frac{1}{\lambda}\int \mathrm{d}^d x\, |\Delta(x)|^2 - \mathrm{i}\hbar \,\mathrm{Tr}\ln\begin{pmatrix} p_0 - \xi(\mathbf{p}) & -\Delta(x) \\ -\Delta^*(x) & p_0 + \xi(\mathbf{p}) \end{pmatrix}, \tag{8.10}$$

with $\xi(\mathbf{k}) = \hbar^2\mathbf{k}^2/2m - \mu$. As before, we adopt the convention that the momentum operator $p_\mu = \mathrm{i}\hbar\partial_\mu$ acts on all the fields to its right, while the ordinary derivative ∂_μ acts only on the next field to its right.

The effective action $\Gamma[\Delta_c^*, \Delta_c]$ is given by the right side of Eq. (8.10) with the pair field Δ replaced with its average, satisfying the field equation

$$\frac{\delta\Gamma[\Delta_c^*, \Delta_c]}{\delta\Delta_c^*(x)} = 0. \tag{8.11}$$

If this equation yields a solution with $\Delta_c \neq 0$, the global U(1) symmetry (8.2) is spontaneously broken for

$$\Delta_c \to e^{2i\alpha}\Delta_c \neq \Delta_c \tag{8.12}$$

under this transformation. To see if this is the case, we evaluate Eq. (8.11) assuming that the solution $\Delta_c(x) = \bar{\Delta} > 0$ is a constant. The integral over spacetime contained in the trace Tr in the expression (8.10) for the effective action then simply gives a volume factor, while the momentum operators p_μ simply produce factors of $\hbar k_\mu$. The only remaining integrals in the trace Tr are those over frequency and wave vector. By virtue of the relation (7.71), we obtain

$$\frac{1}{\lambda}\bar{\Delta} = i\hbar \,\mathrm{tr} \int \frac{d^d k}{(2\pi)^d} S_F(k) \begin{pmatrix} 0 & 0 \\ -1 & 0 \end{pmatrix}, \tag{8.13}$$

where S_F is the fermion propagator of the theory (8.8),

$$S_F(k) = \begin{pmatrix} \hbar\omega - \xi(\mathbf{k}) & -\bar{\Delta} \\ -\bar{\Delta}^* & \hbar\omega + \xi(\mathbf{k}) \end{pmatrix}^{-1}$$

$$= \frac{1}{\hbar^2\omega^2 - E^2(\mathbf{k}) + i\eta} \begin{pmatrix} [\hbar\omega + \xi(\mathbf{k})]e^{i\omega 0^+} & \bar{\Delta} \\ \bar{\Delta}^* & [\hbar\omega - \xi(\mathbf{k})]e^{-i\omega 0^+} \end{pmatrix}, \tag{8.14}$$

with

$$E(\mathbf{k}) \equiv \sqrt{\xi^2(\mathbf{k}) + |\bar{\Delta}|^2} \tag{8.15}$$

the single-fermion excitation energy. As before, $\eta = 0^+$ is an infinitesimally small positive parameter that is to be set to zero at the end of the calculation, and the exponentials $\exp(\pm i\omega 0^+)$ are the usual convergence factors typical for a nonrelativistic theory, see below Eq. (2.69). For $\bar{\Delta} \to 0$, the upper left entry in the propagator reduces to the correct expression of an ideal Fermi gas, see Eq. (3.14), while the lower right entry gives, because of the use of the two-component spinor (8.3), the complex conjugate of that expression.

Equation (8.13) constitutes the celebrated *BCS gap equation*:

$$-\frac{1}{\lambda} = i\hbar \int \frac{d^d k}{(2\pi)^d} \frac{1}{\hbar^2\omega^2 - E^2(\mathbf{k}) + i\eta}. \tag{8.16}$$

The frequency integral is readily evaluated with the help of contour integration, giving

$$-\frac{1}{\lambda} = \frac{1}{2} \int \frac{d^D k}{(2\pi)^D} \frac{1}{E(\mathbf{k})}. \tag{8.17}$$

The integral on the right diverges in the ultraviolet, where the excitation energy $E(\mathbf{k})$ can be replaced with that of a noninteracting Fermi gas, $|\xi(\mathbf{k})|$. The theory can be rendered finite to this order by introducing the renormalized coupling constant λ_r through

$$\frac{1}{\lambda_r} \equiv \frac{1}{\lambda} + \frac{1}{2}\int \frac{\mathrm{d}^D k}{(2\pi)^D} \frac{1}{|\xi(\mathbf{k})|}. \tag{8.18}$$

Recall that the renormalized coupling constant is the physical one, which is accessible in experiment. The diverging integral in Eq. (8.18) can be regularized by introducing a wave number cutoff Λ. In $D = 3$, Eq. (8.18) then assumes the form

$$\frac{1}{\lambda_r} = \frac{1}{\lambda} + \frac{1}{2\pi^2}\frac{m}{\hbar^2}\Lambda + O\left(\Lambda^0\right), \tag{8.19}$$

where the part of the integral that remains finite in the limit $\Lambda \to \infty$ has been omitted. Two limits can be distinguished here. The *weak-coupling BCS limit* corresponds to taking the bare coupling constant to zero, $\lambda \to 0^-$. In this limit, the first term on the right of Eq. (8.19) dominates so that $1/\lambda_r \to -\infty$. In the opposite limit, obtained by letting $\lambda \to -\infty$, the second term on the right of Eq. (8.19) dominates so that $1/\lambda_r \to +\infty$ now when the cutoff is send to infinity. In this limit, which will be covered in Chap. 10, the two-particle interaction is such that the electrons form tightly bound pairs of mass $2m$.

In dimensional regularization, where divergences proportional to powers of the cutoff never show up, no renormalization is required here. With this kept in mind, we drop the index "r" on λ in the following.

In the weak-coupling limit, the integrand in the gap equation (8.17) is peaked near $\xi(\mathbf{k}) = 0$. The integrals over wave vector can therefore be approximated by

$$\int \frac{\mathrm{d}^3 k}{(2\pi)^3} \approx \nu(0) \int_{-\hbar\omega_D}^{\hbar\omega_D} \mathrm{d}\xi \int \frac{\mathrm{d}\Omega_\mathbf{k}}{4\pi} \tag{8.20}$$

with $\int \mathrm{d}\Omega_\mathbf{k}$ denoting the angle integrals and $\nu(0) = mk_F/2\pi^2\hbar^2$ the density of states per spin degree of freedom at the Fermi surface of the metallic state, and where we specialized to $D = 3$. The integral over ξ in Eq. (8.21) is cut off at ($\hbar$ times) the so-called Debye frequency ω_D. This cutoff is introduced to account for the fact that phonons, which mediate the attractive interaction between the conduction electrons, can at most transfer an energy $\hbar\omega_D$. Since this energy, which is a measure of the inverse lattice spacing, is typically of order 10^2 K, while the Fermi temperature is typically 10^4 K, the interaction is nonzero only in a thin shell around the Fermi surface of the metallic state, and most electrons are not affected. With the approximation (8.20), the zero-temperature gap equation takes the form

$$-\frac{1}{\lambda} = \frac{1}{2}\nu(0) \int_{-\hbar\omega_D}^{\hbar\omega_D} \mathrm{d}\xi \frac{1}{E} = \nu(0) \ln\left(\frac{2\hbar\omega_D}{|\bar{\Delta}|}\right), \tag{8.21}$$

which indeed has a nontrivial solution

$$|\bar{\Delta}| = 2\hbar\omega_{\rm D}\, {\rm e}^{1/\nu(0)\lambda}, \tag{8.22}$$

where it is to be remembered that the coupling constant λ is negative. This shows that, at the absolute zero of temperature, the field Δ describing the Cooper pairs develops a nonzero average, implying that the Cooper pairs have condensed into a single quantum state, and that global U(1) symmetry is spontaneously broken. Moreover, the spectrum (8.15) is gapped, meaning that a minimum energy of $|\bar{\Delta}|$ is required to excite a fermionic particle in the superconducting state. The instability of the conduction electrons towards the formation of Cooper pairs occurs however small the effective attraction between the electrons may be.

Note that these results hinge on the specific form (8.6) of the Hubbard-Stratonovich transformation chosen, with the decoupling of the quartic interaction carried out in the Cooper channel, see Eq. (8.7). The two other possibilities are a decoupling in the *direct channel*, where the analog of Eq. (8.7) reads $\Delta \propto \psi^*_\sigma \psi_\sigma$ with $\sigma =\uparrow$ or $\downarrow$, or in the *exchange channel*, $\Delta \propto \psi^*_\sigma \psi_{\sigma'}$ with $\sigma \neq \sigma'$.

8.3 Anderson-Bogoliubov Mode

In the previous section, the order parameter Δ was assumed to be constant. We now somewhat relax this constraint by allowing the phase of the order parameter to vary in spacetime,

$$\Delta_{\rm c}(x) = |\bar{\Delta}|\, {\rm e}^{2{\rm i}\varphi(x)}, \tag{8.23}$$

where $|\bar{\Delta}|$ is the constant solution (8.22) of the zero-temperature gap equation. This can be achieved by a U(1) transformation (8.2) with the constant transformation parameter α replaced with a spacetime-dependent phase $\varphi(x)$, see Eq. (8.7). The approximation, where the phase of an order parameter depends on spacetime while the absolute value is constant, is called the *London limit*. The phase field $\varphi(x)$ physically represents the Nambu-Goldstone field emerging from the spontaneously broken global U(1) symmetry. The objective is to derive the effective theory describing this collective excitation known as the *Anderson-Bogoliubov mode*. After the local U(1) transformation, we obtain, instead of Eq. (8.10), the one-fermion-loop effective action

$$\Gamma_1[V_\mu] = -{\rm i}\hbar\, {\rm Tr}\ln \begin{pmatrix} p_0 - V_0 - \xi(\mathbf{p}+\mathbf{V}) & -\bar{\Delta} \\ -\bar{\Delta}^* & p_0 + V_0 + \xi(\mathbf{p}-\mathbf{V}) \end{pmatrix}, \tag{8.24}$$

where $V_\mu \equiv \hbar\partial_\mu\varphi$, playing the role of an Abelian gauge field, is the derivative of the phase of the order parameter. Expanded in a power series, the effective action assumes the form

$$\Gamma_1[V_\mu] = -\mathrm{i}\hbar \operatorname{Tr} \ln \left[S_\mathrm{F}^{-1}(1 - S_\mathrm{F}\Lambda)\right]$$
$$= \mathrm{i}\hbar \operatorname{Tr} \ln(S_\mathrm{F}) + \mathrm{i}\hbar \operatorname{Tr} \sum_{\ell=1}^{\infty} \frac{1}{\ell} (S_\mathrm{F}\Lambda)^\ell, \tag{8.25}$$

where $S_\mathrm{F}(k)$ is the fermion propagator (8.14) and

$$\Lambda \equiv \frac{1}{2m}(\mathbf{p}\cdot\mathbf{V} + \mathbf{V}\cdot\mathbf{p}) + \left(V_0 + \frac{1}{2m}\mathbf{V}^2\right)\tau^3, \tag{8.26}$$

cf. (3.11). The first term at the right of Eq. (8.25) is irrelevant for our present purposes. Below, we consider only the $\ell = 1$ and $\ell = 2$ terms in the series, and ignore derivatives on the field V_μ.

The first term in the power series yields the contribution

$$\Gamma_1^{(\ell=1)} = \mathrm{i}\hbar \operatorname{Tr} S_\mathrm{F}\left[\hbar\frac{\partial\varphi}{\partial t} + \frac{\hbar^2}{2m}(\nabla\varphi)^2\right]\tau^3. \tag{8.27}$$

To illustrate the role of the convergence factors in the propagator (8.14), we evaluate in detail the frequency integral contained in the trace Tr,

$$I \equiv \hbar \int \frac{\mathrm{d}\omega}{2\pi} \frac{(\hbar\omega + \xi)\mathrm{e}^{\mathrm{i}\omega 0^+} - (\hbar\omega - \xi)\mathrm{e}^{-\mathrm{i}\omega 0^+}}{\hbar^2\omega^2 - E^2 + \mathrm{i}\eta}. \tag{8.28}$$

For the terms in the integrand proportional to ξ, the convergence factor is immaterial, and the integration over ω can be readily carried out with the help of contour integration. For the terms proportional to ω, the convergence factors do matter, as we now demonstrate. To this end, we break the first term apart into two terms with simple denominators

$$\hbar \int \frac{\mathrm{d}\omega}{2\pi} \frac{\hbar\omega \mathrm{e}^{\mathrm{i}\omega 0^+}}{\hbar^2\omega^2 - E^2 + \mathrm{i}\eta} = \frac{1}{2}\frac{\hbar}{E} \int \frac{\mathrm{d}\omega}{2\pi} \left(\frac{\hbar\omega \mathrm{e}^{\mathrm{i}\omega 0^+}}{\hbar\omega + E - \mathrm{i}\eta} - \frac{\hbar\omega \mathrm{e}^{\mathrm{i}\omega 0^+}}{\hbar\omega - E + \mathrm{i}\eta}\right), \tag{8.29}$$

where $E > 0$. Because of the convergence factor $\exp(\mathrm{i}\omega 0^+)$, the contour in the complex ω plane has to be closed in the upper half where only the first term on the right has a pole—that is, only that term contributes, yielding a factor $\frac{1}{2}\mathrm{i}$. The other term in the integral I proportional to ω can be treated similarly. Added together, the integral gives

$$I = \mathrm{i}(1 - \xi/E). \tag{8.30}$$

The effective action (8.27) therefore leads to the Lagrangian density

$$\mathcal{L}_1^{(\ell=1)} = -\bar{n}\left[\hbar\partial_t\varphi + \frac{\hbar^2}{2m}(\nabla\varphi)^2\right], \tag{8.31}$$

where $\bar{n}$ is the average electron density,

$$\bar{n} = \int \frac{\mathrm{d}^D k}{(2\pi)^D} \left(1 - \frac{\xi(\mathbf{k})}{E(\mathbf{k})}\right). \tag{8.32}$$

In the weak-coupling limit, where $|\bar{\Delta}| << \mu$, the combination $\xi(\mathbf{k})/E(\mathbf{k})$ reduces to the sign function $\mathrm{sgn}[\xi(\mathbf{k})]$, and the expression for $\bar{n}$ assumes the same form as for an ideal Fermi gas.

We continue with the second term in the power series (8.25). In evaluating this term, we use the approximation (8.20) with the range of the final integral extended to $(-\infty, \infty)$. This is justified by the rapid convergence of the integral. After some algebra, and using the integral (3.43) with $d = 1$, i.e.,

$$\hbar \int \frac{\mathrm{d}\omega}{2\pi} \frac{1}{(\hbar^2\omega^2 - E^2 + \mathrm{i}\eta)^l} = \mathrm{i} \frac{(-1)^l}{\sqrt{4\pi}} \frac{\Gamma(l - \frac{1}{2})}{\Gamma(l)} \frac{1}{(E^2)^{l-1/2}} \tag{8.33}$$

as well as its Euclidean counterpart

$$\int \mathrm{d}\xi \frac{1}{(E^2)^{l+1/2}} = \sqrt{4\pi} \frac{\Gamma(l)}{\Gamma(l + \frac{1}{2})} \frac{1}{|\bar{\Delta}|^{2l}}, \tag{8.34}$$

with $\Gamma(n)$ the gamma function, we obtain

$$\mathcal{L}_1^{(\ell=2)} = \nu(0) \left[\hbar\partial_t\varphi + \frac{\hbar^2}{2m}(\nabla\varphi)^2\right]^2. \tag{8.35}$$

The contributions (8.31) and (8.35),

$$\mathcal{L}_1^{(\ell=1,2)} = -\bar{n}\left[\hbar\partial_t\varphi + \frac{\hbar^2}{2m}(\nabla\varphi)^2\right] + \nu(0)\left[\hbar\partial_t\varphi + \frac{\hbar^2}{2m}(\nabla\varphi)^2\right]^2, \tag{8.36}$$

constitute the effective theory governing the Nambu-Goldstone field emerging from the spontaneously broken global U(1) symmetry at low frequencies and long wave lengths. It is of precisely the same form as the effective theory (4.121) describing the collective excitations in a superfluid, see Sec. 4.9. The fact that the phase of the order parameter appears in a combination which is invariant under Galilei transformations was to be expected for the microscopic theory (8.1) enjoys this invariance. The spectrum of the gapless Anderson-Bogoliubov mode that follows from the Lagrangian density (8.36) is linear with a speed of propagation given by $c_\mathrm{s} = v_\mathrm{F}/\sqrt{3}$, where $v_\mathrm{F} = \hbar k_\mathrm{F}/m$ is the Fermi velocity. It is mentioned again that the Nambu-Goldstone field appears in the effective theory only in combination with derivatives. For stable equilibrium configurations, the phase φ of the order parameter is a constant, and the system is rigid, preferring to have the same value of the phase throughout the sample. In the following sections, various physical properties of superconductors are derived from the classical field theory (8.36).

8.4 Linear Response

Up to this point, we considered neutral superconductors. As a first step towards understanding the electromagnetic properties of charged superconductors we in this section study their response to weak electromagnetic perturbations at the absolute zero of temperature. We thereto couple the effective theory (8.36) minimally to a background field A_c^μ by letting $\partial^\mu \varphi \to \partial^\mu \varphi - (e/\hbar c) A_c^\mu$, with the four-vector A^μ introduced in Eq. (2.180). Since the linear effect of an electromagnetic perturbation is to change the phase of the order parameter and not its magnitude, the complete linear response is obtained in the London limit which we consider. The linear term in Eq. (8.36) is a total derivative and not important for our present purposes. As only terms at most quadratic in the Nambu-Goldstone fields are relevant to linear response, also the higher-order terms can be ignored. The pertinent terms in the Lagrangian density thus read

$$\mathcal{L}_{\rm eff} = \nu(0) \left[(\hbar \partial_t \varphi - e\Phi_c)^2 - c_s^2 \left(\hbar \nabla \varphi + \frac{e}{c} \mathbf{A}_c \right)^2 \right], \tag{8.37}$$

which is of a form first studied by Stückelberg in the context of particle physics. The Coulomb interaction between the electrons will be accounted for shortly.

The expressions for the electric charge and current densities follow from applying Eq. (2.183) to the effective theory (8.37). Explicitly, the induced charge and current densities read

$$j_e^0 = -2\nu(0)\,(\hbar \partial_t \varphi - e\Phi_c) \tag{8.38}$$

and

$$\mathbf{j}_e = \frac{\bar{n}}{m} \left(\hbar \nabla \varphi + \frac{e}{c} \mathbf{A}_c \right). \tag{8.39}$$

The Nambu-Goldstone field can be eliminated from these expressions through its field equation,

$$(\partial_t^2 - c_s^2 \nabla^2) \varphi = \frac{e}{\hbar c} \left(c \partial_t \Phi_c + c_s^2 \nabla \cdot \mathbf{A}_c \right). \tag{8.40}$$

As the theory (8.37) is quadratic, this is tantamount to integrating out this field, with the results

$$j_e^0(k) = -\frac{e\bar{n}}{mc^2} \frac{c^2 \mathbf{k}^2 \Phi_c(k) - c\omega \mathbf{k} \cdot \mathbf{A}_c(k)}{\omega^2 - c_s^2 \mathbf{k}^2} \tag{8.41}$$

and

$$\mathbf{j}_e(k) = \frac{e\bar{n}}{mc} \left[\mathbf{A}_c(k) + \frac{c_s^2 \mathbf{k}\mathbf{k} \cdot \mathbf{A}_c(k) - c\omega \mathbf{k} \Phi_c(k)}{\omega^2 - c_s^2 \mathbf{k}^2} \right] \tag{8.42}$$

after Fourier transforming. These expressions, which are explicitly gauge invariant, satisfy charge conservation $\partial_t j_e^0 + \nabla \cdot \mathbf{j}_e = 0$. A resonance is seen to occur in the response to the external fields when their frequency and wave length are tuned such that the Anderson-Bogoliubov mode is excited, i.e., when $\omega^2 = c_s^2 \mathbf{k}^2$.

In the static limit ($\omega \to 0$), Eq. (8.42) reduces to the gauge-invariant *London equation*

$$j_e^i(\mathbf{k}) = \frac{e\bar{n}}{mc}\left(\delta^{ij} - \frac{k^i k^j}{\mathbf{k}^2}\right) A_c^j(\mathbf{k}). \tag{8.43}$$

This result can also be derived directly from the effective theory (8.37) by integrating out the Nambu-Goldstone field. This gives, in the static limit, the following Lagrangian governing the vector potential

$$L_L = -\frac{1}{8\pi}\frac{1}{\lambda_L^2}\int \frac{d^D k}{(2\pi)^D} A_c^i(-\mathbf{k})\left(\delta^{ij} - \frac{k^i k^j}{\mathbf{k}^2}\right) A_c^j(\mathbf{k}), \tag{8.44}$$

where λ_L is a characteristic length scale, known as the London penetration depth, which in $D = 3$ reads

$$\lambda_L^2 \equiv \frac{mc^2}{4\pi e^2 \bar{n}}. \tag{8.45}$$

The result (8.43) now follows by using Eq. (2.183) with $q = -e$, which translates into

$$\frac{e}{c}\mathbf{j}_e(\mathbf{k}) = -\frac{\delta L_L}{\delta \mathbf{A}_c(-\mathbf{k})}. \tag{8.46}$$

Note that a diamagnetic term of the form (8.44), which can be represented in coordinate space as

$$\begin{aligned} L_L &= -\frac{1}{8\pi}\frac{1}{\lambda_L^2}\int d^D x\, A_c^i(\mathbf{x})\left(\delta_{ij} - \frac{\partial_i \partial_j}{\nabla^2}\right) A_c^j(\mathbf{x}) \\ &= -\frac{1}{8\pi}\frac{1}{\lambda_L^2}\int d^D x\, (\nabla \times \mathbf{A}_c) \cdot \frac{1}{\nabla^2}(\nabla \times \mathbf{A}_c), \end{aligned} \tag{8.47}$$

also arises in a normal metal, see Eq. (6.54). There, however, it is canceled by a second contribution, see Eq. (6.62), while in a superconductor it is not.

Comparison of the Lagrangian (8.44) with minus the magnetic energy $-(1/8\pi)\int d^D x\, (\nabla \times \mathbf{A}_c)^2$, or

$$L_M = -\frac{1}{8\pi}\int \frac{d^D k}{(2\pi)^D} A_c^i(-\mathbf{k})\mathbf{k}^2\left(\delta^{ij} - \frac{k^i k^j}{\mathbf{k}^2}\right) A_c^j(\mathbf{k}), \tag{8.48}$$

shows that the range of variations in the vector potential becomes finite in the superconducting state. Put differently, the field $\mathbf{A}_c$ acquires a mass. This implies that a small magnetic field applied to a superconducting material is exponentially screened from the interior of the sample within the characteristic length λ_L. This property of a superconductor to expel magnetic fields from its interior was experimentally discovered by Meissner and Ochsenfeld, and is usually referred to as the *Meissner effect*. Splitting the vector potential $\mathbf{A}(\mathbf{k})$ into a longitudinal and a transverse part, $\mathbf{A}(\mathbf{k}) = \mathbf{A}^{\|}(\mathbf{k}) + \mathbf{A}^{\perp}(\mathbf{k})$, with $\mathbf{k} \times \mathbf{A}^{\|}(\mathbf{k}) = 0$ and $\mathbf{k} \cdot \mathbf{A}^{\perp}(\mathbf{k}) = 0$, i.e., explicitly

$$\mathbf{A}^{\perp}(\mathbf{k}) = \mathbf{A}(\mathbf{k}) - \frac{\mathbf{k}\mathbf{k}}{\mathbf{k}^2} \cdot \mathbf{A}(\mathbf{k}), \tag{8.49}$$

we can cast the sum of the two Lagrangians in the compact form

$$L_{\rm LM} = -\frac{1}{8\pi} \int \frac{\mathrm{d}^D k}{(2\pi)^D} \mathbf{A}_{\rm c}^{\perp}(-\mathbf{k}) \cdot (\lambda_{\rm L}^{-2} + \mathbf{k}^2) \mathbf{A}_{\rm c}^{\perp}(\mathbf{k}). \tag{8.50}$$

The Nambu-Goldstone field, which in a neutral superconductor represents a sound mode, is seen to modify the magnetic Lagrangian by endowing the vector potential with a gauge-invariant mass term. The mechanism by which a Nambu-Goldstone field loses its independent significance when coupled to a gauge theory, and produces a mass term for the photon is generally known as the *Anderson-Higgs mechanism*

8.5 Flux Quantization

Consider a piece of superconducting material with a hole in it. When the system is in its ground state, the combination $\hbar\nabla\varphi + (e/c)\mathbf{A}$ will by Eq. (8.37), with $\mathbf{A}_{\rm c}$ replaced with $\mathbf{A}$, be zero well inside the sample and away from the hole—that is, the vector potential becomes a pure gauge. Since the field 2φ introduced in Eq. (8.23) is a phase, it is defined modulo 2π. When the hole is circled once by going around it along a closed contour Γ sufficiently far away from the hole, φ may therefore have changed a multiple of π upon returning to the starting point,

$$\oint_\Gamma \mathrm{d}\mathbf{x} \cdot \nabla\varphi = -w\pi \tag{8.51}$$

with $w = 0, \pm 1, \pm 2, \cdots$ an integer. Since $\hbar\nabla\varphi + (e/c)\mathbf{A} = 0$ everywhere along the loop, this implies that

$$\oint_\Gamma \mathrm{d}\mathbf{x} \cdot \mathbf{A} = w\pi\hbar c/e. \tag{8.52}$$

The left side of this equation denotes the magnetic flux Φ through the area S bounded by the closed contour Γ

$$\Phi = \int_S \mathrm{d}\mathbf{S} \cdot (\nabla \times \mathbf{A}) = w\Phi_0 \tag{8.53}$$

so that the magnetic flux is quantized in units of $\Phi_0 \equiv \pi\hbar c/e$. In Eq. (8.53), d**S** denotes a surface element of S. Note that in a superconductor, the unit of magnetic flux is half of that we encountered before, see Eq. (5.28). This is because the condensate consists of Cooper pairs which have a charge twice that of an electron.

A nonzero magnetic flux through the hole must be supported by an electric current flowing in a layer of thickness λ_L around it. The quantization of flux then implies that this current cannot decay smoothly, but only in jumps at which the flux drops by a multiple of Φ_0. Electric currents in superconductors, which usually are referred to as *supercurrents*, are apparently not affected by impurity scattering and other processes that lead to a finite electric resistance in ordinary metals.

8.6 Plasma Mode

The presence of the gapless Anderson-Bogoliubov mode in a neutral superconductor is clearly reflected in the linear response expressions (8.41) and (8.42). This changes qualitatively when the Coulomb potential between the electrons is included. To account for this long-range interaction, we add the kinetic term $(\nabla\Phi)^2/8\pi$ to the effective theory (8.37), see Sec. 6.5. The relevant part then reads

$$\mathcal{L}_{\rm eff} = \nu(0)\,(\hbar\partial_t\varphi - e\Phi)^2 - \frac{\bar{n}}{2m}\,(\hbar\nabla\varphi)^2 + \frac{1}{8\pi}(\nabla\Phi)^2, \tag{8.54}$$

where the background fields are put to zero for the time being. To see what form the instantaneous Coulomb repulsion takes in the effective theory, we integrate out the scalar potential Φ, thereby reversing the steps in Sec. 6.5. This gives as effective action in Fourier space

$$\Gamma = \int \frac{\mathrm{d}^d k}{(2\pi)^d}\varphi(-k)\left(\nu(0)\frac{\mathbf{k}^2}{\omega_{\rm p}^2/c_{\rm s}^2 + \mathbf{k}^2}\hbar^2\omega^2 - \frac{\bar{n}}{2m}\hbar^2\mathbf{k}^2\right)\varphi(k), \tag{8.55}$$

which reduces to the expression for a neutral superconductor in the limit $e \to 0$, where the plasma frequency $\omega_{\rm p}$ introduced in Eq. (6.126) tends to zero. It follows that the long-range Coulomb interaction drastically changes the spectrum of the Anderson-Bogoliubov mode from being gapless, $\omega^2(\mathbf{k}) = c_{\rm s}^2\mathbf{k}^2$, in the absence of the Coulomb interaction, to

$$\omega^2(\mathbf{k}) = \omega_{\rm p}^2 + c_{\rm s}^2\mathbf{k}^2, \tag{8.56}$$

where the plasma frequency $\omega_{\rm p}$ denotes the lowest attainable frequency. In other words, the Coulomb repulsion between electrons turns the gapless sound mode into a gapped plasma mode. Physically, this gap arises due to charge screening.

To study the linear response of the system with the Coulomb interaction included, the derivatives in the denominator in Eq. (8.55) can be neglected so that

$$\mathcal{L}_{\rm eff} = -\frac{1}{8\pi e^2}\frac{\hbar^2}{e^2}\partial_t\varphi\nabla^2\partial_t\varphi - \frac{\hbar^2}{2m}\bar{n}\,(\nabla\varphi)^2 . \tag{8.57}$$

After coupling to the background fields $A^\mu_{\rm c}$, the Lagrangian density assumes the form

$$\mathcal{L}_{\rm eff} = -\frac{1}{8\pi e^2}\,(\hbar\partial_t\varphi - e\Phi_{\rm c})\,\nabla^2\,(\hbar\partial_t\varphi - e\Phi_{\rm c}) - \frac{\bar{n}}{2m}\left(\hbar\nabla\varphi + \frac{e}{c}\mathbf{A}_{\rm c}\right)^2, \tag{8.58}$$

which replaces the linear response Lagrangian (8.37). While the expression (8.39) for the induced current density remains unchanged, Eq. (8.38) is supplanted by

$$j^0_{\rm e} = \frac{1}{4\pi e^2}\nabla^2\,(\hbar\partial_t\varphi - e\Phi_{\rm c})\,. \tag{8.59}$$

The Nambu-Goldstone field can again be eliminated from these expressions through its field equation, which instead of Eq. (8.40) now reads

$$(\partial_t^2 + \omega_{\rm p}^2)\nabla^2\varphi = \frac{e}{\hbar c}\left(c\partial_t\nabla^2\Phi_{\rm c} - \omega_{\rm p}^2\nabla\cdot\mathbf{A}_{\rm c}\right), \tag{8.60}$$

giving

$$j^0_{\rm e}(k) = -\frac{1}{4\pi e c^2}\frac{\omega_{\rm p}^2}{\omega^2-\omega_{\rm p}^2}\left[c^2\mathbf{k}^2\Phi_{\rm c}(k) - c\omega\mathbf{k}\cdot\mathbf{A}_{\rm c}(k)\right] \tag{8.61}$$

and

$$\mathbf{j}_{\rm e}(k) = \frac{e\bar{n}}{mc}\left(\mathbf{A}_{\rm c}(k) + \frac{\omega_{\rm p}^2\mathbf{k}\mathbf{k}\cdot\mathbf{A}_{\rm c}(k)/\mathbf{k}^2 - c\omega\mathbf{k}\Phi_{\rm c}(k)}{\omega^2-\omega_{\rm p}^2}\right) \tag{8.62}$$

after Fourier transforming. Resonances in the response to the background fields now occur at long wave lengths when the frequency of the external fields coincides with the plasma frequency, i.e., when the plasma mode is excited.

8.7 Josephson Effect

The identification (4.118) of the chemical potential in the effective theory of a superfluid also holds for a neutral superconductor, i.e., $\mu(x) = -\hbar\partial_t\varphi(x)$ if the nonlinear term is ignored. For a charged superconductor, this translates into

$$\mu(x) = -\hbar\partial_t\varphi(x) + e\Phi(x) \tag{8.63}$$

after minimal substitution. Taking the gradient, we obtain the linear Euler equation, cf. Eq. (1.164)

$$\nabla\mu = -\partial_t\mathbf{p}_{\rm s} - e\mathbf{E}, \tag{8.64}$$

with

$$\mathbf{p}_s \equiv \hbar\nabla\varphi + \frac{e}{c}\mathbf{A} \tag{8.65}$$

the superfluid momentum.

Consider two pieces of superconducting material connected over a small area through a barrier consisting of, say, an insulator. The junction must be thin enough so that charges can tunnel through. Such a coupling of two superconductors is referred to as a *weak link* or *Josephson junction*. The energy of the junction is a function of the gauge-invariant phase difference $\Delta_{\mathbf{A}}\varphi$,

$$\Delta_{\mathbf{A}}\varphi \equiv \int_\Gamma \mathbf{dx}\cdot\left(\nabla\varphi + \frac{e}{\hbar c}\mathbf{A}\right) = \Delta\varphi + \frac{e}{\hbar c}\int_\Gamma \mathbf{dx}\cdot\mathbf{A} \tag{8.66}$$

where the line integral is along a path Γ through the junction connecting the two superconductors. The additional term assures that the variable $\Delta_{\mathbf{A}}\varphi$ is gauge invariant and therefore physical. As the current through the junction is obtained by taking the functional derivative of the junction energy with respect to $\mathbf{A}$, it too is a function f of $\Delta_{\mathbf{A}}\varphi$, satisfying

$$f(\Delta_{\mathbf{A}}\varphi) = f(\Delta_{\mathbf{A}}\varphi + \pi) \tag{8.67}$$

for φ is a phase variable with periodicity π.

Applying Eq. (8.63) to both superconductors, we conclude that the time dependence of $\Delta_{\mathbf{A}}\varphi$ is determined by

$$\begin{aligned}\partial_t\Delta_{\mathbf{A}}\varphi &= -\frac{1}{\hbar}\Delta\mu + \frac{e}{\hbar}\Delta\Phi + \frac{e}{\hbar c}\int_\Gamma \mathbf{dx}\cdot\partial_t\mathbf{A} \\ &= -\frac{1}{\hbar}\int_\Gamma \mathbf{dx}\cdot(\nabla\mu + e\mathbf{E}) \\ &= -\frac{1}{\hbar}\int_\Gamma \mathbf{dx}\cdot\nabla\mu_e,\end{aligned} \tag{8.68}$$

where we introduced the *electrochemical potential* μ_e in the last line. The line integral on the right equals ($-e$ times) the voltage U applied across the junction so that

$$\partial_t\Delta_{\mathbf{A}}\varphi = \frac{e}{\hbar}U. \tag{8.69}$$

If the voltage is maintained constant across the junction, this equation can be integrated to yield

$$\Delta_{\mathbf{A}}\varphi(t) = \frac{eU}{\hbar}t + \text{const.} \tag{8.70}$$

Together with the periodicity (8.67) of the current flowing through the junction, this implies that the current oscillates with an angular frequency $\omega_J = 2eU/\hbar$. This constitutes the celebrated *ac Josephson effect*. The double charge $2e$ appearing in this expression for the frequency reflects the fact that Cooper pairs tunnel through the junction.

8.8 Effective Potential

For a constant order parameter field, the effective action (8.10) reduces to an effective potential density $\mathcal{V}_{\text{eff}}$ as the integral over spacetime produces merely a volume factor, $\Gamma = -\int \mathrm{d}^d x \mathcal{V}_{\text{eff}}$. The effective potential can be evaluated in closed form for the momentum operators p_μ become simply replaced with $\hbar k_\mu$, where k_μ is the loop integration variable. Writing

$$\begin{pmatrix} \hbar\omega - \xi(\mathbf{k}) & -\bar{\Delta} \\ -\bar{\Delta}^* & \hbar\omega + \xi(\mathbf{k}) \end{pmatrix} = S_{\text{F},0}^{-1}(k)\left[1 - S_{\text{F},0}(k)\begin{pmatrix} 0 & \bar{\Delta} \\ \bar{\Delta}^* & 0 \end{pmatrix}\right], \tag{8.71}$$

where $S_{\text{F},0}(k)$ is the propagator (8.14) with $\Delta = 0$, and splitting the logarithm in two, we recognize, after expanding the second logarithm in a Taylor series, the form

$$\begin{aligned} \mathcal{V}_{\text{eff}} = &-\frac{1}{\lambda}|\bar{\Delta}|^2 + \mathrm{i}\hbar \int \frac{\mathrm{d}^d k}{(2\pi)^d} \ln \begin{pmatrix} \hbar\omega - \xi(\mathbf{k}) & 0 \\ 0 & \hbar\omega + \xi(\mathbf{k}) \end{pmatrix} \\ &+ \mathrm{i}\hbar \int \frac{\mathrm{d}^d k}{(2\pi)^d} \ln\left(1 - \frac{|\bar{\Delta}|^2}{\hbar^2\omega^2 - \xi^2(\mathbf{k})}\right), \end{aligned} \tag{8.72}$$

where an irrelevant additive constant has been dropped. The integrals over the loop frequency ω are best evaluated by first taking derivatives of the integrals with respect to ξ (first integral) and $|\bar{\Delta}|^2$ (second integral). With the $\mathrm{i}\eta$ prescription applied and the convergence factors included, the integrals can be handled as in Eq. (8.28) to yield

$$\mathcal{V}_{\text{eff}} = -\frac{1}{\lambda}|\bar{\Delta}|^2 - \int \frac{\mathrm{d}^D k}{(2\pi)^D} \left[E(\mathbf{k}) - \xi(\mathbf{k})\right], \tag{8.73}$$

where the single-fermion excitation energy $E(\mathbf{k})$ was introduced in Eq. (8.15). For $\bar{\Delta} \to 0$, Eq. (8.73) reduces to

$$\begin{aligned} \mathcal{V}_{\text{eff}} &= -\int \frac{\mathrm{d}^D k}{(2\pi)^D} \left[|\xi(\mathbf{k})| - \xi(\mathbf{k})\right] \\ &= 2\int \frac{\mathrm{d}^D k}{(2\pi)^D} \xi(\mathbf{k})\theta[-\xi(\mathbf{k})] \end{aligned} \tag{8.74}$$

appropriate for a noninteracting Fermi gas, see Eq. (6.10).

The remaining integrals in Eq. (8.73) can be evaluated in the weak-coupling limit by using the approximation (8.20), with the result for $D = 3$

$$\mathcal{V}_{\text{eff}} = -\frac{1}{\lambda}|\bar{\Delta}|^2 - \nu(0)|\bar{\Delta}|^2\left[\frac{1}{2} + \ln\left(\frac{2\hbar\omega_{\text{D}}}{|\bar{\Delta}|}\right)\right], \tag{8.75}$$

valid for $|\bar{\Delta}| << \hbar\omega_{\text{D}}$, and where an irrelevant term independent of $|\bar{\Delta}|$ has been dropped. In the mean-field approximation, where fluctuations are ignored and the order parameter is given by Eq. (8.22), the first and last term on the right cancel, and the effective potential reduces to the simple form

$$\mathcal{V}_{\text{eff}} = -\frac{\nu(0)}{2}|\bar{\Delta}|^2. \tag{8.76}$$

This physically denotes the *condensation energy* density of a conventional superconductor at the absolute zero of temperature.

8.9 Effective Action

If the order parameter varies in space and time, the effective action can no longer be evaluated in closed form. An important approximation follows from assuming that Δ is small. In that case, the effective action can be expanded in powers of the order parameter. We carry out this so-called *Landau expansion* up to fourth order. Apart from a constant term, the decomposition (8.71) leads to

$$\begin{aligned}\Gamma &= \frac{1}{\lambda}\int \mathrm{d}^d x|\Delta|^2 - \mathrm{i}\hbar\,\mathrm{Tr}\ln\left[1 - S_{\mathrm{F},0}\begin{pmatrix} 0 & \Delta \\ \Delta^* & 0 \end{pmatrix}\right] \\ &= \frac{1}{\lambda}\int \mathrm{d}^d x|\Delta|^2 + \mathrm{i}\hbar\,\mathrm{Tr}\sum_{\ell=1}^{\infty}\frac{1}{\ell}\left[S_{\mathrm{F},0}\begin{pmatrix} 0 & \Delta \\ \Delta^* & 0 \end{pmatrix}\right]^{\ell} \\ &\equiv \frac{1}{\lambda}\int \mathrm{d}^d x|\Delta|^2 + \sum_{\ell=1}^{\infty}\Gamma_1^{(\ell)},\end{aligned} \tag{8.77}$$

where for notational convenience we dropped the index "c" on the pair field. In the $\ell = 1$ term of the expansion, the trace over the 2×2 matrix immediately yields zero,

$$\Gamma_1^{(\ell=1)} = \mathrm{i}\hbar\,\mathrm{Tr}\begin{pmatrix} 0 & [p_0 - \xi(\mathbf{p})]^{-1}\,\Delta \\ [p_0 + \xi(\mathbf{p})]^{-1}\,\Delta^* & 0 \end{pmatrix} = 0. \tag{8.78}$$

In a similar fashion, all terms $\Gamma_1^{(\ell)}$ with ℓ odd give zero in accordance with the global U(1) symmetry that forbids such terms. The second term in the Landau expansion gives

$$\Gamma_1^{(\ell=2)} = \frac{\mathrm{i}}{2}\hbar\,\mathrm{Tr}\begin{pmatrix} 0 & [p_0 - \xi(\mathbf{p})]^{-1}\,\Delta(x) \\ [p_0 + \xi(\mathbf{p})]^{-1}\,\Delta^*(x) & 0 \end{pmatrix}^2 \tag{8.79}$$

or

$$\Gamma_1^{(\ell=2)} = \mathrm{i}\hbar\,\mathrm{Tr}\frac{1}{p_0 + \xi(\mathbf{p})}\Delta^*(x)\frac{1}{p_0 - \xi(\mathbf{p})}\Delta(x), \tag{8.80}$$

where we recall the definition of the momentum operator p_μ as operating on everything that appears to the right. Applying the derivative expansion rules outlined in Sec. 3.1, we can write more explicitly

$$\Gamma_1^{(\ell=2)} = \mathrm{i}\hbar \int \mathrm{d}^d x \int \frac{\mathrm{d}^d k}{(2\pi)^d} \frac{1}{\hbar\omega + \xi(\mathbf{k})} \frac{1}{\hbar\omega - \mathrm{i}\hbar\partial_t - \xi(\mathbf{k} + \mathrm{i}\nabla)} \Delta^*(x)\Delta(x), \quad (8.81)$$

where, by convention, the derivative ∂_μ operates only on $\Delta^*(x)$.

Derivatives in the fourth-order term are irrelevant for most applications and will be ignored here. That term then simply becomes

$$\begin{aligned}\Gamma_1^{(\ell=4)} &= \frac{\mathrm{i}}{4}\hbar \int \mathrm{d}^d x \int \frac{\mathrm{d}^d k}{(2\pi)^d} \begin{pmatrix} 0 & [\hbar\omega - \xi(\mathbf{k})]^{-1}\,\Delta \\ [\hbar\omega + \xi(\mathbf{k})]^{-1}\,\Delta^* & 0 \end{pmatrix}^4 \\ &= \frac{\mathrm{i}}{2}\hbar \int \mathrm{d}^d x \int \frac{\mathrm{d}^d k}{(2\pi)^d} \frac{1}{[\hbar^2\omega^2 - \xi^2(\mathbf{k})]^2} |\Delta(x)|^4. \end{aligned} \quad (8.82)$$

The expansion up to fourth order thus reads

$$\Gamma = \frac{1}{\lambda} \int \mathrm{d}^d x |\Delta|^2 + \Gamma_1^{(\ell=2)} + \Gamma_1^{(\ell=4)}, \quad (8.83)$$

with $\Gamma_1^{(\ell=2)}$ and $\Gamma_1^{(\ell=4)}$ given by Eq. (8.81) and (8.82), respectively. The validity of this expansion rests on the assumption that the order parameter is small. One regime where this is satisfied is in the vicinity of the transition temperature T_c where superconductivity sets in. This regime is the subject of the next section.

8.10 Ginzburg-Landau Theory

To access the region around T_c, temperature must be included in the formalism. We do so by applying the substitution (2.43) to the effective action (8.83). This yields the Euclidean Lagrangian density, defined through $S_E = \int \mathrm{d}\tau \mathrm{d}^D x \, \mathcal{L}_E$, see Eq. (2.39),

$$\begin{aligned}\mathcal{L}_{\text{eff}}^{\text{E}} &= -\frac{1}{\lambda}|\Delta|^2 + \frac{1}{\beta}\sum_n \int \frac{\mathrm{d}^D k}{(2\pi)^D} \frac{1}{-\mathrm{i}\hbar\omega_n - \xi(\mathbf{k})} \frac{1}{-\mathrm{i}\hbar\omega_n - \hbar\partial_\tau + \xi(\mathbf{k}+\mathrm{i}\nabla)} \Delta^*\Delta \\ &+ \frac{1}{2}\frac{1}{\beta}\sum_n \int \frac{\mathrm{d}^D k}{(2\pi)^D} \frac{1}{[\hbar^2\omega_n^2 + \xi^2(\mathbf{k})]^2} |\Delta|^4, \end{aligned} \quad (8.84)$$

where the sum is over the fermionic Matsubara frequencies $\omega_n = \pi(2n+1)/\hbar\beta$. Because the integrands in Eq. (8.84) are peaked near $\xi(\mathbf{k}) = 0$ in the weak-coupling limit, wave numbers $|\mathbf{k}|$ can be replaced with the Fermi wave number k_F, and the integrals can be approximated as in Eq. (8.20) with $\omega_D \to \infty$ with only a small error. The remaining integrations are then readily carried out with the help of the following integrals, which can be checked using contour integration,

$$\int \mathrm{d}\xi \frac{1}{(\hbar\omega_n^2 + \xi^2)^2} = \frac{\pi}{2|\hbar\omega_n|^3}, \tag{8.85}$$

and

$$\int \mathrm{d}\xi \frac{1}{(-\mathrm{i}\hbar\omega_n - \xi)} \frac{1}{(-\mathrm{i}\hbar\omega_n + \xi - r)} = \frac{-\pi}{\hbar|\omega_n| - \mathrm{i}\,\mathrm{sgn}(\omega_n)\, r/2}, \tag{8.86}$$

with r real. This gives for the effective Lagrangian (8.84)

$$\mathcal{L}_{\mathrm{eff}}^{\mathrm{E}} = -\frac{1}{\lambda}|\Delta|^2 - 2\pi\nu(0)\frac{1}{\beta}\int \frac{\mathrm{d}\Omega_{\mathbf{k}}}{\Omega_D}\sum_{n=0}^{\infty}$$
$$\times \left[\mathrm{Re}\left(\frac{1}{\hbar\omega_n - \mathrm{i}(\hbar\partial_\tau + \hbar^2\nabla^2/2m - \mathrm{i}\hbar v_{\mathrm{F}}\hat{\mathbf{k}}\cdot\nabla)/2}\right)\Delta^*\Delta - \frac{1}{4\hbar^3\omega_n^3}|\Delta|^4\right] \tag{8.87}$$

with $v_{\mathrm{F}} = \hbar k_{\mathrm{F}}/m$ the Fermi velocity, and $\hat{\mathbf{k}}$ the unit vector in the direction of the wave vector $\mathbf{k}$. The real-part function $\mathrm{Re}(x)$ arises from rewriting the sum over the Matsubara frequencies as

$$\sum_{n=-\infty}^{\infty} \frac{1}{\hbar|\omega_n| - \mathrm{i}\,\mathrm{sgn}(\omega_n)\, r/2} = \sum_{n=0}^{\infty}\frac{1}{\hbar\omega_n - \mathrm{i}r/2} + \sum_{n=0}^{\infty}\frac{1}{\hbar\omega_n + \mathrm{i}r/2}$$
$$= 2\,\mathrm{Re}\sum_{n=0}^{\infty}\frac{1}{\hbar\omega_n - \mathrm{i}r/2}. \tag{8.88}$$

The quadratic term without derivatives involves the diverging frequency sum $\sum_n 1/\omega_n$ which must be regularized. We do so by introducing the cutoff n_{max} which is related to the Debye frequency, i.e., the physical cutoff, through

$$\omega_{\mathrm{D}} = \pi(2n_{\mathrm{max}} + 1)/\hbar\beta. \tag{8.89}$$

The temperature T_{c} where superconductivity sets in is determined by the condition that the coefficient of this quadratic term changes sign. With the definition

$$\alpha(T) \equiv -\frac{1}{\lambda} - 2\pi\nu(0)\frac{1}{\hbar\beta}\sum_{n=0}^{n_{\mathrm{max}}}\frac{1}{\omega_n}, \tag{8.90}$$

the condition reads $\alpha(T_{\mathrm{c}}) = 0$, or

$$2\nu(0)\sum_{n=0}^{n_{\mathrm{max}}}\frac{1}{2n+1} = -\frac{1}{\lambda}. \tag{8.91}$$

The sum appearing here can be evaluated in closed form in terms of the digamma function

$$2\sum_{k=0}^{l-1} \frac{1}{2k+1} = -\psi\left(\tfrac{1}{2}\right) + \psi\left(\tfrac{1}{2}+l\right), \tag{8.92}$$

where

$$\psi\left(\tfrac{1}{2}\right) = -\gamma - 2\ln(2), \tag{8.93}$$

with $\gamma = 0.577\cdots$ Euler's constant, and $\psi(\frac{1}{2}+l) = \ln(l) + O(l^{-2})$. This yields for large $n_{\rm max}$

$$\nu(0)[\gamma + \ln(2n_{\rm max}+2) + \ln(2)] = -\frac{1}{\lambda}, \tag{8.94}$$

or

$$2\mathrm{e}^{\gamma}(2n_{\rm max}+1) = \mathrm{e}^{-1/\nu(0)\lambda}, \tag{8.95}$$

using the approximation $2n_{\rm max} + 2 \approx 2n_{\rm max} + 1$. By virtue of the relation (8.89) between the Debye frequency $\omega_{\rm D}$ and $n_{\rm max}$ at $T = T_{\rm c}$, Eq. (8.95) yields the BCS expression for the transition temperature:

$$\begin{aligned} k_{\rm B}T_{\rm c} &= \frac{2}{\pi}\,\mathrm{e}^{\gamma}\hbar\omega_{\rm D}\,\mathrm{e}^{1/\nu(0)\lambda} \\ &= \frac{1}{\pi}\,\mathrm{e}^{\gamma}\,|\bar{\Delta}| \approx 0.567\,|\bar{\Delta}|, \end{aligned} \tag{8.96}$$

where $\bar{\Delta}$ is the solution (8.22) of the zero-temperature gap equation. The BCS prediction $2|\bar{\Delta}| \approx 3.52\, k_{\rm B}T_{\rm c}$, relating the zero-temperature energy gap to the transition temperature, is one of the hallmarks of a weak-coupling superconductor. The first line in Eq. (8.96) provides an explanation of the *isotope effect*, which stands for the experimental observation that the transition temperature of different isotopes of a given metallic element varies as $T_{\rm c} \propto M^{-1/2}$, with M the ionic mass, for $\omega_{\rm D} \propto M^{-1/2}$.

With the explicit expression for $T_{\rm c}$, the coefficient (8.90) can be cast into a more convenient form. Going through the same steps leading to Eq. (8.96), we obtain the expression

$$\frac{2}{\pi}\,\mathrm{e}^{\gamma}\frac{\hbar\omega_{\rm D}}{k_{\rm B}T} = \exp\left\{-\frac{1}{\nu(0)}\left[\alpha(T) + \frac{1}{\lambda}\right]\right\}, \tag{8.97}$$

which yields, after using Eq. (8.96),

$$\alpha(T) = \nu(0)\ln(T/T_{\rm c}) \approx \nu(0)\,(T/T_{\rm c} - 1)\,. \tag{8.98}$$

This expression shows that when the transition temperature is approached from above, the coefficient α becomes negative, indicating the appearance of a Cooper pair condensate.

By introducing the cutoff of the phonon spectrum, i.e., the Debye energy, we regularized the divergent part in Eq. (8.87). The remaining sums in that expression are readily evaluated with the help of the summation formula ($k = 1, 2\cdots$)

$$\sum_{n=0}^{\infty} \frac{1}{\omega_n^k} = \left(\frac{\hbar\beta}{\pi}\right)^k \left(1 - 2^{-k}\right) \zeta(k), \tag{8.99}$$

with $\zeta(k)$ the Riemann zeta function, and the expression

$$\psi(z) = -\gamma - \frac{1}{z} + \sum_{m=1}^{\infty} \left(\frac{1}{m} - \frac{1}{m+z}\right) \tag{8.100}$$

for the digamma function. Near the critical point, we thus obtain as effective theory

$$\begin{aligned} \mathcal{L}_{\rm eff} = \nu(0) \Bigg\{ & - \int \frac{{\rm d}\Omega_{\mathbf{k}}}{\Omega_D} \, {\rm Re}\, \psi \left[\frac{1}{2} - \frac{\rm i}{4\pi k_{\rm B} T} \left(-{\rm i}\hbar \frac{\partial}{\partial t} + \frac{\hbar^2}{2m} \nabla^2 - {\rm i}\hbar v_{\rm F} \hat{\mathbf{k}} \cdot \nabla\right)\right] \Delta^* \Delta \\ & + \left[\ln\left(\frac{T_{\rm c}}{T}\right) + \psi\left(\frac{1}{2}\right)\right] |\Delta|^2 - \frac{7\zeta(3)}{16\pi^2} \frac{1}{k_{\rm B}^2 T_{\rm c}^2} |\Delta|^4 \Bigg\}, \end{aligned} \tag{8.101}$$

where we returned to Minkowski spacetime by replacing τ with it and using the relation (2.39). Expanding up to second order in derivatives with the help of the series expansion of the digamma function,

$$\psi\left(\tfrac{1}{2} + x\right) = \psi\left(\tfrac{1}{2}\right) + \frac{\pi^2}{2} x - 7\zeta(3) x^2 + O\left(x^3\right), \tag{8.102}$$

we arrive at the celebrated *Ginzburg-Landau theory*

$$\mathcal{L}_{\rm GL} = \nu(0) \left[\xi_0^2 \Delta^* \nabla^2 \Delta + \left(1 - \frac{T}{T_{\rm c}}\right) |\Delta|^2 - \frac{7\zeta(3)}{16\pi^2} \frac{1}{k_{\rm B}^2 T_{\rm c}^2} |\Delta|^4\right]. \tag{8.103}$$

Here, $\xi_0 \propto \hbar v_{\rm F}/k_{\rm B} T_{\rm c}$ is a length scale known as the *coherence length*. Specifically,

$$\xi_0^2 \equiv \frac{7\zeta(3)}{48\pi^2} \frac{\hbar^2 v_{\rm F}^2}{k_{\rm B}^2 T_{\rm c}^2}. \tag{8.104}$$

In deriving Eq. (8.103), use is made of the angular averages in $D = 3$

$$\int \frac{{\rm d}\Omega_{\mathbf{k}}}{4\pi} \hat{k}^i \hat{k}^j = \frac{1}{3} \delta^{ij}, \quad \int \frac{{\rm d}\Omega_{\mathbf{k}}}{4\pi} \hat{k}^i = 0, \tag{8.105}$$

and $\ln(T_{\rm c}/T)$ is expanded around $T_{\rm c}$. Moreover, it is assumed that ∇^2 yields a real quantity when acting on the order parameter, i.e., ${\rm Re}\, \nabla^2 \Delta = \nabla^2 \Delta$. The leading terms in time derivatives are

$$\mathcal{L}_{\rm TD} = \nu(0) \Delta^* \left(-\frac{\pi}{8} \frac{\hbar}{k_{\rm B} T_{\rm c}} {\rm Re}\, \partial_t + \frac{7\zeta(3)}{16\pi^2} \frac{1}{k_{\rm B}^2 T_{\rm c}^2} \hbar^2 \partial_t^2\right) \Delta, \tag{8.106}$$

assuming that $\mathrm{Re}\,\partial_t^2\Delta = \partial_t^2\Delta$. The derivative expansion is valid only in the limit of low frequencies $\omega << k_B T_c/\hbar$ and small wave numbers $|\mathbf{k}| << 1/\xi(T)$, where $\xi(T)$ is the finite-temperature counterpart of the coherence length. (8.104) at the absolute zero of temperature,

$$\xi^2(T) \equiv \frac{\xi_0^2}{|1 - T/T_c|}, \tag{8.107}$$

valid near T_c. For higher frequencies and larger wave numbers, the full dependence as contained in the argument of the digamma function in Eq. (8.101) must be retained.

The dynamics described by the time-dependent Ginzburg-Landau theory is dissipative. This is in contrast to the Gross-Pitaevskii theory (4.40), governing the dynamics of a Bose-Einstein condensate, where the coefficient in front of the time derivative is purely imaginary. For a pure superconductor below the transition temperature, where the Fourier transform of $\Delta(x)$ involves the exponential $\exp(-i\omega t)$, the time derivative ∂_t yields a purely imaginary frequency $-i\omega$ so that the term linear in time derivatives in Eq. (8.106) drops out.

The situation is different just above T_c. Due to thermal fluctuations, small superconducting regions with a nonzero order parameter briefly appear in the metallic state before decaying again. Being a dissipative phenomenon, ∂_t gives a real number when acting on the order parameter so that the linear term in time derivatives in Eq. (8.106) now survives. Specifically, it leads to the diffusion equation

$$-\partial_t \Delta = \frac{1}{\tau_\Delta}\left[1 - \xi^2(T)\nabla^2\right]\Delta \tag{8.108}$$

if the nonlinear term in Eq. (8.103) is ignored. This linear equation shows that just above the transform temperature, where this time-dependent theory applies, the order parameter relaxes exponentially towards its zero equilibrium value with

$$\tau_\Delta \equiv \frac{\pi\hbar}{8k_B}\frac{1}{T - T_c}, \qquad T > T_c \tag{8.109}$$

the relaxation time of the uniform order parameter. The short wave-length modes decay faster than the uniform ($\mathbf{k} = 0$) mode with relaxation rate

$$\frac{1}{\tau_{\mathbf{k}}} \equiv \frac{1}{\tau_\Delta}\left[1 + \xi^2(T)\mathbf{k}^2\right]. \tag{8.110}$$

Physically, Eq. (8.108) reflects the continuum of fermionic excitations into which a Cooper pair can decay. Right at the transition temperature, the diffusion equation (8.108) reduces to

$$-\partial_t \Delta = -\frac{8}{\pi}\frac{k_B T_c}{\hbar}\xi_0^2\nabla^2\Delta. \tag{8.111}$$

The spectrum $E_\Delta(\mathbf{k})$ of the pair field for $T > T_c$ that follows from Eqs. (8.103) and (8.106),

$$E_\Delta(\mathbf{k}) = -\mathrm{i}\frac{8}{\pi}k_B T_c \left[\left(\frac{T}{T_c} - 1\right) + \xi_0^2 \mathbf{k}^2\right], \tag{8.112}$$

is purely dissipative. Other cases where the time-dependent Ginzburg-Landau theory is applicable will be discussed in Sec. 8.15. In the following sections, first the time-independent theory will be considered.

8.11 Condensation Energy

If the condensate is stationary and uniform in space, the minimization of the Ginzburg-Landau theory (8.103) with respect to Δ^* yields as temperature dependence of the order parameter,

$$\bar{\Delta}(T) = \left(\frac{8\pi^2}{7\zeta(3)}\right)^{1/2} k_B T_c \left(1 - \frac{T}{T_c}\right)^{1/2} \approx 3.1\, k_B T_c \left(1 - \frac{T}{T_c}\right)^{1/2}, \tag{8.113}$$

where, without loss of generality, it is assumed that $\bar{\Delta}(T)$ is real. It shows that the order parameter vanishes smoothly when the transition temperature is approached from below.

The result (8.113) can also be extracted from the gap equation (8.16) for it follows from minimizing the same expression, albeit before evaluating the integrals. At finite temperature, the gap equation assumes the form

$$-\frac{1}{\lambda} = \frac{1}{\beta}\int \frac{\mathrm{d}^3k}{(2\pi)^3} \sum_n \frac{1}{\hbar^2\omega_n^2 + \xi^2(\mathbf{k}) + \bar{\Delta}^2(T)} \tag{8.114}$$

by virtue of the substitution rule (2.43). The sum is readily evaluated using the fermionic counterpart of Eq. (2.69),

$$\frac{1}{\beta}\sum_n \frac{1}{-\mathrm{i}\hbar\omega_n + x} = -f(x) + \frac{1}{2}, \tag{8.115}$$

with $f(x)$ the Fermi-Dirac distribution function

$$f(x) = \frac{1}{\mathrm{e}^{\beta x} + 1}, \tag{8.116}$$

or its regularized form with the vacuum term $(\frac{1}{2})$ omitted. The resulting equation,

$$-\frac{1}{\lambda} = \frac{1}{\beta}\int \frac{\mathrm{d}^3k}{(2\pi)^3} \frac{1}{2E(\mathbf{k})} \tanh\left[\tfrac{1}{2}\beta E(\mathbf{k})\right], \tag{8.117}$$

with $E^2(\mathbf{k}) = \xi^2(\mathbf{k}) + \bar{\Delta}^2(T)$, is valid for all temperatures. However, the full temperature dependence of the gap $\bar{\Delta}(T)$ cannot be obtained from it in closed form. Near the critical point, Eq. (8.114) can be expanded in powers of $\bar{\Delta}$ as

$$-\frac{1}{\lambda} = \frac{1}{\beta}\int\frac{d^3k}{(2\pi)^3}\sum_n\left[\frac{1}{\hbar^2\omega_n^2+\xi^2(\mathbf{k})} - \frac{1}{[\hbar^2\omega_n^2+\xi^2(\mathbf{k})]^2}\bar{\Delta}^2(T) + O(\bar{\Delta}^4)\right]$$

$$= -\frac{1}{\lambda} + \nu(0)\left(1-\frac{T}{T_c}\right) - \nu(0)\frac{7\zeta(3)}{8\pi^2k_B^2T_c^2}\bar{\Delta}^2(T) + O(\bar{\Delta}^4), \qquad (8.118)$$

where the last step follows from the integrals (8.85) and (8.86) with $r = 0$, and from the summation formula (8.99). We also regularized the ultraviolet-divergent term $1/[\hbar^2\omega_n^2+\xi^2(\mathbf{k})]$ in Eq. (8.118) as before in Sec. 8.10 by using Eq. (8.90) with Eq. (8.98). The temperature dependence of $\bar{\Delta}(T)$ that follows from Eq. (8.118) is the same as derived in Eq. (8.113).

Substitution of the solution (8.113) back into the Ginzburg-Landau theory (8.103) yields the condensation energy density

$$\mathcal{V}_{\text{eff}} = -\frac{4\pi^2\nu(0)}{7\zeta(3)}k_B^2T_c^2\left(1-\frac{T}{T_c}\right)^2$$

$$= -\frac{1}{2}\nu(0)\bar{\Delta}^2(T)\left(1-\frac{T}{T_c}\right), \qquad (8.119)$$

which is negative just below T_c where this expression applies. Comparison with the zero-temperature result (8.76) shows that Eq. (8.119) can be obtained from the zero-temperature result by simply replacing $\bar{\Delta}^2$ with $\bar{\Delta}^2(T)(1-T/T_c)$.

A formal expression for the condensation energy for arbitrary $T < T_c$ can be obtained as follows. The derivative of the effective potential (8.73), which is a function of $\bar{\Delta}^2$, with respect to the coupling constant λ, yields

$$\frac{\partial\mathcal{V}_{\text{eff}}}{\partial\lambda} = \frac{1}{\lambda^2}\bar{\Delta}^2(T), \qquad (8.120)$$

valid for all temperatures. As $\lambda \to 0^-$, the energy gap tends to zero. Integrating Eq. (8.120) with respect to the coupling constant from 0 to λ, we therefore obtain the difference between the ground-state energy in the superconducting and metallic state ($\bar{\Delta} = 0$) at the same temperature,

$$\mathcal{V}_{\text{eff}} = \int_0^\lambda\frac{d\lambda'}{\lambda'^2}\bar{\Delta}^2(T). \qquad (8.121)$$

At zero temperature, $\bar{\Delta}$ is determined by Eq. (8.22), and the integration can be carried out to yield the condensation energy (8.76) at the absolute zero of temperature.

Linear response theory studied in Sec. 8.4 showed that when a superconductor at the absolute zero of temperature is placed in a small magnetic field it expels the field completely from its interior. This Meissner effect is an important physical characteristic of a superconductor not only at zero temperature, but for all temperatures below T_c. To understand this, note that it requires an energy of $H^2/8\pi$ per unit volume to resist the magnetic pressure. If the applied magnetic field H is increased at a given temperature below T_c, it will reach a critical value H_c where the superconductor undergoes a first-order phase transition to the metallic state. At this point, the energy gained by remaining in the superconducting state is not sufficient to keep the field out, and the condensation energy density will make a jump

$$\frac{1}{8\pi} H_c^2 = -\mathcal{V}_{\text{eff}} \ (> 0). \tag{8.122}$$

By Eq. (8.119),

$$H_c = \left(\frac{32\pi^3 \nu(0)}{7\zeta(3)} \right)^{\frac{1}{2}} k_B \left(T_c - T \right) \tag{8.123}$$

close to T_c. This critical field, which smoothly tends to zero when the transition temperature at zero field is approached from below, is known as the *thermodynamic critical field*.

8.12 Ginzburg Region

The Ginzburg-Landau theory, which is obtained by expanding the effective action in a power series about $\Delta = 0$, is valid near the transition temperature where Δ is small. However, when the temperature is too close to T_c the theory breaks down. In most applications of the Ginzburg-Landau theory (8.103), fluctuations in the order parameter are ignored—that is, the theory is a mean-field theory derived from applying the saddle-point approximation to the functional integral (8.9). The regime where mean-field theory breaks down and where fluctuations in Δ are to be included is called the *Ginzburg region*. These thermal fluctuations are a classical phenomenon characteristic of all systems undergoing a smooth phase transition. Being an equilibrium phenomenon, time can be ignored altogether and only the time-independent Ginzburg-Landau Lagrangian (8.103) need be considered with the associated finite-temperature action (divided by $\hbar$),

$$S_{\text{GL}} \equiv \frac{1}{\hbar} \int_0^{\hbar\beta} \mathrm{d}\tau \int \mathrm{d}^D x \mathcal{L}_{\text{GL}}^{\text{E}}, \tag{8.124}$$

with $\mathcal{L}_{\text{E}}$ the Euclidean Lagrangian density. The size of the Ginzburg region can be estimated by requiring that the corrections to mean-field theory are small. To find the dimensionless perturbation parameter of the theory, we rescale the fields and integration variables as follows:

$$\mathbf{x}' \equiv \frac{\mathbf{x}}{\xi(T)}, \quad \psi \equiv \left[\frac{\nu(0)}{k_{\mathrm{B}}T_{\mathrm{c}}}\xi^3(T)\left(1-\frac{T}{T_{\mathrm{c}}}\right)\right]^{1/2} \Delta, \tag{8.125}$$

$$\frac{1}{g} \equiv \frac{1}{3}\hbar^2 v_{\mathrm{F}}^2 \frac{\nu(0)}{k_{\mathrm{B}}T_{\mathrm{c}}}\xi(T)\left(1-\frac{T}{T_{\mathrm{c}}}\right). \tag{8.126}$$

In terms of these dimensionless variables, the action assumes the simple form

$$S_{\mathrm{GL}} = \int \mathrm{d}^3x' \left(|\nabla'\psi|^2 - |\psi|^2 + g|\psi|^4\right), \tag{8.127}$$

where $g \propto |1 - T/T_{\mathrm{c}}|^{-1/2}$ is the dimensionless coupling constant. Mean-field theory breaks down if $g \geq 1$. From the steps leading to Eq. (8.127), it follows that the exponent of $|1 - T/T_{\mathrm{c}}|$ crucially depends on the number of space dimensions. In D dimensions one finds instead $g \propto |1 - T/T_{\mathrm{c}}|^{(D-4)/2}$. This exponent indicates that for $D > 4$, the dimensionless coupling constant tends to zero when the transition temperature is approached, implying that a perturbative expansion around the mean-field solution becomes increasingly accurate. On the other hand, for $D < 4$, $g \to \infty$ as the transition temperature is approached, and perturbation theory breaks down close to the transition temperature. In this Ginzburg region, mean-field theory gives the wrong critical behavior.

For the case at hand, the condition $g \geq 1$ can be cast in terms of the coherence length (8.104), the transition temperature, and the jump in the heat capacity Δc_V at the critical point as

$$|1 - T/T_{\mathrm{c}}| \leq \frac{1}{4\xi_0^6(\Delta c_V/k_{\mathrm{B}})^2}. \tag{8.128}$$

Here,

$$\Delta c_V = \frac{8\pi^2\nu(0)}{7\zeta(3)}k_{\mathrm{B}}^2 T_{\mathrm{c}}, \tag{8.129}$$

which follows from the condensation energy (8.119) by using the thermodynamic relation

$$c_V = -\frac{T}{V}\frac{\partial^2\Omega}{\partial T^2}, \tag{8.130}$$

with $\Omega = V\mathcal{V}_{\mathrm{eff}}$ and V the volume of the system. Since Eq. (8.119) represents only the condensation energy, inserting it into Eq. (8.130) yields the jump in the heat capacity. For the metallic state, the heat capacity c_V^{n} reads

$$c_V^{\mathrm{n}} = \frac{2\pi^2}{3}\nu(0)k_{\mathrm{B}}^2 T \tag{8.131}$$

so that

$$\frac{\Delta c_V}{c_V^{\rm n}} = \frac{12}{7\xi(3)} \approx 1.43. \tag{8.132}$$

With ξ_0 being of order the size of a Cooper pair $\approx 10^2$ nm, $T_{\rm c}$ being of order several Kelvin, and Δc_V being of order several Joule/(K cm)^3, it follows that the Ginzburg region is very small for a conventional BCS superconductor,

$$|1 - T/T_{\rm c}| \approx 10^{-8} - 10^{-12}. \tag{8.133}$$

In other words, thermal fluctuations in the order parameter become important only very close to the critical point, and the Ginzburg-Landau theory, although a mean-field theory, applies in a broad region around $T_{\rm c}$.

8.13 Magnetic Vortices

An external magnetic field larger than the critical field $H_{\rm c}$, if penetrating the superconductor at all, it does so by forming a number of thin filaments carrying quantized magnetic flux. To study these *magnetic vortices*, we minimally couple the Ginzburg-Landau theory (8.103) to a magnetic vector potential $\mathbf{A}(\mathbf{x})$. This leads to the Hamiltonian

$$\begin{aligned} H_{\rm GL} = \nu(0) \int {\rm d}^3 x \left[\xi_0^2 \left| \left(\nabla + {\rm i}\frac{2e}{\hbar c}\mathbf{A} \right) \Delta \right|^2 - \left(1 - \frac{T}{T_{\rm c}} \right) |\Delta|^2 + \frac{3\xi_0^2}{\hbar^2 v_{\rm F}^2} |\Delta|^4 \right] \\ + \frac{1}{8\pi} \int {\rm d}^3 x \, (\nabla \times \mathbf{A})^2, \end{aligned} \tag{8.134}$$

where the charge $q = -2e(< 0)$ is the double charge of a Cooper pair described by the pair field Δ. The last term represents the usual magnetic energy. We seek a solution of a form similar to the vortex solution (1.152) in the Gross-Pitaevskii theory with winding number w,

$$\Delta(\mathbf{x}) = \bar{\Delta}(T) f(\rho)\, {\rm e}^{-{\rm i}w\theta} \tag{8.135}$$

in cylindrical coordinates (ρ, θ, z). Here, $\bar{\Delta}(T)$, given by Eq. (8.113), minimizes the energy of the uniform system. The minus sign in the exponential is chosen so that the magnetic flux carried by the vortex is oriented in the positive z-direction for $w > 0$, see Sec. 8.5. This *Ansatz*, in which the Nambu-Goldstone field behaves as the real-space azimuthal angle θ, assumes that the solution is cylindrically symmetric with the vortex line defining the symmetry axis. The vector potential is taken to be of the form $\mathbf{A}(\mathbf{x}) = A^\theta(\rho)\mathbf{e}_\theta$ so that it produces a magnetic induction in the direction of the vortex axis. The London limit corresponds to setting $f(\rho)$

to unity in Eq. (8.135). With this *Ansatz*, the first and last term of the Ginzburg-Landau Hamiltonian assume the form

$$\left|\left(\nabla + \mathrm{i}\frac{2e}{\hbar c}\mathbf{A}\right)\Delta\right|^2 = \bar{\Delta}^2(T)\left[\left(\frac{\mathrm{d}f}{\mathrm{d}\rho}\right)^2 + \frac{4e^2\mathbf{Q}^2}{\hbar^2c^2}f^2\right] \tag{8.136}$$

$$(\nabla \times \mathbf{A})^2 = \left[\frac{1}{\rho}\frac{\mathrm{d}}{\mathrm{d}\rho}\left(\rho A^\theta\right)\right]^2, \tag{8.137}$$

respectively, where we introduced the vector

$$\mathbf{Q}(\rho) \equiv \mathbf{A}(\rho) - \frac{\hbar c}{2e}\nabla\theta \tag{8.138}$$

which has only one nonzero component, *viz.*

$$Q^\theta = A^\theta - w\frac{\hbar c}{2e}\frac{1}{\rho}. \tag{8.139}$$

The Euler-Lagrange equations for the fields $f(\rho)$ and $A^\theta(\rho)$ can for $\rho \neq 0$ be cast in the form

$$\xi^2(T)\left(\frac{\mathrm{d}^2}{\mathrm{d}\rho^2} + \frac{1}{\rho}\frac{\mathrm{d}}{\mathrm{d}\rho} - \frac{4e^2\mathbf{Q}^2}{\hbar^2c^2}\right)f + f - f^3 = 0, \tag{8.140}$$

$$\left(\frac{\mathrm{d}^2}{\mathrm{d}\rho^2} + \frac{1}{\rho}\frac{\mathrm{d}}{\mathrm{d}\rho} - \frac{1}{\rho^2}\right)Q^\theta - \frac{1}{\lambda_\mathrm{L}^2(T)}f^2Q^\theta = 0, \tag{8.141}$$

where $\xi(T)$ is the temperature-dependent coherence length (8.107), and where we introduced the finite-temperature counterpart of the zero-temperature London penetration depth (8.45) defined through

$$\lambda_\mathrm{L}^2(T) \equiv \frac{1}{2}\frac{\lambda_\mathrm{L}^2}{1 - T/T_\mathrm{c}}, \tag{8.142}$$

valid near T_c. This length scale, which is the inverse square root of the coefficient of the term $\mathbf{A}^2/8\pi$ in the Ginzburg-Landau Hamiltonian (8.134) with $f(\rho)$ set to unity, specifies the distance over which the magnetic induction varies. In deriving Eq. (8.141), it is used that for $\rho \neq 0$, $\nabla \times \mathbf{A} = \nabla \times \mathbf{Q}$, and

$$\left(\frac{\mathrm{d}^2}{\mathrm{d}\rho^2} + \frac{1}{\rho}\frac{\mathrm{d}}{\mathrm{d}\rho} - \frac{1}{\rho^2}\right)\frac{1}{\rho} = 0. \tag{8.143}$$

Also the factor of ρ arising from the measure $\int \mathrm{d}^3x$ when expressed in cylindrical coordinates is taken into account.

The field equations describing a straight magnetic vortex form a set of coupled nonlinear ordinary differential equations. They simplify for ρ small compared to both the correlation length $\xi(T)$ and the penetration depth $\lambda_\mathrm{L}(T)$ to

$$\left(\frac{\mathrm{d}^2}{\mathrm{d}\rho^2} + \frac{1}{\rho}\frac{\mathrm{d}}{\mathrm{d}\rho} - \frac{4e^2\mathbf{Q}^2}{\hbar^2c^2}\right)f = 0, \tag{8.144}$$

$$\left(\frac{\mathrm{d}^2}{\mathrm{d}\rho^2} + \frac{1}{\rho}\frac{\mathrm{d}}{\mathrm{d}\rho} - \frac{1}{\rho^2}\right)Q^\theta = 0. \tag{8.145}$$

The latter differential equation is solved by

$$Q^\theta(\rho) = \frac{1}{2}B^z\rho - w\frac{\hbar c}{2e}\frac{1}{\rho}, \tag{8.146}$$

where the coefficient of the first term is chosen such that it yields the magnetic induction along the vortex axis,

$$B^z = \frac{1}{\rho}\frac{\mathrm{d}}{\mathrm{d}\rho}(\rho Q^\theta). \tag{8.147}$$

With this solution for $Q^\theta(\rho)$, Eq. (8.144) shows that $f(\rho)$ and, thus, the order parameter vanishes for $\rho \to 0$ as

$$f(\rho) \sim \rho^{|w|} \tag{8.148}$$

with a power of ρ determined by the winding number—that is, a magnetic vortex possesses a core of radius $\rho \sim \xi(T)$ in which the order parameter decreases from its bulk value to zero at the vortex axis.

In the London limit, where $f(\rho) = 1$, the field equations also simplify in that they decouple. In that limit, Eq. (8.141) can be solved exactly, giving

$$Q^\theta(\rho) = -\frac{w\Phi_0}{2\pi\lambda_\mathrm{L}(T)}K_1\left(\frac{\rho}{\lambda_\mathrm{L}(T)}\right), \tag{8.149}$$

with K_1 the first-order modified Bessel function of the second kind which for small x has the series expansion $K_1(x) = 1/x + \cdots$, and for large x falls off exponentially,

$$K_1(x) \sim \sqrt{\frac{\pi}{2x}}\,\mathrm{e}^{-x}. \tag{8.150}$$

The coefficient of the solution with $\Phi_0 = \pi\hbar c/e$ has been chosen such that for $\rho \to 0$, the vector potential $A^\theta(\rho)$ tends to zero. The magnetic induction follows as:

$$B^z = \frac{w\Phi_0}{2\pi\lambda_\mathrm{L}^2(T)}K_0\left(\frac{\rho}{\lambda_\mathrm{L}(T)}\right), \tag{8.151}$$

where it is used that $K_1'(x) + K_1(x)/x = -K_0(x)$. For $\rho > \lambda_\mathrm{L}(T)$, the field falls off exponentially,

$$B^z(\rho) \sim \frac{w\Phi_0}{\sqrt{8\pi\lambda_\mathrm{L}^3(T)\rho}}\,\mathrm{e}^{-\rho/\lambda_\mathrm{L}(T)}, \tag{8.152}$$

implying that the persistent currents around the vortex core, which generate this induction field, tend to zero at a distance $\rho \sim \lambda_\mathrm{L}(T)$ from the vortex axis. For $\rho < \lambda_\mathrm{L}(T)$, the magnetic induction associated with a straight vortex behaves logarithmically,

$$B^z(\rho) = -\frac{w\Phi_0}{2\pi\lambda_\mathrm{L}^2(T)}\ln\left(\frac{\rho}{\lambda_\mathrm{L}(T)}\right) + \cdots. \tag{8.153}$$

The divergence for $\rho \to 0$ is an artifact of the London limit, where $f(\rho) = 1$. In reality, as was shown above, $f(\rho)$ tends to zero in the vortex core, and the logarithm is cut off at $\rho \sim \xi(T)$.

Outside the London limit and for $\rho < \lambda_{\,\mathrm{L}}(T)$, where $Q^{\theta} = -w\hbar c/2e\rho$, the field equation (8.140) becomes identical to the differential equation (1.153) describing a straight vortex in the Gross-Pitaevskii theory,

$$\xi^2(T)\left(\frac{\mathrm{d}^2}{\mathrm{d}\rho^2} + \frac{1}{\rho}\frac{\mathrm{d}}{\mathrm{d}\rho} - \frac{w^2}{\rho^2}\right) f + f - f^3 = 0. \tag{8.154}$$

For $\lambda_{\mathrm{L}}(T) > \rho > \xi\ \ (T)$, $f(\rho)$ tends to the asymptotic value of a spatially uniform system as in Eq. (1.155).

We next study the vortex from a distance larger than the correlation length $\xi(T)$ and the penetration depth $\lambda_{\mathrm{L}}(T)$. In that regime, a perturbative expansion around $f(\rho) = 1$ and $Q^{\theta}(\rho) = 0$ can be carried out. Setting $f(\rho) = 1 - \eta(\rho)$ with $\eta(\rho) > 0$ and linearizing the field equation (8.140), we obtain

$$\left(\frac{\mathrm{d}^2}{\mathrm{d}\rho^2} + \frac{1}{\rho}\frac{\mathrm{d}}{\mathrm{d}\rho} - \frac{2}{\xi^2(T)}\right)\eta = 0, \tag{8.155}$$

where it is to be noted that variations in the modulus of the pair field are characterized by the length scale $\xi(T)/\sqrt{2}$, which is shorter than the coherence length $\xi(T)$. The solution which tends to zero for $\rho \to \infty$ is

$$\eta(\rho) = qK_0\left(\frac{\rho}{\xi(T)/\sqrt{2}}\right), \tag{8.156}$$

with $q > 0$ a dimensionless constant which has to be determined numerically by matching this solution of the linearized theory to that of the full theory. To physically interpret this solution, note that $(1/2\pi)K_0$ is the Green function of the so-called modified Helmholtz operator $(\nabla^2 - \mu^2)$ in two dimensions. Specifically,

$$-(\nabla_\perp^2 - \mu^2)\frac{1}{2\pi}K_0(\mu\rho) = \delta(\mathbf{x}_\perp), \tag{8.157}$$

where $\mathbf{x}_\perp = (x^1, x^2, 0)$ denotes the position vector perpendicular to the vortex axis, or in Fourier space

$$G(\mathbf{x}_\perp) \equiv \int \frac{\mathrm{d}^2k}{(2\pi)^2}\frac{1}{\mathbf{k}^2 + \mu^2}\mathrm{e}^{\mathrm{i}\mathbf{k}\cdot\mathbf{x}_\perp} = \frac{1}{2\pi}K_0(\mu\rho). \tag{8.158}$$

The asymptotic solution (8.156) can therefore be thought of as produced by a point source $q\delta(\mathbf{x}_\perp)$ at the origin of the two-dimensional plane perpendicular to the vortex axis,

$$\left(\nabla_\perp^2 - \frac{2}{\xi^2(T)}\right)\eta = -2\pi q\delta(\mathbf{x}_\perp). \tag{8.159}$$

The asymptotic solution of Eq. (8.141) is of the form (8.149), but now with a coefficient that has to be determined numerically,

$$Q^\theta(\rho) = -\frac{m}{2\pi}\frac{\Phi_0}{\lambda_L(T)}K_1\left(\frac{\rho}{\lambda_L(T)}\right). \tag{8.160}$$

The sign of m specifies the orientation of the vortex. For $m > 0$, the vortex is oriented in the positive z-direction while for $m < 0$, it is oriented in the negative z-direction. The coefficient cannot be fixed by the behavior at the vortex center as in Eq. (8.149) because this is outside the realm where the asymptotic solution is valid. The prefactor $\Phi_0/\lambda_L(T)$ in Eq. (8.160) is included so that m is a dimensionless constant. The asymptotic solution (8.160) can be thought of as produced by a pointlike magnetic dipole located at the vortex center,

$$\left(\nabla_\perp^2 - \frac{1}{\lambda_L^2(T)}\right)\mathbf{Q} = -m\,\Phi_0\mathbf{e}_z \times \nabla\delta(\mathbf{x}_\perp), \tag{8.161}$$

where $\mathbf{e}_z$ is the unit vector in the direction of the vortex axis, and use is made of the identity $K_1(x) = -K_0'(x)$.

The linearized field equations (8.159) and (8.161) are the Euler-Lagrange equations of the Hamiltonian

$$\begin{aligned} H_{\rm GL}^{(2)} &= \frac{\hbar^2}{2m}\bar{n}(1 - T/T_c)L\int d^2x_\perp\left[\frac{1}{2}(\nabla_\perp\eta)^2 + \frac{1}{\xi^2(T)}\eta^2 - 2\pi q\delta(\mathbf{x}_\perp)\eta\right] \qquad (8.162)\\ &+\frac{1}{4\pi}L\int d^2x_\perp\left\{\frac{1}{2}(\nabla_\perp\times\mathbf{Q})^2 + \frac{1}{2}\frac{1}{\lambda_L^2(T)}\mathbf{Q}^2 - m\,\Phi_0\left[\mathbf{e}_z\times\nabla\delta(\mathbf{x}_\perp)\right]\cdot\mathbf{Q}\right\}\end{aligned}$$

with L the length of the vortex. This result can also be obtained by expanding the Ginzburg-Landau theory (8.134) with the *Ansatz* (8.135) and $f = 1 - \eta$ up to quadratic order in η and $\mathbf{Q}$, and by including the sources. The equivalence of the Euler-Lagrange equation for $\mathbf{Q}$ and Eq. (8.161) follows from noting that a massive vector field is automatically divergence-free, as can be verified by taking the divergence of the Euler-Lagrange equation. In the London limit, $\eta = 0$ and only the second line in Eq. (8.162) survives.

The Hamiltonian (8.162) can be used to estimate the vortex energy by integrating out the scalar field η and the vector potential $\mathbf{A}$, or $\mathbf{Q}$. Since the linearized theory is at most quadratic in these fields, the integrals are Gaussian. In the London limit, where the first line in Eq. (8.162) drops out and $m = w$, we thus obtain

$$\begin{aligned} H_{\rm GL}^{(2)} &= -\frac{w^2\Phi_0^2}{8\pi}L\int d^2x_\perp d^2x'_\perp\left[\mathbf{e}_z\times\nabla\delta(\mathbf{x}_\perp)\right]\cdot G_A(\mathbf{x}_\perp - \mathbf{x}'_\perp)\left[\mathbf{e}_z\times\nabla'\delta(\mathbf{x}'_\perp)\right]\\ &= -\frac{w^2\Phi_0^2}{8\pi}L\int d^2x_\perp d^2x'_\perp\,\nabla_\perp\delta(\mathbf{x}_\perp)\cdot G_A(\mathbf{x}_\perp - \mathbf{x}'_\perp)\nabla'_\perp\delta(\mathbf{x}'_\perp)\\ &= \frac{w^2\Phi_0^2}{8\pi\lambda_L^2(T)}L\int d^2x_\perp d^2x'_\perp\delta(\mathbf{x}_\perp)G_A(\mathbf{x}_\perp - \mathbf{x}'_\perp)\delta(\mathbf{x}'_\perp), \qquad (8.163)\end{aligned}$$

with $G_A(\mathbf{x}_\perp)$ the Green function (8.158) with μ replaced with $\lambda_{\rm L}^{-1}(T)$. The last line follows (ignoring an irrelevant diverging constant) by integration by parts. We arrive in this way at the energy per unit length of the vortex

$$\varepsilon_L = \left(\frac{w\Phi_0}{4\pi\lambda_{\rm L}(T)}\right)^2 \ln\left(\frac{\lambda_{\rm L}(T)}{\xi(T)}\right), \tag{8.164}$$

to logarithmic accuracy. The logarithmic divergence of $K_0[r/\lambda_{\rm L}(T)]$ for $\rho \to 0$ has been cut off at $\rho \sim \xi(T)$. This is justified by noting that the London limit corresponds to taking $\xi(T) \to 0$, which implies that the vortex core shrinks to zero in this limit. By reintroducing $\xi(T)$ in Eq. (8.164) by hand, the finite core size is accounted for. Equation (8.164) shows that the energy of a vortex is proportional to the winding number squared. A configuration of given topological charge $w_{\rm tot}$ with singly quantized vortices is therefore energetically favorable over one with multiply quantized vortices leading to the same overall charge $w_{\rm tot}$. For this reason we take $|w| = 1$ in the following.

The vortex energy (8.164) defines a critical field as follows. When the superconductor is placed in a uniform external magnetic field H, the Hamiltonian (8.163) must be augmented by the term $-\Phi_0 LH$, representing the interaction of the vortex line, carrying one magnetic flux quantum Φ_0, with the external field. This field penetrates the sample only if the additional term overcomes the vortex energy, i.e., when

$$H > \frac{\varepsilon_L}{\Phi_0} \equiv H_{\rm c_1}. \tag{8.165}$$

Below the critical field $H_{\rm c_1}$, the sample is in the Meissner state where the external field is excluded from the interior of the superconductor. Above $H_{\rm c_1}$, a superconducting material to which the London limit applies is in the so-called *mixed state* where the applied field pierces the superconductor through a number of thin filaments in the form of magnetic vortices carrying quantized magnetic flux.

8.14 Mixed State

The linearized theory (8.162) describes a straight vortex line along the z axis. By including additional sources, it is readily generalized to describe multiple straight vortex lines. Repeating the steps leading to Eq. (8.163), we obtain

$$\frac{H_{\rm GL}^{(2)}}{L} = \frac{\Phi_0^2}{8\pi\lambda_{\rm L}^2(T)}\left[-\sum_{\alpha,\beta} q_\alpha q_\beta G_\eta(\mathbf{x}_\alpha - \mathbf{x}_\beta) + \sum_{\alpha,\beta} m_\alpha m_\beta G_A(\mathbf{x}_\alpha - \mathbf{x}_\beta)\right], \tag{8.166}$$

where the sum extends over the vortices with charges q_α and m_α centered at $\mathbf{x}_\alpha$ in the x^1–x^2 plane. The coefficient in front of the scalar part in Eq. (8.162) has been rewritten in terms of the combination $\Phi_0/\lambda_L(T)$, which is independent of the electric charge. The result (8.166) reveals two types of interactions acting between vortices: one associated with the scalar field η and the other associated with the magnetic vector potential. In the London limit, the scalar interaction is ignored. The two interactión potentials are both given by the Green function (8.158) with μ replaced with $\sqrt{2}/\xi(T)$ for the scalar interaction and with $1/\lambda_L(T)$ for the vector interaction. Both potentials are short range. The field η always leads to an attractive interaction between vortices independent of their orientation. This is because the charges q_α are all positive. The magnetic vector field, on the other hand, can lead to either a repulsive or an attractive interaction, depending on the relative orientation of the two vortices. The magnetic interaction is repulsive between two parallel vortices, for which $m_\alpha/m_\beta > 0$, while it is attractive between antiparallel vortices, for which $m_\alpha/m_\beta < 0$. The vortices in a superconducting sample placed in a uniform external magnetic field orient themselves along the field direction. The two interactions between vortices then compete, with the one having the longest range dominating. In other words, two different types of superconductors can be identified, depending on whether the ratio of the two length scales,

$$\frac{\lambda_L(T)}{\xi(T)/\sqrt{2}} \equiv \sqrt{2}\kappa, \tag{8.167}$$

is smaller or larger than unity. Here, κ is the so-called *Ginzburg-Landau parameter*, which close to T_c assumes the form

$$\kappa = \left(\frac{\pi}{7\zeta(3)\nu(0)}\right)^{1/2} \frac{3ck_B T_c}{e\hbar v_F^2} \tag{8.168}$$

independent of temperature.

For so-called type-I superconductors, $\kappa < 1/\sqrt{2}$, and the screening length of the repulsive magnetic interaction is smaller than the coherence length $\xi(T)$, which defines the vortex core radius—that is, the magnetic repulsion is practically zero outside the vortex core so that vortices in type-I superconductors experience only an attractive force. They will consequently coalesce, with the result that the entire sample becomes normal. For type-II superconductors, $\kappa > 1/\sqrt{2}$ so that in these materials the magnetic repulsion dominates, and isolated vortices may exist. This conclusion can also be reached by a more standard argument as follows.

Let H_{c_2} denote the upper critical magnetic field below which a type-II metal becomes superconducting. In the vicinity of this critical field, the order parameter is small and the fourth-order term in the Ginzburg-Landau theory (8.134) can be neglected. The Euler-Lagrange equation for Δ then becomes

$$-\left(\nabla + \mathrm{i}\frac{2e}{\hbar c}\mathbf{A}\right)^2 \Delta = \frac{1}{\xi^2(T)}\Delta, \tag{8.169}$$

with $\xi(T)$ the temperature-dependent coherence length (8.107), and $\mathbf{A}$ the vector potential associated with the uniform external magnetic field, $\mathbf{H} = \nabla \times \mathbf{A}$. The curl of the vector potential yields the applied magnetic field because screening effects due to supercurrents, which are proportional to $|\Delta|^2$, lead to higher-order terms which are dropped in the linear approximation. By multiplying this equation with $\hbar^2/2M$, it takes the form of the time-independent Schrödinger equation describing a particle of charge $-2e(<0)$, mass M, and energy $\hbar^2/2M\xi^2(T)$ in a uniform applied magnetic field. The energy of the lowest Landau level is, cf. Eq. (3.129)

$$\frac{\hbar^2}{2M\xi^2(T)} = \frac{e\hbar}{Mc}H. \tag{8.170}$$

This defines the highest field where superconductivity can occur, i.e., the critical magnetic field H_{c_2},

$$H_{c_2} = \frac{\hbar c}{2e}\frac{1}{\xi^2(T)} \tag{8.171}$$

expressed in terms of the coherence length $\xi(T)$. Using the Ginzburg-Landau parameter (8.168) and the thermodynamic critical field (8.123), we obtain,

$$H_{c_2} = \sqrt{2}\kappa H_c. \tag{8.172}$$

In this way, the same conclusion is reached as before, namely that if $\kappa > 1/\sqrt{2}$, the upper critical field H_{c_2} is lager than the critical field H_c below which the entire sample becomes superconducting, making a mixed state with magnetic vortex lines possible.

When the applied field increases starting at H_{c_1}, more and more vortices carrying one unit of magnetic flux start to penetrate a type-II superconducting sample. Because of the dominating repulsive magnetic interaction, the vortices are driven apart. This is called to an halt by the finite extent of the system. The vortices then order themselves in a regular lattice, which turns out to be a triangular lattice, so as to minimize the repulsive interaction. This picture that the so-called *Abrikosov flux lattice* results from the repulsive magnetic interaction is confirmed by the fact that the lower and upper critical fields H_{c_1} and H_{c_2}, marking the boundaries of the mixed state, move closer when the value of the Ginzburg-Landau parameter κ becomes smaller. With $\xi(T)$ kept fixed, a smaller κ implies a shorter penetration depth, and therefore a stronger screening of the repulsive magnetic interaction.

The 2D vortex density of an Abrikosov flux lattice is given by $n_\otimes = 2/(\sqrt{3}a^2)$, with a the lattice spacing. The magnetic flux $\Phi = \int \mathrm{d}^2x B^z$ through the lattice, which is given by the number of vortices $N_\otimes$ multiplied by the fundamental flux unit Φ_0 carried by a single vortex, increases with the applied field. At the upper critical field H_{c_2}, the magnetic induction B becomes homogeneous and saturates the applied field, i.e., no further vortices are nucleated. The maximum vortex density $n_{\otimes,\max}$ is

$$n_{\otimes,\max} = \frac{H_{c_2}}{\Phi_0} = \frac{1}{2\pi}\frac{1}{\xi^2(T)}, \tag{8.173}$$

by Eq. (8.171). Not only becomes the Abrikosov flux lattice denser when the applied field increases, also the cylindrical cross-section of the normal-conducting cores smoothly deforms into a hexagonal one. Precisely at H_{c_2}, the magnetic vortices are densely packed, and the superconducting cell borders are squeezed to zero thickness. In this way, the mixed state turns into a homogeneous, metallic state. The area that can be assigned to a single vortex is $S_\otimes = 1/n_\otimes = 2\pi\xi^2(T)$ so that $N_\otimes S_\otimes$ covers the whole surface S perpendicular to the applied field.

In closing this section, we briefly mention the existence of superconducting materials for which the Ginzburg-Landau parameter is close to the critical value $\kappa = 1/\sqrt{2}$ separating type-II from type-I superconductors. For such materials, which are known as type-II/1 superconductors, the screening of the two interactions are of the same order of magnitude. As a result, type-II/1 superconductors exhibit a remarkable experimental phenomenon. If the lower critical field H_{c_1} is crossed from below, a whole flux lattice jumps in instead of one vortex after the other as for deep type-II superconductors. This is because the attractive scalar interaction in these materials dominates at long distance, while the repulsive magnetic interaction dominates at short distances. The lattice spacing of a flux lattice in such superconductors corresponds to the minimum in the potential.

8.15 Gapless Superconductivity

The discussion so far has concentrated mostly on systems in thermal equilibrium. The present section deals with time-dependent non-equilibrium phenomena. This can lead to quite unexpected behavior, such as gapless superconductivity where electrons at the Fermi surface of the metallic state face, in contrast to the usual case, no energy gap. To appreciate the relevance of the notion of time in the description of superconductivity, note that the conventional BCS theory is invariant under time reversal. The paired electrons have momenta $\hbar\mathbf{k}$ and $-\hbar\mathbf{k}$, where the latter may be obtained from the former by changing the time direction of the electron. Because of the invariance under time reversal the energies $\xi(\mathbf{k})$ and $\xi(-\mathbf{k})$ are degenerate by Kramers' theorem. The question arises what happens when time reversal symmetry is broken. Such a breakdown may, for example, be caused by impurity atoms which have a magnetic moment. Experimentally, it is well established that the effect of magnetic impurities is dramatic: even a small amount is sufficient to destroy conventional superconductivity entirely. Physically, such scatterers, which act differently on spin up and spin down electrons, lift the time-reversal degeneracy, and thereby impede the pairing of two electrons of opposite spin. On the other hand, impurities not disrupting time-reversal symmetry only slightly affect the superconducting properties. This result is known as Anderson's theorem. The reason is that, by Kramers' theorem, the electron energy levels remain doubly degenerate so that time-reversed states can still form Cooper pairs. This will, in particular, lead to the same transition temperature as for a pure superconductor.

The relation between breaking of time-reversal symmetry and gapless superconductivity may be clarified by a simple example. Consider a current-carrying state. All wave vectors are then shifted in the same way, $\pm\mathbf{k} \to \pm\mathbf{k} + \mathbf{q}$. As

$$\xi(\mathbf{k}) \to \xi(\mathbf{k}+\mathbf{q}), \quad \xi(-\mathbf{k}) \to \xi(-\mathbf{k}+\mathbf{q}), \tag{8.174}$$

the common drift momentum $\hbar\mathbf{q}$ lifts the degeneracy of $\xi(\mathbf{k})$ and $\xi(-\mathbf{k})$ which has been exact because of time-reversal symmetry. Instead of the propagator (8.14), we now have

$$S_{\mathrm{F}}^{\mathbf{q}}(k) = \begin{pmatrix} \hbar\omega - \xi(\mathbf{k}+\mathbf{q}) & -\bar{\Delta} \\ -\bar{\Delta}^* & \hbar\omega + \xi(-\mathbf{k}+\mathbf{q}) \end{pmatrix}^{-1}, \tag{8.175}$$

where the order parameter is taken to be a constant $\bar{\Delta}$. The corresponding spectrum is

$$E(\mathbf{k}) = \frac{\hbar^2}{m}\mathbf{k}\cdot\mathbf{q} \pm \sqrt{\xi^2(\mathbf{k}) + |\bar{\Delta}|^2}, \tag{8.176}$$

assuming $|\mathbf{q}| << k_{\mathrm{F}}$. It becomes gapless at the Fermi surface of the metallic state for

$$\hbar\hat{\mathbf{k}}\cdot\mathbf{q} = \mp\frac{|\bar{\Delta}|}{v_{\mathrm{F}}}, \tag{8.177}$$

with $\hat{\mathbf{k}}$ denoting the unit vector $\hat{\mathbf{k}} = \mathbf{k}/|\mathbf{k}|$. Although the global U(1) symmetry is still spontaneously broken for $\bar{\Delta} \neq 0$, there exist gapless fermionic excitations just as in the metallic state.

Since perturbations that break time-reversal symmetry split up Cooper pairs, their main effect is to impart a finite lifetime to the pairs. The time-dependence of the pair field then becomes

$$\Delta(t) \propto \mathrm{e}^{-t/\tau_K}, \tag{8.178}$$

with τ_K the average lifetime, so that for $t > \tau_K$, the order parameter is exponentially suppressed.

To see how the transition temperature is affected by pair-breaking perturbations, consider a superconductor close to the critical point where it undergoes a smooth transition to the metallic state. Sufficiently close to this point, the order parameter is small, and governed by the Lagrangian (8.101) with the nonlinear term omitted. For definiteness, we restrict ourselves for the moment to cases where the order parameter Δ is uniform in space—a condition met, for example, in superconducting grains small compared to the coherence length. The linear field equation obtained from the Lagrangian (8.101) then reads

$$\left[\ln\left(\frac{T_c}{T}\right)+\psi\left(\frac{1}{2}\right)-\operatorname{Re}\psi\left(\frac{1}{2}-\frac{\hbar\partial_t}{4\pi k_B T}\right)\right]\Delta=0, \tag{8.179}$$

where T_c is the BCS transition temperature (8.96). With the time dependence (8.178) of the order parameter, Eq. (8.179) turns into an implicit equation,

$$\ln\left(\frac{T_c}{T_t}\right)+\psi\left(\frac{1}{2}\right)-\psi\left(\frac{1}{2}+\frac{\hbar}{4\pi\tau_K k_B T_t}\right)=0, \tag{8.180}$$

for the true transition temperature T_t in the presence of a pair-breaking perturbation characterized by the relaxation time τ_K. Using the expansion (8.102) of the digamma function, we find that for weak pair breaking, the depression of T_c is linear in $1/\tau_K$,

$$k_B\left(T_c-T_t\right)=\frac{\pi\hbar}{8\tau_K}. \tag{8.181}$$

On the other hand, using the asymptotic behavior for $x\to\infty$,

$$\psi(x)\sim\ln(x), \tag{8.182}$$

we find that superconductivity is completely destroyed, i.e., the transition temperature T_t becomes zero for

$$\frac{\hbar}{\tau_K}=\frac{\pi}{e^{\gamma}}k_B T_c\approx 1.76\,k_B T_c, \tag{8.183}$$

where use is made of Eq. (8.93).

If magnetic impurities are responsible for the pair breaking, the order parameter is described by a diffusion equation of the type

$$-\frac{\partial}{\partial t}\Delta=-D\left(\nabla+i\frac{2e}{\hbar c}\mathbf{A}\right)^2\Delta+\cdots, \tag{8.184}$$

rather than by the diffusion equation (8.111) of a pure superconductor. In Eq. (8.184), $D=\frac{1}{3}\tau_F v_F^2$ is the diffusion constant and τ_F is the relaxation time of electrons near the Fermi surface of the metallic state due to impurity scattering. Moreover, the constant term $\propto\Delta$, which vanishes at the true transition temperature, and the nonlinear terms are suppressed in Eq. (8.184), as in Eq. (8.111).

Comparison of the two diffusion equations shows that the coefficient $\gamma \equiv \nu(0)\xi_0^2$ of the kinetic term $-|\nabla\Delta|^2$ in the Lagrangian (8.103) describing a pure superconductor must be replaced with

$$\gamma_{\boxdot} = \frac{\pi\hbar\nu(0)D}{8k_{\rm B}T_{\rm c}} \tag{8.185}$$

for a superconductor with impurities. This result can be checked explicitly by repeating the derivation of Eq. (8.103) using the electron propagator (6.33) instead of the propagator of the pure system. As a result of this substitution, the coherence length. (8.107) of the pure system is replaced with

$$\xi_{\boxdot}^2(T) = \frac{\pi\hbar D}{8k_{\rm B}T_{\rm c}} \frac{1}{|1 - T/T_{\rm c}|}, \tag{8.186}$$

in the case of impurities.

As an application, consider a small alloy (so that Δ is uniform) in an external magnetic field $\mathbf{H}$ near the transition temperature (so that nonlinear terms can be ignored). From the diffusion equation (8.184) with $\nabla = 0$ it follows that the lifetime τ_K of Cooper pairs introduced in Eq. (8.178) is given by

$$\frac{1}{\tau_K} = \frac{4e^2}{\hbar^2 c^2} D\,\langle \mathbf{A}^2 \rangle, \tag{8.187}$$

where the average is over the sample volume. For a small spherical sample of radius R in a uniform field, the vector potential can be written as

$$\mathbf{A} = \frac{1}{2}\mathbf{x} \times \mathbf{H}. \tag{8.188}$$

The applied magnetic field appears here because screening effects due to supercurrents, which are proportional to $|\Delta|^2$, lead to higher-order terms which are dropped in the linear approximation. The spatial average is

$$\langle \mathbf{A}^2 \rangle = \frac{1}{10}\mathbf{H}^2 R^2. \tag{8.189}$$

With τ_K being determined, the implicit equation (8.180) yields the value of the upper critical field $H_{\rm c_2}$ for this sample as a function of temperature.

If the order parameter varies in space, $1/\tau_K$ is the lowest eigenvalue of the equation

$$-D\left(\nabla + {\rm i}\frac{2e}{\hbar c}\mathbf{A}\right)^2 \Delta = \frac{1}{\tau_K}\Delta, \tag{8.190}$$

valid at the true transition temperature where the constant term $\propto \Delta$ vanishes. This equation, which is obtained by ignoring the nonlinear term, is of the form of the time-independent Schrödinger equation describing a particle of mass $\hbar/2D$ and charge $-2e$ in an external magnetic field, cf. Eq. (8.169). The lowest eigenvalue is given by

$$\frac{1}{\tau_K} = \frac{2e}{\hbar c} D H_{c_2}. \tag{8.191}$$

Equation (8.180) then yields as implicit equation for the upper critical field $H_{c_2}(T)$

$$\ln\left(\frac{T_c}{T}\right) + \psi\left(\frac{1}{2}\right) - \psi\left(\frac{1}{2} + \frac{D}{2\pi k_B T}\frac{eH_{c_2}}{c}\right) = 0. \tag{8.192}$$

If, in analogy with the Ginzburg-Landau parameter κ (8.172), the parameter $\kappa_\square$ is defined through

$$\sqrt{2}\kappa_\square \equiv \frac{H_{c_2}}{H_c}, \tag{8.193}$$

with H_c the thermodynamic field (8.123), Eq. (8.181) with τ_K given by Eq. (8.191) leads to the expression

$$\kappa_\square = \left(\frac{7\zeta(3)}{\pi\nu(0)}\right)^{1/2} \frac{c}{2\pi^2 e D}, \tag{8.194}$$

near the transition temperature.

Notes

(i) The history of superconductivity is described in [Hoddeson *et al.* (1992)] and [Schrieffer and Tinkham (1999)]. For a nontechnical introduction to the subject, see [Ginzburg and Andryushin (1994)].

(ii) There exist a host of review articles, monographs and textbooks on the subject from all possible angles and at all possible levels. See, for example, [London (1950); de Gennes (1989); Saint-James *et al.* (1969); Abrikosov (1988)], and [Crisan (1989)]. The subject is covered in the textbooks [Abrikosov *et al.* (1963); Fetter and Walecka (1971)], and [Weinberg (1996)]. For a compact presentation, see [Kopnin (2002)]. For a presentation highlighting the spontaneously symmetry breakdown of global U(1) phase invariance (in neutral BCS systems), see [Weinberg (1986)].

(iii) The exclusion of an applied magnetic field from the interior of a superconducting piece of material was first observed by Meissner and Ochsenfeld (1933).

(iv) The successful phenomenological theory of superconductivity was formulated by Ginzburg and Landau (1950).

(v) The groundbraking theory of superconductivity by Bardeen, Cooper, and Schrieffer was presented in [Bardeen *et al.* (1957)].

(vi) The Ginzburg-Landau equations were first derived from the microscopic BCS theory by Gorkov (1959a).

(vii) The presence of a gapless Nambu-Goldstone mode in a neutral BCS system and the role of this collective mode in rendering a charged superconductor gauge invariant was pointed out in [Anderson (1958); Bogoliubov (1958a,b); Bogoliubov *et al.* (1958)], and in [Nambu (1960)]. Nambu and Jona-Lasinio (1961) subsequently introduced the concept of spontaneously broken symmetries into particle physics, which now forms one of the pillars of the standard model of particle physics.

(viii) Josephson proposed the effect now bearing his name in [Josephson (1962)]. For an overview of his work on the subject, see [Josephson (1969)].

(ix) The effective theory (8.36) of a neutral superconductor was proposed independently by Kemoklidze and Pitaevskii (1966) and by Greiter *et al.* (1989). Its derivation from the microscopic BCS theory was given in [Schakel (1990)].

(x) The linear response theory of superconductors covered in Secs. 8.4 and 8.6 is based on [Ambegaokar and Kadanoff (1961)].

(xi) A theory of the form (8.37), describing a massive vector field, was first considered by Stückelberg (1938) in the context of particle physics.

(xii) The quantization of the magnetic flux trapped in a superconducting ring was predicted by London (1950), who also set forth arguments on the persistence of electric currents in a superconductor.

(xiii) The time-dependent Ginzburg-Landau theory was derived by Abrahams and Tsuneto (1966) and by Schmid (1966). For a discussion of its applicability, see [Cyrot (1973)]

(xiv) The observation that impurities which respect time-reversal symmetry only slightly affect the superconducting properties is due to Anderson (1959).

(xv) For an overview of gapless superconductivity with references to original papers, see [de Gennes (1989)].

(xvi) Equation (8.180) was first derived by Abrikosov and Gorkov (1960) in their study of the effect of magnetic impurities on superconductivity by means of the Green function formalism. The result (8.194) for a disordered superconductor is due to Gorkov (1959b).

(xvii) The theory of the mixed state in type-II superconductors, featuring magnetic vortices, is due to Abrikosov (1957). The Abrikosov vortex solution was introduced into the Abelian Higgs model of particle physics, which is equivalent to the Ginzburg-Landau theory, by Nielsen and Olesen (1973). The mixed state is well covered in the textbooks [de Gennes (1989); Saint-James *et al.* (1969)], and [Abrikosov (1988)]. Our presentation is based also on [Bettencourt and Rivers (1995); Piette *et al.* (1995)], and [Speight (1997)].

(xviii) The explanation that the peculiar behavior of type-II/1 superconductors originates from the attractive interaction between flux lines, is due to Kragelöh (1970), see also [Auer and Ullmaier (1973)].

Chapter 9

Duality

This chapter discusses superconducting films in thermal equilibrium. The first part is an extension of Chap. 5 to charged systems. As in that chapter, it proves expedient to describe the system in terms of dual variables, where the vortices are no longer represented by singular objects. By doing so, we enter the realm of bosonization, chiral anomalies, index theorems, and instantons in two dimensions, provided one of the spatial dimensions is pictured as representing Euclidean time. The chapter closes with a modern description of the celebrated Peierls instability in quasi one-dimensional conductors at the absolute zero of temperature, and with a discussion of charge fractionalization that emerges in this condensed-matter system.

9.1 Superconducting Films

To describe a superconducting film in thermal equilibrium, we consider the static part of the zero-temperature Lagrangian density (8.37), and write the corresponding Hamiltonian density in the general form

$$\mathcal{H} = \frac{1}{2}\rho_s \mathbf{v}_s^2 + \frac{1}{8\pi}(\nabla \times \mathbf{A})^2. \tag{9.1}$$

As in our description of a superfluid film in Sec. 5.5, the coefficient $\rho_s = \rho_s(T)$, known as the *superconducting mass density*, generalizes the zero-temperature result $m\bar{n}$ to finite temperatures. This response function specifies what fraction of the electrons takes part in the supercurrent flow. Despite not all electrons are paired, the result $\rho_s(0) = m\bar{n}$ implies that, at zero temperature, all electrons participate in the supercurrent flow. Furthermore, $\mathbf{v}_s$ is the so-called superconducting velocity

$$\mathbf{v}_s \equiv \frac{\hbar}{m}\left(\nabla\varphi + \frac{e}{\hbar c}\mathbf{A}\right), \tag{9.2}$$

where it is recalled that $q = -e(< 0)$ and m are the charge and mass of an electron, and φ is *half* the phase of the superconducting bulk order parameter, see Eq. (8.23). We also included the magnetic energy density in the Hamiltonian (9.1). For the time being, we ignore vortices. The partition function of the system then reads

$$Z = \int \mathrm{D}\varphi \int \mathrm{D}\mathbf{A} \exp\left(-\beta \int \mathrm{d}^2 x \mathcal{H}\right), \tag{9.3}$$

with $\beta = 1/k_\mathrm{B}T$ the inverse temperature. As in Sec. 8.4, the functional integral over φ is readily evaluated with the result, cf. (8.44)

$$\begin{aligned} Z &= \int \mathrm{D}\mathbf{A} \exp\left\{-\frac{\beta}{8\pi} \int \mathrm{d}^2 x \left[\frac{1}{\lambda_\mathrm{L}^2} A^i \left(\delta_{ij} - \frac{\partial_i \partial_j}{\nabla^2}\right) A^j + (\nabla \times \mathbf{A})^2\right]\right\} \\ &= \int \mathrm{D}\mathbf{A} \exp\left[-\frac{\beta}{8\pi} \int \mathrm{d}^2 x (\nabla \times \mathbf{A}) \cdot \left(1 + \frac{\lambda_\mathrm{L}^{-2}}{\nabla^2}\right) (\nabla \times \mathbf{A})\right], \end{aligned} \tag{9.4}$$

where the last term, with $1/\lambda_\mathrm{L}^2 \equiv 4\pi e^2 \rho_\mathrm{s}/m^2 c^2$, is the gauge-invariant, albeit non-local mass term for the vector potential generated by the Anderson-Higgs mechanism. The number of degrees of freedom does not change in the process. This can be seen by noting that a *gapless* vector potential represents no physical degrees of freedom in two dimensions. In Minkowski spacetime, this is readily understood by recognizing that there is no transverse direction in 1 + 1 dimensions. Before the Anderson-Higgs mechanism took place, the system therefore contains only a single physical degree of freedom described by φ. This equals the number of degrees of freedom of a *gapped* vector potential featuring in Eq. (9.4).

We next introduce an auxiliary field $\tilde{h}$ to linearize the magnetic energy term in Eq. (9.4),

$$\exp\left[-\frac{\beta}{8\pi} \int \mathrm{d}^2 x (\nabla \times \mathbf{A})^2\right] = \int \mathrm{D}\tilde{h} \exp\left\{-\frac{1}{4\pi} \int \mathrm{d}^2 x \left[\frac{1}{2\beta} \tilde{h}^2 + \mathrm{i}\tilde{h}(\nabla \times \mathbf{A})\right]\right\}, \tag{9.5}$$

and integrate out the vector potential. Recall that in two space dimensions, the curl of a vector is a scalar: $\nabla \times \mathbf{A} = \epsilon^{ij} \partial_i A^j$. The result (9.5) is a manifestly gauge-invariant expression for the partition function in terms of a massive scalar field $\tilde{h}$, representing the single degree of freedom contained in the theory:

$$Z = \int \mathrm{D}\tilde{h} \exp\left\{-\frac{1}{8\pi\beta} \int \mathrm{d}^2 x \left[\lambda_\mathrm{L}^2 (\nabla \tilde{h})^2 + \tilde{h}^2\right]\right\}. \tag{9.6}$$

In deriving this, the invariance under gauge transformations is used to pick the transverse, or Coulomb gauge, in which the vector potential satisfies the condition $\nabla \cdot \mathbf{A} = 0$. This alternative, but completely equivalent description of a superfluid film is referred to as the *dual theory*. Note the appearance of $1/\beta$ rather than β

in the exponential, meaning that the temperature gets inverted when going from the original to the dual theory. To understand the physical significance of the field $\tilde{h}$ featuring in the dual theory, we note from Eq. (9.5) that it satisfies the Euler-Lagrange equation

$$\tilde{h} = -\mathrm{i}\beta\nabla \times \mathbf{A}. \tag{9.7}$$

That is, the fluctuating field $\tilde{h}$ may be thought of representing the local magnetic induction, which is a scalar in two space dimensions. Equation (9.6) shows that the magnetic induction has a finite penetration depth λ_{L}. In contrast to the original description where the functional integral runs over the vector potential, the integration variable in the dual theory (9.6) is the physical field.

We next include vortices in the description. Since the penetration depth λ_{L} provides the system with an infrared cutoff, a single magnetic vortex in the charged theory has a finite energy. Vortices, which are pointlike in 2D at scales large compared to their core size, can therefore be thermally activated in type-II superconductors. This is different from the superfluid phase of the neutral model, where the absence of an infrared cutoff permits only tightly bound vortex-antivortex pairs to exist. We expect, accordingly, the superconducting phase to describe a plasma of vortices of unit winding number. Vortices with higher winding number carry a much higher energy and can be safely ignored, see Eq. (8.163). The partition function now reads

$$Z = \sum_{N=0}^{\infty} \sum_{w_\alpha = \pm 1} \frac{z^N}{N!} \prod_\alpha \int \frac{\mathrm{d}^2 x_\alpha}{a^2} \int \mathrm{D}\varphi \int \mathrm{D}\mathbf{A} \exp\left(-\beta \int \mathrm{d}^2 x \mathcal{H}\right), \tag{9.8}$$

where z is the fugacity, i.e., the Boltzmann factor associated with the vortex core energy. The length scale a of order the vortex core size is included for dimensional reasons. The velocity appearing in the Hamiltonian (9.1) now includes the Villain potential

$$\mathbf{v}_{\mathrm{s}} = \frac{\hbar}{m}\left(\nabla\varphi + \frac{e}{\hbar c}\mathbf{A}\right) - \mathbf{A}^{\mathrm{V}}, \tag{9.9}$$

which satisfies the condition, cf. Eq. (5.1)

$$\nabla \times \mathbf{A}^{\mathrm{V}}(\mathbf{x}) = -\sum_\alpha \kappa_\alpha \delta(\mathbf{x} - \mathbf{x}_\alpha), \tag{9.10}$$

where the sum extends over all vortices centered at $\mathbf{x}_\alpha$ with winding number $w_\alpha = \pm 1$. Because the phase φ has periodicity π, the circulation of the vortices in a superconductor is quantized in units of $\pi\hbar/m$, i.e., $\kappa_\alpha = \pm\pi\hbar/m$ in Eq. (9.10).

The Villain potential $\mathbf{A}^{\mathrm{V}}$ can be shifted from the first to the second term in the Hamiltonian (9.1) by applying the transformation $\mathbf{A} \to \mathbf{A}' \equiv \mathbf{A} + (mc/e)\mathbf{A}^{\mathrm{V}}$. This results in the shift

$$\nabla \times \mathbf{A} \to \nabla \times \mathbf{A}' = \nabla \times \mathbf{A} + B^{\mathrm{V}}, \tag{9.11}$$

with

$$B^{\mathrm{V}} \equiv -\Phi_0 \sum_\alpha w_\alpha\, \delta(\mathbf{x} - \mathbf{x}_\alpha) \tag{9.12}$$

representing the magnetic flux density associated with the vortices. Here, $\Phi_0 = \pi\hbar/e$ is the elementary flux quantum in a superconductor. Repeating the steps of the previous paragraph, we now obtain instead of Eq. (9.6) as expression for the partition function

$$Z = \sum_{N=0}^{\infty} \sum_{w_\alpha = \pm 1} \frac{z^N}{N!} \prod_\alpha \int \frac{\mathrm{d}^2 x_\alpha}{a^2} \int \mathrm{D}\tilde{h} \times \exp\left\{ -\frac{1}{8\pi\beta} \int \mathrm{d}^2 x \left[\lambda_{\mathrm{L}}^2 (\nabla \tilde{h})^2 + \tilde{h}^2 \right] - \frac{\mathrm{i}}{4\pi} \int \mathrm{d}^2 x B^{\mathrm{V}} \tilde{h} \right\}, \tag{9.13}$$

where $\tilde{h}$ represents the local magnetic induction[1] h

$$\tilde{h} = -\mathrm{i}\beta(\nabla \times \mathbf{A} + B^{\mathrm{V}}) = -\mathrm{i}\beta h. \tag{9.14}$$

The field equation for $\tilde{h}$ obtained from Eq. (9.13) yields for the local magnetic induction:

$$-\lambda_{\mathrm{L}}^2 \nabla^2 h + h = B^{\mathrm{V}}, \tag{9.15}$$

which is the familiar equation in the presence of magnetic vortices.

The last term in (9.13) shows that the coupling constant g with which a magnetic vortex couples to the fluctuating $\tilde{h}$ field is given by the ratio of the elementary flux quantum (contained in the definition of B^{V}) to the penetration depth $1/\lambda_{\mathrm{L}}$ (resulting from rescaling $\tilde{h} \to \tilde{h}/\lambda_{\mathrm{L}}$ so that the coefficient of the gradient term becomes unity),

$$g = \frac{\Phi_0}{\lambda_{\mathrm{L}}} = \frac{2\pi\hbar}{mc} \sqrt{\pi \rho_{\mathrm{s}}}. \tag{9.16}$$

Note that this coupling constant is independent of the electric charge e, and is determined by the Compton wave length $\lambda_{\mathrm{C}} = 2\pi\hbar/mc$ of the electron and the superconducting mass density. By the identity

$$\begin{aligned} &\sum_{N=0}^{\infty} \sum_{w_\alpha = \pm 1} \frac{z^N}{N!} \prod_\alpha \int \frac{\mathrm{d}^2 x_\alpha}{a^2} \mathrm{e}^{\mathrm{i}(\Phi_0/4\pi) \sum_\alpha \tilde{h}(x_\alpha)} \\ &\quad = \sum_{N=0}^{\infty} \frac{(z/a^2)^N}{N!} \left[\int \mathrm{d}^2 x \left(\mathrm{e}^{\mathrm{i}(\Phi_0/4\pi)\tilde{h}(x)} + \mathrm{e}^{-\mathrm{i}(\Phi_0/4\pi)\tilde{h}(x)} \right) \right]^N \\ &\quad = \exp\left\{ 2\frac{z}{a^2} \int \mathrm{d}^2 x \cos[(\Phi_0/4\pi)\tilde{h}(x)] \right\}, \end{aligned} \tag{9.17}$$

[1] We here adopt the standard, but somewhat confusing notation in the theory of superconductivity.

we arrive at the partition function of the massive sine-Gordon model

$$Z = \int \mathrm{D}\tilde{h} \exp\left(-\int \mathrm{d}^2 x \left\{\frac{1}{8\pi\beta}\left[\lambda_{\mathrm{L}}^2(\nabla\tilde{h})^2 + \tilde{h}^2\right] - 2\frac{z}{a^2}\cos\left[(\Phi_0/4\pi)\tilde{h}\right]\right\}\right). \quad (9.18)$$

This expression constitutes the dual formulation of a two-dimensional superconductor. The magnetic vortices of unit winding number $w_\alpha = \pm 1$ turn the theory (9.6) without mutual interactions into an interacting one.

The final form (9.18) demonstrates the rationales for going over to a dual theory. First, it is a formulation directly in terms of a physical field representing the local magnetic induction. There is no redundancy in this description and therefore no gauge invariance. Second, the magnetic vortices are accounted for in a nonsingular fashion. This is different from the original formulation of a 2D superconductor where the local magnetic induction is the curl of an unphysical gauge potential **A**, and where the magnetic vortices appear as singular objects.

Up to this point we have discussed a genuine 2D superconductor. As a model to describe superconducting films this is, however, not adequate. The reason is that the magnetic interaction between the vortices takes place mostly not through the film but through free space surrounding the film where the vector potential is long range. A charged superconducting film is markedly different from a superfluid film. The interaction between the vortices there is mediated by a gapless phase field which is confined to the film so that a genuine 2D theory gives a satisfactory description of a superfluid film. Note that this nonmagnetic interaction is also present in a superconductor, but is irrelevant for type-II materials which we are considering.

To arrive at a more realistic description of magnetic vortices in a film of thickness W, the result (9.15) must be adjusted. First, the induction field now also depends on the third coordinate x^3. Moreover, to account for the fact that both the vortices and the screening currents, which produce the second term in (9.15), are confined to the plane, a Dirac delta function $W\delta(x^3)$ must be included in the second and last term, giving

$$-\nabla^2 h(\mathbf{x}_\perp, x^3) + \frac{1}{\lambda_\perp}\delta(x^3) h(\mathbf{x}_\perp, x^3) = \frac{1}{\lambda_\perp}\delta(x^3) B^{\mathrm{V}}(\mathbf{x}_\perp). \quad (9.19)$$

Here, $1/\lambda_\perp \equiv W/\lambda_{\mathrm{L}}^2$ is the relevant inverse length scale, $\mathbf{x}_\perp$ denotes the coordinates in the plane, and h is the component of the local induction field perpendicular to the film.

To be definite, we consider a single magnetic point vortex of winding number $w = -1$ located at the origin. As in bulk superconductors, magnetic vortices with negative winding numbers produce a magnetic flux in the positive third direction, see Eq. (8.135). The field equation (9.19) can be best solved by going over to Fourier space. With

$$h(\mathbf{k}_\perp, x^3 = 0) = \int \frac{dk^3}{2\pi} h(\mathbf{k}_\perp, k^3), \tag{9.20}$$

it can then be cast in the form

$$(\mathbf{k}_\perp^2 + k_3^2) h(\mathbf{k}_\perp, k^3) + \frac{1}{\lambda_\perp} h(\mathbf{k}_\perp, x^3 = 0) = \frac{\Phi_0}{\lambda_\perp}, \tag{9.21}$$

or, after integrating over k^3, in the form

$$h(\mathbf{k}_\perp, x^3 = 0) = \frac{\Phi_0}{1 + 2\lambda_\perp |\mathbf{k}_\perp|} h(\mathbf{x}_\perp, x^3 = 0). \tag{9.22}$$

In coordinate space, this gives

$$\begin{aligned} h(\mathbf{x}_\perp, 0) &= \int \frac{d^2 k_\perp}{(2\pi)^2} e^{i\mathbf{k}_\perp \cdot \mathbf{x}_\perp} h(\mathbf{k}_\perp, x^3 = 0) \\ &= \frac{\Phi_0}{2\pi} \int_0^\infty dk_\perp \frac{k_\perp}{1 + 2\lambda_\perp k_\perp} J_0(k_\perp |\mathbf{x}_\perp|), \end{aligned} \tag{9.23}$$

with J_0 the zeroth-order Bessel function of the first kind. At small distances from the vortex core ($\lambda_\perp k_\perp >> 1$)

$$h(\mathbf{x}_\perp, 0) \sim \frac{\Phi_0}{4\pi\lambda_\perp |\mathbf{x}_\perp|}, \tag{9.24}$$

while far away ($\lambda_\perp k_\perp << 1$)

$$h(\mathbf{x}_\perp, 0) \sim \frac{\Phi_0 \lambda_\perp}{\pi |\mathbf{x}_\perp|^3}. \tag{9.25}$$

This last equation shows that the field does not exponentially decay as would be the case in a genuine 2D system. The reason for the long range is that most of the magnetic interaction takes place in free space outside the film where the vector potential is long range.

If, as is often the case, the relevant penetration length $\lambda_\perp$ is much larger than the sample size, it can be effectively set to infinity. In this limit, the effect of the magnetic interaction, as can be seen from Eq. (9.24), diminishes and the vortices behave as in a superfluid film. A superconducting film may therefore undergo a Berezinskii-Kosterlitz-Thouless transition at a temperature $T_{\rm BKT}$ characterized by the unbinding of vortex-antivortex pairs. The transition temperature $T_{\rm BKT}$ is well below the bulk temperature T_c where the Cooper pairs form so that the energy gap of the fermions remains finite at the critical point. For temperatures $T_{\rm BKT} \leq T \leq T_c$, there is a plasma of magnetic vortices which disorder the superconducting state. At the critical temperature $T = T_{\rm BKT}$ vortices and antivortices bind into pairs and algebraic long-range order sets in.

9.2 Bosonization

The two dual models (9.6) and (9.18) of the preceding section, describing a 2D superconductor without and with vortices, are reminiscent of the bosonized massless and massive *Schwinger model*, respectively. The massless Schwinger model describes gapless fermions interacting with an electric field in one space and one time dimension. It is defined by the Lagrangian density

$$\mathcal{L} = \bar{\psi}(\mathrm{i}\partial\!\!\!/ + eA\!\!\!/)\psi - \frac{1}{4}F_{\mu\nu}F^{\mu\nu}, \tag{9.26}$$

consisting of a massless Dirac theory minimally coupled to a gauge field through a charge $q = -e(< 0)$. Here, $A\!\!\!/ \equiv A_\mu\gamma^\mu$ with γ^μ the Dirac matrices which obey the algebra (3.25). Moreover, ψ is a two-component Grassmann field describing the fermions, and $\bar{\psi} \equiv \psi^\dagger\beta$, with $\beta \equiv \gamma^0$. A possible representation of the Dirac matrices in $d = 2$ is given in Eq. (3.26). In compliance with standard practice in particle physics, we use natural units here with $\hbar = c = 1$, and adopt instead of Gaussian, Heaviside-Lorentz units where the coefficient of the Maxwell term, with $F^{\mu\nu} = \partial^\mu A^\nu - \partial^\nu A^\mu$, is $\frac{1}{4}$, and Gauss' law reads $\partial_1 E^1 = \rho$, with ρ the electric charge density. Notice that electrodynamics in two spacetime dimensions does not involve a photon because there are no transverse spatial directions. There is also no magnetic field, $\frac{1}{4}F_{\mu\nu}F^{\mu\nu} = -\frac{1}{2}E_1^2$, with $E^1 = F^{10}$ the electric field, and the only interaction present is just the Coulomb interaction. As will we demonstrated in Sec. 9.8 below, gapless fermions described by the massless Dirac theory naturally emerge in the context of quasi one-dimensional condensed matter systems.

Since the Schwinger model is bilinear in ψ, the functional integral over the Grassmann fields in the partition function,

$$Z = \int \mathrm{D}\bar{\psi}\mathrm{D}\psi \int \mathrm{D}A_\mu\, \mathrm{e}^{\mathrm{i}\int \mathrm{d}^2x\,\mathcal{L}}, \tag{9.27}$$

can be formally evaluated to yield the Matthews-Salam determinant

$$\int \mathrm{D}\bar{\psi}\mathrm{D}\psi \exp\left[\mathrm{i}\int \mathrm{d}^2x\,\bar{\psi}(\mathrm{i}\partial\!\!\!/ + eA\!\!\!/)\psi\right] = \mathrm{Det}(p\!\!\!/ + eA\!\!\!/), \tag{9.28}$$

which is a one-fermion-loop result. The determinant can be evaluated in closed form using, for example, the derivative expansion presented in Sec. 3.1. To this end, we for the moment assume that A^μ is a background field and introduce the effective action $\Gamma_1[A^\mu]$ through

$$\mathrm{Det}(p\!\!\!/ + eA\!\!\!/) = \mathrm{e}^{\mathrm{i}\Gamma_1[A^\mu]} \tag{9.29}$$

and expand it as

$$\begin{aligned}\Gamma_1[A^\mu] &= -\mathrm{i}\,\mathrm{Tr}\ln(p\!\!\!/ + eA\!\!\!/)\\ &= -\mathrm{i}\,\mathrm{Tr}\ln(p\!\!\!/) + \mathrm{i}\,\mathrm{Tr}\sum_{\ell=1}^{\infty}\frac{1}{\ell}\left(-eS_{\mathrm{F}}A\!\!\!/\right)^\ell,\end{aligned} \tag{9.30}$$

where S_F is the massless fermion propagator, cf. Eq. (2.140),

$$S_F(x-x') = -i\,\langle \psi(x)\bar{\psi}(x')\rangle\,, \tag{9.31}$$

satisfying $i\partial\!\!\!/ S_F(x) = \delta(x)$. Explicitly,

$$\begin{aligned} S_F(x-x') &= \int \frac{d^2k}{(2\pi)^2}\frac{k\!\!\!/}{k^2+i\eta}e^{-ik\cdot(x-x')} \\ &= i\partial\!\!\!/ \int \frac{d^2k}{(2\pi)^2}\frac{1}{k^2+i\eta}e^{-ik\cdot(x-x')} \\ &= -\frac{1}{2\pi}\partial\!\!\!/ \ln\left(\sqrt{-(x-x')^2}\right) = -\frac{1}{2\pi}\frac{x\!\!\!/ - x\!\!\!/'}{(x-x')^2}, \end{aligned} \tag{9.32}$$

with η the usual infinitesimal switching parameter which, as discussed in Sec. 2.4, is required for causality. In the third line, the Minkowski counterpart of Eq. (5.11) is used. It turns out that only the second term in the series (9.30) is nonzero. This is special to two dimensions. That term gives

$$\Gamma_1^{(\ell=2)} = i\frac{e^2}{2}\,\mathrm{Tr}\,\frac{p\!\!\!/}{p^2}A\!\!\!/\frac{p\!\!\!/}{p^2}A\!\!\!/, \tag{9.33}$$

which by applying the derivative expansion rules of Sec. 3.1 takes the form

$$\Gamma_1^{(\ell=2)} = i\frac{e^2}{2}\int d^2x \frac{d^2k}{(2\pi)^2}\,\mathrm{tr}\,\gamma^\mu\gamma^\nu\gamma^\lambda\gamma^\rho\frac{k_\mu}{k^2}\frac{k_\lambda - i\partial_\lambda}{(k-i\partial)^2}A_\nu A_\rho, \tag{9.34}$$

where, as throughout, the derivatives operate only on the first field to the right. Equation (9.34) combines the contributions of two terms which, by the cyclic property of the trace, are equal. Power counting shows that the momentum integral diverges. To regularize it, we consider the theory in $d = 2-\varepsilon$ dimensions, and take ε to zero at the end of the calculation. The integral is conveniently evaluated with the help of the Feynman parametrization

$$\frac{1}{ab} = \int_0^1 d\alpha\,\frac{1}{[\alpha a + (1-\alpha)b]^2}, \tag{9.35}$$

giving with $q_\lambda \equiv i\partial_\lambda$

$$\begin{aligned} I_{\mu\lambda} &\equiv \int \frac{d^dk}{(2\pi)^d}\frac{k_\mu}{k^2+i\eta}\frac{k_\lambda - q_\lambda}{(k-q)^2+i\eta} \\ &= \int_0^1 d\alpha \int \frac{d^dk}{(2\pi)^d}\frac{k_\mu(k_\lambda - q_\lambda)}{[k^2 - 2(1-\alpha)k\cdot q + (1-\alpha)q^2 + i\eta]^2} \\ &= \int_0^1 d\alpha \int \frac{d^dk}{(2\pi)^d}\frac{[k_\mu + (1-\alpha)q_\mu](k_\lambda - \alpha q_\lambda)}{[k^2 + \alpha(1-\alpha)q^2 + i\eta]^2}, \end{aligned} \tag{9.36}$$

where in the last step the integration variable k^μ is shifted to $k^\mu + (1-\alpha)q^\mu$. Using the basic formulas (3.43) and (3.44) of dimensional regularization, we arrive at

$$I_{\mu\lambda} = -\frac{\mathrm{i}}{2}\frac{\Gamma(\varepsilon/2)}{(4\pi)^{1-\varepsilon/2}}\int_0^1 \mathrm{d}\alpha\left(\eta_{\mu\lambda} - \varepsilon\frac{\partial_\mu\partial_\lambda}{\partial^2}\right)\frac{1}{[\alpha(1-\alpha)\partial^2]^{\varepsilon/2}}. \tag{9.37}$$

The first term on the right diverges as $\sim 1/\varepsilon$ in the limit $\varepsilon \to 0$, while the second remains finite. By the Dirac algebra (3.25), we have

$$\operatorname{tr}\gamma^\mu\gamma^\nu = d\eta^{\mu\nu}, \quad \gamma^\mu\gamma_\mu = d. \tag{9.38}$$

With these rules, the traces over the Dirac matrices in Eq. (9.34) are readily evaluated, giving

$$\begin{aligned}\eta_{\mu\lambda}\operatorname{tr}\gamma^\mu\gamma^\nu\gamma^\lambda\gamma^\rho &= \eta_{\mu\lambda}\operatorname{tr}(-\gamma^\nu\gamma^\mu + 2\eta^{\mu\nu})\gamma^\lambda\gamma^\rho \\ &= \varepsilon\operatorname{tr}\gamma^\nu\gamma^\rho\end{aligned} \tag{9.39}$$

and

$$\begin{aligned}h_{\mu\lambda}\operatorname{tr}\gamma^\mu\gamma^\nu\gamma^\lambda\gamma^\rho &= h_{\mu\lambda}\operatorname{tr}\left(-\gamma^\nu\eta^{\mu\lambda}\gamma^\rho + 2\eta^{\mu\nu}\gamma^\lambda\gamma^\rho\right) \\ &= (2-\varepsilon)\left(2h^{\nu\rho} - h^\mu{}_\mu\eta^{\nu\rho}\right)\end{aligned} \tag{9.40}$$

for any $h_{\mu\lambda}$ symmetric in its indices. Note that the trace (9.39) yields zero when evaluated directly in $d = 2$. The prefactor precisely cancels the diverging $1/\varepsilon$ arising from the momentum integration. Carrying out the remaining integral over the Feynman parameter and putting the results together, we find for the Matthews-Salam determinant (9.28)

$$\frac{\operatorname{Det}(\not{p} + e\not{A})}{\operatorname{Det}(\not{p})} = \exp\left[\frac{\mathrm{i}e^2}{2\pi}\int \mathrm{d}^2x A_\mu\left(\eta^{\mu\nu} - \frac{\partial^\mu\partial^\nu}{\partial^2}\right)A_\nu\right], \tag{9.41}$$

or in manifestly gauge-invariant form

$$\begin{aligned}\frac{\operatorname{Det}(\not{p} + e\not{A})}{\operatorname{Det}(\not{p})} &= \exp\left(-\frac{\mathrm{i}e^2}{4\pi}\int \mathrm{d}^2x\, F_{\mu\nu}\frac{1}{\partial^2}F^{\mu\nu}\right) \\ &= \exp\left(\frac{\mathrm{i}e^2}{2\pi}\int \mathrm{d}^2x\, E^1\frac{1}{\partial^2}E^1\right).\end{aligned} \tag{9.42}$$

Because of the $1/\partial^2$, these expressions are nonlocal. As in Eq. (8.47), the precise meaning of the notation used here follows by going over to Fourier space where, for example,

$$\int \mathrm{d}^2x A_\mu(x)\left(\eta^{\mu\nu} - \frac{\partial^\mu\partial^\nu}{\partial^2}\right)A_\nu(x) = \int\frac{\mathrm{d}^2k}{(2\pi)^2}A_\mu(-k)\left(\eta^{\mu\nu} - \frac{k^\mu k^\nu}{k^2}\right)A_\nu(k). \tag{9.43}$$

The partition function (9.27) can now be written as

$$Z = \operatorname{Det}(\not{p})\int \mathrm{D}A^\mu \exp\left\{\mathrm{i}\int \mathrm{d}^2x\left[\frac{1}{2}m_A^2 A^\mu\left(\eta_{\mu\nu} - \frac{\partial_\mu\partial_\nu}{\partial^2}\right)A^\nu - \frac{1}{4}F_{\mu\nu}F^{\mu\nu}\right]\right\} \tag{9.44}$$

or in manifestly gauge-invariant form as

$$Z = \mathrm{Det}(\not{p}) \int \mathrm{D}A^{\mu} \exp\left[\frac{\mathrm{i}}{2}\int \mathrm{d}^2 x E^1 \left(1 + \frac{m_A^2}{\partial^2}\right) E^1\right] \tag{9.45}$$

with $m_A^2 \equiv e^2/\pi$. In natural units where $\hbar = c = 1$, the electric charge e has the dimension of mass. Equation (9.44) is the Minkowski analog of the partition function (9.4) describing a superconducting film without vortices. It shows that integrating out the massless fermions from the Schwinger model produces a gauge-invariant mass term for the gauge field—that is, we again witness the generation of mass for a gauge field, known as the *Schwinger mechanism* in this guise.

The partition function (9.4) could be equivalently represented by the massive scalar field theory (9.6). The same holds for the Schwinger model. Rather than linearizing the Maxwell term in Eq. (9.44), which would be the analog of the Hubbard-Stratonovich transformation (9.5), we represent the right side of Eq. (9.41) as a functional integral over a scalar field ϕ as

$$\int \mathrm{D}\bar{\psi}\mathrm{D}\psi \exp\left[\mathrm{i}\int \mathrm{d}^2 x\, \bar{\psi}(\mathrm{i}\not{\partial} + e\not{A})\psi\right] = \mathrm{Det}(\not{p}) \int \mathrm{D}\phi \exp\left\{\mathrm{i}\int \mathrm{d}^2 x \left[\frac{1}{2}(\partial_\mu\phi)^2 + \frac{e}{\sqrt{\pi}}\epsilon_{\mu\nu}A^\mu \partial^\nu \phi\right]\right\}, \tag{9.46}$$

where $\epsilon_{\mu\nu}$ is the antisymmetric Levi-Civita tensor in two dimensions, with $\epsilon_{01} = 1$. The scalar field obeys the Euler-Lagrange equation

$$\partial^2 \phi = -\frac{e}{\sqrt{\pi}}\epsilon_{\mu\nu}\partial^\mu A^\nu = \frac{e}{\sqrt{\pi}}E^1. \tag{9.47}$$

Comparing term by term in Eq. (9.46), we arrive at the correspondence

$$\begin{aligned} \bar{\psi}\mathrm{i}\not{\partial}\psi &\,\hat{=}\, \frac{1}{2}(\partial_\mu\phi)^2 \\ \bar{\psi}\gamma^\mu\psi &\,\hat{=}\, \frac{1}{\sqrt{\pi}}\epsilon^{\mu\nu}\partial_\nu\phi, \end{aligned} \tag{9.48}$$

where, by Eq. (2.183), $\bar{\psi}\gamma^\mu\psi = j_e^\mu$ is the electric current density. These rules connect a fermion theory on the left to a boson theory on the right, i.e., the fermion theory is *bosonized*.

The integration over the gauge field can now be easily carried out, say in the Lorentz gauge $\partial_\mu A^\mu = 0$, to yield the bosonized form of the massless Schwinger model

$$Z = \mathrm{Det}(\not{p}) \int \mathrm{D}\phi \exp\left\{\frac{\mathrm{i}}{2}\int \mathrm{d}^2 x \left[(\partial_\mu\phi)^2 - m_A^2\phi^2\right]\right\}, \tag{9.49}$$

which is a massive scalar theory with a decoupled massless free fermion theory represented by the first factor on the right. The mass term of the boson field represents the instantaneous Coulomb interaction in the Schwinger model. More specifically, it denotes the potential energy U of all the charges integrated over time,

$$\begin{aligned}\int \mathrm{d}t\, U &= \frac{1}{2}\int \mathrm{d}t\,\mathrm{d}x^1\mathrm{d}x'^1\, j_e^0(t,x^1)V(x^1-x'^1)j_e^0(t,x'^1)\\ &= -\frac{e^2}{4}\int \mathrm{d}t\,\mathrm{d}x^1\,\mathrm{d}x'^1\,\bar\psi(t,x^1)\gamma^0\psi(t,x^1)|x^1-x'^1|\bar\psi(t,x'^1)\gamma^0\psi(t,x'^1),\end{aligned} \tag{9.50}$$

with $V(x^1)$ the one-dimensional Coulomb potential

$$V(x^1) = -\frac{e^2}{2}|x^1| \tag{9.51}$$

and where the particle number density $\psi^\dagger\psi$ is written as $\bar\psi\gamma^0\psi$. The assertion can be checked by applying the bosonization rule (9.48) and integrating by parts twice

$$\begin{aligned}\int \mathrm{d}t\, U &= -\frac{e^2}{4\pi}\int \mathrm{d}t\,\mathrm{d}x^1\,\mathrm{d}x'^1\,\partial_1\phi(t,x^1)|x^1-x'^1|\partial_1'\phi(t,x'^1)\\ &= \frac{e^2}{4\pi}\int \mathrm{d}t\,\mathrm{d}x^1\,\mathrm{d}x'^1\,\phi(t,x^1)\epsilon(x^1-x'^1)\partial_1'\phi(t,x'^1)\\ &= \frac{e^2}{2\pi}\int \mathrm{d}t\,\mathrm{d}x^1\,\mathrm{d}x'^1\,\phi(t,x^1)\delta(x^1-x'^1)\phi(t,x'^1)\\ &= \frac{m_A^2}{2}\int \mathrm{d}^2x\,\phi^2(x),\end{aligned} \tag{9.52}$$

where $\epsilon(x) \equiv -\theta(-x)+\theta(x)$.

The scalar part in Eq. (9.49) forms the Minkowski analog of the dual theory (9.6) of a superconducting film without taking into account vortices.

9.3 Chiral Anomaly

To understand why the scalar field ϕ in Eq. (9.49) is massive, note that the massless Schwinger model possesses in addition to the local U(1) gauge symmetry

$$\psi(x) \to \mathrm{e}^{\mathrm{i}\alpha(x)}\psi(x), \quad A_\mu(x) \to A_\mu(x) + \frac{1}{e}\partial_\mu\alpha(x) \tag{9.53}$$

also a *global* U(1) symmetry. Under so-called chiral transformations

$$\psi \to \mathrm{e}^{\mathrm{i}\beta\gamma^5}\psi, \tag{9.54}$$

where β is the transformation parameter and $\gamma^5 \equiv \gamma^0\gamma^1$ the matrix which anticommutes with the other γ matrices, see Eq. (3.27). In the representation (3.26) of the

Dirac matrices, γ^5 is diagonal, $\gamma^5 = -\tau^3$. The Noether current associated with the chiral symmetry reads

$$j^5_\mu = \bar{\psi}\gamma_\mu\gamma^5\psi. \tag{9.55}$$

At the classical level, this current is conserved. But, as we will see now, this is no longer true in the quantum theory. The symmetry is said to become anomalous when quantum fluctuations are included.

To obtain the *chiral anomaly*, we couple the Schwinger model (9.26) to an external vector field V^μ by including the source term

$$\mathcal{L}_\mathrm{I} = -j^5_\mu V^\mu, \tag{9.56}$$

and calculate the effective action in a derivative expansion up to linear order in V^μ by integrating out the fermionic degrees of freedom. The chiral current then follows by taking the derivative of the effective action with respect to this source. The relevant term in the expansion of the effective action is the counterpart of the contribution (9.33),

$$\Gamma_1^{(\ell=2)} = -\mathrm{i}e\,\mathrm{Tr}\,\frac{\not{p}}{p^2}\not{A}\frac{\not{p}}{p^2}\not{V}\gamma^5. \tag{9.57}$$

While the momentum integrals are the same as before, the product of Dirac matrices now contains an extra γ^5. The counterparts of Eqs. (9.39) and (9.40) read in $d = 2 - \varepsilon$ spacetime dimensions with ε small, cf. (3.37)

$$\eta_{\mu\lambda}\,\mathrm{tr}\,\gamma^\mu\gamma^\nu\gamma^\lambda\gamma^\rho\gamma^5 = \varepsilon\,\mathrm{tr}\,\gamma^\nu\gamma^\rho\gamma^5 = -2\varepsilon\epsilon^{\nu\rho}, \tag{9.58}$$

with $\epsilon_{01} = -\epsilon^{01} = 1$, and

$$h_{\mu\lambda}\,\mathrm{tr}\,\gamma^\mu\gamma^\nu\gamma^\lambda\gamma^\rho\gamma^5 = 2\left(\epsilon^{\nu\rho}h_\mu{}^\mu - 2\epsilon^{\mu\rho}h^\nu{}_\mu\right) \tag{9.59}$$

by Eq. (3.39). We arrive in this way at the gauge-invariant expression

$$\Gamma_1^{(\ell=2)} = \frac{e}{\pi}\int \mathrm{d}^2x\left(\epsilon^{\mu\nu} + \epsilon^{\nu\lambda}\frac{\partial_\lambda\partial^\mu}{\partial^2}\right)A_\mu V_\nu. \tag{9.60}$$

The chiral current can be readily read off, and the chiral anomaly

$$\left\langle \partial^\mu j^5_\mu \right\rangle = \frac{m_A}{\sqrt{\pi}}\epsilon_{\mu\nu}\partial^\mu A^\nu \tag{9.61}$$

follows, showing that once quantum fluctuations are included, the chiral current is no longer conserved. Since the chiral anomaly is encoded in the one-fermion-loop contribution (9.60) to the effective action, the corresponding $A_\mu V_\nu$ proper vertex is said to be anomaly-induced.

9.4 Fermion Condensate

We next show that the fermions form a condensate in the Schwinger model. Consider the 4-point correlation function

$$G^{(4)}(x) \equiv \langle \bar\psi\psi(x)\bar\psi\psi(0)\rangle . \tag{9.62}$$

For large spacelike separation ($\sqrt{-x^2} \to \infty$), the correlation function factorizes as

$$\langle \bar\psi\psi(x)\bar\psi\psi(0)\rangle \to \langle \bar\psi\psi(x)\rangle \langle \bar\psi\psi(0)\rangle . \tag{9.63}$$

This property, which is known as the *cluster decomposition property*, is a direct consequence of locality. A finite value of $G^{(4)}(x)$ in this limit implies a nonzero average $\langle\bar\psi\psi\rangle \neq 0$ and signals the presence of a fermion condensate.

Calculations in the Schwinger model generally simplify because the fermionic and gauge degrees of freedom can be decoupled. This splitting is special to two dimensions where a vector field has a unique decomposition into a longitudinal and transverse part as

$$A^\mu = \frac{1}{e}\partial^\mu\vartheta - \frac{1}{e}\epsilon^{\mu\nu}\partial_\nu\varphi, \tag{9.64}$$

with ϑ and φ two scalar fields. The coefficients are chosen for convenience. With the parametrization

$$\psi = \mathrm{e}^{\mathrm{i}\vartheta+\mathrm{i}\varphi\gamma^5}\chi, \quad \bar\psi = \bar\chi\,\mathrm{e}^{-\mathrm{i}\vartheta+\mathrm{i}\varphi\gamma^5} \tag{9.65}$$

of the fermion fields, the interaction term in the Lagrangian (9.26) cancels by the relation (3.39), $\gamma^\mu\gamma^5 = \epsilon^{\mu\nu}\gamma_\nu$. The classical theory thus separates into two pieces

$$\mathcal{L} = \bar\chi\mathrm{i}\partial\!\!\!/\chi - \frac{1}{4}F_{\mu\nu}F^{\mu\nu}, \tag{9.66}$$

describing a free fermion and a free boson theory. This splitting persists at the quantum level. Specifically, with the so-called Hodge decomposition (9.64) of the gauge field, the partition function (9.44) assumes the form

$$Z = \int \mathrm{D}\bar\chi\mathrm{D}\chi \int \mathrm{D}A_\mu \exp\left\{\mathrm{i}\int \mathrm{d}^2x\left[\bar\chi\mathrm{i}\partial\!\!\!/\chi - \frac{1}{2\pi}\left(\partial_\mu\varphi\right)^2 + \frac{1}{2e^2}\left(\partial^2\varphi\right)^2\right]\right\}, \tag{9.67}$$

where the first factor at the right side of Eq. (9.44) has been written as a functional integral over the Grassmann fields $\bar\chi$ and χ. The scalar field ϑ is seen to completely decouple from the theory. Note that the lowest-order term $(\partial_\mu\varphi)^2$ has the wrong sign, and the higher-order derivative term is needed to stabilize the theory. Applying the rules spelled out in Sec. 2.4, we find as boson propagator

$$\begin{aligned}\Delta_\mathrm{F}(x) &= \pi \int \frac{\mathrm{d}^2k}{(2\pi)^2}\left(\frac{1}{k^2 - m_A^2} - \frac{1}{k^2}\right)\mathrm{e}^{-ik\cdot x} \\ &= -\frac{\mathrm{i}}{2}\left[K_0\left(m_A\sqrt{-x^2}\right) + \ln\left(\tfrac{1}{2}\mathrm{e}^\gamma m_A\sqrt{-x^2}\right)\right].\end{aligned} \tag{9.68}$$

The coefficient in the argument of the logarithm is fixed by requiring that the propagator becomes zero in the massless limit as follows from the first line in this equation. Note that for spacelike vectors, $-x^2$ is positive.

The correlation function $G^{(4)}(x)$ becomes with the parametrization (9.65)

$$G^{(4)}(x) = \left\langle \bar{\chi}(x) e^{2i\varphi(x)\gamma^5} \chi(x) \bar{\chi}(0) e^{2i\varphi(0)\gamma^5} \chi(0) \right\rangle. \tag{9.69}$$

The evaluation of this expression is facilitated by noting that in the representation (3.26) only mixed terms appear in the Lagrangian,

$$\bar{\chi} i\partial\!\!\!/ \chi = \bar{\chi}_{\rm L}(i\partial_t + i\partial_1)\chi_{\rm R} + \bar{\chi}_{\rm R}(i\partial_t - i\partial_1)\chi_{\rm L}, \tag{9.70}$$

where we introduced the notation

$$\chi \equiv \begin{pmatrix} \chi_{\rm L} \\ \chi_{\rm R} \end{pmatrix}, \quad \bar{\chi} \equiv (\bar{\chi}_{\rm L}, \bar{\chi}_{\rm R}), \tag{9.71}$$

with the subscripts "L" and "R" standing for left and right moving, respectively. The two components $\chi_{\rm L}$ and $\chi_{\rm R}$ represent fermions of negative and positive chirality, respectively,

$$\gamma^5 \chi = \begin{pmatrix} -\chi_{\rm L} \\ \chi_{\rm R} \end{pmatrix}, \tag{9.72}$$

and

$$\bar{\chi} e^{2i\varphi\gamma^5} \chi = e^{-2i\varphi} \bar{\chi}_{\rm L}\chi_{\rm L} + e^{2i\varphi} \bar{\chi}_{\rm R}\chi_{\rm R}. \tag{9.73}$$

Given the special form of the Lagrangian, only the cross terms in the correlation function contribute,

$$\begin{aligned} G^{(4)}(x) = &\left\langle e^{-2i[\varphi(x)-\varphi(0)]} \bar{\chi}_{\rm L}(x)\chi_{\rm L}(x)\bar{\chi}_{\rm R}(0)\chi_{\rm R}(0) \right\rangle \\ &+ \left\langle e^{2i[\varphi(x)-\varphi(0)]} \bar{\chi}_{\rm R}(x)\chi_{\rm R}(x)\bar{\chi}_{\rm L}(0)\chi_{\rm L}(0) \right\rangle. \end{aligned} \tag{9.74}$$

This can be still further simplified by noting that odd powers of $\varphi(x) - \varphi(0)$ give zero upon integration so that $e^{-2i[\varphi(x)-\varphi(0)]}$ in the first term can be replaced with $e^{+2i[\varphi(x)-\varphi(0)]}$, and $G^{(4)}(x)$ can be written as

$$\begin{aligned} G^{(4)}(x) &= \left\langle e^{2i[\varphi(x)-\varphi(0)]} \bar{\chi}(x)\chi(x)\bar{\chi}(0)\chi(0) \right\rangle \\ &= \operatorname{tr} \left\langle e^{2i[\varphi(x)-\varphi(0)]} \chi(0)\bar{\chi}(x)\chi(x)\bar{\chi}(0) \right\rangle \\ &= -\left\langle e^{2i[\varphi(x)-\varphi(0)]} \right\rangle \operatorname{tr} S_{\rm F}(-x) S_{\rm F}(x). \end{aligned} \tag{9.75}$$

In the second line of Eq. (9.75), we introduced the trace and used its cyclic property, while the last line, with $S_{\rm F}$ denoting the massless fermion propagator (9.32), follows because the fermion part of the theory is free. Given that also the boson part is free, the remaining average in Eq. (9.75) is readily evaluated, see Eq. (2.107). With the propagator (9.68), this gives

$$\left\langle \mathrm{e}^{2\mathrm{i}[\varphi(x)-\varphi(0)]} \right\rangle = \mathrm{e}^{4\mathrm{i}[\Delta_{\mathrm{F}}(x)-\Delta_{\mathrm{F}}(0)]}$$
$$= -\frac{\mathrm{e}^{2\gamma}}{4} m_A^2 x^2 \mathrm{e}^{2K_0\left(m_a\sqrt{-x^2}\right)}, \tag{9.76}$$

and the correlation function becomes

$$G^{(4)}(x) = \frac{\mathrm{e}^{2\gamma}}{8\pi^2} m_A^2 \mathrm{e}^{2K_0\left(m_a\sqrt{-x^2}\right)}. \tag{9.77}$$

For large spacelike separation $\sqrt{-x^2} \to \infty$, $G^{(4)}(x)$ tends to a constant, implying that $\bar{\psi}\psi$ develops a nonzero average and the chiral symmetry is spontaneously broken by this fermion condensate. In Sec. 9.6 below, it will be show that $G^{(4)}(x)$ receives an additional contribution that has not been included yet. That contribution turns out to be equal to the one already found so that the condensate is given by

$$\langle\bar{\psi}\psi\rangle = -\frac{\mathrm{e}^{\gamma}}{2\pi} m_A, \tag{9.78}$$

where the minus sign is determined by other means, see Eq. (9.134) below.

By virtue of the identity (3.39), $\gamma^\mu\gamma^5 = \epsilon^{\mu\nu}\gamma_\nu$, the chiral and electric currents are related through

$$j^5_\mu = \epsilon_{\mu\nu} j^\nu_{\mathrm{e}}. \tag{9.79}$$

With the bosonization rule (9.48), we then obtain the correspondence

$$j^5_\mu \mathrel{\hat{=}} -\frac{1}{\sqrt{\pi}}\partial_\mu\phi, \tag{9.80}$$

identifying ϕ as the Nambu-Goldstone field of the spontaneously broken U(1) chiral symmetry. This cannot, however, be the end of the story. Nambu-Goldstone modes are expected to be gapless, but Eq. (9.49) shows that ϕ has acquired a mass. Both observations are reconciled by the observation that the chiral symmetry is no longer an exact symmetry of the bosonized theory. The mass of the would-be Nambu-Goldstone field then indicates by how much the chiral symmetry has become anomalous,

$$\partial^\mu j^5_\mu = -\frac{1}{\sqrt{\pi}}\partial^2\phi. \tag{9.81}$$

On comparing this expressions for the chiral anomaly with Eq. (9.61), we recover the relation (9.47). When the field equation obtained from Eq. (9.49) is used, that relation becomes the Minkowski analog of the relation (9.7) derived in the context of a 2D superconductor,

$$m_A\phi = \epsilon_{\mu\nu}\partial^\mu A^\nu = -E^1. \tag{9.82}$$

9.5 Index Theorem

In two-dimensional Euclidean spacetime, the counterpart of a point vortex studied in the context of a superconducting film is an *instanton*—a point defect in spacetime that instantly appears and disappears again at some space point. We next study their effect. To this end, we go back to the Matthews-Salam determinant (9.41) and evaluate it in the presence of an instanton. We go over to Euclidean spacetime by replacing the time variable $x^0 = t$ with $-\mathrm{i}\tau$. Also $A_0 = \mathrm{i}A_0^{\mathrm{E}}$ in accordance with the substitution $\partial_t = \mathrm{i}\partial_\tau$. In contrast to the convention introduced in Eq. (2.43), we in this and the following two sections set $k^0 = -\mathrm{i}\omega_{\mathrm{E}}$ so as to obtain an Euclidean scalar product $k \cdot x = -(\omega_{\mathrm{E}}\tau + \mathbf{k} \cdot \mathbf{x})$. To arrive at the appropriate Dirac algebra given by Eq. (3.25) with the metric tensor $\eta^{\mu\nu}$ in Minkowski spacetime replaced with $-\delta^{\mu\nu}$, we set $\gamma^0 = -\mathrm{i}\gamma^0_{\mathrm{E}}$. With these definitions, all the gamma matrices become anti-Hermitian, $\gamma^{\mu\dagger}_{\mathrm{E}} = -\gamma^\mu_{\mathrm{E}}$. We keep the definition $\gamma^5 = \gamma^0\gamma^1 = -\mathrm{i}\gamma^0_{\mathrm{E}}\gamma^1$ so that also in Euclidean spacetime γ^5 is Hermitian $\gamma^{5\dagger} = \gamma^5$ and $(\gamma^5)^2 = \gamma^5$. For convenience, we include a short dictionary for going from Minkowski to Euclidean spacetime, complementing the rule (2.39)

$$t = -\mathrm{i}\tau, \quad \gamma^0 = -\mathrm{i}\gamma^0_{\mathrm{E}}, \quad j^0 = -\mathrm{i}j^0_{\mathrm{E}}, \quad A_0 = \mathrm{i}A_0^{\mathrm{E}},$$
$$\bar\psi\psi = -\mathrm{i}\bar\psi_{\mathrm{E}}\psi, \quad \bar\psi\gamma^0\psi = -\bar\psi_{\mathrm{E}}\gamma^0_{\mathrm{E}}\psi, \quad \bar\psi\gamma^i\psi = -\mathrm{i}\bar\psi_{\mathrm{E}}\gamma^i\psi, \quad \bar\psi\mathrm{i}\partial\!\!\!/\psi = \bar\psi_{\mathrm{E}}\partial\!\!\!/_{\mathrm{E}}\psi,$$
$$F_{0i} = \mathrm{i}F^{\mathrm{E}}_{0i}, \quad F_{\mu\nu}F^{\mu\nu} = F^{\mathrm{E}}_{\mu\nu}F^{\mu\nu}_{\mathrm{E}}, \quad \epsilon^{\mu\nu}\partial_\mu A_\nu = -\mathrm{i}\epsilon^{\mu\nu}\partial^{\mathrm{E}}_\mu A^{\mathrm{E}}_\nu, \quad j^\mu A_\mu = j^\mu_{\mathrm{E}}A^{\mathrm{E}}_\mu,$$

where it is recalled that $\bar\psi = \psi^\dagger\gamma^0$. In the following, we drop the sub- or superscript "E", trusting that the use of the Euclidean metric $\eta^{\mu\nu} = -\delta^{\mu\nu}$ is understood so that, for example, with $\epsilon_{01} = 1$, $\epsilon^{01} = +1$ in Euclidean spacetime (whereas in Minkowski spacetime $\epsilon^{01} = -1$).

There exists a powerful index theorem due to Atiyah and Singer, relating the zero eigenvalues of the operator $D\!\!\!\!/ \equiv \partial\!\!\!/ - \mathrm{i}eA\!\!\!/$ in Euclidean spacetime, with A^μ considered a background field, to the so-called Pontryagin index

$$Q \equiv \frac{e}{2\pi}\int \mathrm{d}^2x\, \epsilon_{\mu\nu}\partial^\mu A^\nu = \frac{e}{4\pi}\int \mathrm{d}^2x\, \epsilon_{\mu\nu}F^{\mu\nu}. \tag{9.83}$$

Equivalently,

$$Q = \frac{e}{2\pi}\oint \mathrm{d}x_\mu A^\mu, \tag{9.84}$$

where the integral is over a circle at infinity, showing that Q denotes the winding number of the field configuration. The theorem states that the index of the operator $D\!\!\!\!/$, defined by the number n_+ of zero-eigenvalue modes of positive chirality minus the number n_- of zero-eigenvalue modes of negative chirality, is given by minus the Pontryagin index,

$$n_+ - n_- = -Q. \tag{9.85}$$

To prove it, note that that the operator $\not{D}$ is Hermitian $\not{D} = \not{D}^\dagger$ in Euclidean spacetime with respect to the scalar product

$$\begin{aligned}\langle\varphi_1,\varphi_2\rangle &\equiv \int \mathrm{d}^2x\,\varphi_1^\dagger(x)\varphi_2(x)\\ &= \int \mathrm{d}^2x\,\varphi_{1,\alpha}^*(x)\varphi_{2,\alpha}(x),\end{aligned} \tag{9.86}$$

where $\alpha = 1, 2$ is the spinor index. Its eigenvalues

$$\not{D}\varphi_n(x) = \lambda_n\varphi_n(x) \tag{9.87}$$

are therefore real. Since $\{\gamma^\mu,\gamma^5\} = 0$, nonzero eigenvalues come in pairs: if $\lambda_n \neq 0$ is an eigenvalue of the eigenstate φ_n, $-\lambda_n$ is also an eigenvalue,

$$\not{D}(\gamma^5\varphi_n)(x) = -\lambda_n(\gamma^5\varphi_n)(x) \tag{9.88}$$

of the eigenstate $\gamma^5\varphi_n$. For $\lambda_n \neq 0$, $\lambda_n \neq -\lambda_n$ and the two eigenstates are orthogonal $\varphi_n \perp \gamma^5\varphi_n$. For zero eigenvalues, the two eigenvalue equations become degenerate,

$$\not{D}\varphi_n^0(x) = 0, \quad \not{D}(\gamma^5\varphi_n^0)(x) = 0. \tag{9.89}$$

They can be combined

$$\not{D}(1 \pm \gamma^5)\varphi_n^0(x) = 0 \tag{9.90}$$

so that the zero-eigenvalue states are simultaneously eigenstates of γ^5 with eigenvalues ± 1,

$$\gamma^5\left(1 \pm \gamma^5\right)\varphi_n^0(x) = \pm(1 \pm \gamma^5)\varphi_n^0(x). \tag{9.91}$$

Now consider the trace

$$\mathrm{Tr}\,\gamma^5 = \sum_n \int \mathrm{d}^2x\,\varphi_n^\dagger(x)\gamma^5\varphi_n(x). \tag{9.92}$$

Since φ_n and $\gamma^5\varphi_n$ are orthogonal for $\lambda_n \neq 0$, only zero modes contribute to the sum, and

$$\mathrm{Tr}\,\gamma^5 = n_+ - n_-, \tag{9.93}$$

assuming that the eigenstates are properly normalized.

To compute the trace explicitly, it must be regularized. Fujikawa proposed to choose the regularization

$$\mathrm{Tr}\,\gamma^5 = \lim_{s\to 0}\mathrm{Tr}\,\gamma^5\,\mathrm{e}^{-s\not{D}^2}. \tag{9.94}$$

Using plane waves $\langle x|k\rangle \equiv \mathrm{e}^{-\mathrm{i}k\cdot x}$ as basis to evaluate the trace, we then obtain, cf. Eq. (2.62),

$$\begin{aligned}\mathrm{Tr}\,\gamma^5 &= \lim_{s\to 0}\int\frac{\mathrm{d}^2k}{(2\pi)^2}\,\mathrm{tr}\langle k|\gamma^5\,\mathrm{e}^{-s\not{D}^2}|k\rangle \\ &= \lim_{s\to 0}\int\mathrm{d}^2x\int\frac{\mathrm{d}^2k}{(2\pi)^2}\gamma^5\exp\left\{-s\left[-(k_\mu+eA_\mu)^2+\mathrm{i}e\partial_\mu A^\mu-\tfrac{1}{2}\mathrm{i}e\gamma^\mu\gamma^\nu F_{\mu\nu}\right]\right\},\end{aligned} \tag{9.95}$$

after inserting the unit operator $\int\mathrm{d}^dx|x\rangle\langle x|$. It is recalled that with the Euclidean metric $\eta^{\mu\nu}=-\delta^{\mu\nu}$, $k\cdot x=-(\omega\tau+\mathbf{k}\cdot\mathbf{x})$, *etc.* In deriving Eq. (9.95), use is made of the identity

$$\begin{aligned}\not{D}^2 &= \frac{1}{2}\left(\{\gamma^\mu,\gamma^\nu\}+[\gamma^\mu,\gamma^\nu]\right)D_\mu D_\nu \\ &= D^2-\frac{\mathrm{i}}{2}e\gamma^\mu\gamma^\nu F_{\mu\nu},\end{aligned} \tag{9.96}$$

and of the property that for arbitrary function $f(D^2)$,

$$\int\frac{\mathrm{d}^2k}{(2\pi)^2}\,\mathrm{e}^{\mathrm{i}k\cdot x}f\left(D^2\right)\mathrm{e}^{-\mathrm{i}k\cdot x}=\int\frac{\mathrm{d}^2k}{(2\pi)^2}f\left[-(k_\mu+eA_\mu)^2+\mathrm{i}e\partial_\mu A^\mu\right], \tag{9.97}$$

by the derivative expansion rules spelled out in Sec. 3.1. The momentum integral in Eq. (9.95) is a simple Gaussian and gives

$$\mathrm{Tr}\,\gamma^5=\lim_{s\to 0}\frac{1}{4\pi s}\int\mathrm{d}^2x\,\mathrm{tr}\,\gamma^5\exp\left[-\mathrm{i}es\left(\partial_\mu A^\mu-\tfrac{1}{2}\gamma^\mu\gamma^\nu F_{\mu\nu}\right)\right]. \tag{9.98}$$

Expanding the integrand in s, we notice that the only contribution surviving the limit $s\to 0$ is

$$\mathrm{Tr}\,\gamma^5=-\frac{e}{4\pi}\int\mathrm{d}^2x\,\epsilon_{\mu\nu}F^{\mu\nu}, \tag{9.99}$$

where the Euclidean counterpart of Eq. (9.58) is used to evaluate the trace over the Dirac matrices,

$$\mathrm{tr}\,\gamma^\mu\gamma^\nu\gamma^5=2\mathrm{i}\epsilon^{\mu\nu},\qquad \eta^{\mu\nu}=-\delta^{\mu\nu}, \tag{9.100}$$

with $\epsilon_{01}=+\epsilon^{01}=1$ in Euclidean spacetime. The right side of Eq. (9.99) is precisely minus the Pontryagin index (9.83), which, as already mentioned, denotes the winding number of an instanton. With Eq. (9.93) we thus have proved the index theorem (9.85).

According to the Atiyah-Singer theorem, instantons are always accompanied by fermion zero-eigenvalue modes. As the determinant of an operator is the product of its eigenvalues, it follows that the determinant (9.41) vanishes when zero modes are present. In other words, field configurations with instantons do not contribute to the partition function of the massless Schwinger model.

The zero-eigenvalue mode of the operator $\not{D}$ accompanying an instanton of unit winding number is readily constructed starting from the *Ansatz*

$$A^\mu = -\frac{f(r)}{er}\mathbf{e}_\theta = \frac{f(r)}{er}\left(\frac{x^1}{r}, -\frac{\tau}{r}\right), \tag{9.101}$$

where $r \equiv \sqrt{\tau^2 + x_1^2}$ and $\theta \equiv \arctan(x^1/\tau)$, with corresponding unit vector $\mathbf{e}_\theta$, denote polar coordinates in the τ–x^1 plane. The function $f(r)$ is such that it tends to unity for $r \to \infty$ and vanishes for $r \to 0$. The zero-eigenvalue equation,

$$\gamma^\mu(\partial_\mu - \mathrm{i}eA_\mu)\psi = 0, \tag{9.102}$$

then gives for the first component ψ_1 of the spinor ψ

$$\left[(\partial_\tau - \mathrm{i}\partial_1) + \frac{f(r)}{r}\frac{\tau - \mathrm{i}x^1}{r}\right]\psi_1 = 0. \tag{9.103}$$

With the assumption that $\psi_1 = \psi_1(r)$, this equation reduces to

$$\frac{\tau - \mathrm{i}x^1}{r}\left(\frac{\mathrm{d}}{\mathrm{d}r} + \frac{f(r)}{r}\right)\psi_1(r) = 0, \tag{9.104}$$

which yields the solution

$$\psi_1(r) = \exp\left(-\int_0^r \mathrm{d}r' \frac{f(r')}{r'}\right). \tag{9.105}$$

Given the constraints on the function $f(r)$, this zero-eigenvalue mode is regular at $r = 0$ and falls to zero as $r \to \infty$.

9.6 θ Vacuum

In contrast to the partition function, observable averages generally do receive contributions from field configurations with nonzero winding number. Let, using Dirac's bra-ket notation, $|w\rangle$ denote the vacuum sector belonging to the homotopy class with winding number w, and $\hat{Q}$ the topological charge operator,

$$\hat{Q}|w\rangle = w|w\rangle. \tag{9.106}$$

Consider the operator $\hat{G}$ which increases the topological charge of a state by one,

$$\hat{G}|w\rangle = |w + 1\rangle. \tag{9.107}$$

The operator is said to generate "large gauge transformations" as opposed to "small" ones which do not change the homotopy class and can be smoothly deformed to the identity. It leaves the Hamiltonian or energy operator of the system invariant, and so both operators can be diagonalized simultaneously. The physical vacuum is an eigenstate of the Hamiltonian and should therefore also be an eigenstate of $\hat{G}$. By construction, none of the vacuum sectors $|w\rangle$ are by themselves an eigenstate of $\hat{G}$, but the superposition

$$|\theta\rangle \equiv \sum_{w=-\infty}^{\infty} e^{iw\theta}|w\rangle, \tag{9.108}$$

characterized by the free parameter $0 \leq \theta < 2\pi$, is. Indeed,

$$\begin{aligned}\hat{G}|\theta\rangle &= \sum_{w=-\infty}^{\infty} e^{iw\theta}\hat{G}|w\rangle = \sum_{w} e^{iw\theta}|w+1\rangle \\ &= e^{-i\theta}\sum_{w} e^{i(w+1)\theta}|w+1\rangle = e^{-i\theta}|\theta\rangle.\end{aligned} \tag{9.109}$$

Since the operator $\hat{G}$ generates gauge transformations, albeit "large" ones, a physical state can at most change by a phase factor under the action of this operator—that is, $\hat{G}$ is a unitary operator. This observation is taken into account in the *Ansatz* (9.108). Different θ vacua are orthogonal

$$\begin{aligned}\langle\theta|\theta'\rangle &= \sum_{w,w'} e^{i(w'\theta'-w\theta)}\langle w|w'\rangle \\ &= \sum_{w} e^{iw(\theta'-\theta)} = 2\pi\delta(\theta-\theta')\end{aligned} \tag{9.110}$$

where it is used that $\langle w|w'\rangle = \delta_{w,w'}$. Similarly, for any gauge-invariant operator $\hat{O}$, the transition amplitude between two θ vacua is

$$\begin{aligned}\langle\theta|\hat{O}|\theta'\rangle &= \sum_{w,w'} e^{i(w'\theta'-w\theta)}\langle w|\hat{O}|w'\rangle \\ &= \sum_{w'}\sum_{Q} e^{i(w'\theta'-w'\theta-Q\theta)}\langle w'+Q|\hat{O}|w'\rangle \\ &= 2\pi\delta(\theta-\theta')\sum_{Q} e^{-iQ\theta}\langle Q|\hat{O}|0\rangle,\end{aligned} \tag{9.111}$$

where in the last step it is used that the transition amplitude $\langle w'+Q|\hat{O}|w'\rangle$ between states with winding number w' and $w'+Q$ depends only on Q, while the sum over w' resulted in the delta function. It follows that a gauge-invariant operator cannot generate transitions between different θ vacua so that these are completely decoupled. Writing the amplitude as a functional integral, we arrive at

$$\langle\theta|\hat{O}|\theta'\rangle = 2\pi\delta(\theta-\theta')\sum_{Q} e^{-iQ\theta}\int [D\bar{\psi}D\psi DA^{\mu}]_Q\, O\, e^{-\int d^2x\,\mathcal{L}}, \tag{9.112}$$

where the Lagrangian density $\mathcal{L}$ is the Euclidean counterpart (with the metric $\eta^{\mu\nu} = -\delta^{\mu\nu}$) of Eq. (9.26),

$$\mathcal{L} = i\bar{\psi}(i\partial\!\!\!/ + eA\!\!\!/)\psi + \frac{1}{4}F_{\mu\nu}F^{\mu\nu}. \tag{9.113}$$

The integration in each of the terms in Eq. (9.112) is restricted to field configurations belonging to a specific homotopy class with winding number Q. This is indicated by the subscript on the integration measure. The sum $\sum_Q$ implies that all possible field configurations are to be included irrespective of the homotopy class they belong to. Using the expression (9.83) for Q in terms of the gauge field, we can absorb the prefactor into the action to obtain

$$\langle\theta|\hat{O}|\theta'\rangle = 2\pi\delta(\theta-\theta')\langle O\rangle_\theta, \tag{9.114}$$

where $\langle O\rangle_\theta$ denotes the average

$$\langle O\rangle_\theta \equiv \int [\mathrm{D}\bar{\psi}\mathrm{D}\psi\mathrm{D}A^\mu]_{\mathrm{all}\,Q}\, O \exp\left[-\int \mathrm{d}^2x\left(\mathcal{L}+\mathrm{i}\frac{e\theta}{4\pi}\epsilon_{\mu\nu}F^{\mu\nu}\right)\right]. \tag{9.115}$$

This average differs from the perturbative average in that, first, the θ term is included in the action and, second, field configurations from all homotopy classes are to be integrated over and not just those belonging to the topologically trivial class with zero winding number.

Because the extra term, with θ a constant, is a total divergence, it does not affect the classical field equations. In fact, we already encountered a similar term in Eq. (9.60). If we set $V_\rho = -\frac{1}{2}\partial_\rho\theta(x)$ there, integrate by parts, and take $\theta(x)=\theta$, we recover the θ term (in Minkowski spacetime). The interaction term (9.56) through which the external vector field V_ρ was coupled to the Schwinger model becomes with this choice

$$\begin{aligned}\mathcal{L}_\mathrm{I} &= -j^5_\mu V^\mu \\ &= \frac{1}{2}j^5_\mu\partial^\mu\theta(x) = -\frac{1}{2}\epsilon_{\mu\nu}j^\mu_e\partial^\nu\theta(x),\end{aligned} \tag{9.116}$$

where in the last step Eq. (9.79) was substituted. The right side represents the usual (as opposed to the chiral) coupling to an external vector potential $eA^\mu = -\frac{1}{2}\epsilon_{\mu\nu}\partial^\nu\theta(x)$, which corresponds to setting $\vartheta = 0, \varphi = \frac{1}{2}\theta$ in the Hodge decomposition (9.64). This coupling term can be canceled by going over to the chirally rotated field χ introduced in Eq. (9.65). And, hence, the θ term can be accounted for by simply including a phase factor in the fermion fields,

$$\psi \to \mathrm{e}^{\frac{1}{2}\mathrm{i}\gamma^5\theta}\psi, \quad \bar{\psi} \to \bar{\psi}\,\mathrm{e}^{\frac{1}{2}\mathrm{i}\gamma^5\theta}. \tag{9.117}$$

With this observation, we return to the calculation of the 4-point correlation function (9.62), which reads more explicitly

$$G^{(4)}(x) = \langle 0|\bar{\psi}\psi(x)\bar{\psi}\psi(0)|0\rangle, \tag{9.118}$$

with

$$|0\rangle = \int_0^{2\pi}\frac{\mathrm{d}\theta}{2\pi}|\theta\rangle \tag{9.119}$$

the perturbative vacuum with $w = 0$. The θ vacuum, which constitutes the true vacuum of the massless Schwinger model, is required for the cluster decomposition property. To show this we first insert a complete set of eigenstates in the correlation function. In the limit of large spacelike separation, only vacuum states $|\theta\rangle$ survive, and

$$\begin{aligned} G^{(4)}(x) &\to \int_0^{2\pi} \frac{\mathrm{d}\theta}{2\pi} \langle 0|\bar{\psi}\psi(x)|\theta\rangle \langle\theta|\bar{\psi}\psi(0)|0\rangle \\ &= \int_0^{2\pi} \frac{\mathrm{d}\theta}{2\pi}\frac{\mathrm{d}\theta_1}{2\pi}\frac{\mathrm{d}\theta_2}{2\pi} \langle\theta_1|\bar{\psi}\psi(x)|\theta\rangle \langle\theta|\bar{\psi}\psi(0)|\theta_2\rangle, \end{aligned} \tag{9.120}$$

where all possible θ vacua are included as intermediate state. In the perturbative vacuum, $\bar{\psi}\psi$ has zero average. In the θ vacuum, however, cf. Eq. (9.114)

$$\langle\theta|\bar{\psi}\psi|\theta'\rangle = 2\pi\delta(\theta-\theta')\,\langle\bar{\psi}\psi\rangle_\theta, \tag{9.121}$$

with $\langle\cdots\rangle_\theta$ the average introduced in Eq. (9.115). The main difference with the perturbative average is that the θ vacuum average includes contributions from all topological sectors, and not just from the trivial one with zero winding number. In fact, a closer inspection reveals that the one-instanton sectors with winding number $Q = \pm 1$ are responsible for the formation of the fermion condensate, i.e., for a nonzero average $\langle\bar{\psi}\psi\rangle_\theta$.

To calculate $\langle\bar{\psi}\psi\rangle_\theta$, we use the observation that including the θ term amounts to a chiral U(1) transformation (9.117) under which the product $\bar{\psi}\psi$ transforms as

$$\bar{\psi}\psi \to \bar{\psi}\mathrm{e}^{\mathrm{i}\gamma^5\theta}\psi. \tag{9.122}$$

This implies that $\langle\bar{\psi}\psi\rangle_\theta$ has the following θ dependence

$$\langle\bar{\psi}\psi\rangle_\theta = \langle\bar{\psi}\psi\rangle_{\theta=0}\cos\theta. \tag{9.123}$$

Inserting this in Eq. (9.120) with Eq. (9.121), we find that for large spacelike separation

$$G^{(4)}(x) \to \int_0^{2\pi} \frac{\mathrm{d}\theta}{2\pi}\cos^2\theta \left|\langle\bar{\psi}\psi\rangle_{\theta=0}\right|^2 = \frac{1}{2}\left|\langle\bar{\psi}\psi\rangle_{\theta=0}\right|^2. \tag{9.124}$$

With the perturbative result (9.77) for $G^{(4)}$ substituted on the left side, the correct fermion condensate (9.78) follows from this equation, where it is now understood that the average in that equation is given by Eq. (9.115) with $\theta = 0$. It is somewhat fortuitous that, by calculating $G^{(4)}$, the formation of a fermion condensate, a nonperturbative result, could be obtained in perturbation theory. Note, however, that the perturbative expression (9.77) by itself violates the cluster decomposition property. In general, only the full correlation functions, with the contributions from all topological sectors included, enjoy this property.

In the case of the correlation function $G^{(4)}(x)$, a closer inspection reveals that, in addition to the topologically trivial sector, only the sectors with winding number $Q = \pm 2$ contribute to the asymptotic behavior of the correlation function, and in the limit of large spacelike separation

$$\begin{aligned}\langle 0|\bar{\psi}\psi(x)\bar{\psi}\psi(0)|0\rangle_{Q=2} &= \langle 0|\bar{\psi}\psi(x)\bar{\psi}\psi(0)|0\rangle_{Q=-2} \\ &= \frac{1}{2}\langle 0|\bar{\psi}\psi(x)\bar{\psi}\psi(0)|0\rangle_{Q=0}\,.\end{aligned} \tag{9.125}$$

With these contributions added, the cluster decomposition property is satisfied.

9.7 Massive Schwinger Model

We next turn to the massive Schwinger model in Euclidean spacetime, obtained by including a mass term in the Lagrangian (9.113)

$$\mathcal{L} = \mathrm{i}\bar{\psi}(\mathrm{i}\partial\!\!\!/ - m + eA\!\!\!/)\psi + \frac{1}{4}F_{\mu\nu}^2. \tag{9.126}$$

An important difference with the massless model is that the mass term explicitly breaks chiral symmetry. This implies that even in the absence of the anomaly, a *massive* bosonized theory is expected. Moreover, because the massive Dirac operator has no zero-eigenvalue modes, the index theorem is of no relevance and instanton configurations contribute to the partition function. To account for such configurations, we note that the interaction term $\mathcal{L}_\mathrm{I} = -e j_\mathrm{e}^\mu A_\mu$, with $j_\mathrm{e}^\mu = -\mathrm{i}\bar{\psi}\gamma^\mu\psi$ the current density in Euclidean spacetime, becomes in the bosonized theory

$$\mathcal{L} = \mathrm{i}\frac{e}{\sqrt{\pi}}\phi\epsilon^{\mu\nu}\partial_\mu A_\nu \tag{9.127}$$

after a partial integration. Here, use is made of the bosonization rule (9.48), which has the same form in Euclidean spacetime. The partition function of the bosonized theory then reads, cf. Eq. (9.13),

$$Z = \sum_{N=0}^{\infty}\sum_{w_\alpha=\pm 1}\frac{\xi^N}{N!}\prod_\alpha\int \mathrm{d}^2x_\alpha\int \mathrm{D}\phi\exp\left\{-\int\mathrm{d}^2x\left[-\tfrac{1}{2}(\partial_\mu\phi)^2 + \tfrac{1}{2}m_A^2\phi^2 + \mathrm{i}m_A\tilde{F}^\mathrm{V}\phi\right]\right\} \tag{9.128}$$

where $\xi \equiv \mathrm{e}^{-S_\mathrm{c}}/a^2$ with S_c the instanton core action and a a length scale of order the instanton core size which is needed for dimensional reasons. The source $\tilde{F}^\mathrm{V}$ describes the instantons,

$$\tilde{F}^\mathrm{V} = -\Phi_0\sum_\alpha w_\alpha\,\delta(x - x_\alpha), \tag{9.129}$$

where the sum extends over all instanton locations x_α, and $w_\alpha = \pm 1$ is the winding number of the αth instanton. Because instantons of higher winding number carry much higher energy, it is justified to consider only those of unit winding number $w_\alpha = \pm 1$. The notation for the source used here alludes to the convention that $\tilde{F}^{\rm V} \equiv \frac{1}{2}\epsilon_{\mu\nu}F^{\mu\nu}$ denotes the so-called dual field strength, which physically represents the electric field in $d = 2$. The elementary flux quantum in the Schwinger model is $\Phi_0 = 2\pi/e$ in natural units, which is twice the value for a superconductor. By the identity (9.17) cast in the form

$$\sum_{N=0}^{\infty} \sum_{w_\alpha=\pm 1} \frac{\xi^N}{N!} \prod_\alpha \int \mathrm{d}^2 x_\alpha \mathrm{e}^{\mathrm{i}g \sum_\alpha \phi(x_\alpha)} = \sum_{N=0}^{\infty} \frac{\xi^N}{N!} \left[\int \mathrm{d}^2 x \left(\mathrm{e}^{+\mathrm{i}g\phi(x)} + \mathrm{e}^{-\mathrm{i}g\phi(x)}\right)\right]^N$$
$$= \exp\left\{2\xi \int \mathrm{d}^2 x \cos[g\phi(x)]\right\}, \tag{9.130}$$

the partition function then becomes the analog of Eq. (9.18),

$$Z = \int \mathrm{D}\phi \exp\left\{-\int \mathrm{d}^2 x \left[-\tfrac{1}{2}(\partial_\mu \phi)^2 + \tfrac{1}{2}m_A^2 \phi^2 - 2\xi \cos(g\phi)\right]\right\}, \tag{9.131}$$

with $g \equiv \Phi_0 m_A$ as in Eq. (9.16).

The inclusion of a mass term in the Schwinger model is seen to result in the cosine interaction term in the bosonized theory—that is, we have the correspondence in Euclidean (E) spacetime

$$\mathrm{i}m\bar{\psi}_{\mathrm{E}}\psi \mathrel{\hat{=}} 2\xi \cos\left(\sqrt{4\pi}\phi\right), \tag{9.132}$$

where it is used that

$$g = \Phi_0 m_A = \sqrt{4\pi} \tag{9.133}$$

in the Schwinger model. Instanton contributions, which were suppressed in the massless model, thus generate an interaction term in the bosonized formulation of the massive Schwinger model. Equation (9.132), which takes the form

$$m\bar{\psi}\psi \mathrel{\hat{=}} -2\xi \cos\left(\sqrt{4\pi}\phi\right) \tag{9.134}$$

in Minkowski spacetime, constitutes the third and final bosonization rule complementing the two earlier rules given in Eq. (9.48).

Note that even in the absence of the anomaly, in which case the mass term in Eq. (9.131) is absent, the bosonized theory would still be massive for the expansion of the cosine term produces a term quadratic in ϕ.

9.8 Peierls Instability

Relativisticlike fermions described by the Dirac theory naturally emerge in the context of quasi one-dimensional and also in higher-dimensional condensed matter systems. Consider the simple Hamiltonian

$$H_0 = -t \sum_j \left(a^\dagger_{j,\sigma} a_{j+1,\sigma} + a^\dagger_{j+1,\sigma} a_{j,\sigma} \right), \tag{9.135}$$

describing the hopping of mutually noninteracting electrons between nearest-neighbor atoms on a one-dimensional lattice. Because of the unique ordering of sites along a line, the notation used in Eq. (7.1) for arbitrary dimension has been simplified to $x^1 = ja$, with a the lattice spacing and j an integer labeling the position of the atoms. The parameter t represents the matrix element for the hopping of an electron from one atom at site j to a neighboring atom. The operator $a^\dagger_{j\sigma}$ ($a_{j\sigma}$) creates (annihilates) a conduction electron of spin $\sigma(=\uparrow, \downarrow)$ at the location of the jth atom. In writing the Hamiltonian (9.135), it is assumed that the electrons are well localized so that a tight-binding approximation is justified. The sum $\sum_j$ extends over all lattice sites. The kinetic energy is given by Eq. (7.9) with $D = 1$,

$$\epsilon(k^1) = -2t \cos(k^1 a). \tag{9.136}$$

We consider the system at half-filling, where the number of electrons equals the number of atoms, and the Fermi wave number is given by $k_{\rm F} = \pi/2a$. At low energy and small momentum, only electrons near the Fermi points are relevant. To study these, we introduce the operators $a^{\rm e}$, $a^{\rm o}$ through

$$\begin{aligned} a_{2j} &= {\rm i}^{2j} a^{\rm e}_{2j} \\ a_{2j+1} &= {\rm i}^{2j+1} a^{\rm o}_{2j+1}, \end{aligned} \tag{9.137}$$

for even and odd sites, respectively. Here and in the following, we suppress spin indices. A sum over these hidden indices is always implied. The powers of the imaginary unit i in Eq. (9.137) represent the factor $\exp({\rm i}k_{\rm F}x^1)$, with $x^1 = 2ja$ for even sites and $x^1 = (2j+1)a$ for odd sites, respectively. In terms of the new operators, the tight-binding Hamiltonian (9.135) assumes the form

$$H_0 = -{\rm i}t \sum_j \left[\left(a^{\rm e}_{2j} - a^{\rm e\dagger}_{2j+2} \right) a^{\rm o}_{2j+1} + a^{\rm o\dagger}_{2j+1} \left(a^{\rm e}_{2j+2} - a^{\rm e}_{2j} \right) \right]. \tag{9.138}$$

With the derivative $\partial_1 a^{\rm o}$ defined by

$$\partial_1 a^{\rm e}_{2j+2} = \lim_{a \to 0} \frac{a^{\rm e}_{2j+2} - a^{\rm e}_{2j}}{2a}, \tag{9.139}$$

and a similar definition for $\partial_1 a^{\rm o}$, we obtain in the limit $a \to 0$

$$H_0 = 2ta \int {\rm d}x^1 \psi^\dagger \alpha(-{\rm i}\partial_1)\psi, \tag{9.140}$$

where the two-component spinor ψ contains the even-site and odd-site species of fermions,

$$\psi = \begin{pmatrix} \psi_1 \\ \psi_2 \end{pmatrix} \equiv \frac{1}{\sqrt{2a}} \begin{pmatrix} a^{\rm e} \\ a^{\rm o} \end{pmatrix}, \tag{9.141}$$

with $\alpha = \tau^1$ the first Pauli matrix. Regarding the dimensionful factor $\sqrt{2a}$, we follow the convention (7.17) with the *proviso* that the lattice spacing is $2a$ between even or odd sites. In deriving Eq. (9.140), the sum over the lattice sites is replaced with an integral: $2a \sum_j \rightarrow \int \mathrm{d}x^1$. The right side of Eq. (9.140) is of the form of a massless Dirac Hamiltonian. It describes Bloch electrons with a gapless spectrum that is linear in the crystal momentum k^1

$$\epsilon(k^1) = v_{\rm F}|k^1|, \tag{9.142}$$

where $v_{\rm F} = 2ta$ is the Fermi velocity. This linear spectrum corresponds to expanding the full spectrum (9.136) to linear order around the Fermi momentum, with k^1 in Eq. (9.142) denoting the momentum relative to $+k_{\rm F}$ or $-k_{\rm F}$. The Lagrangian density associated with the Hamiltonian (9.140) reads

$$\mathcal{L}_0 = \mathrm{i}\psi^\dagger \partial_t \psi - v_{\rm F}\psi^\dagger \alpha(-\mathrm{i}\partial_1)\psi. \tag{9.143}$$

With the definitions

$$\gamma^0 \equiv \beta = \tau^3, \quad \gamma^1 \equiv \beta\alpha = \mathrm{i}\tau^2, \tag{9.144}$$

and the Fermi velocity $v_{\rm F}$ set to unity, $\mathcal{L}_0$ takes the form (9.26) with $e = 0$, albeit with a different representation of the Dirac algebra than in Eq. (3.26). The present representation naturally follows from splitting the Hamiltonian into two parts, corresponding to even and odd sites, and introducing the corresponding spinor (9.141).

The lattice Hamiltonian (9.135), a first step towards describing a one-dimensional metal, completely ignores the electron-phonon interaction. This coupling is, however, in one space dimension crucial. A famous argument due to Peierls shows that a one-dimensional metal with a partially filled conduction band is unstable towards a periodic distortion of the linear lattice which opens an energy gap at the Fermi points. As a result, the one-dimensional system at the absolute zero of temperature is not a metal, but an insulator instead. Fröhlich showed that a periodic lattice distortion is accompanied by a modulation of the electron number density with the same periodicity—a so-called *charge density wave*.

With the phonons included, the Lagrangian density describing the system becomes instead of (9.143)

$$\mathcal{L} = \bar{\psi}\mathrm{i}\partial\!\!\!/\psi - g\bar{\psi}\sigma\psi + \frac{1}{2}(\partial_t\sigma)^2 - \frac{1}{2}c_{\rm s}^2(\partial_1\sigma)^2 - \mathcal{V}(\sigma). \tag{9.145}$$

The phonons are described by a real scalar field[2] σ measuring the displacement of the atoms from their equilibrium positions. In Eq. (9.145), c_s denotes the sound velocity, while $\mathcal{V}(\sigma)$ denotes the potential density, which is an even function of the field, $\mathcal{V}(\sigma) = \mathcal{V}(-\sigma)$,

$$\mathcal{V}(\sigma) = \frac{1}{2}\mu^2\sigma^2 + O(\sigma^4), \tag{9.146}$$

with $\mu^2 > 0$ so that the phonons are gapped. These optical phonons are coupled to the electrons through a Yukawa interaction. This coupling violates chiral symmetry, and reduces it to the discrete Z_2 symmetry

$$\sigma \to -\sigma, \quad \psi \to e^{i(\pi/2)\gamma^5}\psi = i\gamma^5\psi, \tag{9.147}$$

where the first form of the fermion transformation brings out that it is a chiral rotation by a discrete angle, see Eq. (9.54).

Suppose the phonons condense and $\langle\sigma\rangle = v$. As a result, the Z_2 symmetry is spontaneously broken, and the electrons acquire a mass $m \equiv gv$. Translated back to the microscopic tight-binding model of electrons hopping on the one-dimensional lattice, a mass term corresponds to a periodic modulation of the electron number density with a period twice the original lattice spacing,

$$H_V = m\sum_j (-1)^j \hat{n}_j, \tag{9.148}$$

where $\hat{n}_j = a_j^\dagger a_j$ is the electron-number operator at site j. Indeed, in the continuum limit

$$\begin{aligned} m\sum_j (-1)^j \hat{n}_j &= m\sum_j \left(a^{e\dagger}_{2j} a^{e}_{2j} - a^{o\dagger}_{2j+1} a^{o}_{2j+1}\right) \\ &\to m\int dx^1 \psi^\dagger \beta\psi, \end{aligned} \tag{9.149}$$

with ψ the spinor (9.141) and $\beta = \tau^3$ the diagonal Pauli matrix, see Eq. (9.144). Since the electrons are tightly bound to the lattice sites, the charge modulation must result from a periodic distortion of the lattice with neighboring atoms moved in opposite directions by the condensed phonons. And, as predicted by Peierls, an energy gap opens at the Fermi points for the gapless spectrum (9.142) is replaced with

$$E^2(k^1) = v_F^2(k^1)^2 + m^2. \tag{9.150}$$

The metallic properties of the free model are now lost. In condensed matter, m is referred to as the Peierls energy gap, which is typically of order 100 K.

[2]Previously, in Sec. 3.4, this field was denoted by ϕ, which in this chapter already denotes the scalar field featuring in the bosonization rules (9.48).

For the phonons to condense, the coefficient of the σ^2 term, which classically is positive ($\mu^2 > 0$), must turn negative. This sign change comes about when quantum corrections are included. Consider the effective action in a background σ field

$$\Gamma_1[\sigma] = -\mathrm{i}\,\mathrm{Tr}\ln(\not{p} - g\sigma) \tag{9.151}$$

obtained by integrating out the fermions. For a constant field $\sigma(x) = \bar{\sigma}$, Eq. (9.151) yields by the identity

$$\mathrm{Tr}\ln(\not{p} - g\sigma) = \mathrm{Tr}\ln\left[\gamma^5(\not{p} - g\sigma)\gamma^5\right] = \mathrm{Tr}\ln\left[-(\not{p} + g\sigma)\right], \tag{9.152}$$

the one-loop effective potential density

$$\mathcal{V}_1 = \mathrm{i}\int \frac{\mathrm{d}^2k}{(2\pi)^2}\ln(k^2 - M^2), \tag{9.153}$$

up to an irrelevant additive constant and $M^2 \equiv g^2\bar{\sigma}^2$. The integral is best calculated by extending it to d dimensions and first taking the derivative with respect to M^2,

$$\begin{aligned}\frac{\partial}{\partial M^2}\mathcal{V}_1 &= -\mathrm{i}\int \frac{\mathrm{d}^dk}{(2\pi)^d}\frac{1}{k^2 - M^2 - \mathrm{i}\eta} \\ &= -\frac{1}{(4\pi)^{d/2}}\frac{\Gamma(1-d/2)}{(M^2)^{1-d/2}}\end{aligned} \tag{9.154}$$

by the basic formula (3.43). Up to an irrelevant additive constant, this gives

$$\mathcal{V}_1 = -\frac{2/d}{(4\pi)^{d/2}}\frac{\Gamma(1-d/2)}{(M^2)^{-d/2}}. \tag{9.155}$$

In the limit $d \to 2$, the right side diverges as $M^2/2\pi(d-2)$. This divergence, which is caused by the linearization of the original model around the Fermi points, can be absorbed by renormalizing the bare mass parameter μ. Up to logarithm accuracy, the one-loop contribution to the effective potential then reads

$$\mathcal{V}_1 = \frac{1}{4\pi}g^2\sigma^2\ln\left(g^2\sigma^2/\Lambda^2\right), \tag{9.156}$$

where Λ is an ultraviolet scale which arises for dimensional reasons. For small σ, the quantum induced term $\propto \sigma^2\ln\sigma^2$ dominates the term $\frac{1}{2}\mu_r^2\sigma^2$ so that the minimum of the effective potential $\mathcal{V}_{\mathrm{eff}}(\sigma) = \mathcal{V}(\sigma) + \mathcal{V}_1(\sigma)$ occurs at $\sigma \neq 0$ however small g may be. Specifically,

$$\langle\sigma\rangle^2 = v^2 = \frac{\Lambda^2}{g^2}\mathrm{e}^{-2\pi\mu_r^2/g^2} \tag{9.157}$$

at weak coupling, and the phonons condense.

9.9 Fractional Charge

Given that the discrete Z_2 symmetry is spontaneously broken, we expect on account of the classification of topological defects given in Sec. 1.9, *kinks* or *domain walls* to emerge as topologically stable solutions of the theory. Such static solutions, characterized by the zeroth homotopy group $\pi_0(Z_2) = Z_2$, assume different asymptotic values $\pm v$ for $x^1 \to -\infty$ and $x^1 \to +\infty$. At the core of a kink, $\sigma(x^1) = 0$, and the position-dependent fermion mass $g\sigma(x^1)$ vanishes. To study these defects, the fermion sector of the model (9.145) is best bosonized

$$\mathcal{L} = \frac{1}{2}(\partial_\mu \phi)^2 + \frac{1}{2}(\partial_t \sigma)^2 - \frac{1}{2}c_s^2(\partial_1 \sigma)^2 - \mathcal{V}(\sigma, \phi), \tag{9.158}$$

where

$$\mathcal{V}(\sigma, \phi) \equiv \mathcal{V}(\sigma) + g\xi\sigma \cos\left(\sqrt{4\pi}\phi\right). \tag{9.159}$$

Asymptotically, the fields satisfy the static, uniform field equations, $\sin\left(\sqrt{4\pi}\phi\right) = 0$ and

$$\mathcal{V}'(\sigma) + g\xi \cos\left(\sqrt{4\pi}\phi\right) = 0. \tag{9.160}$$

Solutions with $\sigma(x^1)$ tending to the same asymptotic value for $x^1 \to -\infty$ as for $x^1 \to +\infty$ must have $\cos\left[\sqrt{4\pi}\phi(-\infty)\right] = \cos\left[\sqrt{4\pi}\phi(+\infty)\right]$ so that the field equation (9.160) assumes the same form in both limits. For such solutions, $\Delta\phi \equiv \phi(+\infty) - \phi(-\infty) = \pm\sqrt{\pi}$, and by Eq. (9.48), they carry one unit of fermion number charge

$$\begin{aligned} Q_e &\equiv \int \mathrm{d}x^1 j_e^0 = \frac{1}{\sqrt{\pi}} \int \mathrm{d}x^1 \partial_1 \phi \\ &= \frac{\Delta\phi}{\sqrt{\pi}} = \pm 1. \end{aligned} \tag{9.161}$$

A solution of this form can be thought of as representing an (anti)fermion of the original fermionic formulation (9.145). That is to say, fermions emerge as solitons of self-interacting boson fields.

Solutions with $\sigma(x^1)$ tending to opposite asymptotic values for $x^1 \to -\infty$ and $x^1 \to +\infty$ must have $\cos\left[\sqrt{4\pi}\phi(-\infty)\right] = -\cos\left[\sqrt{4\pi}\phi(+\infty)\right]$. This follows from noting that since $\mathcal{V}(\sigma)$ is even in σ, the derivative $\mathcal{V}'(\sigma)$ is odd. For such kink solutions, or domain walls $\Delta\phi = \pm\frac{1}{2}\sqrt{\pi}$, implying that they carry half a unit of fermion number charge. This remarkable finding provides an example of *charge fractionalization*. In various other condensed matter systems, topological defects are found to have peculiar quantum numbers associated with them. This phenomenon is not restricted to two dimensions. Also in higher dimensions, where bosonization does not apply, field configurations describing solitons in purely bosonic theories may carry quantum numbers usually associated with fermions, or even fractional charges.

The above results can also be obtained directly in the original fermionic formulation (9.145) by including a source term $j_e^\mu A_\mu$, and evaluating the resulting effective action in a derivative expansion to linear order in A^μ for a σ field varying in space. The fractional fermion number can be traced back to the existence of a fermion zero-energy mode. As in the Schwinger model, there exists an index theorem that assures that the Dirac equation has such a solution in the presence of a topologically nontrivial static background, i.e., a kink. The role of the γ^5 operator in the Schwinger model, which transforms an eigenstate φ_n of the operator $\not{D}$ with eigenvalue $\lambda_n \neq 0$ into an eigenstate $\gamma^5\varphi_n$ with eigenvalue $-\lambda_n$, is here played by charge conjugation. The main difference between the two index theorems is that in the Schwinger model, the time dimension is essential whereas here, for a static kink, it is irrelevant. That is to say, the number of relevant dimensions is reduced from two to one, and the zero-eigenvalue mode corresponds to a zero-energy mode here.

The zero-energy solution is localized on the wall separating the two asymptotic regions. Indeed, for this mid-gap state, the Dirac equation $(\mathrm{i}\gamma^1\partial_1 - g\sigma)\psi = 0$ reduces to

$$\mathrm{i}\partial_1\psi_2 = g\sigma\psi_1, \quad \mathrm{i}\partial_1\psi_1 = -g\sigma\psi_2 \tag{9.162}$$

in the representation (9.144) of the Dirac matrices, yielding for the kink with $\phi(x^1 \to \pm\infty) \to \pm v$ the normalizable solution

$$\psi_2(x^1) = \exp\left[-g\int_0^{x^1} \mathrm{d}x'^1\sigma(x'^1)\right], \quad \psi_1 = -\mathrm{i}\psi_2. \tag{9.163}$$

This zero mode decreases exponentially with distance from the domain wall, and whence is localized on it. Because this mode has zero energy, the ground state in the presence of a kink is doubly degenerate. Let $|\pm\rangle$ denote the ground state with the zero mode (un)occupied, then

$$\langle+|\hat{Q}_\mathrm{e}|+\rangle = \langle-|\hat{Q}_\mathrm{e}|-\rangle + 1, \tag{9.164}$$

where the last term on the right denotes the charge of the zero mode, assuming that this state is properly normalized to unity. Since under charge conjugation $Q_\mathrm{e} \to -Q_\mathrm{e}$, it follows that

$$\langle-|\hat{Q}_\mathrm{e}|-\rangle = -\langle+|\hat{Q}_\mathrm{e}|+\rangle. \tag{9.165}$$

When combined, these equations yield

$$\langle\pm|\hat{Q}_\mathrm{e}|\pm\rangle = \pm\frac{1}{2}, \tag{9.166}$$

showing that the ground state in the presence of a kink appears as a degenerate doublet with fermion number $\pm\frac{1}{2}$ depending on whether the zero mode is occupied or not. The two ground states are charge conjugates of each other.

A third way to derive the fractional charge carried by the kink is to consider the generalized model (3.24) and corresponding Goldstone-Wilczek current $\langle j^\mu \rangle$ given in Eq. (3.46). Because the second field ϕ_2 in the extended model must be taken to zero, the phase θ introduced in Eq. (3.30) only takes the values 0 or π (mod 2π). For a kink $|\Delta\theta| = \pi$ so that $\int \mathrm{d}x^1 \langle j^0 \rangle = \pm\frac{1}{2}$ by Eq. (3.46).

Notes

(i) Vortices in superconducting films were first discussed by Pearl (1964, 1965). See also the textbook [de Gennes (1989)].

(ii) The first experiment to study the possibility of a Berezinskii-Kosterlitz-Thouless transition in superconducting films was reported in [Beasley *et al.* (1979)].

(iii) Instantons in the (1+1)-dimensional Higgs model, which forms the Minkowski counterpart of the Ginzburg-Landau theory of type-II superconductors in two space dimensions discussed in Sec. 9.2, were first studied by Callan *et al.* (1977). We follow the approach used in [Schaposnik (1978)], which originates from work on the pure compact Abelian gauge theory in 2+1 dimensions by Polyakov (1977), see also [Polyakov (1987)].

(iv) Quantum electrodynamics of charged massless fermions in 1+1 spacetime dimensions, known as the Schwinger model, was first analyzed in [Schwinger (1962)]. One of various other important papers on the subject is [Lowenstein and Swieca (1971)].

(v) The close connection between the Schwinger mechanism and the Meissner effect in superconductors (Higgs mechanism) was pointed out by Anderson (1963).

(vi) The bosonization of the Schwinger model presented in Sec. 9.2 is based on [Roskies and Schaposnik (1981); Naón (1985); Schaposnik (1986); Nielsen and Wirzba (1987); Das and Karev (1987)], and [Jolicoeur and Le Guillou (1993)]. The interpretation that the Nambu-Goldstone mode acquires a gap because chiral symmetry is explicitly broken in the Schwinger model is due to Nielsen and Schroer (1977b).

(vii) The determinant obtained after the functional integrations over the anticommuting Grassmann fields are carried out was first considered by Matthews and Salam (1955).

(viii) The derivation of the chiral anomaly using the derivative expansion method was given by Das and Karev (1987), see also [Das (1999)]. For a review on anomalies, see [Jackiw (1985)].

(ix) Anomalies, index theorems, fermionic zero-eigenvalue modes, and θ vacua are covered in [Coleman (1979)], which is reprinted in [Coleman (1988)], and in [Rajaraman (1982); Polyakov (1987)]. In [Dittrich and Reuter (1986)], many technical details concerning these topics are worked out.

(x) The powerful regularization method used in establishing the index theorem in Sec. 9.5 is due to Fujikawa (1979, 1980), see also [Fujikawa and Suzuki (2004)]. The theorem was originally proved in this context by Nielsen and Schroer (1977a).

(xi) The fermion condensate in the Schwinger model was first calculated by Baaquie (1982). The argument leading to Eq. (9.124) is due to Bodwin and Kovacs (1987). The importance of the topological sectors with nonzero winding number was shown in [Jayewardena (1988)] and [Adam (1994)].

(xii) A key reference on the massive Schwinger model is [Coleman *et al*. (1975)].

(xiii) The instability of a linear chain towards a periodic distortion of the lattice accompanied by the opening of an energy gap in the fermionic spectrum was pointed out by Peierls (1955).

(xiv) For a review on fractional charges in condensed matter physics, see [Krive and Rozhavskii (1987)].

(xv) The presence of solitons carrying fractional fermion number in the model (9.145) was uncovered by Jackiw and Rebbi (1976), and in its condensed-matter counterpart, where it describes polyacetylene, by Su *et al*. (1979, 1980). The ground state of polyacetylene can be pictured as a chain molecule with alternating single and double bonds. The carbon nuclei linked by double bonds are held closer than those linked by single bonds. For lucid presentations, see [Wilczek (2002)] and [Schrieffer (2004)].

Chapter 10

From BCS to BEC

This chapter considers the pairing theory, whose weak-coupling BCS limit with loosely bound Cooper pairs was covered in Chap. 8, in the limit where the pairs are tightly bound. This so-called *composite boson limit*, although studied already in the 1960ies, has become an exciting research area only in the past decade or so with possible applications in the context of superconductor-insulator transitions. Very recently, the crossover from the weak-coupling BCS limit to the composite boson limit has been experimentally realized in ultracold Fermi gases.

10.1 Composite Boson Limit

For studying the composite boson limit, it proves prudent to swap the coupling constant λ in the gap equation (8.16) for a more convenient parameter, namely the binding energy ϵ_a of a fermion pair in vacuum. Both parameters characterize the strength of the fermion-fermion interaction. To establish the connection between the two, consider the time-independent Schrödinger equation for the problem at hand. In reduced coordinates, it reads

$$\left[-\frac{\hbar^2}{m}\nabla^2 + \lambda\,\delta(\mathbf{x})\right]\psi(\mathbf{x}) = -\epsilon_a\psi(\mathbf{x}), \tag{10.1}$$

where the reduced mass is $m/2$ and the delta-function potential, with $\lambda < 0$, represents the attractive local fermion-fermion interaction $\mathcal{L}_\mathrm{I} \equiv -\lambda\psi_\uparrow^*(x)\psi_\downarrow^*(x)\psi_\downarrow(x)\psi_\uparrow(x)$ in Eq. (8.1). We stress that this is a two-particle problem in vacuum and not the famous Cooper problem of two interacting fermions on top of a filled Fermi sea. The equation is readily solved in the Fourier representation, yielding, cf. Eq. (9.22)

$$\psi(\mathbf{k}) = -\frac{\lambda}{\hbar^2\mathbf{k}^2/m + \epsilon_a}\psi(\mathbf{x} = 0), \tag{10.2}$$

or

$$-\frac{1}{\lambda} = \int \frac{\mathrm{d}^D k}{(2\pi)^D} \frac{1}{\hbar^2 \mathbf{k}^2/m + \epsilon_a}. \tag{10.3}$$

The integral diverges in the ultraviolet for $D \geq 2$. Being of the type (3.43), the integral gives

$$-\frac{1}{\lambda} = \frac{\Gamma(1 - D/2)}{(4\pi)^{D/2}} \left(\frac{m}{\hbar^2}\right)^{D/2} \frac{1}{\epsilon_a^{1-D/2}} \tag{10.4}$$

in dimensional regularization, thus providing an explicit relation between the coupling constant λ and the binding energy ϵ_a.

In the regularization scheme with a large wave-number cutoff Λ, the divergence is absorbed by defining a renormalized coupling constant λ_r as in Eq. (8.19) through

$$\begin{aligned} D = 2: \quad & \frac{1}{\lambda_r} = \frac{1}{\lambda} + \frac{1}{2\pi} \frac{m}{\hbar^2} \ln(\Lambda/\kappa), \\ D = 3: \quad & \frac{1}{\lambda_r} = \frac{1}{\lambda} + \frac{1}{2\pi^2} \frac{m}{\hbar^2} \Lambda, \end{aligned} \tag{10.5}$$

where κ is an arbitrary scale of dimension inverse length on which the renormalized coupling constant depends in two space dimensions, $\lambda_r = \lambda_r(\kappa)$. In this way, the bound-state equation (10.3) assumes the form

$$\begin{aligned} D = 2: \quad & \epsilon_a = \frac{\hbar^2}{m} \kappa^2 \exp\left(4\pi \frac{\hbar^2}{m} \frac{1}{\lambda_r}\right), \\ D = 3: \quad & \epsilon_a = 16\pi^2 \left(\frac{\hbar^2}{m}\right)^3 \frac{1}{\lambda_r^2}. \end{aligned} \tag{10.6}$$

Because positive powers of the cutoff do not show up in dimensional regularization, the coupling constant in three space dimensions is not renormalized in that scheme, and Eq. (10.4) with $D = 3$ immediately yields the above result. The three-dimensional relation can be equivalently cast in the form

$$\epsilon_a = \frac{\hbar^2}{ma^2}, \tag{10.7}$$

with a the s-wave scattering length introduced in Eq. (6.17), $\lambda_r \equiv 4\pi\hbar^2 a/m$. For an attractive interaction, $\lambda_r, a < 0$. The expressions (10.6) and (10.7) allow for replacing the coupling constant λ_r or scattering length a with the binding energy ϵ_a.

When the bound-state equation (10.3) is substituted in the gap equation (8.17) with $E(\mathbf{k}) = \sqrt{\xi^2(\mathbf{k}) + |\bar{\Delta}|^2}$, the latter assumes the form

$$\int \frac{\mathrm{d}^D k}{(2\pi)^D} \frac{1}{\hbar^2 \mathbf{k}^2/m + \epsilon_a} = \frac{1}{2} \int \frac{\mathrm{d}^D k}{(2\pi)^D} \frac{1}{E(\mathbf{k})}, \tag{10.8}$$

where the divergences in the integrals on the left and the right precisely cancel. By inspection, it follows that, in addition to the BCS solution, this equation has a second solution given by

$$\bar{\Delta} \to 0, \quad \mu \to -\frac{1}{2}\epsilon_a. \tag{10.9}$$

It is important to note that, in contrast to the weak-coupling limit, the chemical potential characterizing the ensemble of fermions is *negative* here. As a result, the spectrum $E(\mathbf{k})$ of the elementary fermionic excitations becomes fully gapped even in the limit $|\Delta| \to 0$,

$$\xi(\mathbf{k}) \approx \epsilon(\mathbf{k}) + \frac{1}{2}\epsilon_a. \tag{10.10}$$

This is a hallmark of the composite boson limit. To appreciate the physical significance of the specific value found for the chemical potential in this limit, observe that the spectrum $E_{\bowtie}(\mathbf{q})$ of the two-fermion bound state measured relative to the pair chemical potential 2μ reads

$$E_{\bowtie}(\mathbf{q}) = \frac{\hbar^2}{4m}\mathbf{q}^2 - \mu_{\bowtie}, \tag{10.11}$$

where $\mu_{\bowtie}$ is defined as $\mu_{\bowtie} \equiv \epsilon_a + 2\mu$ and may be understood as the chemical potential characterizing the ensemble of composite bosons. The negative value for μ found in Eq. (10.9) is precisely the condition for Bose-Einstein condensation of an ideal gas of composite bosons in the $\mathbf{q} = 0$ state.

Including quadratic terms in $\bar{\Delta}$, we obtain as solution to Eq. (10.8)

$$\mu = -\frac{1}{2}\epsilon_a + (1 - D/4)\frac{|\bar{\Delta}|^2}{\epsilon_a}, \tag{10.12}$$

which leads to the chemical potential of the composite bosons

$$\mu_{\bowtie} = (2 - D/2)\frac{|\bar{\Delta}|^2}{\epsilon_a}, \tag{10.13}$$

characterizing the now interacting Bose gas. These conclusions can be explicitly checked in two space dimensions ($D = 2$), where the integral on the right of the gap equation becomes elementary. Introducing a wave number cutoff, we obtain as solution

$$\epsilon_a = \sqrt{\mu^2 + |\bar{\Delta}|^2} - \mu, \tag{10.14}$$

or

$$\mu = \frac{\epsilon_a^2 - |\bar{\Delta}|^2}{2\epsilon_a}, \tag{10.15}$$

suppressing terms that tend to zero when the cutoff is send to infinity. Note that these expressions are exact so that no higher-order terms in $|\bar{\Delta}|$ contribute to μ and $\mu_{\bowtie}$ in $D = 2$.

10.2 Ultracold Fermi Gases

Shortly after Bose-Einstein condensation had been achieved in ultracold dilute Bose gases of alkali atoms in magneto-optical traps, also fermionic alkali atoms have been studied. After successfully creating a degenerate Fermi sea in an ultracold dilute Fermi gas, Jin and collaborators at JILA, University of Colorado at Boulder, concentrated on tuning the interatomic interaction by using so-called magnetic *Feshbach resonances*. The ability to control the interactions between atoms is unique to experiments on trapped dilute gases, unmatched by other condensed matter experiments. Using this handle, the group produced in 2003 for the first time a BCS-like condensate of weakly bound pairs of fermionic atoms, and subsequently managed to realize the intriguing crossover from the weak-coupling BCS limit to the strong-coupling BEC limit, where the fermionic atoms are tightly bound in pairs and form a Bose-Einstein condensate.

At a Feshbach resonance, a new bound state emerges and, as a result, the scattering length characterizing the interatomic interaction diverges and changes sign. This can be elucidated by considering the simple example of two identical particles of mass m interacting through a three-dimensional spherical well of radius r_0 and depth V_0, parametrized as

$$V_0 = -\frac{\hbar^2}{m} k_0^2. \tag{10.16}$$

We seek a solution of the time-independent Schrödinger equation for zero energy and zero angular momentum. Writing the wave function as $\psi(\mathbf{x}) = u(r)/r$ with $r \equiv |\mathbf{x}|$ the interparticle distance, we obtain the two differential equations

$$\begin{aligned} \frac{d^2 u}{dr^2} + k_0^2 u = 0, \quad & r < r_0 \\ \frac{d^2 u}{dr^2} = 0, \quad & r > r_0 \end{aligned} \tag{10.17}$$

with the solution

$$u(r) = \begin{cases} C \sin(k_0 r), & r < r_0 \\ 1 - r/a, & r > r_0 \end{cases}, \tag{10.18}$$

where the integration constant a physically denotes the scattering length characterizing the potential. As always, the two constants C and a are fixed by the matching conditions at the edge of the potential, giving

$$a = r_0 \left(1 - \frac{\tan(k_0 r_0)}{k_0 r_0} \right), \quad C = -\frac{1}{k_0 a \cos(k_0 r_0)}. \tag{10.19}$$

When $k_0 r_0$ is an odd multiple of $\frac{1}{2}\pi$, the equation for C breaks down, while that for the scattering length gives $|a| = \infty$ so that $C = 1$ by Eq. (10.18). It follows

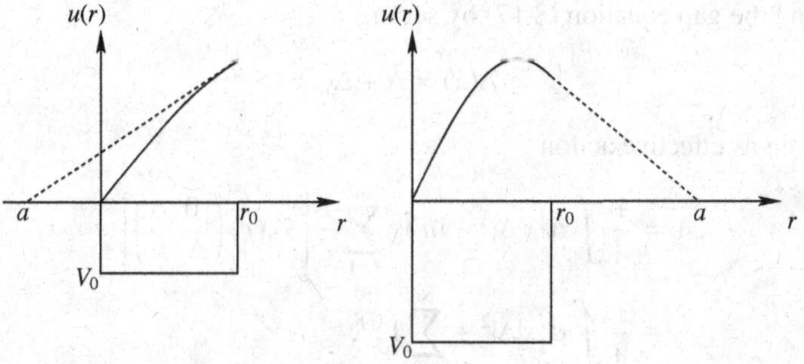

Fig. 10.1 The scattering length a for the three-dimensional spherical well of radius r_0 and depth V_0, without (left panel: $a < 0$) and with a bound state present (right panel: $a > 0$).

that the scattering length changes sign from negative for $k_0 r_0 < \frac{1}{2}\pi$ to positive for $k_0 r_0 > \frac{1}{2}\pi$, see Fig. 10.1. Physically, a new bound state, i.e., a state with negative energy arises at this value of $k_0 r_0$.

In experiments on ultracold Fermi gases, the presence of two two-particle channels is exploited. In addition to a triplet channel of two unbound atoms with parallel spins, there also exists a singlet bound state of two atoms with opposite spins. An applied magnetic field shifts the energy of the triplet state relative to that of the bound state. When, by tuning the applied field, the energy levels get close, relatively small changes in field strength can have dramatic effects on the interaction between the atoms. If the bound state has the lowest energy, the fermions tend to form pairs, which can subsequently undergo Bose-Einstein condensation. In this regime, the scattering length is positive so that the effective interatomic interaction is repulsive. If, on the other hand, the triplet state has the lowest energy, the scattering length is negative, and the effective interatomic interaction is attractive. Although isolated pairs are unstable, in the presence of a Fermi surface, two fermions can still form a loosely bound Cooper pair in this BCS regime.

10.3 Derivative Expansion

In Chap. 8, the one-loop-fermion effective action (8.10) has been evaluated in two limiting cases. Namely, in the London limit, where only the phase of the pair field Δ is assumed to depend on spacetime, and in the limit where $|\Delta|$ is assumed to be small. To study the composite boson limit, we consider the more general case and, following Sec. 3.1, expand the effective action about the constant value $\bar{\Delta}$

satisfying the gap equation (8.17) by setting

$$\Delta(x) = \bar{\Delta} + \tilde{\Delta}(x) \tag{10.20}$$

and obtain as effective action

$$\begin{aligned}\Gamma &= \frac{1}{\lambda}\int \mathrm{d}^d x |\Delta|^2 + \mathrm{i}\hbar\,\mathrm{Tr}\sum_{\ell=1}^{\infty}\frac{1}{\ell}\left[S_{\mathrm{F}}(p)\begin{pmatrix}0 & \tilde{\Delta}\\ \tilde{\Delta}^* & 0\end{pmatrix}\right]^{\ell} \\ &\equiv \frac{1}{\lambda}\int \mathrm{d}^d x|\Delta|^2 + \sum_{\ell=1}^{\infty}\Gamma_1^{(\ell)},\end{aligned} \tag{10.21}$$

where S_{F} is the fermion propagator (8.14). Using the derivative expansion method as in Sec. 3.4, we find that the one-fermion-loop induces the following terms quadratic in the field $\tilde{\Delta}$

$$\begin{aligned}\Gamma_1^{(\ell=2)} = \frac{\mathrm{i}}{2}\hbar\,\mathrm{Tr}&\frac{1}{p_0^2 - E^2(\mathbf{p})}\frac{1}{(p_0 + \mathrm{i}\hbar\partial_t)^2 - E^2(\mathbf{p}+\mathrm{i}\hbar\nabla)} \\ &\times\Big\{\bar{\Delta}^2\,\tilde{\Delta}^*\tilde{\Delta}^* + [p_0 + \xi(\mathbf{p})][p_0 + \mathrm{i}\hbar\partial_t - \xi(\mathbf{p}+\mathrm{i}\hbar\nabla)]\tilde{\Delta}\tilde{\Delta}^* \\ &\quad + \bar{\Delta}^{*2}\tilde{\Delta}\tilde{\Delta} + [p_0 - \xi(\mathbf{p})][p_0 + \mathrm{i}\hbar\partial_t + \xi(\mathbf{p}+\mathrm{i}\hbar\nabla)]\tilde{\Delta}^*\tilde{\Delta}\Big\},\end{aligned} \tag{10.22}$$

where it is recalled that, by convention, the ordinary derivatives ∂_μ only act on the first field to their right. Ignoring for the moment derivatives in this expression, we obtain after carrying out the integral over the loop frequency ω and using the gap equation (8.17) as one-loop effective potential

$$\mathcal{V}_1^{(\ell=2)} = \frac{1}{8}\int\frac{\mathrm{d}^D k}{(2\pi)^D}\frac{1}{E^3(\mathbf{k})}\left(\bar{\Delta}^2\,\tilde{\Delta}^{*2} + \bar{\Delta}^{*2}\tilde{\Delta}^2 + 2|\bar{\Delta}|^2|\tilde{\Delta}|^2\right). \tag{10.23}$$

With the approximation (10.10), the integral over the loop wave vectors becomes elementary, yielding

$$\int\frac{\mathrm{d}^D k}{(2\pi)^D}\frac{1}{E^3(\mathbf{k})} = \frac{4\Gamma(3 - D/2)}{(4\pi)^{D/2}}\left(\frac{m}{\hbar^2}\right)^{D/2}\epsilon_a^{D/2-3} \tag{10.24}$$

in dimensional regularization.

We next include the derivatives in Eq. (10.22). For simplicity, we set $\bar{\Delta}$ to zero in these contributions. The frequency integrals are then readily evaluated, with the result

$$\begin{aligned}\mathcal{L}_1^{(\ell=2)} &= -\frac{1}{2}\int\frac{\mathrm{d}^D k}{(2\pi)^D}\frac{1}{\mathrm{i}\hbar\partial_t - 2\xi(\mathbf{k}) + \hbar^2\nabla^2/4m}\tilde{\Delta}\tilde{\Delta}^* \\ &\quad - \frac{1}{2}\int\frac{\mathrm{d}^D k}{(2\pi)^D}\frac{1}{-\mathrm{i}\hbar\partial_t - 2\xi(\mathbf{k}) + \hbar^2\nabla^2/4m}\tilde{\Delta}^*\tilde{\Delta} - \mathcal{V}_1^{(\ell=2)}\end{aligned} \tag{10.25}$$

with $\xi(\mathbf{k}) = \hbar^2\mathbf{k}^2/2m - \mu$ as always and

$$\int \frac{\mathrm{d}^D k}{(2\pi)^D} \frac{1}{\mathrm{i}\hbar\partial_t - 2\xi(\mathbf{k}) + \hbar^2\nabla^2/4m} =$$
$$-\frac{\Gamma(1-D/2)}{(4\pi)^{D/2}}\left(\frac{m}{\hbar^2}\right)^{D/2} \epsilon_a^{D/2-1} - \frac{\Gamma(2-D/2)}{(4\pi)^{D/2}}\left(\frac{m}{\hbar^2}\right)^{D/2} \epsilon_a^{D/2-2}\left(\mathrm{i}\hbar\partial_t + \frac{\hbar^2\nabla^2}{4m}\right), \tag{10.26}$$

after expanding in derivatives. When combined with the tree term $|\tilde{\Delta}|^2/\lambda$, the first term at the right of Eq (10.26) precisely leads to the renormalization (8.18) of the coupling constant, giving $|\tilde{\Delta}|^2/\lambda_r$. The second term on the right gives, on adding the contribution (10.23), the result

$$\mathcal{L}_1^{(\ell=2)} = \frac{1}{2}\frac{\Gamma(2-D/2)}{(4\pi)^{D/2}}\left(\frac{m}{\hbar^2}\right)^{D/2} \epsilon_a^{D/2-2}\, \Psi^\dagger K(\mathrm{i}\partial)\Psi, \qquad \Psi \equiv \begin{pmatrix}\tilde{\Delta} \\ \tilde{\Delta}^*\end{pmatrix}, \tag{10.27}$$

where the differential operator $K(\mathrm{i}\partial)$ stands for the 2×2 matrix

$$K(\mathrm{i}\partial) =$$
$$\begin{pmatrix} \mathrm{i}\hbar\partial_t + \hbar^2\nabla^2/4m - (2-D/2)|\bar{\Delta}|^2/\epsilon_a & -(2-D/2)\bar{\Delta}^2/\epsilon_a \\ -(2-D/2)\bar{\Delta}^{*2}/\epsilon_a & -\mathrm{i}\hbar\partial_t + \hbar^2\nabla^2/4m - (2-D/2)|\bar{\Delta}|^2/\epsilon_a \end{pmatrix}. \tag{10.28}$$

This effective theory, obtained after integrating out the fermionic degrees of freedom from the theory of superconductivity in the composite boson limit, is nothing but the Lagrangian (4.60) of the Bogoliubov theory, describing a gas of repulsively interacting bosons. From this comparison it follows that the composite bosons have, as expected, a mass $m_{\bowtie} = 2m$ twice the fermion mass m, and a small chemical potential given by Eq. (10.13). It also follows that the number density of composite bosons condensed in the ground state of a uniform system reads

$$\bar{n}_{\bowtie,0} = \frac{\Gamma(2-D/2)}{(4\pi)^{D/2}}\left(\frac{m}{\hbar^2}\right)^{D/2} \epsilon_a^{D/2-2}|\bar{\Delta}|^2, \tag{10.29}$$

and that the repulsive interaction $g_{\bowtie}$ between the composite bosons is given by

$$g_{\bowtie} = \frac{(4\pi)^{D/2}}{4}\frac{2-D/2}{\Gamma(2-D/2)}\left(\frac{\hbar^2}{m}\right)^{D/2} \epsilon_a^{1-D/2}. \tag{10.30}$$

Both these parameters generally depend on the binding energy ϵ_a, which is used in place of the coupling constant λ to characterize the fermion-fermion interaction. This implies that the *attractive* interaction responsible for the binding of fermions in pairs translates into a *repulsive* interaction between the composite bosons. Remarkably, in $D = 2$, $g_{\bowtie}$ is independent of ϵ_a and depends just on atomic constants,

$$g_{\bowtie} = 1/2\nu(0) \tag{10.31}$$

with $\nu(0) = m/2\pi\hbar^2$ the density of states per spin degree of freedom at the Fermi "surface" of the metallic state in two space dimensions, see Eq. (6.60).

For $D = 2$, the general expression (10.29) reduces to

$$\bar{n}_{\bowtie,0} = \frac{1}{4\pi}\frac{m}{\hbar^2}\frac{|\bar{\Delta}|^2}{\epsilon_a}. \tag{10.32}$$

For the fermion number density $\bar{n}$, we obtain from the particle number equation (8.32) in $D = 2$ the expression

$$\bar{n} = \frac{1}{2\pi}\frac{m}{\hbar^2}\left(\sqrt{\mu^2 + |\bar{\Delta}|^2} + \mu\right), \tag{10.33}$$

with $\bar{n} = k_\mathrm{F}^2/2\pi$. For small $|\bar{\Delta}|$,

$$\bar{n} = \frac{1}{4\pi}\frac{m}{\hbar^2}\frac{|\bar{\Delta}|^2}{|\mu|} + O(|\bar{\Delta}|^4) = \frac{1}{2\pi}\frac{m}{\hbar^2}\frac{|\bar{\Delta}|^2}{\epsilon_a} + O(|\bar{\Delta}|^4) \tag{10.34}$$

so that $\bar{n} \approx 2\bar{n}_{\bowtie,0}$, showing that in a first approximation all fermions reside in the condensate (as tightly bound pairs).

The derivative expansion carried out in this section can be repeated close to the transition temperature T_c as in Sec. 8.10, where the pair field Δ vanishes. It turns out that exactly the same effective theory arises as obtained here at the absolute zero of temperature. The reason is that in the composite boson limit, the transition temperature T_c is much smaller than the dissociation temperature $k_\mathrm{B}T_{\bowtie} \approx \epsilon_a$ at which the tightly bound fermion pairs are broken up by thermal fluctuations. Hence, for all temperatures in the range $T \leq T_\mathrm{c} << T_{\bowtie}$ the system is effectively in the zero-temperature regime.

10.4 Quantum Phase Transitions

Apart from being realized in ultracold Fermi gases, the composite boson limit of the pairing theory may also be relevant in describing superconductor-insulator transitions at very low temperatures. Such a transition is an example of a *quantum phase transition*. Quantum critical phenomena differ from conventional critical phenomena occurring at finite temperatures in that quantum rather than thermal fluctuations drive the transition. Since a quantum phase transition takes place at the absolute zero of temperature, a parameter other than the temperature need be tuned to trigger the transition.

Although irrelevant for thermal phase transitions in equilibrium, time needs to be included when describing quantum phase transitions. In addition to a diverging correlation length ξ, quantum phase transitions are therefore also characterized by a diverging time scale ξ_t. They indicate, respectively, the distance and time period over which the system fluctuates coherently. The two are related, with the diverging correlation time scaling with the diverging correlation length as

$$\xi_t \sim \xi^z, \tag{10.35}$$

where z is the so-called *dynamic exponent*. It is a measure for the asymmetry between the time and space directions close to the critical point. The dynamic exponent is to be added to the set of critical exponents used to characterize a thermal phase transition. As thermal phase transitions in equilibrium are specified by two independent exponents, quantum phase transitions require three independent exponents.

The traditional scaling theory of continuous thermal phase transitions in equilibrium is readily extended to include the time dimension because the relation (10.35) implies the presence of only one independent diverging scale. Let $\delta \equiv K - K_c$, with K the tuning parameter, measure the distance from the critical value K_c. A physical observable O at the absolute zero of temperature depends on K as well as on other variables, such as an external field, energy, or momentum. These other variables are collectively denoted by Γ. According to scaling theory, O can in the critical region close to the critical point be written as

$$O(\Gamma, K) = \xi^{d_O} O(\hat{\Gamma}), \qquad (T = 0), \tag{10.36}$$

where d_O defines the scaling dimension of the observable O, $\xi \sim |\delta|^{-\nu}$, with ν the correlation length exponent, and $\hat{\Gamma}$ is obtained from Γ by rescaling it with factors of the correlation length so that $\hat{\Gamma}$ is independent of that scale. To be specific, if an external field scales as $\Gamma \sim \xi^{d_\Gamma}$, the rescaled field is defined as $\hat{\Gamma} \equiv \xi^{-d_\Gamma}\Gamma$. The right side of Eq. (10.36) does not explicitly depend on K, but only implicitly through ξ.

The data of an observable O as a function of Γ obtained for different values of the tuning parameter K can be collapsed onto a single curve if instead of $O(\Gamma, K)$, the rescaled quantity $|\delta|^{\nu d_O} O(\Gamma, K)$ is plotted as a function not of Γ, but of $\hat{\Gamma}$. Indeed, by Eq. (10.36), the combination $|\delta|^{\nu d_O} O(\Gamma, K)$ depends only on $\hat{\Gamma}$ and is thus independent of the distance from the critical point. Data collapse is often used to determine the location of the critical point as well as the values of the critical exponents by varying these parameters until an optimal collapse is achieved.

Since a physical system is always at some finite temperature, the scaling law (10.36) must be extended to account for this. This is achieved by going over to imaginary time $\tau = it$, with τ restricted to the interval $0 \leq \tau \leq \hbar\beta$. The temporal dimension thus becomes of finite extent. The behavior at finite temperature is still controlled by the quantum critical point, provided the correlation time satisfies $\xi_t < \hbar\beta$. If this condition is fulfilled, the system will not note the finite extent of the time dimension at all. Now, sufficiently close to the critical point, this condition breaks down no matter how small T, and ($\hbar/k_B$ times) the inverse temperature replaces ξ_t as relevant time scale. This is in fact what makes quantum phase transitions experimentally accessible. For the zero-temperature scaling (10.36) is then replaced with the finite-size scaling

$$O(\Gamma, K, T) = T^{-d_O/z} O(\hat{\Gamma}_T, \xi_t T), \qquad (T \neq 0), \tag{10.37}$$

where, instead of using the correlation length $\xi \sim \xi_t^{1/z}$ to remove the scale from quantities, the inverse temperature is used: $\hat{\Gamma}_T \equiv T^{d_\Gamma/z}\Gamma$.

10.5 Superconductor-Insulator Transition

In Sec. 10.2, it was shown that the composite boson limit of the pairing theory is described by the zero-temperature Bogoliubov theory of repulsively interacting tightly bound fermion pairs. That theory is equivalent to the phase-only theory (4.121). To be applicable to superconductor-insulator transitions, the theory must account for impurities. This can be done as in Sec. 4.13, where it was shown that including a moderate amount of impurities in the weak-coupling Bogoliubov theory leads to localization without destroying the superfluid state completely. A superconductor-insulator transition is a result of the interplay between superconductivity and localization, with the superconducting state giving way to an insulating state when localization effects start to dominate. In the composite boson limit, the transition to the insulating state is a result not of the unbinding of electron pairs, but rather of the quenching of the condensate of composite bosons. In other words, electron pairs exist on both sides of the transition, with the insulating state characterized by localized composite bosons. Such a state, which is a result not only of repulsive interactions, as in a Mott insulator, but also of (Anderson) localization, is called an *Anderson-Mott insulator*.

As a warm-up exercise, we consider the two terms in the zero-temperature effective theory (4.121) quadratic in the Nambu-Goldstone field φ and write them in the most general form

$$\mathcal{L}_{\rm eff} = \frac{1}{2}\bar{n}^2 \kappa \hbar^2 (\partial_t \varphi)^2 - \frac{1}{2}\frac{\hbar^2}{m^2}\rho_{\rm s}(\nabla\varphi)^2. \tag{10.38}$$

The coefficient $\rho_{\rm s}$ is the superfluid mass density, which is a response function and in general does not equal $m\bar{n}$. The other coefficient features the compressibility introduced in Eq. (2.202) which is related to the (0, 0)-component of the polarization tensor Π^{00} through

$$\lim_{|\mathbf{k}|\to 0} \Pi^{00}(0, \mathbf{k}) = \bar{n}^2 \kappa, \tag{10.39}$$

where, as is typical for response functions, the frequency transfer is put to zero before the wave vector transfer **k** is. Equation (10.38) leads to the general expression for the speed of sound

$$c_s^2 = \frac{\rho_s}{m^2\bar{n}^2\kappa}, \tag{10.40}$$

which reduces to the less general expression (2.203) if ρ_s is replaced with $m\bar{n}$.

Simple dimensional analysis shows that the effective Lagrangian density scales near a smooth phase transition as

$$\mathcal{L}_{\text{eff}} \sim \xi^{-(D+z)}, \tag{10.41}$$

while

$$(\nabla\varphi)^2 \sim \xi^{-2}, \quad (\partial_t\varphi)^2 \sim \xi_t^{-2} \sim \xi^{-2z}. \tag{10.42}$$

Combining these hyperscaling arguments, and noting that the mass parameter is inessential with regards to critical behavior, we arrive at the scaling laws for the two coefficients appearing in the effective theory (10.38):

$$\rho_s \sim \xi^{-(D+z-2)}, \quad \kappa \sim \xi^{-(D-z)}. \tag{10.43}$$

The first conclusion is consistent with the universal jump (5.96) predicted for the Berezinskii-Kosterlitz-Thouless phase transition, which corresponds to taking $z = 0$ and $D = 2$.

In the presence of impurities it is reasonable to assume that the compressibility stays finite at the quantum critical point, implying $z = D$. This remarkably simple argument thus predicts an exact and nontrivial value for the dynamic exponent.

Without impurities, the dynamic exponent is $z = 2$. This agrees with the observation that in a nonrelativistic theory one time derivative appears in combination with two space derivatives, $i\hbar\partial_t + \hbar^2\nabla^2/2m$. This last argument is, however, to be treated with care when applied to the phase-only theory (4.121). In that theory, the time and space derivatives appear in a symmetrical form, yet z is in general not unity, as we just saw. The difference is that in the effective theory, the relative coefficient c^2 scales according to Eq. (10.40) with the scaling laws (10.43) as

$$c_s^2 \sim \xi^{2(1-z)}, \tag{10.44}$$

while the relative coefficient m in the microscopic theory does not scale.

We proceed to include long-range interactions in the effective theory, cf. Sec. 8.6. A case of particular interest is the three-dimensional Coulomb potential

$$V(\mathbf{x}) = \frac{q^2}{r}, \tag{10.45}$$

($r \equiv |\mathbf{x}|$) whose Fourier transform in D space dimensions reads

$$V(\mathbf{k}) = 2^{D-1}\pi^{(D-1)/2}\Gamma[\tfrac{1}{2}(D-1)]\frac{q^2}{|\mathbf{k}|^{D-1}}. \tag{10.46}$$

Here, q stands for the electric charge, which in the case of Cooper pairs is twice the electron charge. The rationale for using the 3D Coulomb potential, even when considering charges confined to move in a lower dimensional space, is that the electrostatic interaction remains three-dimensional. To appreciate under which circumstances the Coulomb interaction becomes important, note that for dimensional reasons $1/|\mathbf{x}| \sim k_\mathrm{F}$ and $\bar{n} \sim k_\mathrm{F}^D$, where k_F is the Fermi wave number. The ratio of the Coulomb interaction energy to the Fermi energy $\epsilon_\mathrm{F} = \hbar^2 k_\mathrm{F}^2/2m$ is therefore proportional to $\bar{n}^{-1/D}$. This means that the lower the electron number density $\bar{n}$ is, the more important the Coulomb interaction becomes.

The effective theory with the Coulomb interaction included becomes after passing over to the Fourier representation, cf. Eq. (8.57)

$$\mathcal{L}_\mathrm{eff} = \frac{\hbar^2}{2}\varphi(-k)\left(\frac{\omega^2}{V(\mathbf{k})} - \frac{\rho_\mathrm{s}}{m^2}\mathbf{k}^2\right)\varphi(k) \tag{10.47}$$

or

$$\mathcal{L}_\mathrm{eff} = \frac{\hbar^2}{2}\varphi(-k)\left(\frac{1}{q'^2}|\mathbf{k}|^{D-1}\omega^2 - \frac{\rho_\mathrm{s}}{m^2}\mathbf{k}^2\right)\varphi(k), \tag{10.48}$$

where q' appearing in this theory governing a plasma mode is a redefined charge parameter,

$$q'^2 \equiv 2^{D-1}\pi^{(D-1)/2}\Gamma[\tfrac{1}{2}(D-1)]q^2. \tag{10.49}$$

The charge is connected to the (0, 0)-component of the polarization tensor (10.39) via

$$q'^2 = \lim_{|\mathbf{k}|\to 0}\frac{|\mathbf{k}|^{d-1}}{\Pi^{00}(0,\mathbf{k})}. \tag{10.50}$$

A simple hyperscaling argument like the one given above for the case without Coulomb interaction shows that near a smooth transition, the charge scales as

$$q'^2 \sim \xi^{1-z}, \tag{10.51}$$

independent of the number of space dimensions D.

In experiments on charged systems, instead of the superfluid mass density, usually the conductivity σ is measured. To see the relation between the two, we introduce a vector potential in the effective theory by replacing $\nabla\varphi$ with $\nabla\varphi - (q/\hbar c)\mathbf{A}$ in Eq. (10.38). The term in the action quadratic in $\mathbf{A}$ then becomes in momentum space

$$S_\sigma = -\frac{1}{2}\frac{q^2}{m^2c^2}\int\frac{\mathrm{d}^d k}{(2\pi)^d}\mathbf{A}(-k)\cdot\rho_\mathrm{s}(k)\mathbf{A}(k). \tag{10.52}$$

By Eq. (2.183), the electric current density,

$$\frac{q}{c}\mathbf{j}_{\rm e}(k) = \frac{\delta S_\sigma}{\delta \mathbf{A}(-k)} \tag{10.53}$$

obtained from this action can be written as

$$q\mathbf{j}_{\rm e}(k) = \sigma(k)\mathbf{E}(k), \tag{10.54}$$

see Eq. (6.40), with the conductivity

$$\sigma(k) = \mathrm{i}\frac{q^2}{m^2}\frac{\rho_{\rm s}(k)}{\omega} \tag{10.55}$$

essentially given by the superfluid mass density. With the scaling laws (10.43) and (10.51), it then follows from this equation that the conductivity scales as

$$\sigma \sim \xi^{3-(D+z)}. \tag{10.56}$$

In the presence of random impurities, the charge is expected to be finite at the transition so that $z = 1$ by Eq. (10.51). This is again an exact result, which replaces the value $z = D$ of a disordered system without Coulomb interaction. For $z = 1$, the conductivity scales as

$$\sigma \sim \xi^{2-D}, \tag{10.57}$$

implying that in two space dimensions, the conductivity is a marginal operator which remains finite at the quantum critical point.

Notes

(i) The composite boson limit of the pairing theory was first considered by Eagles (1969). Key theoretical papers on the subject include [Leggett (1980); Nozières and Schmitt-Rink (1985); Randeria *et al.* (1990); Drechsler and Zwerger (1992); Sá de Melo *et al.* (1993)], and [Marini *et al.* (1998)]. In these references, also the crossover between the weak-coupling BCS limit and the strong-coupling composite boson limit is studied. An alternative derivation of the effective theory (10.27) using T-matrix theory can be found in [Haussmann (1993)].

(ii) The first degenerate Fermi gas was produced by DeMarco and Jin (1999) in a trapped dilute gas of ^{40}K atoms using evaporative cooling. Bose-Einstein condensation of tightly bound fermion pairs was first observed by Greiner *et al.* (2003); Jochim *et al.* (2003), and Zwierlein *et al.* (2003). The experimental realization of condensation in the BCS-BEC crossover regime was first reported in [Regal *et al.* (2004)] and [Zwierlein *et al.* (2004)].

(iii) For authoritative introductions to scaling theory applied to thermal phase transitions in equilibrium, and references to the original literature, see [Fisher (1983, 1998)] and [Stanley (1999)].

(iv) One of the earliest and influential studies on quantum critical phenomena is by Hertz (1976). For introductions to quantum phase transitions, see [Continentino (1994)] and [Sondhi *et al.* (1997)].

(v) Section 10.5 is based on [Fisher and Fisher (1988); Fisher *et al.* (1990); Fisher (1990)], and [Girvin *et al.* (1992)].

(vi) For a review of the Anderson-Mott transition, see [Belitz and Kirkpatrick (1994)].

(vii) The scaling (10.57) is due to Wegner (1976).

(viii) Experimental support for the presence of a quantum critical point in disordered superconducting films and the prediction, based on the phase-only theory, that the conductivity remains finite at this point was reported in [Hebard and Paalanen (1990)]. For an assessment of the experimental status of the phase-only theory, see [Liu and Goldman (1994)] and [Goldman and Marković (1998)]. For a paper arguing against the applicability of this theory, see [Ramakrishnan (1989)]. A recent critical analysis of the experimental data on quantum phase transitions in two-dimensional systems can be found in [Shangina and Dolgopolov (2003)].

Chapter 11

Superfluid ^{3}He

In this chapter, the BCS theory, featuring Cooper pairing in the relative s-wave orbital state, is extended to p-wave pairing to describe superfluid ^{3}He. The resulting order parameter transforms as a vector under rotations both in spin and orbital space. The superfluid ^{3}He phases are characterized by the different ways the rotational and global phase symmetries are spontaneously broken. The resulting symmetry breaking patterns are intricate and give rise to a host of surprising physical properties displayed by the superfluid ^{3}He phases. The anisotropic superfluid ^{3}He-A phase features a topological line defect in momentum space. As a result of this remarkable property, the energy gap in the fermionic excitation spectrum vanishes at two nodes on the Fermi surface of the normal liquid state. At these so-called boojums, the spectrum of the fermionic excitations vanishes linearly, as for massless relativistic fermions. The presence of gapless fermionic excitations, or zero modes implies that even at zero temperature, superfluid ^{3}He-A has a normal component. They are at the heart of various special effects such as a chiral anomaly and the Callan-Harvey effect usually associated with relativistic theories. Marking the locations on the Fermi surface of the normal liquid state where two energy levels cross, the boojums are diabolic points, which produce a geometric phase that has direct physical consequences.

11.1 P-State Triplet Pairing

A common characteristic of superfluids is the spontaneously broken U(1) symmetry of global phase transformations. Often, as in superfluid ^{4}He and in neutral superconductors with Cooper pairs forming a spin singlet in a relative s-wave orbital state, this is the only symmetry broken. In such systems, the rotational symmetries in spin and orbital space remain symmetries of the new ground state. In superfluid ^{3}He, on the other hand, these symmetries are, in addition to the global

U(1) symmetry, spontaneously broken. Provided the small dipole-dipole interaction is neglected, the spin rotations (S) and spatial, or orbital rotations (L) can be considered as independent symmetries so that the relevant symmetry group is

$$G = \mathrm{SO}^S(3) \times \mathrm{SO}^L(3) \times \mathrm{U}^N(1). \tag{11.1}$$

The global U(1) group of phase transformations, which is generated by the particle number N, has been given a superscript N. The conventional description of superfluid ^{3}He involves an order parameter $A^{\alpha i}$, with $\alpha = 1, 2, 3$ the spin and $i = 1, 2, 3$ the orbital index. This complex matrix acquires a phase factor under U(1) phase transformations and, as will be shown shortly, transforms as a vector both under rotations in spin and orbital space. In contrast, the order parameter of a superconductor with Cooper pairing in the relative s-wave orbital state is invariant under these rotation groups so that rotational symmetry is unbroken there.

Superfluid ^{3}He can be modeled by a Lagrangian of the BCS-type (8.1) with the interaction term

$$\mathcal{L}_{\mathrm{BCS}} = -\lambda \psi_\uparrow^* \psi_\downarrow^* \psi_\downarrow \psi_\uparrow = -\frac{\lambda}{2} \sum_{\sigma,\tau=\uparrow,\downarrow} \psi_\sigma^* \psi_\tau^* \psi_\tau \psi_\sigma \tag{11.2}$$

replaced with

$$\mathcal{L}_{^3\mathrm{He}} = -\frac{\lambda}{2} \sum_{\sigma,\tau=\uparrow,\downarrow} \psi_\sigma^* (-\mathrm{i}\check{\partial}_i) \psi_\tau^* \psi_\tau (-\mathrm{i}\check{\partial}_i) \psi_\sigma, \tag{11.3}$$

appropriate for pairing in the relative p-wave orbital state. Here, ψ_σ are Grassmann fields describing the ^{3}He atoms with spin $\sigma = \uparrow, \downarrow$, $\check{\nabla}$ is the right minus left gradient operator, $\check{\nabla} \equiv (\nabla - \overleftarrow{\nabla})/2k_{\mathrm{F}}$ normalized by twice the Fermi wave number k_{F} for convenience. The coupling constant is negative $\lambda < 0$ so that the effective interatomic interaction is attractive. Since the fields anticommute, the terms with $\sigma = \tau$ vanish identically in the BCS interaction term (11.2), while they need not in Eq. (11.3) because of the gradient operators.

As in BCS theory, this theory is linearized by introducing auxiliary fields Δ and $\Delta^\dagger$ through the functional identity, cf. Eq. (8.6)

$$\begin{aligned} &\exp\left\{-\frac{\mathrm{i}}{2\hbar}\lambda \int \mathrm{d}^d x \left[\psi_\sigma^*(-\mathrm{i}\check{\partial}_i)\psi_\tau^*\right]\left[\psi_\tau(-\mathrm{i}\check{\partial}_i)\psi_\sigma\right]\right\} = \int \mathrm{D}\Delta^\dagger \mathrm{D}\Delta \\ &\times \exp\left(-\frac{\mathrm{i}}{2\hbar}\int \mathrm{d}^d x \left\{\Delta_{\sigma\tau}^{*i}\left[\psi_\tau(-\mathrm{i}\check{\partial}_i)\psi_\sigma\right] + \left[\psi_\sigma^*(-\mathrm{i}\check{\partial}_i)\psi_\tau^*\right]\Delta_{\tau\sigma}^i - \frac{1}{\lambda}\Delta_{\sigma\tau}^{*i}\Delta_{\tau\sigma}^i\right\}\right). \end{aligned} \tag{11.4}$$

The Euler-Lagrange equation for Δ

$$\Delta_{\sigma\tau}^i = \lambda \psi_\sigma (-\mathrm{i}\check{\partial}_i) \psi_\tau \tag{11.5}$$

shows that it physically describes pairs of ^{3}He atoms. The specific form is to be compared to Eq. (8.7) appropriate for pairing in a relative s-wave orbital state. Because of the derivative at the right side of Eq. (11.5), the pair field $\Delta^i_{\sigma\tau}$ describes a spin triplet in a relative p-wave orbital state. Indeed, $\Delta^i_{\sigma\tau}$ is symmetric in its spin indices, $\Delta^i_{\sigma\tau} = \Delta^i_{\tau\sigma}$, as appropriate for a spin triplet, and the index i indicates that the field is a vector under spatial rotations, as appropriate for a p-wave orbital state. The partition function Z of the theory at the absolute zero of temperature can then be represented by the functional integral

$$Z = \int \mathrm{D}\Psi^\dagger \mathrm{D}\Psi \int \mathrm{D}\Delta^\dagger \mathrm{D}\Delta \exp\left\{\frac{\mathrm{i}}{2\hbar}\int \mathrm{d}^d x\left[\Psi^\dagger K(\mathrm{i}\partial)\Psi + \frac{1}{\lambda}\Delta^{*i}_{\sigma\tau}\Delta^i_{\tau\sigma}\right]\right\} \tag{11.6}$$

with $K(\mathrm{i}\partial)$ the differential operator

$$K(\mathrm{i}\partial) = \begin{pmatrix} \mathrm{i}\hbar\partial_t + \hbar^2\nabla^2/2m + \mu & -(1/2k_\mathrm{F})[\mathbf{\Delta}\cdot(-\mathrm{i}\nabla) + \mathrm{i}\overleftarrow{\nabla}\cdot\mathbf{\Delta}]g^\dagger \\ -(1/2k_\mathrm{F})g[\mathbf{\Delta}^\dagger\cdot(-\mathrm{i}\nabla) + \mathrm{i}\overleftarrow{\nabla}\cdot\mathbf{\Delta}^\dagger] & \mathrm{i}\hbar\partial_t - \hbar^2\nabla^2/2m - \mu \end{pmatrix}. \tag{11.7}$$

Here, g is the so-called metric spinor

$$g \equiv \mathrm{i}\sigma^2 = \begin{pmatrix} 0 & 1 \\ -1 & 0 \end{pmatrix}, \tag{11.8}$$

and Ψ stands for the four-component Nambu multiplet

$$\Psi \equiv \begin{pmatrix} \psi \\ g\psi^* \end{pmatrix}, \qquad \psi \equiv \begin{pmatrix} \psi_\uparrow \\ \psi_\downarrow \end{pmatrix}. \tag{11.9}$$

The reason for introducing the spinor g in this multiplet is that the symmetric matrices Δ^i ($i = 1, 2, 3$) may be expressed as a linear combination of the symmetric matrices $\sigma^\alpha g$ with σ^α ($\alpha = 1, 2, 3$) the Pauli matrices:

$$\Delta^i_{\sigma\tau} = A^{\alpha i}(\sigma^\alpha g)_{\sigma\tau}, \tag{11.10}$$

where $A^{\alpha i}$ are expansion coefficients, and

$$\Delta^i g^\dagger = A^{\alpha i}\sigma^\alpha. \tag{11.11}$$

The three matrices $\sigma^\alpha g$ transform in the same way as Δ^i under spin rotations,

$$\sigma^\alpha g \to \sigma'^\alpha g = U(\boldsymbol{\theta}^S)\,\sigma^\alpha g\, U^\mathrm{T}(\boldsymbol{\theta}^S), \tag{11.12}$$

with $U(\boldsymbol{\theta}^S)$ the 2×2 spin rotation matrix

$$U(\boldsymbol{\theta}^S) \equiv \exp\left(\mathrm{i}\boldsymbol{\theta}^S \cdot \frac{\boldsymbol{\sigma}}{2}\right) \tag{11.13}$$

and $\boldsymbol{\theta}^S$ the rotation vector, specifying the direction of the rotation axis and the rotation angle $|\boldsymbol{\theta}^S|$. Moreover, $\frac{1}{2}\boldsymbol{\sigma}$, satisfying the algebra

$$\left[\frac{\sigma^\alpha}{2}, \frac{\sigma^\beta}{2}\right] = \mathrm{i}\epsilon^{\alpha\beta\gamma}\frac{\sigma^\gamma}{2}, \tag{11.14}$$

denotes the $SO^S(3)$ generators in the spinor representation. The transformation property (11.12) follows from the fact that the Pauli matrices transform unitarily under spin rotations,

$$\sigma^\alpha \to \sigma'^\alpha = U(\boldsymbol{\theta}^S)\,\sigma^\alpha U^\dagger(\boldsymbol{\theta}^S), \tag{11.15}$$

and from the identity $\sigma^2\sigma^\alpha = -\sigma^{*\alpha}\sigma^2$, leading to

$$gU(\boldsymbol{\theta}^S) = U^*(\boldsymbol{\theta}^S)g. \tag{11.16}$$

The transformation properties of the expansion coefficients $A^{\alpha i}$ in Eq. (11.10) follow, of course, from those of $\Delta^i_{\sigma\tau}$. Specifically, the spin rotation (11.12) induces the transformation of the coefficients $A^{\alpha i}$:

$$A^{\alpha i}\sigma^\alpha g \to A^{\alpha i}\,U(\boldsymbol{\theta}^S)\,\sigma^\alpha g\,U^{\mathrm{T}}(\boldsymbol{\theta}^S) \equiv A'^{\alpha' i}\sigma^{\alpha'} g. \tag{11.17}$$

To proceed, we use the identities

$$\begin{aligned} U(\boldsymbol{\theta}^S)\,\sigma^\alpha g\,U^{\mathrm{T}}(\boldsymbol{\theta}^S) &= U(\boldsymbol{\theta}^S)\,\sigma^\alpha U^\dagger(\boldsymbol{\theta}^S)g \\ &= R^{\alpha\alpha'}(-\boldsymbol{\theta}^S)\,\sigma^{\alpha'} g, \end{aligned} \tag{11.18}$$

where in the first step Eq. (11.16) is used, and $R(\boldsymbol{\theta}^S)$ is the 3×3 orthogonal rotation matrix

$$R(\boldsymbol{\theta}^S) \equiv \exp\left(\mathrm{i}\boldsymbol{\theta}^S \cdot \mathbf{t}\right), \tag{11.19}$$

with $\mathbf{t}$ denoting the $SO^S(3)$ generators in the vector representation,

$$(t^\alpha)^{\beta\gamma} = -\mathrm{i}\epsilon^{\alpha\beta\gamma}, \tag{11.20}$$

satisfying the algebra (11.14),

$$\left[t^\alpha, t^\beta\right] = \mathrm{i}\epsilon^{\alpha\beta\gamma}t^\gamma. \tag{11.21}$$

The last step in Eq. (11.18) follows from the algebra (11.14) written in the form

$$\left[\boldsymbol{\theta}^S \cdot \frac{\boldsymbol{\sigma}}{2}, \sigma^\alpha\right] = -(\boldsymbol{\theta}^S \cdot \mathbf{t})^{\alpha\beta}\,\sigma^\beta. \tag{11.22}$$

From Eqs. (11.17) and (11.18) we finally conclude that $A^{\alpha i}$ transforms as a vector with respect to its spin index,

$$A'^{\alpha' i} = R^{\alpha'\alpha}(\boldsymbol{\theta}^S)\,A^{\alpha i}. \tag{11.23}$$

By Eq. (11.5), $A^{\alpha i}$ transforms under orbital rotations in the same way as ∂_i, that is also as a vector, while the U(1) group of phase transformations

$$\psi_\sigma(x) \to \psi'_\sigma(x) = \mathrm{e}^{\mathrm{i}\bar{\alpha}}\psi_\sigma(x) \tag{11.24}$$

acts as a multiplicative factor on the order parameter. The complete transformation property of $A^{\alpha i}$ under the symmetry group (11.1) is therefore:

$$A \to A' = \mathrm{e}^{2\mathrm{i}\tilde{\alpha}} R(\boldsymbol{\theta}^S) A R^{\mathrm{T}}(\boldsymbol{\theta}^L), \tag{11.25}$$

or more explicitly,

$$A^{\alpha i} \to A'^{\alpha' i'} = \mathrm{e}^{2\mathrm{i}\tilde{\alpha}} R^{\alpha' \alpha}(\boldsymbol{\theta}^S) A^{\alpha i} R^{i' i}(\boldsymbol{\theta}^L), \tag{11.26}$$

with $R(\boldsymbol{\theta}^L)$ the orthogonal 3×3 orbital rotation matrix given by Eq. (11.19) with S replaced with L.

To physically interpret the order parameter $A^{\alpha i}$, we recall from standard quantum mechanics that spin rotations are generated by the spin operator $\hat{\mathbf{S}}$ and, similarly, orbital rotations by the intrinsic orbital angular momentum operator $\hat{\mathbf{L}}$. That is to say, by Eq. (11.25)

$$\hat{S}^\alpha A^{\beta i} = \hbar \, (t^\alpha)^{\beta\gamma} A^{\gamma i}, \tag{11.27}$$

$$\hat{L}^i A^{\beta j} = \hbar \, (t^i)^{jk} A^{\beta k}, \tag{11.28}$$

with t^α the generators (11.20) of the $\mathrm{SO}^S(3)$ group in the vector representation, and a similar expression for the generators t^i. To proceed, we go over to a spherical basis in both spin and orbital space. In three dimensions, this basis is defined by the three unit vectors

$$\mathbf{e}_+ \equiv -\frac{1}{\sqrt{2}}(\mathbf{e}_1 + \mathrm{i}\mathbf{e}_2), \quad \mathbf{e}_0 \equiv \mathbf{e}_3; \quad \mathbf{e}_- \equiv \frac{1}{\sqrt{2}}(\mathbf{e}_1 - \mathrm{i}\mathbf{e}_2), \tag{11.29}$$

with $\mathbf{e}_i$ $(i = 1, 2, 3)$ the usual Cartesian unit vectors. An arbitrary vector $\mathbf{v}$,

$$\mathbf{v} = \sum_{i=1}^{3} v^i \mathbf{e}_i, \tag{11.30}$$

can be re-expressed in terms of the spherical basis as

$$\mathbf{v} = \sum_{m=-1}^{1} v^m \mathbf{e}_m, \tag{11.31}$$

with $v^m = \mathbf{e}_m^* \cdot \mathbf{v}$. The change of basis can be summarized by the transformation matrix

$$P \equiv \frac{1}{\sqrt{2}} \begin{pmatrix} -1 & -\mathrm{i} & 0 \\ 0 & 0 & \sqrt{2} \\ 1 & -\mathrm{i} & 0 \end{pmatrix} \tag{11.32}$$

as

$$v^m = (P^*)^{mi} v^i, \tag{11.33}$$

With respect to the spherical basis in both spin and orbital space, the order parameter assumes the form

$$a^{m_S m_L} = (P^*)^{m_S \alpha} A^{\alpha i} (P^*)^{m_L i}, \tag{11.34}$$

and

$$\hat{S}^\alpha a^{m_S m_L} = \hbar (M^\alpha)^{m_S m'_S} a^{m'_S, m_L} \tag{11.35}$$
$$\hat{L}^i a^{m_S, m_L} = \hbar (M^i)^{m_L m'_L} a^{m_S, m'_L}, \tag{11.36}$$

with

$$M^\alpha \equiv P^* t^\alpha P^{\mathrm{T}} \tag{11.37}$$
$$M^i \equiv P^* t^i P^{\mathrm{T}}. \tag{11.38}$$

That is, the generators t^α and t^i undergo a unitary transformation in passing from the Cartesian to the spherical bases. Explicitly,

$$M^1 = \frac{1}{\sqrt{2}} \begin{pmatrix} & 1 & \\ 1 & & 1 \\ & 1 & \end{pmatrix}, \quad M^2 = \frac{1}{\sqrt{2}} \begin{pmatrix} & -\mathrm{i} & \\ \mathrm{i} & & -\mathrm{i} \\ & \mathrm{i} & \end{pmatrix}, \quad M^3 = \begin{pmatrix} 1 & & \\ & 0 & \\ & & -1 \end{pmatrix}, \tag{11.39}$$

which is the standard matrix representation of the angular momentum operator in a spherical basis, and

$$\begin{aligned} \hat{S}^3 a^{m_S m_L} &= \hbar m_S a^{m_S m_L} \\ \hat{L}^3 a^{m_S m_L} &= \hbar m_L a^{m_S m_L}, \end{aligned} \tag{11.40}$$

with $m_S, m_L = 1, 0, -1$, revealing that the entries $a^{m_S m_L}$ of the order parameter in the spherical bases have definite magnetic quantum numbers. This facilitates identifying the various superfluid ^{3}He states, as will be shown in the next section. The spin- and orbital-quantization axes need not point in the same direction so that $\hat{S}^3$ and $\hat{L}^3$ in Eq. (11.40) can refer to different axes.

11.2 Phase Diagram

Figure 11.1 gives the phase diagram of ^{3}He in the presence of an applied magnetic field at ultra low temperatures. At pressures below about 34 bar (3.4 MPa), ^{3}He remains liquid all the way down to the absolute zero of temperature, becoming superfluid below 2.5 mK. In the absence of an applied field, two superfluid phases exist, the so-called A and B phases. Depending on the pressure P, the normal Fermi liquid may undergo a phase transition into the B phase either directly ($P <$ 21 bar) or via the A phase ($P >$ 21 bar). The transitions from the normal to the A and B phases are smooth, while the transition from the A to the B phase

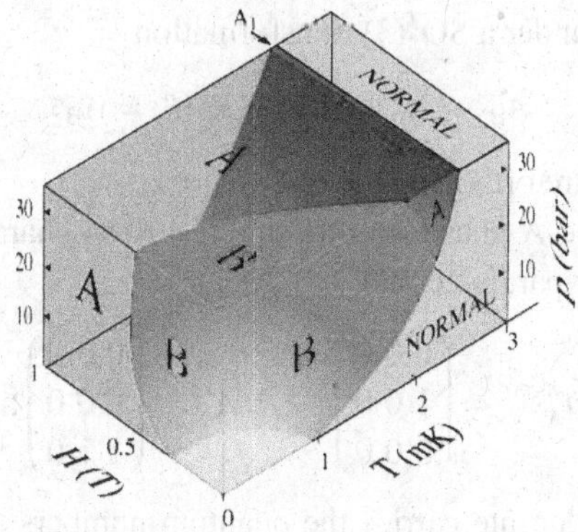

Fig. 11.1 Phase diagram of liquid ^{3}He in the mK temperature regime. (Figure reprinted with kind permission of M. Krusius, Low Temperature Laboratory, Helsinki University of Technology.)

is first order. If an external magnetic field H is applied to the system, a third superfluid phase, the so-called A_1 phase, arises between the normal liquid and the superfluid ^{3}He-A phase, occupying only a small part of the phase diagram. Both the transitions from the normal to the A_1 state and from the A_1 to the A state are smooth. With increasing magnetic field, the regime of the superfluid A phase increases at the expense of the B phase. In each of the three superfluid phases, the symmetry of the system is broken differently.

In the superfluid ^{3}He-B phase, the symmetry G is spontaneously broken down to the diagonal subgroup $SO^J(3)$ of the product $SO^S(3) \times SO^L(3)$,

$$G \to H_B = SO^J(3), \tag{11.41}$$

where J refers to the total angular momentum operator, $\hat{\mathbf{J}} \equiv \hat{\mathbf{S}} + \hat{\mathbf{L}}$. In its simplest form, the order parameter reads

$$a_B^{m_S m_L} = \begin{pmatrix} 0 & 0 & -1 \\ 0 & 1 & 0 \\ -1 & 0 & 0 \end{pmatrix}, \qquad A_B^{\alpha i} = \begin{pmatrix} 1 & 0 & 0 \\ 0 & 1 & 0 \\ 0 & 0 & 1 \end{pmatrix}. \tag{11.42}$$

The representation in the spherical bases on the left shows that pairing takes place in the state with zero total angular momentum $m_J \equiv m_S + m_L = 0$. Like a superfluid with pairing in the s-state, superfluid ^{3}He-B is isotropic, having no preferred direction. In the Cartesian bases, the order parameter is simply the unit matrix. From this it follows that although spin and orbital rotations by themselves are no longer symmetries of this superfluid state, the combined action of a spin rotation followed by an orbital rotation with the same rotation vector, $\boldsymbol{\theta}^L = \boldsymbol{\theta}^S = \boldsymbol{\theta}$, is still a symmetry. Indeed, under a SO^J(3) transformation

$$A_B \to A'_B = R(\boldsymbol{\theta})A_B R^T(\boldsymbol{\theta}) = A_B \tag{11.43}$$

for rotation matrices are orthogonal $R(\boldsymbol{\theta})R^T(\boldsymbol{\theta}) = 1$.

The superfluid ^{3}He-A phase is in its simplest form characterized by the suitably normalized order parameter

$$a_A^{m_S m_L} = \begin{pmatrix} 0 & 0 & 0 \\ 1 & 0 & 0 \\ 0 & 0 & 0 \end{pmatrix}, \qquad A_A^{\alpha i} = \begin{pmatrix} 0 & 0 & 0 \\ 0 & 0 & 0 \\ 1 & i & 0 \end{pmatrix}, \tag{11.44}$$

showing that the condensate carries the quantum numbers $m_S = 0, m_L = 1$. In this state, the symmetry is spontaneously broken down to a product of two U(1) groups

$$G \to H_A = U^S(1) \times U^{L-\frac{1}{2}N}(1). \tag{11.45}$$

The residual $U^S(1)$ symmetry shows that the SO^S(3) group of spin rotations is spontaneously broken down to rotations about the spin-quantization axis as in a ferro- or antiferromagnet. In the absence of an outside agent, this axis settles on an arbitrary but fixed direction throughout the system, assuming the system is uniform. Given this preferred direction, superfluid ^{3}He-A is an example of an *anisotropic* superfluid. The second residual symmetry group $U^{L-\frac{1}{2}N}(1)$ comprises both a rotation about the orbital-quantization axis and a $U^N(1)$ phase transformation. In other words, in this superfluid state, the SO^L(3) group of orbital rotations is spontaneously broken down to rotations about the preferred axis combined with a simultaneous global phase transformation. This is a direct consequence of the fact that pairing takes place in a state with $m_L = 1$ as can be seen as follows. Consider the combined action of a rotation by an angle θ^S about the preferred axis in spin space, a rotation by an angle θ^L about the preferred axis in orbital space, and a phase transformation. Under this action, the matrix elements $a^{m_S m_L}$ of the order parameter in the spherical bases pick up the phase factor

$$a^{m_S m_L} \to a'^{m_S m_L} = e^{i(\theta^S m_S + \theta^L m_L + 2\alpha)} a^{m_S m_L}. \tag{11.46}$$

The elements with nonzero magnetic quantum numbers are not invariant under rotations. However, a nontrivial realization of rotational invariance may be achieved by allowing a phase transformation to compensate. Specifically, for the case at hand with $m_S = 0, m_L = 1$, the choice $\alpha = -\frac{1}{2}\theta^L$ leads to a generalized invariance of orbital rotations about the preferred axis denoted by $U^{L-\frac{1}{2}N}(1)$. As remarked before, the directions of the spin- and orbital-quantization axes need not coincide.

Finally, the superfluid ^{3}He-A$_1$ phase is characterized by the order parameter

$$a_{\rm A_1}^{m_S m_L} = \begin{pmatrix} 1 & 0 & 0 \\ 0 & 0 & 0 \\ 0 & 0 & 0 \end{pmatrix}, \qquad A_{\rm A_1}^{\alpha i} = \begin{pmatrix} 1 & {\rm i} & 0 \\ {\rm i} & -1 & 0 \\ 0 & 0 & 0 \end{pmatrix}, \tag{11.47}$$

representing a condensate with quantum numbers $m_S = m_L = 1$. As in the superfluid ^{3}He-A state, the symmetry in this state is spontaneously broken down to a product of two U(1) groups

$$G \to H_{\rm A_1} = {\rm U}^{S-\frac{1}{2}N}(1) \times {\rm U}^{L-\frac{1}{2}N}(1). \tag{11.48}$$

Now, pure rotations about the preferred axes in both spin and orbital space are not symmetry operations unless compensated by a phase transformation.

Even more exotic superfluid ^{3}He states show up in the core of topological defects, generated, for example, by setting the system in rotation (see Sec. 11.7 below). A systematic classification of all possible states, based on representation group theory, can be given, but is outside the scope of this textbook.

Now that the forms of the order parameter for the uniform ^{3}He-B, A, and A$_1$ states have been specified, the collective Nambu-Goldstone modes emerging from these ground states can be studied by subjecting the constant order parameters to a general symmetry transformation G. Being a symmetry transformation, the transformed order parameter also describes the ground state, with the Nambu-Goldstone fields parametrizing the ground-state manifold, see Sec. 1.5.

11.3 Fermionic Excitations

The spectrum of the elementary fermionic excitations in the superfluid ^{3}He states is determined by the one-fermion, or *Bogoliubov* Hamiltonian

$$H = \begin{pmatrix} \xi(\mathbf{k}) & \boldsymbol{\sigma} \cdot A \cdot \check{\mathbf{k}} \\ \boldsymbol{\sigma} \cdot A^* \cdot \check{\mathbf{k}} & -\xi(\mathbf{k}) \end{pmatrix} \tag{11.49}$$

which can be read off from the expression (11.6) for the partition function at the absolute zero of temperature by assuming a constant order parameter Δ^i and using Eq. (11.11). Here, $\check{\mathbf{k}}$ denotes the abbreviation $\check{\mathbf{k}} \equiv \mathbf{k}/k_{\rm F}$.

For the superfluid ^{3}He-B state, with $A^{\alpha i} = \bar{\Delta}_{\rm B}\delta^{\alpha i}$, where we included a complex parameter $\bar{\Delta}_{\rm B}$ of dimension energy, the general expression (11.49) reduces to

$$H_{\rm B} = \begin{pmatrix} \xi(\mathbf{k}) & \bar{\Delta}_{\rm B}\boldsymbol{\sigma} \cdot \check{\mathbf{k}} \\ \bar{\Delta}_{\rm B}^*\boldsymbol{\sigma} \cdot \check{\mathbf{k}} & -\xi(\mathbf{k}) \end{pmatrix}, \tag{11.50}$$

showing that an effective spin-orbit coupling emerges in this superfluid state. The spectrum $E_\mathrm{B}(\mathbf{k})$ of the elementary fermionic excitations then follows as

$$E_\mathrm{B}^2(\mathbf{k}) = \xi^2(\mathbf{k}) + |\bar{\Delta}_\mathrm{B}|^2\check{\mathbf{k}}^2, \tag{11.51}$$

which is isotropic in momentum space. The isotropy is closely related to the residual $\mathrm{SO}^J(3)$ symmetry. In the weak-coupling limit, where the chemical potential can be approximated by that of a noninteracting Fermi gas $\mu \approx \hbar^2 k_\mathrm{F}^2/2m$, and $|\bar{\Delta}_\mathrm{B}| << \mu$, the spectrum becomes similar for small energies to that of a BCS superconductor,

$$E_\mathrm{B}^2(\mathbf{k}) \approx \hbar^2 v_\mathrm{F}^2(|\mathbf{k}| - k_\mathrm{F})^2 + |\bar{\Delta}_\mathrm{B}|^2. \tag{11.52}$$

The minimum of energy $|\bar{\Delta}_\mathrm{B}|$ is reached for momenta on the Fermi surface of the normal liquid state, $|\mathbf{k}| = k_\mathrm{F}$.

In the opposite limit, where the ^{3}He atoms become tightly bound in pairs and form a Bose-Einstein condensate, the chemical potential is negative so that the minimum of energy is now reached for $|\mathbf{k}| = 0$, and

$$E_\mathrm{B}^2(\mathbf{k}) \approx \mu^2 + |\bar{\Delta}_\mathrm{B}|^2\check{\mathbf{k}}^2 \tag{11.53}$$

for small $\mathbf{k}$. This spectrum is of a relativistic form, describing a massive Dirac fermion of rest energy $-\mu > 0$ moving in vacuum with the speed of light c given by $|\bar{\Delta}_\mathrm{B}|/\hbar k_\mathrm{F}$. Indeed, in this limit, the Bogoliubov Hamiltonian (11.50) assumes the form of a Dirac Hamiltonian, cf. (3.119)

$$H_\mathrm{B} = c\boldsymbol{\alpha}\cdot\mathbf{p} - \mu\beta, \tag{11.54}$$

with $\boldsymbol{\alpha}$ and β the 4×4 matrices introduced in Eq. (3.120). Without loss of generality, the gap parameter is here assumed to be real.

The order parameter of the superfluid ^{3}He-A state, $A^{\alpha i} = \bar{\Delta}_\mathrm{A}\delta^{\alpha 3}(\delta^{i1} + \mathrm{i}\delta^{i2})$, can be written in the more general form

$$A^{\alpha i} = \bar{\Delta}_\mathrm{A} d^\alpha(e_1^i + \mathrm{i}\, e_2^i), \tag{11.55}$$

where we again included a gap parameter $\bar{\Delta}_\mathrm{A}$. Here, $\mathbf{v}_1, \mathbf{v}_2, \mathbf{d} \equiv \mathbf{v}_3 = \mathbf{v}_1 \times \mathbf{v}_2$ and $\mathbf{e}_1, \mathbf{e}_2, \mathbf{l} \equiv \mathbf{e}_1 \times \mathbf{e}_2$ are two sets of orthonormal vectors in spin and orbital space, respectively, with the so-called orbital vector $\mathbf{l}$ denoting the orbital-quantization axis, and $\mathbf{d}$ denoting the spin-quantization axis. Equation (11.55) reduces to the simple form when the two orthonormal sets are given the standard orientation. The two quantization axes need not point in the same direction. However, the relatively weak dipole-dipole interaction of the form $(\mathbf{d}\cdot\mathbf{l})^2$ tends to align the two parallel or antiparallel to each other. In the presence of an external magnetic field

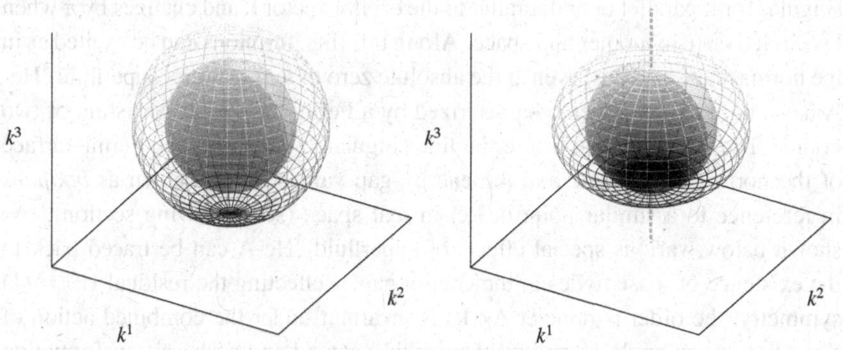

Fig. 11.2 The energy gap in the superfluid ^{3}He-B (left) and ^{3}He-A (right) phase

H, the **d** axis tends to orient itself perpendicular to **H**. With the choice (11.55) for $A^{\alpha i}$, the general expression (11.49) assumes the form

$$H_{\mathrm{A}} = \begin{pmatrix} \xi(\mathbf{k}) & \bar{\Delta}_{\mathrm{A}}\boldsymbol{\sigma}\cdot\mathbf{d}(\mathbf{e}_1\cdot\check{\mathbf{k}} + \mathrm{i}\,\mathbf{e}_2\cdot\check{\mathbf{k}}) \\ \bar{\Delta}_{\mathrm{A}}^*\boldsymbol{\sigma}\cdot\mathbf{d}(\mathbf{e}_1\cdot\check{\mathbf{k}} - \mathrm{i}\,\mathbf{e}_2\cdot\check{\mathbf{k}}) & -\xi(\mathbf{k}) \end{pmatrix}, \tag{11.56}$$

or with Eq. (3.120)

$$H_{\mathrm{A}} = \beta\xi(\mathbf{k}) + \Delta_{\mathrm{A}}\left(\boldsymbol{\alpha}\cdot\mathbf{d}\,\mathbf{e}_1\cdot\check{\mathbf{k}} + \mathrm{i}\beta\boldsymbol{\alpha}\cdot\mathbf{d}\,\mathbf{e}_2\cdot\check{\mathbf{k}}\right) \tag{11.57}$$

if the gap parameter $\bar{\Delta}_{\mathrm{A}}$ is taken to be real. This gives as spectrum for the fermionic excitations in the superfluid ^{3}He-A state

$$\begin{aligned} E_{\mathrm{A}}^2(\mathbf{k}) &= \xi^2(\mathbf{k}) + |\bar{\Delta}_{\mathrm{A}}|^2\left[(\mathbf{e}_1\cdot\check{\mathbf{k}})^2 + (\mathbf{e}_2\cdot\check{\mathbf{k}})^2\right] \\ &= \xi^2(\mathbf{k}) + |\bar{\Delta}_{\mathrm{A}}|^2(\mathbf{l}\times\check{\mathbf{k}})^2. \end{aligned} \tag{11.58}$$

It has the remarkable property that in the weak-coupling limit, where $\mu \approx \hbar^2 k_{\mathrm{F}}^2/2m$ and $|\bar{\Delta}_{\mathrm{A}}| << \mu$, it becomes gapless at the two points $\mathbf{k} = \pm k_{\mathrm{F}}\mathbf{l}$, see Fig. 11.2. All matrix elements of the Bogoliubov Hamiltonian (11.56) vanish in the weak-coupling limit for these values of the momentum. In fact, writing

$$\begin{aligned} \Delta_{\mathrm{A}}(\mathbf{k}) &\equiv \bar{\Delta}_{\mathrm{A}}\left(\mathbf{e}_1\cdot\check{\mathbf{k}} + \mathrm{i}\,\mathbf{e}_2\cdot\check{\mathbf{k}}\right) \\ &\approx |\bar{\Delta}_{\mathrm{A}}|\sqrt{1 - (\mathbf{l}\cdot\check{\mathbf{k}})^2}\,\mathrm{e}^{2\mathrm{i}\varphi_{\mathrm{A}}(\mathbf{k})}, \end{aligned} \tag{11.59}$$

where $\varphi_{\mathrm{A}}(\mathbf{k})$ is the **k**-dependent phase

$$\varphi_{\mathrm{A}}(\mathbf{k}) \equiv \varphi + \frac{1}{2}\arctan\left(\frac{\mathbf{e}_2\cdot\check{\mathbf{k}}}{\mathbf{e}_1\cdot\check{\mathbf{k}}}\right), \tag{11.60}$$

with φ (half) the usual phase of the order parameter, $\bar{\Delta}_{\mathrm{A}} = |\bar{\Delta}_{\mathrm{A}}|\mathrm{e}^{\mathrm{i}2\varphi}$, we see that ^{3}He-A exhibits a topological line defect in momentum space as the phase $\varphi_{\mathrm{A}}(\mathbf{k})$ is singular for $\mathbf{k}$ parallel or antiparallel to the orbital vector $\mathbf{l}$, and changes by π when $\mathbf{l}$ is circled once in momentum space. Along this line, fermions can be excited as in the normal state—that is, even at the absolute zero of temperature, superfluid ^{3}He-A has a normal component characterized by a Fermi "surface" consisting of two points. These two points, where the line singularity intersects the Fermi surface of the normal liquid state and the energy gap vanishes, are known as *boojums* in reference to a similar point defect in real space (see following section). As shown below, various special effects of superfluid ^{3}He-A can be traced back to the existence of these nodes in the energy gap. Reflecting the residual $\mathrm{U}^{L-\frac{1}{2}N}(1)$ symmetry, the order parameter $\Delta_\mathrm{A}(\mathbf{k})$ is invariant under the combined action of a rotation by an angle θ^L around the orbital vector $\mathbf{l}$ and a phase transformation $\Delta_\mathrm{A}(\mathbf{k}) \to \Delta'_\mathrm{A}(\mathbf{k}) = \Delta_\mathrm{A}(\mathbf{k})\mathrm{e}^{-\mathrm{i}\theta^L}$, i.e., $\varphi \to \varphi' = \varphi - \frac{1}{2}\theta^L$.

The elementary fermionic excitations in the ^{3}He-A_1 and other superfluid states can be studied in the same way as was done here for ^{3}He-B and ^{3}He-A.

11.4 Boojums

Leaving the superfluid ^{3}He-A state unchanged, the combined symmetry transformation, consisting of a rotation by an angle θ^L about the orbital-quantization axis $\mathbf{l}$ followed by a phase transformation with $\varphi = -\frac{1}{2}\theta^L$ undoing the rotation again, can even be carried out locally with the transformation parameter changing from one point in spacetime to another. Although this superfluid state is trivially invariant under these local transformations, physical observables must respect this dynamically generated local $\mathrm{U}^{L-\frac{1}{2}N}(1)$ invariance. This observation can be used to quickly derive expressions describing anisotropic superfluid ^{3}He-A, starting from their isotropic counterparts by imposing this local symmetry.

Consider, for example, the expression (8.65) for the superfluid momentum in a neutral ($e = 0$) BCS system, $\mathbf{p}_s = \hbar\nabla\varphi$. As it stands, this expression is not invariant under local $\mathrm{U}^{L-\frac{1}{2}N}(1)$ transformations and therefore needs to be extended to describe superfluid ^{3}He-A. Let the *Dreibein* $\mathbf{e}_1, \mathbf{e}_2, \mathbf{l} \equiv \mathbf{e}_1 \times \mathbf{e}_2$ vary in space. Imposing invariance under local $\mathrm{U}^{L-\frac{1}{2}N}(1)$ transformations, we arrive at the expression

$$p_\mathrm{s}^i = \hbar(\partial_i\varphi + \tfrac{1}{2}\mathbf{e}_1 \cdot \partial_i\mathbf{e}_2) \tag{11.61}$$

which contains an extra term, playing a role similar to the Villain potential $\mathbf{A}^\mathrm{V}$ introduced in Eq. (1.158) to describe a real-space vortex in an ideal classical fluid or superfluid ^{4}He. In fact, the extra term here originates from the line defect in momentum space as can be seen by writing $\mathbf{p}_s$ in terms of the phase (11.60) as

$$p_s^i = \hbar\partial_i \int \frac{d\Omega_{\mathbf{k}}}{4\pi} \varphi_A(\mathbf{k}), \tag{11.62}$$

with $\mathbf{k}$ restricted to the Fermi surface of the normal liquid state, $|\mathbf{k}| = k_F$. In deriving this, it is used that

$$\partial_i \varphi_A(\mathbf{k}) = \partial_i \varphi + \frac{1}{2} \frac{\mathbf{e}_1 \cdot \check{\mathbf{k}}\, \partial_i \mathbf{e}_2 \cdot \check{\mathbf{k}} - \mathbf{e}_2 \cdot \check{\mathbf{k}}\, \partial_i \mathbf{e}_1 \cdot \check{\mathbf{k}}}{(\mathbf{e}_1 \cdot \check{\mathbf{k}})^2 + (\mathbf{e}_2 \cdot \check{\mathbf{k}})^2}, \tag{11.63}$$

and

$$\int \frac{d\Omega_{\mathbf{k}}}{4\pi} \frac{k_1^2}{k_1^2 + k_2^2} = \frac{1}{2}. \tag{11.64}$$

Since $\mathbf{e}_1$ and $\mathbf{e}_2$ are orthogonal, $\mathbf{e}_1 \cdot \partial_i \mathbf{e}_2 = -\mathbf{e}_2 \cdot \partial_i \mathbf{e}_1$. Because of the extra term, the superflow is no longer a potential flow and the curl of the superfluid momentum not necessarily vanishes,

$$\begin{aligned}
(\nabla \times \mathbf{p}_s)^i &= \frac{\hbar}{2} \epsilon^{ijk} \int \frac{d\Omega_{\mathbf{k}}}{4\pi} \left[\partial_j, \partial_k\right] \varphi_A(\mathbf{k}) \\
&= \frac{\hbar}{8} \epsilon^{ijk} \sum_{q=\pm 1} \int d\Omega_{\mathbf{k}} \delta(\hat{\mathbf{k}} - q\mathbf{l})\, \mathbf{l} \cdot \left(\partial_j \mathbf{l} \times \partial_k \mathbf{l}\right) \\
&= \frac{\hbar}{4} \epsilon^{ijk} \mathbf{l} \cdot \left(\partial_j \mathbf{l} \times \partial_k \mathbf{l}\right),
\end{aligned} \tag{11.65}$$

with $\hat{\mathbf{k}}$ the unit vector $\hat{\mathbf{k}} \equiv \mathbf{k}/|\mathbf{k}|$, and where use is made of the vector identity (3.100), i.e., $(\mathbf{a} \times \mathbf{b}) \cdot (\mathbf{c} \times \mathbf{d}) = \mathbf{a} \cdot \mathbf{c}\, \mathbf{b} \cdot \mathbf{d} - \mathbf{a} \cdot \mathbf{d}\, \mathbf{b} \cdot \mathbf{c}$, and

$$\partial_i \mathbf{e}_1 \cdot \mathbf{l}\, \partial_j \mathbf{e}_2 \cdot \mathbf{l} = \partial_i \mathbf{e}_1 \cdot \partial_j \mathbf{e}_2. \tag{11.66}$$

Equation (11.65) constitutes the celebrated *Mermin-Ho relation*, which can be equivalently presented as

$$\frac{2m}{\hbar} \mathbf{l} \cdot (\nabla \times \mathbf{v}_s) = \frac{1}{2} \epsilon^{ijk} l^i\, \mathbf{l} \cdot \left(\partial_j \mathbf{l} \times \partial_k \mathbf{l}\right), \tag{11.67}$$

with $\mathbf{v}_s \equiv \mathbf{p}_s/m$ the superfluid velocity. The derivation given here, which is based on the identity

$$[\partial_u, \partial_v] \arctan\left(\frac{v}{u}\right) = 2\pi\delta(u)\delta(v), \tag{11.68}$$

with $\partial_u = \partial/\partial u$, $\partial_v = \partial/\partial v$, and [,] the commutator, shows that the Mermin-Ho relation has its origin in the presence of the line defect in momentum space. The identity (11.68) can be readily checked by integrating the left and right sides over a surface in the uv plane which includes the origin, and using Stoke's theorem to convert the surface integral on the left into a line integral around a closed path bounding the surface with arctan(v/u) the polar angle.

When integrated over a closed surface S, the right side of the Mermin-Ho relation (11.65) yields $2\pi\hbar$ times an integer w_2, telling how often the two-sphere S^2 parametrized by the orbital-quantization axis $\mathbf{l}$ is covered in the process. In other words, w_2 is the winding number of the mapping of a two-sphere $S^2_\mathbf{x}$ in real space, into which the closed surface can be smoothly deformed, onto the internal two-sphere S^2. It is recalled here that the second homotopy group $\pi_2(S^2) = Z$. Rather than proving this in general, we take as an example $\mathbf{l} = \hat{\mathbf{r}}$, with $\hat{\mathbf{r}}$ the unit vector in the radial direction, so that the winding number should we unity. And indeed, with θ and φ denoting the azimuthal and polar angle, respectively, and $\mathrm{d}\mathbf{S} = \mathrm{d}\varphi\mathrm{d}\theta \sin\theta\,\hat{\mathbf{r}}$,

$$\begin{aligned} w_2 \equiv \frac{1}{8\pi}\int_{S^2_\mathbf{x}} \mathrm{d}S^i \epsilon^{ijk}\mathbf{l}\cdot\left(\partial_j\mathbf{l}\times\partial_k\mathbf{l}\right) &= \frac{1}{8\pi}\int_{S^2_\mathbf{x}} \mathrm{d}S^i \epsilon^{ijk}\cos\theta\,[\partial_j,\partial_k]\varphi \\ &= \frac{1}{2}\int_{S^2_\mathbf{x}} \mathrm{d}S^3\cos\theta\,\delta(x^1)\delta(x^2) = 1 \end{aligned} \tag{11.69}$$

for $[\partial_j,\partial_k]\varphi = \epsilon^{jk3}2\pi\delta(x^1)\delta(x^2)$ by Eq. (11.68). Topologically stable configurations where the orbital-quantization axis $\mathbf{l}$ varies in space are called *textures*. Since the left side of the Mermin-Ho relation (11.65) is the curl of a vector, it must have singularities on the surface S to give the required result when integrated over that surface,

$$2\pi\hbar w_2 = \int_S \mathrm{d}\mathbf{S}\cdot(\nabla\times\mathbf{p}_\mathrm{s}) = m\oint \mathrm{d}\mathbf{x}\cdot\mathbf{v}_\mathrm{s}. \tag{11.70}$$

Here, $\mathrm{d}\mathbf{S}$ denotes a surface element of S, and the line integral encloses only the singular points of $\mathbf{v}_\mathrm{s}$ on the surface. Configurations with nonzero winding number w_2 therefore carry $2w_2$ units of circulation $\kappa_0 \equiv \pi\hbar/m \approx 0.067\,\mathrm{mm}^2/\mathrm{s}$,

$$\kappa = \oint \mathrm{d}\mathbf{x}\cdot\mathbf{v}_\mathrm{s} = 2w_2\kappa_0. \tag{11.71}$$

Note that the expression for the circulation quantum is half that for superfluid ^{4}He because the mass $2m$ of a pair of ^{3}He atoms enters here.

To explicitly construct a solution with nonzero winding number, consider a spherical vessel containing a drop of superfluid ^{3}He-A. The orbital-quantization axis $\mathbf{l}$ has the tendency to orient itself perpendicular to the boundary of the vessel. As *Ansatz* satisfying this boundary condition, we take for the orbital-quantization axis the *hedgehog* configuration $\mathbf{l} = \hat{\mathbf{r}}$, where we have chosen the center of the vessel as the origin of the coordinate system. A possible choice of the two other unit vectors spanning the *Dreibein* is given by $\mathbf{e}_1 = \mathbf{e}_\varphi$ and $\mathbf{e}_2 = \mathbf{e}_\theta$, with $\mathbf{e}_\varphi$ and $\mathbf{e}_\theta$

the remaining two unit vectors of the spherical coordinate system. The associated Villain potential $A^{V,i} = (\kappa_0/2\pi)\mathbf{e}_1 \cdot \partial_i \mathbf{e}_2$ is of the Dirac type, cf. Eq. (3.80),

$$\mathbf{A}^V = -\frac{\kappa_0}{2\pi}\frac{\cos\theta}{r\sin\theta}\mathbf{e}_\varphi, \tag{11.72}$$

with two Dirac strings at $\theta = 0$ and $\theta = \pi$, respectively, each carrying one unit of circulation κ_0,

$$\nabla \times \mathbf{A}^V = \frac{\kappa_0}{2\pi}\frac{\hat{\mathbf{r}}}{r^2} - \kappa_0 \cos\theta\, \delta(x^1)\delta(x^2)\mathbf{e}_3. \tag{11.73}$$

Here, $\mathbf{e}_3$ is the unit vector in the third direction, and

$$\cos\theta\delta(x^1)\delta(x^2) = \pm\delta(x^1)\delta(x^2) \tag{11.74}$$

for the string along the positive and negative third axis, respectively. Unlike for magnetic monopoles, these strings are physical, being genuine vortex lines. The *Ansatz* therefore describes a point singularity at the center of the vessel at which two vortex lines terminate, see Fig. 11.3. This configuration is metastable. Because of the finite line tension, the energy can be lowered by reducing the length of the two vortices at the cost of some additional bending energy. Under a perturbation, the endpoints of the vortex lines on the surface start to move towards each other, while at the same time the hedgehog moves towards the surface, until the vortex lines have completely shrunk to zero, leaving behind a point singularity on the surface. Mermin dubbed this point defect a "boojum" in reference to a mysterious object in Lewis Carroll's poem *The Hunting of the Snark* that can cause something to "softly and suddenly vanish away".

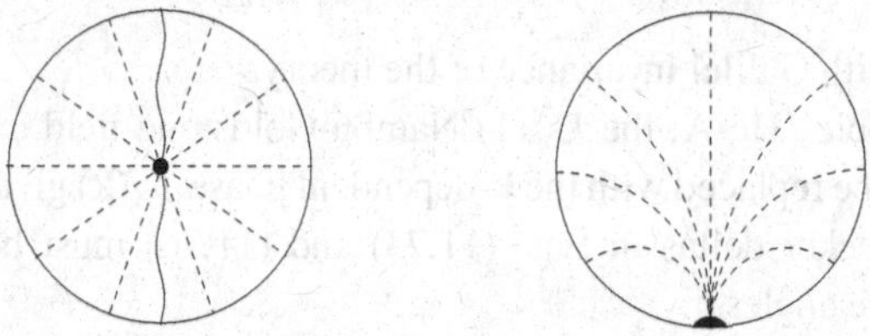

Fig. 11.3 A superfluid ^{3}He-A droplet in a spherical vessel. *Left*: Hedgehog with the orbital-quantization axis pointing everywhere radially outward, accompanied by two line singularities, connecting the hedgehog to the surface. *Right*: Boojum.

11.5 Callan-Harvey Effect

The nonvanishing commutator of derivatives at the boojums in momentum space, which is at the root of the Mermin-Ho relation (11.65), is also at the origin of an apparent violation of a conservation law. Consider first an isotropic fermionic superfluid at the absolute zero of temperature. The mass current density $\mathbf{g}_s$, or, equivalently, the momentum density carried by the superflow, $\mathbf{g}_s = \rho_s \mathbf{v}_s$, is componentwise proportional to the superfluid velocity, cf. Eq. (4.216). At $T = 0$, $\rho_s = mn$ so that the superfluid mass current density can be equivalently written as

$$g_s^i = \hbar n \delta^{ij} \partial_j \varphi, \tag{11.75}$$

with φ the Nambu-Goldstone field arising from the spontaneous breakdown of the global $U^N(1)$ symmetry. In Eq. (11.75), $n = k_F^3/3\pi^2$ is the total fermion number density because, as for bosonic superfluids [see Eq. (4.210)], all the particles participate in the superfluid motion. If the superflow is slow so that quadratic terms in the superfluid velocity $\mathbf{v}_s$ can be ignored, the momentum tensor T_s^{ij} describing the superflow follows from Eq. (1.98) as

$$T_s^{ij} = P\delta^{ij}, \tag{11.76}$$

where P is the pressure of the condensate,

$$P = -\hbar n \partial_t \varphi, \tag{11.77}$$

by Eq. (1.92) with $\rho = \rho_s = mn$ and $\mathbf{v} = \mathbf{v}_s = (\hbar/m)\nabla\varphi$. When combined, Eqs. (11.75) and (11.76) automatically lead to the conservation of the mass current of an isotropic superfluid

$$\partial_t g_s^i + \partial_j T_s^{ij} = \hbar n \delta^{ij} [\partial_t, \partial_j] \varphi = 0, \tag{11.78}$$

in accordance with Galilei invariance of the theory.

For anisotropic ^{3}He-A, the $U^N(1)$ Nambu-Goldstone field φ of an isotropic superfluid must be replaced with the $\mathbf{k}$-dependent phase $\varphi_A(\mathbf{k})$ given in Eq.(11.60), while the Kronecker deltas in Eqs. (11.75) and (11.76) must be replaced with averages over the angles

$$\delta^{ij} \to 3 \int \frac{d\Omega_\mathbf{k}}{4\pi} \hat{k}^i \hat{k}^j. \tag{11.79}$$

The mass current density (11.75) then assumes the form

$$\mathbf{g}_s = 3\hbar n \int \frac{d\Omega_\mathbf{k}}{4\pi} \hat{\mathbf{k}}\hat{\mathbf{k}} \cdot \nabla\varphi_A(\mathbf{k}), \tag{11.80}$$

leading to

$$\mathbf{g}_s = n \left[\mathbf{p}_s + \frac{\hbar}{4} \nabla \times \mathbf{l} - \frac{\hbar}{2} \mathbf{l}\mathbf{l} \cdot (\nabla \times \mathbf{l}) \right], \tag{11.81}$$

with $\mathbf{p}_s$ given by Eq. (11.61). This expression for the superfluid mass current density at the absolute zero of temperature can be readily checked by taking the scalar and vector products of the right sides of the last two equations with the orbital vector **l**, and carrying out the simple integral

$$I^{ijmn} \equiv \int \frac{\mathrm{d}\Omega_{\mathbf{k}}}{4\pi} \frac{k^i k^j k^m k^n}{k_1^2 + k_2^2} \tag{11.82}$$

for the relevant values of the indices $i, j, m, n = 1, 2, 3$, using spherical coordinates. This gives, for example,

$$I^{1122} = \frac{1}{2} I^{1133} = \frac{1}{2} I^{2233} = \frac{1}{3} I^{2222} = \frac{1}{12}. \tag{11.83}$$

Similarly, the expression for the momentum tensor (11.76) with Eq. (11.77) generalizes to

$$T_{\mathrm{s}}^{ij} = -3\hbar n \int \frac{\mathrm{d}\Omega_{\mathbf{k}}}{4\pi} \hat{k}^i \hat{k}^j \partial_t \varphi_{\mathrm{A}}(\mathbf{k}). \tag{11.84}$$

When combined, this gives

$$\begin{aligned} \partial_t g_{\mathrm{s}}^i + \partial_j T_{\mathrm{s}}^{ij} &= 3\hbar n \int \frac{\mathrm{d}\Omega_{\mathbf{k}}}{4\pi} \hat{k}^i \hat{k}^j [\partial_t, \partial_j] \varphi_{\mathrm{A}}(\mathbf{k}) \\ &= \frac{3}{4} \hbar n \sum_{q=\pm 1} \int \mathrm{d}\Omega_{\mathbf{k}} \hat{k}^i \hat{k}^j \delta(\hat{\mathbf{k}} - q\mathbf{l}) \left(\partial_t \mathbf{l} \times \partial_j \mathbf{l} \right) \cdot \mathbf{l} \\ &= -\frac{3}{2} \hbar n l^i \partial_t \mathbf{l} \cdot (\nabla \times \mathbf{l}), \end{aligned} \tag{11.85}$$

where use is made again of the key identity (11.68). Equation (11.85) reveals that the superfluid mass current, or momentum is not conserved, contrary to what is expected from Galilei invariance of the theory. The derivation presented here shows that the cause of this *linear momentum paradox* are sources located at the positions of the boojums in momentum space. At these locations, the liquid is normal. This part of the liquid has not yet been taken into account.

To study the effect of the boojums, consider a texture where the orbital vector **l** varies slowly in space and time $\mathbf{l} = \mathbf{l}(x)$, and define an "induction" field

$$\mathbf{B} \equiv \hbar k_{\mathrm{F}} \nabla \times \mathbf{l}(x), \tag{11.86}$$

giving a measure of the spatial variation of the orientation of the orbital vector throughout the texture. The **B** field is assumed to be constant in space and time. Its direction is taken to define the third direction, $\mathbf{B} = B\mathbf{e}_3$, $B > 0$. For convenience, we also assume that at $t = 0$, the orbital vector at the origin points in the positive 3-direction. By choosing the two other axes appropriately, we can write

$$B = \hbar k_{\mathrm{F}} \partial_1 l^2. \tag{11.87}$$

Because of the variation of the position of the boojums throughout the texture as specified by the **B** field, the momenta of the gapless fermionic excitations, or zero modes at these nodes, $\hbar\mathbf{k}_\pm \equiv \pm\hbar k_\mathrm{F}\mathbf{l}$ change as

$$\hbar\Delta k_\pm^2 = |qB|\Delta x^1, \tag{11.88}$$

where both $\Delta k_\pm^2$ and Δx^1 are defined to be positive, and $q = \pm e$ with

$$e \equiv \frac{1}{k_\mathrm{F}}\mathbf{k}_+ \cdot \mathbf{l} = 1 \tag{11.89}$$

the "electric charge". With this definition, the fermions at the north pole of the Fermi surface have charge $q = +e$, while those at the south pole have charge $q = -e$. As in the Landau problem considered in Sec. 3.8, let L be the extent of the system in the 2-direction. The wave vector in this direction is then quantized as

$$\Delta k^2 = 2\pi l/L \tag{11.90}$$

with l an integer, and a unit change $\Delta k^2 = 2\pi/L$ in momentum space, corresponding to a single state, induces the change in real space

$$\Delta x^1 = 2\pi\hbar/|qB|L \tag{11.91}$$

by Eq. (11.88). Defining the flux Φ through the area S as

$$\Phi \equiv \int_S \mathrm{d}\mathbf{S} \cdot \mathbf{B}, \tag{11.92}$$

we infer from Eq. (11.91) that just one unit of flux ($= 2\pi\hbar/e$) penetrates the (infinitesimal) rectangle of length L and width Δx^1 orientated perpendicular to the **B** field. In other words, Φ counts the number of zero modes that are produced by the texture at the boojums. More specifically, the number of zero modes per unit area perpendicular to **B** induced in this way equals $|qB|/2\pi\hbar$.

We next introduce a constant "electric" field

$$\mathbf{E} \equiv -\hbar k_\mathrm{F}\partial_t \mathbf{l}(x) \tag{11.93}$$

to specify the change of the orbital vector **l** in time. In the presence of such a field, the momenta of the fermions at the boojums change in time according to $\mathrm{d}p_q^3/\mathrm{d}t = q|E|$, with $E < 0$ the component of **E** along the vortex, or string in momentum space. Since the fermion density n in one space dimension is $n = k_\mathrm{F}/2\pi$, the electric field in fact pumps fermions (of charge $q = +e$) through the node in the energy gap at the north pole of the Fermi surface at a rate

$$\frac{\mathrm{d}n_+}{\mathrm{d}t} = \frac{q|E|}{2\pi\hbar} \tag{11.94}$$

per unit length of string—that is, in a time-dependent texture, fermions are created at the north pole. Simultaneously, fermions are annihilated at the south pole at the same rate. This implies that in the presence of a time-dependent $\mathbf{l}$ texture, fermions carrying momentum $-\hbar k_F \mathbf{l}$ disappear from the south pole of the Fermi surface and fermions carrying momentum $+\hbar k_F \mathbf{l}$ appear at the north pole. In the texture, therefore, momentum $\mathbf{P}_n = \int d^3x\, \mathbf{g}_n$ is created out of the Fermi sea at a rate

$$\begin{aligned} \frac{d\mathbf{P}_n}{dt} = \frac{d}{dt}\int d^3x\, \mathbf{g}_n &= V \sum_{q=\pm e} q\hbar k_F \mathbf{l} \frac{q|E|}{2\pi\hbar}\frac{|qB|}{2\pi\hbar} \\ &= -V\frac{1}{2\pi^2\hbar} k_F \mathbf{l}\, \mathbf{E}\cdot\mathbf{B}, \end{aligned} \tag{11.95}$$

where V is the volume of the system and in the last step it is used that $\mathbf{E}$ and $\mathbf{B}$ point in opposite directions, $E < 0$ while $B > 0$. The local form of this expression reads

$$\partial_t g_n^i + \partial_j T_n^{ij} = -\frac{1}{2\pi^2\hbar} k_F l^i \mathbf{E}\cdot\mathbf{B}, \tag{11.96}$$

with T_n^{ij} the momentum tensor of the normal component. Equation (11.96) reveals that the momentum carried by the normal component is not conserved, leading to a chiral-like anomaly with the role of the electromagnetic vector potential $\mathbf{A}$ played by the orbital vector $\hbar k_F \mathbf{l}$. In massless electrodynamics, the anomaly is proportional to E in one space dimension, see Eq. (9.61), and to $\mathbf{E}\cdot\mathbf{B}$ in three dimensions. The three-dimensional result can be derived with the help of the derivative expansion method in the same way as the one-dimensional result in Sec. 9.3.

Comparison of Eq. (11.96) with Eq. (11.85) shows that the right sides of the equations precisely cancel each other. This means that the total momentum, i.e., the momentum carried by the superfluid and the normal component combined is conserved, in accordance with Galilei invariance of the full theory. The two components of the liquid by themselves violate momentum conservation. But through the exchange of particles between the string and the condensate, the anomalies precisely cancel: momentum is flowing into the boojums (normal liquid) from the outside world, i.e., the condensate. This momentum appears in the dimensionally reduced fermion world of the string as a source term, giving rise to the anomaly.

This solution to the linear momentum paradox in superfluid ^{3}He provides a physical example of the mechanism first proposed in the context of particle physics by Callan and Harvey who considered a string in real space rather than in momentum space though.

11.6 Wess-Zumino Term

If the spin degree of freedom is ignored, the Bogoliubov Hamiltonian (11.56) describing superfluid ^{3}He-A takes the form of the generic two-level Hamiltonian (3.71),

$$H = \begin{pmatrix} \xi(\mathbf{k}) & \bar{\Delta}_{\rm A}(\mathbf{e}_1 \cdot \check{\mathbf{k}} + \mathrm{i}\, \mathbf{e}_2 \cdot \check{\mathbf{k}}) \\ \bar{\Delta}_{\rm A}(\mathbf{e}_1 \cdot \check{\mathbf{k}} - \mathrm{i}\, \mathbf{e}_2 \cdot \check{\mathbf{k}}) & -\xi(\mathbf{k}) \end{pmatrix}, \tag{11.97}$$

where the gap parameter $\bar{\Delta}_{\rm A}$ is assumed to be real. Defining

$$\boldsymbol{\lambda} \equiv \lambda\left(\bar{\Delta}_{\rm A}\mathbf{e}_1 \cdot \check{\mathbf{k}}, -\bar{\Delta}_{\rm A}\mathbf{e}_2 \cdot \check{\mathbf{k}}, \xi\right), \tag{11.98}$$

with $\lambda \equiv 2E(\mathbf{k})$ and $E(\mathbf{k})$ the single-fermion energy spectrum (11.58), we can cast the Hamiltonian in the equivalent form (3.78) with

$$\varphi = -\arctan\left(\frac{\mathbf{e}_2 \cdot \check{\mathbf{k}}}{\mathbf{e}_1 \cdot \check{\mathbf{k}}}\right), \tag{11.99}$$

cf. $\varphi_{\rm A}(\mathbf{k})$ introduced in Eq. (11.60), and

$$\cos\theta = \xi/E. \tag{11.100}$$

The Wess-Zumino term (3.97) for the problem at hand becomes

$$S_{\rm WZ} = \frac{\hbar}{2}\int \mathrm{d}^3 x \sum_\sigma \int \frac{\mathrm{d}^3 k}{(2\pi)^3}\, \gamma_{\mathbf{k}} \tag{11.101}$$

with the Berry phase $\gamma_{\mathbf{k}} = \gamma_- = -\gamma_+$, with γ_+ introduced in Eq. (3.101),

$$\gamma_{\mathbf{k}} = \frac{1}{2}\int_0^1 \mathrm{d}u \int \mathrm{d}t\, [\partial_t \cos\theta\, \partial_u \varphi - (t \leftrightarrow u)]. \tag{11.102}$$

We concern ourselves only with negative energies, whence only γ_- is taken into account, and we reintroduced the spin degree of freedom by including the sum $\sum_\sigma$ in Eq. (11.101) which simply produces a factor of 2. With

$$\partial_t \varphi = \frac{\mathbf{e}_2 \cdot \check{\mathbf{k}}\, \partial_t \mathbf{e}_1 \cdot \check{\mathbf{k}} - \mathbf{e}_1 \cdot \check{\mathbf{k}}\, \partial_t \mathbf{e}_2 \cdot \check{\mathbf{k}}}{(\mathbf{e}_1 \cdot \check{\mathbf{k}})^2 + (\mathbf{e}_2 \cdot \check{\mathbf{k}})^2}, \tag{11.103}$$

$\partial_t \cos\theta = \partial_t(\xi/E) = -(\xi/E^2)\partial_t E$, and

$$\partial_t E = \frac{\bar{\Delta}_{\rm A}^2}{E}\left(\mathbf{e}_1 \cdot \check{\mathbf{k}}\, \partial_t \mathbf{e}_1 \cdot \check{\mathbf{k}} + \mathbf{e}_2 \cdot \check{\mathbf{k}}\, \partial_t \mathbf{e}_2 \cdot \check{\mathbf{k}}\right), \tag{11.104}$$

and similar equations for ∂_t replaced with ∂_u, the Berry phase assumes the form

$$\gamma_{\mathbf{k}} = -\frac{\bar{\Delta}_{\rm A}^2}{2}\frac{\xi}{E^3}\int_0^1 \mathrm{d}u \int \mathrm{d}t \left[\partial_u \mathbf{e}_1 \cdot \check{\mathbf{k}}\, \partial_t \mathbf{e}_2 \cdot \check{\mathbf{k}} - (t \leftrightarrow u)\right]. \tag{11.105}$$

This gives for the Wess-Zumino term (11.101)

$$\Gamma_{\rm WZ} = -\int_0^1 \mathrm{d}u \int \mathrm{d}^4x\, L_0[\partial_u \mathbf{e}_1 \cdot \mathbf{l}\, \partial_t \mathbf{e}_2 \cdot \mathbf{l} - (t \leftrightarrow u)], \tag{11.106}$$

or

$$\Gamma_{\rm WZ} = -\int_0^1 \mathrm{d}u \int \mathrm{d}^4x\, L_0 \mathbf{l} \cdot (\partial_u \mathbf{l} \times \partial_t \mathbf{l}) \tag{11.107}$$

where use is made again of the vector identity (3.100) to write

$$\begin{aligned}[\partial_u \mathbf{e}_1 \cdot \mathbf{l}\, \partial_t \mathbf{e}_2 \cdot \mathbf{l} - (t \leftrightarrow u)] &= [\mathbf{e}_1 \cdot \partial_u \mathbf{l}\, \mathbf{e}_2 \cdot \partial_t \mathbf{l} - (t \leftrightarrow u)] \\ &= \mathbf{l} \cdot (\partial_u \mathbf{l} \times \partial_t \mathbf{l}).\end{aligned} \tag{11.108}$$

Here,

$$\begin{aligned} L_0 &\equiv \frac{\hbar}{2} \bar{\Delta}_{\rm A}^2 \int \frac{\mathrm{d}^3k}{(2\pi)^3} (\mathbf{l} \cdot \hat{\mathbf{k}})^2 \frac{\xi}{E^3} \\ &= \frac{\hbar}{2} \bar{\Delta}_{\rm A}^2 \int \frac{\mathrm{d}^3k}{(2\pi)^3} \cos^2\vartheta \frac{\xi}{E^3}, \end{aligned} \tag{11.109}$$

where the orbital vector $\mathbf{l}$ is taken to define the third axis in momentum space and $\hat{\mathbf{k}} \cdot \mathbf{l} = \cos\vartheta$. With $\mathbf{k} \approx k_{\rm F}\hat{\mathbf{k}}$, the energy spectrum (11.58) reads when expressed in terms of this angle $E^2 = \xi^2 + \bar{\Delta}_{\rm A}^2(1 - \cos^2\vartheta)$, and

$$\frac{\partial E}{\partial \cos\vartheta} = -\bar{\Delta}_{\rm A}^2 \frac{\cos\vartheta}{E} \tag{11.110}$$

so that L_0 can be rewritten as

$$\begin{aligned} L_0 &= \frac{\hbar}{2} \int \frac{\mathrm{d}^3k}{(2\pi)^3} \cos^2\vartheta \frac{\xi}{E} \frac{\partial}{\partial E}\left(-\frac{1}{E}\right) \\ &= \frac{\hbar}{2} \int \frac{\mathrm{d}^3k}{(2\pi)^3} \xi \cos\vartheta \frac{\partial}{\partial \cos\vartheta} \frac{1}{E} \\ &= \frac{\hbar}{2} \int \frac{\mathrm{d}^3k}{(2\pi)^3} \left[\mathrm{sgn}(\xi) - \frac{\xi}{E}\right]. \end{aligned} \tag{11.111}$$

The last line is obtained by integrating by parts and using that the integral measure includes a $\int_{-1}^1 \mathrm{d}\cos\vartheta$.

The physical implication of the Wess-Zumino term follows from considering the Euler-Lagrange equation for the orbital vector,

$$L_0 \frac{\partial}{\partial t} \mathbf{l} = \mathbf{l} \times \frac{\delta}{\delta \mathbf{l}} \Gamma'[\mathbf{l}], \tag{11.112}$$

where $\Gamma'[\mathbf{l}]$ is the effective action governing the orbital vector with the Wess-Zumino term omitted, and where it is assumed that L_0 is constant in time. The term on the left originates from the Wess-Zumino term. In deriving that term it is used that $\delta\mathbf{l}\cdot(\partial_u\mathbf{l}\times\partial_t\mathbf{l}) = 0$ so that as in the derivation of the Landau-Lifshitz equation (7.59)

$$\begin{aligned}\delta\Gamma_{\mathrm{WZ}} &= -\int_0^1 \mathrm{d}u \int \mathrm{d}^4x\, L_0[\mathbf{l}\cdot(\partial_u\delta\mathbf{l}\times\partial_t\mathbf{l}) + \mathbf{l}\cdot(\partial_u\mathbf{l}\times\partial_t\delta\mathbf{l})] \\ &= -\int_0^1 \mathrm{d}u \int \mathrm{d}^4x\, L_0\{\partial_u[\mathbf{l}\cdot(\delta\mathbf{l}\times\partial_t\mathbf{l})] + \partial_t[\mathbf{l}\cdot(\partial_u\mathbf{l}\times\delta\mathbf{l})]\} \\ &= \int \mathrm{d}^4x\, L_0\,[\delta\mathbf{l}\cdot(\mathbf{l}\times\partial_t\mathbf{l})], \end{aligned} \tag{11.113}$$

if a total time derivative ignored. In addition, the vector identities $\mathbf{a}\cdot(\mathbf{b}\times\mathbf{c}) = -\mathbf{b}\cdot(\mathbf{a}\times\mathbf{c})$ and $\mathbf{l}\times(\mathbf{l}\times\partial_t\mathbf{l}) = -\partial_t\mathbf{l}$ are used. The Euler-Lagrange equation (11.112) assumes the form of the conservation law for angular momentum provided $\mathbf{L} \equiv L_0\mathbf{l}$ is identified as the average intrinsic angular momentum density.

In the composite boson limit, where all the fermions form tightly bound pairs, the intrinsic angular momentum at the absolute zero of temperature is given by $\hbar N/2$, with N the total fermion number as each pair carries an intrinsic angular momentum $\hbar$ oriented along the orbital vector $\mathbf{l}$. In the weak-coupling limit, this is no longer the case. In terms of the total particle number density n given in Eq. (8.32), L_0 can be written as

$$L_0 = \frac{\hbar}{2}(n - C_0) \tag{11.114}$$

with C_0 the constant

$$C_0 \equiv \sum_\sigma \int \frac{\mathrm{d}^3k}{(2\pi)^3}\,\theta(-\xi), \tag{11.115}$$

which takes the value $C_0 = k_\mathrm{F}^3/3\pi^2$ in the weak-coupling limit, where $\mu = \hbar^2k_\mathrm{F}^2/2m$, and vanishes ($C_0 = 0$) in the composite boson limit, where $\mu < 0$. A nonzero value $C_0 \neq 0$ is thus understood as resulting from the gapless fermionic excitations at the boojums. Previously, the finding that the average intrinsic angular momentum in superfluid ^{3}He-A came out smaller than $\hbar N/2$ was known as the *angular momentum paradox*.

To arrive at an estimate for L_0 in superfluid ^{3}He-A, we return to the original expression (11.109) and approximate it by

$$L_0 \approx \frac{\hbar}{2}\bar{\Delta}_\mathrm{A}^2\nu(0)\int_{-1}^1 \mathrm{d}\cos\vartheta\,\cos^2\vartheta \int_{-\mu}^\infty \mathrm{d}\xi\,\frac{\partial}{\partial\xi}\left(-\frac{1}{E}\right) \approx \frac{\hbar}{8}n\frac{\bar{\Delta}_\mathrm{A}^2}{\mu^2} \tag{11.116}$$

with $\nu(0) = mk_\mathrm{F}/2\pi^2\hbar^2$ the density of states at the Fermi surface of the normal liquid state per spin degree. This approximation is justified in the weak-coupling limit, where $\mu \approx \hbar^2k_\mathrm{F}^2/2m$, as the main contribution to the integral then comes from $\xi \approx 0$. Since $|\bar{\Delta}_\mathrm{A}/|\mu \approx 10^{-3}$ for superfluid ^{3}He-A, the intrinsic angular momentum is six orders of magnitude smaller than in the composite boson limit.

11.7 Topological Defects

Superfluid ^{3}He harbors a host of topologically stable defects that in richness and properties is unmatched to date by other condensed matter systems, although spinor Bose-Einstein condensates are quickly catching up. This section covers only the experimentally most relevant ones, starting with vortices that have been observed in the isotropic superfluid ^{3}He-B state. Such vortices are generated by setting the entire cryostat in rotation as routinely done by, for example, the ROTA group at the Low Temperature Laboratory of the Helsinki University of Technology The core of ^{3}He-B vortices is of order 1 μm and can be considered a macroscopic system by itself which may undergo a phase transition. As shown in Sec. 1.8, in a conventional, isotropic superfluid like superfluid ^{4}He where only the $U^N(1)$ symmetry of global phase transformations is spontaneously broken, the vortex core is in the normal state. This can be understood as follows. The superfluid momentum $\mathbf{p}_s$ for such a superfluid is given by

$$\mathbf{p}_s = \hbar\nabla\varphi, \tag{11.117}$$

with φ the $U^N(1)$ Nambu-Goldstone field, so that the superflow is a potential flow. For a vortex with nonzero winding number w_1, i.e.,

$$\frac{1}{2\pi}\oint \mathrm{d}\mathbf{x}\cdot\nabla\varphi \equiv w_1 \neq 0, \tag{11.118}$$

where the integral is taken over a closed path around the vortex, φ changes by a fixed amount. Consequently, when this contour shrinks down to zero, eventually a region is reached in which $\nabla\varphi$, and thus the kinetic energy would diverge. To avoid this real-space singularity in the vortex core, the $U^N(1)$ symmetry is restored, and the liquid reverts to its normal state there. Superfluid ^{3}He-B possesses in contrast vortices which do no destruct superfluidity in their cores. This is made possible by the nontrivial internal structure of the superfluid ^{3}He phases. The simple argument just given for a conventional, isotropic superfluid namely breaks down if the superflow is not a potential flow. Moreover, some superfluid phases have quantum numbers which are compatible with those of certain vortices in ^{3}He-B. Vortices with a superfluid core escape the vortex singularity in real space by transforming it into momentum space. The energy gap vanishes then only for certain rather than for all momenta on the Fermi surface (of the liquid state) as is the case for real-space singularities.

The key point in understanding which superfluid states may figure as inner vortex core states is their quantum numbers. In superfluid ^{4}He, a centered axisymmetric vortex line of winding number w_1 imparts exactly w_1 units of angular momentum to each atom. Similarly, in superfluid ^{3}He-B at the absolute zero of temperature, such a vortex imparts exactly w_1 units of angular momentum to each pair. The topological charge in ^{3}He-B therefore acts as the angular momentum quantum number of the pairs, and the inner vortex core state must have $m_J = w_1$. From this observation, it immediately follows that an axisymmetric vortex in ^{3}He-B with winding number $w_1 \geq 3$ would always have a normal core, since no superfluid ^{3}He state with $m_J \geq 3$ exists. Furthermore, by recalling that the state with $m_J = 2$ is the A_1 phase, we conclude that this state may figure as inner vortex core state of a doubly quantized ($w_1 = 2$) axisymmetric vortex. In experiment, using NMR, only singly quantized vortices ($w_1 = 1$) are observed in superfluid ^{3}He-B in rotation. The superfluid inner core of an axisymmetric vortex of this type consists of the state with $m_J = 1$. This state, which is called the ϵ phase, is characterized in its simplest form by the order parameter

$$a_\epsilon^{m_S m_L} = \begin{pmatrix} 0 & s & 0 \\ u & 0 & 0 \\ 0 & 0 & 0 \end{pmatrix}, \qquad A_\epsilon^{\alpha i} = \begin{pmatrix} 0 & 0 & u \\ 0 & 0 & iu \\ s & is & 0 \end{pmatrix}, \tag{11.119}$$

where s and u are two complex constants. In this core state, which has the same momentum-space topology as the A phase, with a line singularity piercing the Fermi surface of the normal liquid state, the original symmetry G is spontaneously broken down to a single U(1) group,

$$G \to H_\epsilon = \mathrm{U}^{J-N}(1). \tag{11.120}$$

In equilibrium, these vortices repel each other and form a hexagonal lattice.

By lowering the temperature, the vortex core undergoes a first-order transition to a new core state. From being axisymmetric, the vortex develops a remarkable double-core structure for which only a rotation by the angle π about the vortex axis is still a symmetry operation. Together with the unit element, this forms the discrete subgroup C_2 of the axial U(1) symmetry group. The corresponding core state is the so-called *axiplanar state* in which the residual $\mathrm{U}^{J-N}(1)$ symmetry of the ϵ state is further broken down to the subgroup C_2^{J-N}. The order parameter of the axiplanar state has four nonzero matrix elements, corresponding to $m_J = 1$, as for the ϵ state, and $m_J = -1$. Also the equilibrium vortex lattice changes from being hexagonal for the axisymmetric vortices to centered rectangular for the double-core vortices.

Various types of exotic vortices, including a vortex sheet, have been found in superfluid ^{3}He-A in rotation. For simplicity, we consider the system in a strong enough external magnetic field so that the vector $\mathbf{d}$ is locked in the plane perpendicular to the applied field. The Mermin-Ho relation (11.65) was shown in Eq. (11.71) to relate the texture of $\mathbf{l}$ to the circulation of the superfluid. A texture covering the unit sphere once carries two circulation quanta. If $\mathbf{l}$ is uniform,

$\nabla \times \mathbf{v}_s$ vanishes in the superfluid state and only vortices with a singular core can be created by rotation. Such singular vortices have been experimentally observed at relatively slow rotation velocity of the cryostat. These vortices are singly quantized and have a hard inner and a soft outer core. Within the hard core, which has a radius of order the coherence length $\xi \approx 10 - 100$ nm, the order parameter deviates sharply from its value in bulk ^{3}He-A, and the superfluid momentum is undefined. In the much larger soft core, the orbital vector changes smoothly. Outside this core, $\mathbf{l}$ is practically uniform and locked to $\mathbf{d}$. The size of the soft core is of order the dipolar healing length $\xi_D \approx 10\ \mu$m.

The other types of vortices that have been observed in ^{3}He-A have a smooth core and, in accordance with the Mermin-Ho relation, carry two or even four quanta of circulation when the vector $\mathbf{d}$ is dipole locked to $\mathbf{l}$ and both vary in tandem.

The most common vortex line is doubly quantized and has a core in which the order parameter changes smoothly while the gap parameter $\bar{\Delta}$ remains constant. Neither $\mathbf{v}_s$ nor $\mathbf{l}$ have singularities, see Fig. 11.4 for a cross section of the vortex line. It is of a type first proposed by Anderson and Toulouse, and by Chechetkin (ATC). Following the motion of the tip of the cone, which points in the direction of $\mathbf{l}$, throughout the plane, we seen that it traces out the surface of a two-sphere. At the same time, going around the perimeter of the figure, the cone rotates about its axis twice, showing that this texture carries two circulation quanta. More specifically, the texture consists of two parts: a circular part where $\mathbf{l}$ sweeps the northern hemisphere and a hyperbolic part where it sweeps the southern hemisphere. By the Mermin-Ho relation, each of these parts carries one quantum of circulation. This doubly quantized vortex can therefore be thought of consisting of two so-called Mermin-Ho vortices, which in particle physics are known as *merons* (Greek for "fraction"). In the context of ferro- and antiferromagnets, which share the same spontaneous symmetry breakdown of a SO(3) group down to rotations about a preferred axis as ^{3}He-A, such textures are known as *skyrmions*, which were first introduced by Skyrme in the context of particle physics. Note that as $\mathbf{l}$ sweeps the entire surface of the unit sphere, each of the Fermi points, which are located at the intersections of this direction with the Fermi surface of the liquid state, covers this sphere in momentum space exactly once. These smooth vortices are generated by slowly setting superfluid ^{3}He-A in rotation as they nucleate much easier than vortices with a singular core. They are now also produced in spinor Bose-Einstein condensates. For an axisymmetric ATC vortex, $\mathbf{l}$ can be parametrized as

$$\mathbf{l} = \sin[\eta(\rho)]\mathbf{e}_\rho + \cos[\eta(\rho)]\mathbf{e}_z \qquad (11.121)$$

with (ρ, θ, z) cylindrical coordinates, and where $\eta(\rho)$ changes from $\eta(0) = 0$ at the center of the texture to $\eta(\infty) = \pi$ at spatial infinity. The resulting superfluid

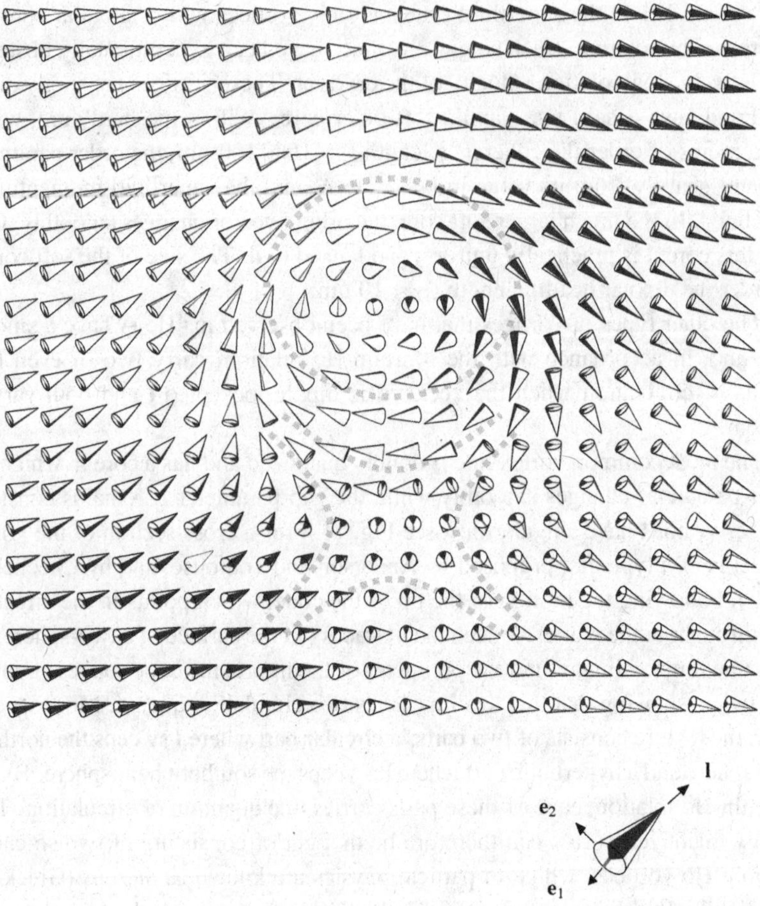

Fig. 11.4 Cross section of a doubly quantized smooth vortex line as observed in superfluid ^{3}He-A in rotation. The plane is perpendicular to the axis of rotation. The cone points in the direction of the orbital vector **l**, while the marks on the cone indicate how it revolves around its axis. [Reprinted from V. B. Eltsov and M. Krusius in: *Topological Defects and the Non-Equilibrium Dynamics of Symmetry Breaking Phase Transitions*, Proceedings of the NATO Advanced Study Institute, February 1999, Les Houches, France, NATO Science Series C **549**, edited by Y. M. Bunkov and H. Godfrin (Kluwer Academic Publishers, Dordrecht, 2000), Fig. 7 on p. 334 with kind permission of Springer Science and Business Media.]

velocity,

$$\mathbf{v}_s = \frac{\hbar}{2m\rho}\{1 - \cos[\eta(\rho)]\}\,\mathbf{e}_\theta, \quad \nabla \times \mathbf{v}_s = \frac{\hbar}{2m\rho}\sin[\eta(\rho)]\frac{\mathrm{d}}{\mathrm{d}\rho}\eta(\rho)\mathbf{e}_z, \qquad (11.122)$$

is regular everywhere, yet the texture carries two quanta of circulation $\oint \mathrm{d}\mathbf{x}\cdot\mathbf{v}_s = 2\kappa_0$, where the line integral is around a loop at spatial infinity.

Notes

(i) For introductions to the theory of superfluid ^{3}He, see [Anderson and Brinkman (1975); Leggett (1978); Mineev (1983)], and [Volovik (1984)]. The first reference, which is reprinted in [Anderson (1984)], also discusses the history of the subject. For a comprehensive coverage with a host of references to the original literature, see [Vollhardt and Wölfle (1990)]. The subject is covered in the the textbooks by Popov [Popov (1983, 1987)]. Volovik (1991, 2001) provides a profound original discussion of the subject and establishes connections between the exotic properties observed in superfluid ^{3}He and phenomena in particle physics and cosmology. This chapter is in part based on [Schakel (1989)].

(ii) The experimentally observed ^{3}He-A, or axial phase was postulated by Anderson and Morel (1961). In [Anderson and Brinkman (1973)], the stability of the Anderson-Morel (AM) phase was argued, which since then has come to be known also as the Anderson-Brinkman-Morel (ABM) phase. The experimentally observed ^{3}He-B, or isotropic phase was postulated by Balian and Werthamer (1963) and is also known as the Balian-Werthamer (BW) phase. The identification of the phases is due to Leggett, see [Leggett (1975)] for a review.

(iii) The Mermin-Ho relation was derived in [Mermin and Ho (1976)]. For an entertaining account of how the word "boojum" found its way into the physics literature, see [Mermin (1981)]. For one of the many applications of the Mermin-Ho relation, see [Blaha (1976)]. An overview of its applications is given in [Mermin (1978)].

(iv) For an early discussion of the linear and the angular momentum paradoxes with references to the original literature, see [Volovik and Mineev (1981)].

(v) The observation that the orbital vector $\mathbf{A} = \hbar k_F \mathbf{l}$ couples as an electromagnetic vector potential to the gapless fermionic excitations in ^{3}He-A is due to Combescot and Dombre (1986).

(vi) The chiral anomaly in superfluid ^{3}He-A was established in [Balatsky *et al.* (1986)] and [Volovik (1986a)]. Our derivation is based on the Landau-levels technique presented in Sec. 3.8 and also borrows from [Witten (1985)].

(vii) The observation that the anomalies in superfluid ^{3}He-A cancel through the Callan-Harvey effect [Callan and Harvey (1985)] is due to Stone and Gaitan (1987).

(viii) Section 11.6 is based on [Volovik (1986b); Balatsky (1987)], and [Goff *et al.* (1989)].

(ix) For lucid accounts of the topological defects realized in superfluid ^{3}He in rotation, see [Hakonen and Lounasmaa (1987); Lounasmaa and Thuneberg (1999)], and [Eltsov and Krusius (2000)]. In these papers, also references are given to the original literature where these defects were proposed.

(x) The continuous ATC vortices observed in superfluid ^{3}He-A are known as skyrmions in particle physics, first introduced in that context by Skyrme (1961).

Chapter 12

Quantum Hall Effect

This final chapter covers the quantum Hall effect which arises in two-dimensional (2D) electron gases at low temperature placed in a strong external magnetic field. Electrons residing at, for example, the interface of two slightly different semiconductors form an effectively two-dimensional system at sufficiently low temperatures where their motion normal to the plane is suppressed. Transport of electrons is then confined to the interface. The magnetic field is applied perpendicular to the plane. When in addition an electric field **E** is applied, say, in the 1-direction, the transverse or *Hall resistance* R_H in the direction perpendicular to both fields is observed to display plateaus as a function of the magnetic field. At these plateaus, the Hall resistance is quantized to a few parts per billion (*sic*!) to

$$R_\mathrm{H} = \frac{1}{\nu}\frac{2\pi\hbar c}{e^2}, \tag{12.1}$$

where ν is either an integer or a simple fraction, see Fig. 12.1. The case where ν is an integer is referred to as the *integral quantum Hall effect* (IQHE), while the case where ν is a simple fraction is referred to as the *fractional quantum Hall effect* (FQHE). Typical experimental conditions for the IQHE is a temperature of order $T \sim 1 - 4$ K and an applied magnetic field of order $H \sim 3 - 16$ T, while the FQHE takes place at considerable lower temperatures $T \sim 20 - 100$ mK and in stronger fields $H \sim 15 - 30$ T. Note that in two space dimensions, resistance R and resistivity ρ have the same dimension. With L^i denoting the extent of the probe in the i-direction ($i = 1, 2$), the two quantities are related through

$$R = \rho L^1/L^2, \quad R_\mathrm{H} = \rho_\mathrm{H} \tag{12.2}$$

for the longitudinal and transverse directions, respectively. In Fig. 12.1, also the longitudinal, or magnetoresistivity ($\rho = \rho_{xx}$) is shown. It is observed to vanish in the regions where the Hall resistance is quantized. These two experimental observations are the defining characteristics of the quantum Hall effect.

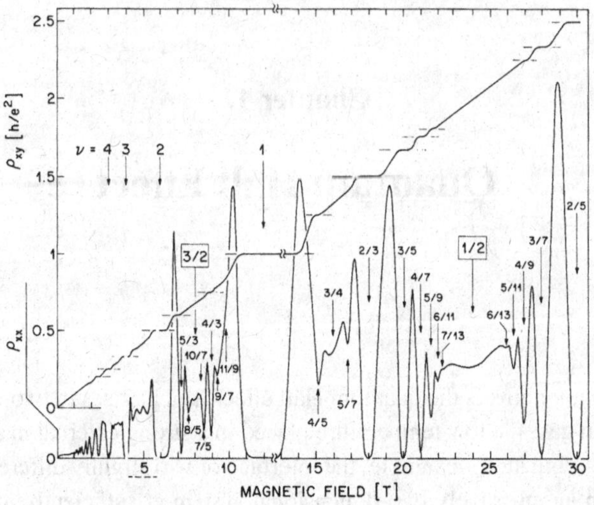

Fig. 12.1 The transverse Hall resistance $\rho_{xy} = R_{\mathrm{H}}$, and the longitudinal magnetoresistivity $\rho_{xx} = \rho$ in the direction of the applied electric field as function of the magnetic field H. Reprinted figure with permission from R. Willett, J. P. Eisenstein, H. L. Stormer, D. C. Tsui, A. C. Gossard, and J. H. English, Phys. Rev. Lett. **59**, 1776 (1987). Copyright 1987 by the American Physical Society.

Below it will be shown that the IQHE can be explained with the help of a more or less free two-dimensional electron gas. The FQHE, on the other hand, is the hallmark of a new condensed-matter state, in which the Coulomb repulsion plays a decisive role. The chapter closes with a discussion of graphene whose fermionic excitations have a gapless linear spectrum. This system features a half-integer quantum Hall effect.

12.1 2D Electron Gas in a Strong Magnetic Field

We start by considering a free electron gas confined to a plane and placed in a strong magnetic field. We assume that the induction field $\mathbf{B} = \nabla \times \mathbf{A}(\mathbf{x})$ acting on the individual electrons coincides with the applied magnetic field $\mathbf{H}$. We assume further that the external field is uniform and applied perpendicular to the plane, and consider for the moment a grand canonical ensemble specified by the chemical potential μ. Below, we will consider the experimentally more relevant case of a canonical ensemble with fixed number of electrons. The zero-temperature partition function Z of the system is given by

$$Z = \int \mathrm{D}\psi^* \mathrm{D}\psi \exp\left(\frac{i}{\hbar} \int \mathrm{d}^3 x \mathcal{L}\right), \tag{12.3}$$

where $\mathcal{L}$ is the Lagrangian density

$$\mathcal{L} = \psi^*(i\hbar\partial_t + \mu - H)\psi, \tag{12.4}$$

with ψ a Grassmann field describing the electrons. It is assumed that the Zeeman splitting between the spin-$\uparrow$ and spin-$\downarrow$ levels is such that the energy levels of the former are too high to be occupied. In other words, the system is assumed to be spin polarized so that the electron spin is irrelevant to the problem. The one-particle Hamiltonian H appearing in Eq. (12.4) reads

$$H = \frac{1}{2m}\left(-i\hbar\nabla + \frac{e}{c}\mathbf{A}\right)^2 \tag{12.5}$$

with $q = -e(< 0)$ and m the charge and mass of the electrons. It describes an electron moving in a plane pierced by an external magnetic field. The energy eigenvalues E_n of this Landau problem,

$$E_n = \hbar\omega_c\left(n + \tfrac{1}{2}\right) \tag{12.6}$$

with $n = 0, 1, 2 \cdots$ labeling the levels and

$$\omega_c \equiv \frac{eB}{mc}, \tag{12.7}$$

the cyclotron frequency ($eB > 0$), are the two-dimensional counterparts of the three-dimensional eigenvalues (3.129). As was shown in Sec. 3.8, the Landau levels have a degeneracy of $|eB|/2\pi\hbar c$ per unit area. The total number of degenerate states per Landau level is therefore given by the number $N_\otimes = \Phi/\Phi_0$ of flux quanta present—that is, every state has associated with it precisely one magnetic flux quantum $\Phi_0 = 2\pi\hbar c/e$. The two-dimensional counterpart of the one-fermion-loop effective Lagrangian density (3.149) reads

$$\mathcal{L}_1 = \frac{eB}{2\pi\hbar c}\sum_{n=0}^{\infty}(\mu - E_n)\theta(\mu - E_n). \tag{12.8}$$

The sum over n is up to the integer i introduced in Eq. (3.151) which satisfies the inequalities

$$\mu - \hbar\omega_c(i + \tfrac{1}{2}) < 0 < \mu - \hbar\omega_c(i - \tfrac{1}{2}). \tag{12.9}$$

Here, we disregard the possibility that Landau levels coincide with the chemical potential, in which case the value of the step function in Eq. (12.8) becomes ambiguous. The integer i physically represents the number of filled Landau levels. The sum in Eq. (12.8) is readily evaluated, with the result

$$\mathcal{L}_1 = \frac{eB}{2\pi\hbar c} i \left(\mu - \tfrac{1}{2} i \hbar\omega_c\right). \tag{12.10}$$

For small fields, the Lagrangian density (12.10) becomes

$$\mathcal{L}_1 = \frac{1}{8\pi}\frac{2m}{\hbar^2}\mu^2 - \frac{1}{8\pi}\left(\frac{2m}{\hbar^2}\right)^{-1}\left(\frac{eB}{\hbar c}\right)^2 \cdots . \tag{12.11}$$

The first term on the right is independent of the magnetic field and corresponds to the free particle contribution in the absence of an applied field, cf. Eq. (6.10)

$$\frac{1}{4\pi}\frac{m}{\hbar^2}\mu^2 = -\int \frac{\mathrm{d}^2 k}{(2\pi)^2}\left[\epsilon(\mathbf{k}) - \mu\right]\theta\left[\mu - \epsilon(\mathbf{k})\right], \tag{12.12}$$

with $\epsilon(\mathbf{k}) = \hbar^2\mathbf{k}^2/2m$ the kinetic energy of the electrons. The second term at the right side of Eq. (12.11) yields the low-field (diamagnetic) susceptibility

$$\chi = \frac{\partial^2 \mathcal{L}_1}{\partial B^2} = -\mu_{\mathrm{B}}^2\,\nu(0), \tag{12.13}$$

with $\mu_{\mathrm{B}} = e\hbar/2mc$ the Bohr magneton, and

$$\nu(0) = \frac{m}{2\pi\hbar^2} \tag{12.14}$$

the density of states per spin degree of freedom at the Fermi "surface" in two space dimensions, see Eq. (6.60).

By taking the derivative of the effective Lagrangian density (12.10) with respect to the chemical potential, we obtain for the average electron number density n

$$n = \frac{\partial \mathcal{L}_1}{\partial \mu} = i\frac{eB}{2\pi\hbar c}. \tag{12.15}$$

This result is easily understood by recalling that the integer i denotes the number of filled Landau levels and that $eB/2\pi\hbar c$ is the degeneracy per level. In general, the ratio ν of the number N of electrons to the number $N_\otimes = \Phi/\Phi_0$ of flux quanta gives the number of filled Landau levels,

$$\nu \equiv \frac{N}{N_\otimes} = \frac{n}{B/\Phi_0}. \tag{12.16}$$

For a grand canonical ensemble of electrons without mutual interactions in an applied field, Eq. (12.15) shows that the filling factor is an integer $\nu = i$.

If, in addition to a magnetic field, also a uniform static electric field $\mathbf{E}$ is applied, an electric current density $\langle \mathbf{j}_{\mathrm{e}} \rangle$ is induced. With the electric field chosen in the 1-direction specified by $A^0 = -cEx^1$, the induced current density assumes the form

$$\langle j_{\mathrm{e}}^2 \rangle = -i\frac{e}{2\pi\hbar}E. \tag{12.17}$$

This result is obtained simply by multiplying the average electron number density (12.15) with the drift velocity $\mathbf{v}_{\mathrm{D}}$, which denotes the velocity of a charged particle moving in a crossed electric and magnetic field for which the Lorentz force $-e[\mathbf{E}+(\mathbf{v}/c)\times\mathbf{B}]$ vanishes, i.e.,

$$\mathbf{v}_D = c\frac{\mathbf{E}\times\mathbf{B}}{B^2} = -c\frac{E}{B}\mathbf{e}_2. \tag{12.18}$$

Here, $(\mathbf{E}\times\mathbf{B})^i = \epsilon^{ij}E^jB$, with $\epsilon^{12} = -\epsilon^{21} = 1$, and $\mathbf{e}_2$ is the unit vector in the 2-direction. It follows that each filled Landau level contributes a factor $e^2/2\pi\hbar$ to the Hall conductivity σ_H,

$$\sigma_H = ie^2/2\pi\hbar, \tag{12.19}$$

which is defined through

$$-e\langle j_e^2\rangle = \sigma_H E. \tag{12.20}$$

It is related to the Hall resistance or resistivity R_H through $\sigma_H = 1/R_H$.

Note that the results (12.15) and (12.17) can be derived from the Lagrangian density

$$\mathcal{L}_{CS} = \frac{1}{2}\vartheta\epsilon^{\mu\nu\lambda}A_\mu\partial_\nu A_\lambda, \tag{12.21}$$

with the parameter $\vartheta = -\nu e^2/2\pi\hbar c^2$ and $\nu = i$ the integer filling factor. Indeed,

$$n = -\frac{c}{e}\frac{\partial\mathcal{L}_{CS}}{\partial A_0} = i\frac{eB}{2\pi\hbar c} \tag{12.22}$$

and with the gauge choice $A^\mu = (-cEx^1, 0, Bx^1)$

$$\langle j_e^i\rangle = -\frac{c}{e}\frac{\partial\mathcal{L}_{CS}}{\partial A_i} = i\frac{eE}{2\pi\hbar}\epsilon^{i1}. \tag{12.23}$$

A term like (12.21) is called a *Chern-Simons term*. Involving the antisymmetric Levi-Civita symbol $\epsilon^{\mu\nu\lambda}$ with $\epsilon_{012} = 1$, it is special to 2+1 dimensions $(\mu, \nu, \lambda = 0, 1, 2)$.

12.2 Many-Particle Wave Function

We continue by studying the wave function describing a free two-dimensional electron gas in a strong applied magnetic field. For later convenience we now choose the symmetric gauge

$$A^i = -\frac{1}{2}B\epsilon^{ij}x_j \tag{12.24}$$

to describe an applied magnetic field in the *negative* third direction. With this gauge choice, the one-particle Hamiltonian (12.5) becomes

$$H = \frac{1}{2m}\left[\left(-i\hbar\partial_1 + \frac{1}{2}\frac{e}{c}Bx^2\right)^2 + \left(-i\hbar\partial_2 - \frac{1}{2}\frac{e}{c}Bx^1\right)^2\right]. \tag{12.25}$$

It proves convenient to introduce complex coordinates

$$z \equiv x^1 + \mathrm{i}x^2, \quad z^* \equiv x^1 - \mathrm{i}x^2 \tag{12.26}$$

for which

$$\partial_z f(x^1, x^2) = (\partial_z x^1)\,\partial_1 f + (\partial_z x^2)\,\partial_2 f = \frac{1}{2}(\partial_1 f - \mathrm{i}\partial_2 f), \tag{12.27}$$

with $\partial_z \equiv \partial/\partial z$ and $f(x^1, x^2)$ an arbitrary function, or

$$\partial_z = \frac{1}{2}(\partial_1 - \mathrm{i}\partial_2), \tag{12.28}$$

and similarly with $\partial_{z^*} \equiv \partial/\partial z^*$

$$\partial_{z^*} = \frac{1}{2}(\partial_1 + \mathrm{i}\partial_2). \tag{12.29}$$

In these complex variables, the one-particle Hamiltonian takes the form

$$H = -2\frac{\hbar^2}{m}\partial_{z^*}\partial_z + \frac{1}{8}m\omega_c^2 z^* z - \frac{1}{2}\omega_c L, \tag{12.30}$$

where ω_c is the cyclotron frequency (12.7) and L the orbital angular momentum operator

$$L = -\mathrm{i}\hbar x^1\partial_2 + \mathrm{i}\hbar x^2\partial_1 = \hbar(z\partial_z - z^*\partial_{z^*}), \tag{12.31}$$

pointing in the positive 3-direction. Following standard quantum mechanics, we introduce creation and annihilation operators to construct the wave functions. Being a two-dimensional problem, two pairs are needed. The choice

$$a^\dagger \equiv \frac{1}{\sqrt{2\ell^2}}\left(-2\ell^2\partial_z + \tfrac{1}{2}z^*\right), \quad a \equiv \frac{1}{\sqrt{2\ell^2}}\left(2\ell^2\partial_{z^*} + \tfrac{1}{2}z\right), \tag{12.32}$$

and

$$b^\dagger \equiv \frac{1}{\sqrt{2\ell^2}}\left(-2\ell^2\partial_{z^*} + \tfrac{1}{2}z\right), \quad b \equiv \frac{1}{\sqrt{2\ell^2}}\left(2\ell^2\partial_z + \tfrac{1}{2}z^*\right), \tag{12.33}$$

with ℓ the magnetic length (5.29) defined through $\ell^2 \equiv \hbar c/eB$, gives two nonzero commutators, *viz.* $[a, a^\dagger] = [b, b^\dagger] = 1$. The Hamiltonian and orbital angular momentum operator read expressed in terms of these operators

$$H = \hbar\omega_c\left(a^\dagger a + \tfrac{1}{2}\right) \tag{12.34}$$

and

$$L = \hbar(b^\dagger b - a^\dagger a), \tag{12.35}$$

respectively. As for a harmonic oscillator, the eigenfunctions are constructed by operating with the creation operators on the ground-state wave function

$$|m, n\rangle = \frac{\left(b^\dagger\right)^m \left(a^\dagger\right)^n}{\sqrt{m!}\,\sqrt{n!}}|0, 0\rangle. \tag{12.36}$$

The resulting one-electron wave functions satisfy the eigenvalue equations

$$H|m,n\rangle = \hbar\omega_c\left(n+\tfrac{1}{2}\right)|m,n\rangle, \quad L|m,n\rangle = \hbar(m-n)|m,n\rangle, \tag{12.37}$$

with $n, m = 0, 1, 2, 3, \cdots$ the eigenvalues of the operator $a^\dagger a$ and $b^\dagger b$, respectively. In the symmetric gauge, the infinite degeneracy of a Landau level is reflected by the fact that the energy eigenvalues are independent of m. If the system is of finite extent, with $N_\otimes$ flux quanta piercing the plane, then m takes the values $m = 0, 1, 2, \cdots, N_\otimes - 1$.

As usual, the explicit form of the ground-state wave function is found by solving the equations

$$a|0,0\rangle = b|0,0\rangle = 0, \tag{12.38}$$

yielding

$$\langle z, z^*|0,0\rangle = \frac{1}{\sqrt{2\pi\ell^2}}\,\mathrm{e}^{-z^*z/4\ell^2}, \tag{12.39}$$

where the normalization is chosen such that a single particle is contained in the entire (two-dimensional) volume,

$$\langle 0,0|0,0\rangle = 1. \tag{12.40}$$

In more detail, the eigenfunctions of the lowest Landau level ($n = 0$) read

$$\langle z, z^*|m,0\rangle = \mathcal{N}_m z^m\,\mathrm{e}^{-z^*z/4\ell^2}, \tag{12.41}$$

with the normalization $\mathcal{N}_m = 1/\sqrt{2^{m+1}\pi m!\,\ell^{2m+2}}$. The magnetic length $\ell = \sqrt{\hbar c/eB}$ is related to the area $S_\otimes$ occupied by a single flux quantum in the following way:

$$S_\otimes B = \Phi_0, \quad \text{or} \quad S_\otimes = 2\pi\ell^2. \tag{12.42}$$

It is instructive to compute the average square radius $z^*z = \mathbf{x}^2$ in the state $|m,0\rangle$. This gives

$$\langle m,0|z^*z|m,0\rangle = 2(m+1)\,\ell^2, \tag{12.43}$$

showing that the state denoted by $|m,0\rangle$ is distributed around a circle of radius $r_m = \sqrt{2(m+1)}\,\ell$. In other words, the average area occupied by a single state is pierced by exactly one flux quantum:

$$B(\pi r_{m+1}^2 - \pi r_m^2) = \Phi_0, \tag{12.44}$$

as it should be, see Fig. 12.2 for an illustration.

From these one-particle functions, the many-particle wave function describing the $\nu = 1$ ground state of electrons without mutual interactions is constructed as follows. Let z_α be the complex coordinate of the αth electron. The wave function

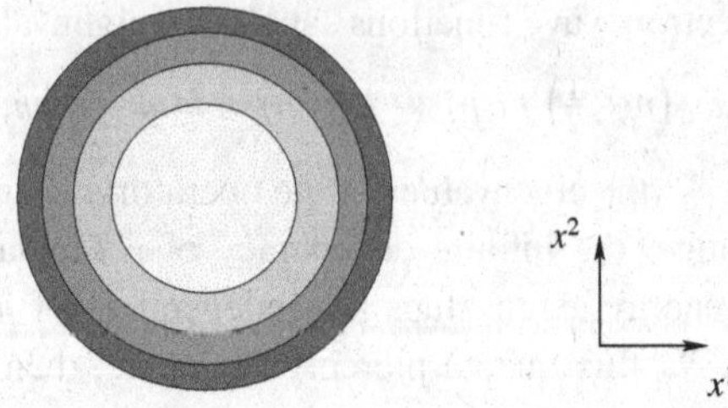

Fig. 12.2 The average areas occupied by the four states $m = 0, 1, 2, 3$ in the lowest Landau level are indicated by different gray scales. Exactly one flux quantum passes through each annulus.

$\psi^{(1)}(z_1, z_2, \cdots, z_N)$ must be antisymmetric under the interchange of two coordinates. This property is implemented by means of a Slater determinant built from the one-particle wave functions in the usual way, giving as many-particle wave function describing the $\nu = 1$ ground state with all the available states in the lowest Landau level occupied by electrons:

$$\psi^{(1)}(z_1, z_2, \cdots, z_N) = \begin{vmatrix} 1 & \cdots & 1 \\ z_1 & \cdots & z_N \\ \vdots & \vdots & \vdots \\ z_1^{N-1} & \cdots & z_N^{N-1} \end{vmatrix} \exp\left(-\sum_\alpha z_\alpha^* z_\alpha / 4\ell^2\right)$$
$$= \prod_{\alpha<\beta}(z_\alpha - z_\beta) \exp\left(-\sum_\alpha z_\alpha^* z_\alpha / 4\ell^2\right), \tag{12.45}$$

up to normalization. For notational convenience, we suppress the complex conjugates of the z's in the argument of the wave function. Note that in accord with the Pauli principle, the wave function vanishes when two coordinates z_α, z_β coincide. Pertaining to electrons without mutual interactions, the energy eigenvalue $E_{\rm tot}$ of this many-particle ground state is simply the sum of the energies of the individual electrons in the ground state, i.e., $E_{\rm tot} = NE_0$.

To obtain the spectrum of the angular momentum operator $L_{\rm tot} = \hbar \sum_\alpha (z_\alpha \partial_{z_\alpha} - z_\alpha^* \partial_{z_\alpha^*})$, we rotate the coordinate system by an angle θ about the third direction, i.e., the direction perpendicular to the plane. Under such a rotation, which is generated by $L_{\rm tot}$, the complex coordinates acquire a phase factor $z_\alpha \to e^{i\theta} z_\alpha$. The rotated ground-state wave function is consequently related to the unrotated one in the following way

$$e^{(i/\hbar)\theta L_{\rm tot}} \psi^{(1)}(z_1, z_2, \cdots, z_N) = \psi^{(1)}\left(e^{i\theta} z_1, e^{i\theta} z_2, \cdots, e^{i\theta} z_N\right) \tag{12.46}$$
$$= e^{i\theta N(N-1)/2} \psi^{(1)}(z_1, z_2, \cdots, z_N),$$

showing that the total angular momentum carried by the filled lowest Landau level is given by ($\hbar$ times)

$$m_{\text{tot}} = \frac{1}{2}N(N-1). \tag{12.47}$$

This result is readily understood by noting that by the Pauli exclusion principle, fermions cannot occupy the same state. If the state with angular momentum m is occupied, the next fermion has to go in the state $m+1$, and so on. And,

$$m_{\text{tot}} = \sum_{m=0}^{N-1} m, \tag{12.48}$$

in agreement with Eq. (12.47). The spacing between the angular momentum eigenvalues is determined by the statistics of the particles involved. The statistics is defined by the phase factor $e^{i\varsigma\pi}$ picked up by the wave function when two identical particles are interchanged. This can be achieved by first taking one particle around the other through an azimuthal angle π, and then translating the two particles (see Fig. 12.6 below for a more detailed description). For fermions, $\varsigma = 1$ and the factor in front of the sum in Eq. (12.48) is unity.

12.3 Fluxons

We next consider a charged particle confined to move in a plane that is pierced by an infinitely thin solenoid, or fluxon carrying a flux $\Phi = \alpha\Phi_0$. Without loss of generality, the location of what we call the fluxon can be taken as the origin. The vector potential describing the fluxon can be chosen to be

$$A^i(\mathbf{x}) = -\frac{\alpha\Phi_0}{2\pi}\frac{\epsilon^{ij}x_j}{\mathbf{x}^2} = -\frac{\alpha\Phi_0}{2\pi}\partial_i\theta, \tag{12.49}$$

where $\mathbf{x}$ denotes the position vector in the plane and θ is the azimuthal angle. The induction field $B = \alpha\Phi_0\delta(\mathbf{x})$ of the fluxon is a delta function in the plane by the identity

$$[\partial_1, \partial_2]\,\theta = 2\pi\delta(\mathbf{x}), \tag{12.50}$$

and is directed in the *negative* third direction. The Hamiltonian describing the system is given by

$$H_\alpha = \frac{\hbar^2}{2m}(-i\nabla - \alpha\nabla\theta)^2\,. \tag{12.51}$$

If ψ_0 is an eigenfunction of the Hamiltonian H_0 with energy E_0, then

$$\psi_\alpha(\mathbf{x}) \equiv e^{i\alpha\theta}\psi_0(\mathbf{x}) \tag{12.52}$$

is an eigenfunction of the Hamiltonian H_α with the same energy E_0. That is to say, inserting a fluxon in the plane does not change the energy spectrum. It changes, however, the spectrum of the angular momentum operator. In terms of the azimuthal angle θ, this operator reads $L = -i\hbar\partial/\partial\theta$, and so if the eigenfunction ψ_0 describes a state of angular momentum $\hbar m$, the eigenfunction ψ_α has the shifted eigenvalue

$$L\psi_\alpha = \hbar(m+\alpha)\psi_\alpha. \tag{12.53}$$

Only when α is an integer, the wave function (12.52) is single-valued away from the origin. We will henceforth restrict ourselves to integer α's. Note that the azimuthal angle θ is undefined at the location of the fluxon so that the wave function is singular there.

When the system is placed in a strong magnetic field, the energy spectrum is again unaffected by the presence of the fluxon at the origin, and is still given by the Landau-level spectrum (12.37):

$$H_\alpha|m,n\rangle_\alpha = \hbar\omega_c\left(n+\tfrac{1}{2}\right)|m,n\rangle_\alpha, \tag{12.54}$$

where the wave function $|m,n\rangle_\alpha$ is obtained from the one in the absence of a fluxon by including the phase factor $e^{i\alpha\theta}$:

$$\langle z,z^*|m,n\rangle_\alpha \equiv e^{i\alpha\theta}\langle z,z^*|m,n\rangle = (z/|z|)^\alpha\langle z,z^*|m,n\rangle, \tag{12.55}$$

with $|z| = \sqrt{z^*z}$. But the angular momentum of the state is again shifted:

$$L|m,n\rangle_\alpha = \hbar(m-n+\alpha)|m,n\rangle_\alpha, \tag{12.56}$$

implying that the fluxon increases the angular momentum of each state by $\hbar\alpha$. If the background magnetic field provides $N_\otimes$ flux quanta so that $m = 0,1,2,\cdots,N_\otimes$, the combination $m+\alpha$ runs from α to $N_\otimes+\alpha$.

Specialized to the lowest Landau level $n = 0$, the lowest angular momentum is not 0 but α, and the eigenfunctions are given by

$$\langle z,z^*|m,0\rangle_\alpha = N_m\frac{z^{m+\alpha}}{|z|^\alpha}e^{-z^*z/4\ell^2}, \tag{12.57}$$

with the normalization $N_m = 1/\sqrt{2^{m+1}\pi m!\,\ell^{2m+2}}$. Since the eigenfunctions $|m,0\rangle$ and $|m,0\rangle_\alpha$ differ only by a phase factor, the average square orbit radius z^*z in the presence of the magnetic fluxon is still $2(m+1)\,\ell^2$, see Eq. (12.43), and nothing changes to Fig. 12.2 when the fluxon is included.

The factor $(z/|z|)^\alpha$ appearing in the wave function (12.57) reminds us of the asymptotic solution of a magnetic vortex in a superconductor discussed in Sec. 8.13, see Eq. (8.135). Close to the vortex core, corresponding to the opposite limit $|z| \to 0$, the solution vanishes, see Eq. (8.148). With this in mind, we are led to study the wave function

$$\langle z, z^*|m, 0\rangle'_\alpha = \mathcal{N}_{m+\alpha} z^{m+\alpha}\, \mathrm{e}^{-z^* z/4\ell^2}, \tag{12.58}$$

where the normalization factor is given by $\mathcal{N}_m$ introduced below Eq. (12.57) with m replaced with $m + \alpha$. The normalization is such that the new wave function is again normalized to unity. It is readily checked that this wave function has the same eigenvalues as the previous one—that is, the energy and angular momentum carried by the state described by the new wave function are the same as those carried by the state described by $|m, 0\rangle_\alpha$. However, in contrast to the previous one, the new wave function is not related to $|m, 0\rangle$ through a simple phase factor, and is not singular at the location of the fluxon, tending to zero there. In a sense, the fluxon has become part of the system, whereas previously it had more the status of the grin of the Cheshire Cat[1]. This is also reflected by the average square radius:

$${}'_\alpha\langle m, 0|z^* z|m, 0\rangle'_\alpha = 2(m + 1 + \alpha)\,\ell^2, \tag{12.59}$$

showing that the area occupied by the fluxon is no longer available to the electron. For example, when the fluxon carries one flux quantum ($\alpha = 1$) so that it occupies an area $S_\otimes = 2\pi\ell^2$, the state m is distributed around a circle with radius $r_m = \sqrt{2(m+2)}\,\ell$, and not around one with radius $\sqrt{2(m+1)}\,\ell$ which would be the case in the absence of the fluxon. In short, an elementary flux quantum at the origin pushes the electron away and forces a given angular momentum state to move to the next higher state, see Fig. 12.3

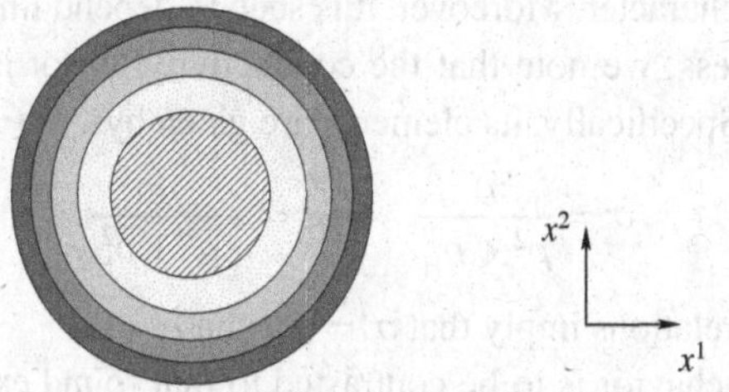

Fig. 12.3 The average areas occupied by the four states $m = 0, 1, 2, 3$ in the lowest Landau level in the presence of a flux quantum at the origin. Exactly one flux quantum passes through each annulus. The same gray scales are used for the states as in the previous figure, Fig. 12.2.

Finally, a fluxon placed at an arbitrary point z_0 in the plane rather than in the origin is described by the wave function

$$\langle z, z^*|m, 0\rangle'_\alpha = \mathcal{N}_{m+\alpha}(z - z_0)^\alpha z^m\, \mathrm{e}^{-z^* z/4\ell^2}, \tag{12.60}$$

[1]This refers to Lewis Carroll's fable *Alice in Wonderland*. The famous line reads: *"Well I've often seen a cat without a grin", thought Alice, "but a grin without a cat! It is the most curious thing, I ever saw in all my life!"*

which reduces to Eq. (12.58) for $z_0 \to 0$.

12.4 IQHE

Before discussing the IQHE, we first consider a *classical* system of charges confined to move in a plane. In a crossed electric and magnetic field, the point particles of charge $q = -e(< 0)$ moving with a velocity $\mathbf{v}$ are accelerated by the Lorentz force $-e[\mathbf{E} + (\mathbf{v}/c) \times \mathbf{B}]$ and decelerated by dissipative processes. By Ohm's law, the current density $\mathbf{j}_e = n\mathbf{v}$, with n the particle number density, is related to the applied fields in the following way, cf. Eq. (6.40),

$$-e\mathbf{j}_e = \sigma_0[\mathbf{E} + (\mathbf{v}/c) \times \mathbf{B}] = \sigma_0[\mathbf{E} + (\mathbf{j}_e/cn) \times \mathbf{B}], \tag{12.61}$$

where σ_0 is the zero-field conductivity. With the magnetic field applied perpendicular to the plane and the electric field laying in the plane, the self-consistent equation is readily solved to yield

$$\begin{pmatrix} E^1 \\ E^2 \end{pmatrix} = \begin{pmatrix} \rho_0 & \rho_H \\ -\rho_H & \rho_0 \end{pmatrix} \begin{pmatrix} -ej_e^1 \\ -ej_e^2 \end{pmatrix} = \begin{pmatrix} \rho_0 & B/ecn \\ -B/ecn & \rho_0 \end{pmatrix} \begin{pmatrix} -ej_e^1 \\ -ej_e^2 \end{pmatrix}, \tag{12.62}$$

where the 2×2 matrix is the resistivity tensor with $\rho_0 = \sigma_0^{-1}$ the zero-field longitudinal resistivity. The Hall resistivity $\rho_H = R_H$ for this classical system, given by the off-diagonal elements of the resistivity tensor, is independent of σ_0, reflecting its nondissipative character. Moreover, it is seen to depend linearly on B.

For completeness, we note that the conductivity tensor is the inverse of the resistivity tensor. Specifically, its elements are given by

$$\sigma = \frac{\rho}{\rho^2 + \rho_H^2}, \quad \sigma_H = -\frac{\rho_H}{\rho^2 + \rho_H^2}. \tag{12.63}$$

For $\rho_H \neq 0$, these relations imply that $\sigma = 0$ when $\rho = 0$.

The classical behavior is to be contrasted to that found experimentally in the IQHE, where the Hall resistance is constant over finite B-intervals. More specifically, the Hall resistivity, or resistance within a plateau takes the values given by Eq. (12.1) with the filling factor ν defined in Eq. (12.16) frozen at integer values. Superficially, this is as for the free electron gas studied in Sec. 12.1. However, that model as it stands is not realistic. The point is that in experiment not the chemical potential is fixed, as assumed in Sec. 12.1, but rather the electron number—that is, QHE systems are to be thought of as isolated islands. Typical electron number densities are $n \sim 10^{11}$ cm^{-2} for the IQHE. To explain the experimental fact that the Hall resistance remains constant when the applied field at fixed density is varied, impurities must be included in the theory. In the presence of random impurities, the single-particle states fall into two classes: localized and extended states. The former are spatially localized states which describe electrons trapped by impurities. In contrast to extended states, localized states do not contribute to charge transport. Extended states are similar to those encountered in the free-electron model without impurities, where the density of states $\nu(\epsilon)$ consists of a chain of delta functions peaked around the energies $E_n = \hbar\omega_c(n + \frac{1}{2})$ of the Landau levels,

$$\nu(\epsilon) = \frac{1}{2\pi\ell^2} \sum_{n=0}^{\infty} \delta(\epsilon - E_n). \tag{12.64}$$

In the limit of vanishing magnetic field, the free-gas density of states becomes constant

$$\lim_{B\to 0} \nu(\epsilon) = \frac{1}{2\pi\ell^2\hbar\omega_c} \int_0^{\infty} \mathrm{d}E\, \delta(\epsilon - E) = \frac{m}{2\pi\hbar^2}, \tag{12.65}$$

in accord with Eq. (12.14). A second effect caused by the impurities is to broaden the Landau levels, see Fig. 12.4. Now, when the Fermi energy is in a region of

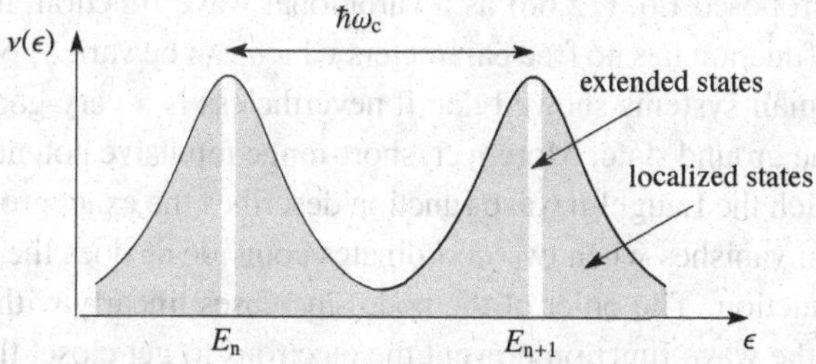

Fig. 12.4 Schematic representation of the broadened density of states due to impurities. Centered around and close to the Landau levels of the pure system are extended states. Further away are localized states. In the pure limit, $\nu(\epsilon)$ reduces to a chain of delta functions.

localized states, only the filled Landau levels below the Fermi energy contribute to the Hall resistance. If the applied magnetic field is slightly changed so that the energies of the Landau levels shift, additional localized states are filled or emptied. But because these states do not carry electric current, the conductivity remains unchanged. This explains the width of the plateaus in the Hall resistance and the fact that the resistance in the direction of the applied field is zero around specific values of the magnetic field. The conclusion that localized states trapped by impurities are the main origin for the stabilization of the IQHE is corroborated by the experimental finding that the plateau widths increase systematically with increasing impurity concentration. When, on the other hand, the Fermi energy is in a region of extended states, i.e., when an impurity-broadened Landau level is being filled, the conductivity in the direction of the electric field is nonzero, and the Hall resistance rises more or less linearly to the next plateau.

12.5 Laughlin Wave Function

The FQHE is characterized by plateaus in the Hall resistance for *fractional* filling factors. This constitutes the hallmark of a new remarkable condensed matter state that originates from the strong Coulomb repulsion.

Taking an imaginative leap, Laughlin proposed the following *Ansatz* for the many-particle ground-state wave function describing the case with filling factor $\nu = 1/(2l+1)$ for l an integer:

$$\psi^{(2l+1)}(z_1, z_2, \cdots, z_N) = \prod_{\alpha<\beta}(z_\alpha - z_\beta)^{2l+1} \exp\left(-\sum_\alpha z_\alpha^* z_\alpha / 4\ell^2\right), \tag{12.66}$$

where a normalization factor is omitted. For l set to zero, corresponding to the filling factor $\nu = 1$, the *Ansatz* reduces to the known result (12.45) of a single filled Landau level. This wave function has a few stunning properties. Although Laughlin initially proposed Eq. (12.66) as a variational wave function, his remarkably simple wave function has no free parameters which can be varied. Numerical simulations on small systems showed that it nevertheless is a very good approximation to the true ground state. Moreover, short-range repulsive potentials have been found for which the Laughlin wave function describes the exact ground state. The wave function vanishes when two coordinates coincide as does the $\nu = 1$ ground-state wave function. The order of the nodes increases linearly with l. Physically, the nodes in the wave function prevent the electrons to get close, the more so the larger l, and in this way minimize the Coulomb repulsion.

Probably the most stunning feature of the Laughlin wave function is that it does not describe ordinary electrons. To demonstrate this, note that interchanging two coordinates z_α and z_β yields a factor $(-1)^{2l+1} = e^{i(2l+1)\pi}$. Since the interchange of two identical quantum particles defines their statistics ς, the objects described by the Laughlin wave function possess a statistics given by the inverse filling factor

$$\varsigma = 1/\nu = 2l + 1. \tag{12.67}$$

Since $(-1)^{2l+1} = -1$, the objects have a fermionlike character, but they are no ordinary electrons for which $l = 0$. The statistics of these "superfermions" also shows up when the coordinate system is rotated by an angle θ around the third direction as in Eq. (12.46). This rotation is generated by the total angular momentum operator $L_{\rm tot}$. When applied to the Laughlin wave function, this gives

$$\begin{aligned} e^{(i/\hbar)\theta L_{\rm tot}}\psi^{(2l+1)}(z_1, z_2, \cdots, z_N) &= \psi^{(2l+1)}\left(e^{i\theta}z_1, e^{i\theta}z_2, \cdots, e^{i\theta}z_N\right) \\ &= e^{i\theta(2l+1)N(N-1)/2}\psi^{(2l+1)}(z_1, z_2, \cdots, z_N) \end{aligned} \tag{12.68}$$

showing that the wave function describes a state of total angular momentum ($\hbar$ times)

$$m_{\text{tot}} = \frac{1}{2}(2l+1)N(N-1). \tag{12.69}$$

This can be written analogously to (12.48) as a sum over the available angular momentum states as

$$m_{\text{tot}} = \varsigma \sum_{m=0}^{N-1} m, \tag{12.70}$$

involving the statistics of the superfermions $\varsigma = (2l+1)$ which determines the spacing between two eigenvalues of the angular momentum operator. The fermionic character of the superfermions is reflected by the fact that an angular momentum state can only accommodate one superfermion.

12.6 Composite Fermions

To understand the nature of superfermions, Jain rewrote the Laughlin wave function (12.66) in the equivalent form

$$\psi^{(2l+1)}(z_1, z_2, \cdots, z_N) = \prod_{\alpha<\beta} (z_\alpha - z_\beta)^{2l} \psi^{(1)}(z_1, z_2, \cdots, z_N), \tag{12.71}$$

where $\psi^{(1)}(z_1, z_2, \cdots, z_N)$ is the ground-state wave function (12.45) describing a lowest Landau level filled by fermions without mutual interactions. This form is closely related to the single-particle wave function (12.60) of the lowest Landau level in the presence of an an infinitely thin fluxon located at z_0 carrying α flux quanta. The important part of that wave function is the prefactor $(z - z_0)^\alpha$. Comparison with the factor $(z_\alpha - z_\beta)^{2l}$ appearing in Eq. (12.71) shows that an electron centered at z_α sees $2l$ flux quanta located at each of the locations z_β of the other electrons—that is, every electron is endowed with a fluxon carrying $2l$ flux quanta. The Laughlin wave function (12.66) therefore describes *composite fermions* carrying electric charge $q = -e(< 0)$ and magnetic flux $2l\Phi_0$. These composite fermions fill the lowest Landau level. Since an electron occupies on average a surface area that is pierced by one flux quantum, a composite fermion, consisting of the original electron and $2l$ flux quanta attached to it, occupies on average an area which is a factor $2l + 1$ larger. As a result, the electronic charge is depleted in the areas around the centers of the original electrons. Through the formation of composite fermions, the system therefore considerably reduces its electrostatic Coulomb energy. The more flux quanta an electron is coerced into acquiring, the further away the other electrons are pushed and the more the electrostatic Coulomb energy is reduced. Given the numerical success of the Laughlin wave function, it is apparently justified to treat the composite fermions in a first approximation as void of mutual interactions.

In acquiring $2l$ flux quanta, the electrons render part of the applied magnetic field $H = B$ ineffective, reducing it in the sample to a value B_{eff} determined by

$$\Phi_{\text{eff}} = \Phi - 2lN\Phi_0. \tag{12.72}$$

Here, $\Phi_{\text{eff}} = \int d^2x B_{\text{eff}}$ is the magnetic flux associated with the effective field felt by the composite fermions, while Φ is the magnetic flux associated with the applied field. To acquire a notion of the effectiveness of this mechanism, consider a typical electron number density $n \sim 10^{11}\ \text{cm}^{-2}$. With $2l$ flux quanta attached to each electron, it follows that a magnetic field of order $2ln\Phi_0 \sim 8l\text{T}$ is made ineffective, which are large fields. The formation of composite fermions harbors therefore two advantages for strongly interacting electrons in a strong magnetic field transform into weakly interacting composite fermions in a weaker effective field B_{eff}.

At the Hall plateaus characterized by a fractional filling factor, the resulting field B_{eff} in the sample is such that the composite fermions fill exactly an integer number i_{eff} of Landau levels. That is to say, the effective filling factor ν_{eff}, defined as the ratio of the number of composite fermions to the number of effective flux quanta Φ_{eff}/Φ_0, is an integer here,

$$\nu_{\text{eff}} = \frac{N}{N_{\otimes,\text{eff}}} = i_{\text{eff}}, \tag{12.73}$$

where the number N of electron equals the number of composite fermions. The Laughlin wave function (12.71) describes the case $\nu_{\text{eff}} = i_{\text{eff}} = 1$. The generalization to higher effective filling factors is obtained by replacing $\psi^{(1)}(z_1, z_2, \cdots, z_N)$ introduced in Eq. (12.45) with the many-particle wave function $\psi^{(1/i_{\text{eff}})}(z_1, z_2, \cdots, z_N)$ describing the ground state with i_{eff} filled Landau levels. Combined with Eq. (12.72), Eq. (12.73) can be rewritten as

$$i_{\text{eff}} = \pm \frac{N}{N_\otimes - 2lN}, \tag{12.74}$$

where the minus sign corresponds to cases where B and B_{eff} point in opposite directions. For the experimentally accessible filling factor $\nu = N/N_\otimes$, whose definition involves the applied field, this gives

$$\nu = \frac{i_{\text{eff}}}{2li_{\text{eff}} \pm 1}. \tag{12.75}$$

This formula predicts sequences of FQHE states emanating from $\nu = 1/2l = \frac{1}{2}, \frac{1}{4}, \cdots$ which are precisely the most dominant fractions with $\nu < 1$ observed in experiments. Upon invoking particle-hole symmetry, which transforms electrons into holes and *vice versa*, and turns ν into $1 - \nu$, one obtains a second series of dominant fractions with $\nu < 1$ found experimentally. FQHE states with $\nu > 1$ can be treated similarly by writing, for example, $\nu = \frac{5}{4} = 1 + \frac{1}{4}$. Experimentally, FQHE states emanating from $\nu = \frac{1}{2}, \frac{3}{2}, \frac{1}{4}, \frac{3}{4}, \frac{5}{4}$ and possibly from $\nu = \frac{3}{8}, \frac{3}{14}$ have been observed. Note that this theory of the FQHE as the IQHE of composite fermions predicts only fractions with odd denominators. With only one exception (see below), this is precisely what is found experimentally.

12.7 Fractional Charge

Laughlin also proposed an *Ansatz* for the wave function of an elementary excitation around the $\nu = 1/(2l + 1)$ ground state:

$$\psi^{(2l+1)}(z_0; z_1, z_2, \cdots, z_N) = \prod_{\alpha=1}^{N} (z_\alpha - z_0)\psi^{(2l+1)}(z_1, z_2, \cdots, z_N). \tag{12.76}$$

Again, there are strong numerical indications that this *Ansatz* correctly describes such an elementary excitation. From the discussion in Sec. 12.3 it follows that an elementary excitation is a fluxon. The angular momentum of the state is readily found to be ($\hbar$ times)

$$m_{\text{tot}} = \frac{1}{2}(2l + 1)N(N - 1) + N, \tag{12.77}$$

which is to be compared to the angular momentum of the ground state $\frac{1}{2}(2l + 1)N(N - 1)$ found in Eq. (12.47). The last term in Eq. (12.77) shows that the inserted fluxon forces all the N composite fermions to move to the next higher angular momentum state. Moreover, since the wave function vanishes when any of the coordinates z_α labeling the position of the composite fermions approaches z_0, an elementary excitation leads to a local depletion of the charge density around z_0. Such an excitation is therefore a hole excitation. Because the surface area $S_\otimes = 2\pi\ell^2$ occupied by a flux quantum is a factor $2l + 1$ smaller than that occupied by an electron, it is expected that also the charge and statistics associated with an elementary excitation is a factor $2l + 1$ smaller than that of an electron. This would imply that an elementary excitation carries fractional charge $q = -e/(2l + 1)$ and fractional statistics $\varsigma = 1/(2l + 1)$. Tunneling and other experiments have been performed that indeed strongly suggest the presence of fractional charges in FQHE states.

12.8 Near $\nu = 1/2$

Figure 12.1 shows that at half filling of the lowest Landau level ($\nu = \frac{1}{2}$), no Hall plateau is present and that the longitudinal resistance does not vanish. If the Jain construction is carried out for this filling factor and two fluxons are attached to every electron, no magnetic field is left over. The composite fermions are therefore expected to move in a free-field environment.

Beautiful experiments have been performed that confirm this prediction, see Fig. 12.5. They involve samples into which holes (antidots) are punched with a diameter of 100 – 200 nm. The antidots form a quadratic lattice with lattice spacing of 500 – 700 nm. In the presence of a small magnetic field, the longitudinal magnetoresistance shows two resonances for both positive and negative fields corresponding to the radius

$$r = v_F/\omega_c \tag{12.78}$$

of a classical charge e circling a single ($= 1^2$) dot (largest field), and four ($= 2^2$) dots (smallest field). The former resonance is more pronounced than the latter. Resonances corresponding to trajectories enclosing nine ($= 3^2$) or more dots are not visible. Since only electrons near the Fermi surface are involved in scattering processes and, consequently, contribute to the resistance, it is the Fermi velocity v_F that appears in the expression for the radius. For comparison, also the results for a sample without holes is included in Fig. 12.5. Apart from Shubnikov-de-Haas oscillations, which we not discuss, nothing special happens there.

Now, in a field close to $B_{\frac{1}{2}}$ corresponding to half filling, the composite fermions are subject to a small effective field $B_{\text{eff}} = B - B_{\frac{1}{2}}$. It is expected that as a function of the effective field, the longitudinal magnetoresistance again shows resonances. This is indeed what is observed in experiment, see Fig. 12.5. In comparing the two cases, the effective field is rescaled by a factor $1/\sqrt{2}$. This is because at half filling the system is spin-polarized involving only electrons with their spins oriented antiparallel to the field, whereas in the small-field case both spin orientations are present. The Fermi sea of the polarized system is therefore twice that of the unpolarized system, and hence the Fermi momentum and velocity are a factor $\sqrt{2}$ larger than in the unpolarized system. By Eq. (12.78), the location of the resonances in the half-filled case are therefore also a factor $\sqrt{2}$ larger. Because of their broadness, only the dominant resonances, corresponding to trajectories around a single dot, are observed at half filling.

These and other experiments impressively demonstrate that composite fermions have properties very different from those of ordinary electrons. Whereas electrons exposed to a very large magnetic field $B_{\frac{1}{2}}$ would orbit on very small cir-

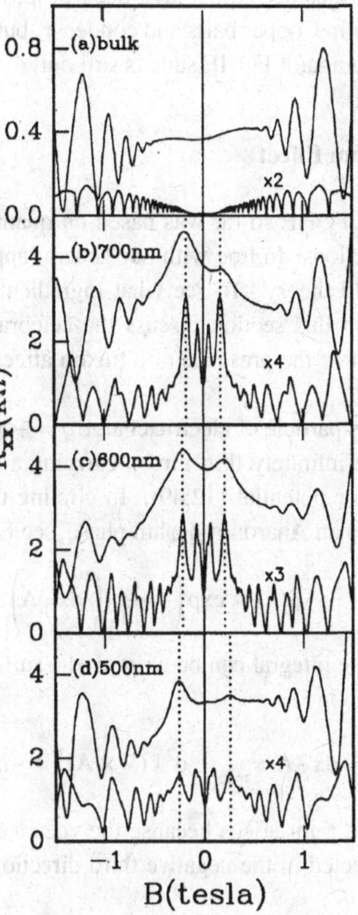

Fig. 12.5 Longitudinal magnetoresistance ($R_{xx} = R$) in the absence (a) and in the presence (b-d) of an antidot lattice as a function of the magnetic field. The length scales specify the lattice spacings. The lower curves pertain to the small-field case, while the upper curves pertain to the half-filled case. Reprinted figure with permission from W. Kang, H. L. Stormer, L. N. Pfeiffer, K. W. Baldwin, and K. W. West, Phys. Rev. Lett. **71**, 3850 (1993). Copyright 1993 by the American Physical Society.

cles, composite fermions, oblivious to this field, move in straight lines. Also their mass is unrelated to the electron mass as it arises solely from interactions.

The state $\nu = \frac{5}{2}$, mirroring the $\nu = \frac{1}{2}$ state in a higher Landau level ($\frac{5}{2} = 2 + \frac{1}{2}$), should behave the same as the $\nu = \frac{1}{2}$ state, but, surprisingly, it does not. Experimentally, the $\nu = \frac{5}{2}$ state, an even-denominator fraction, has all the characteristics of a FQHE state. Various scenarios have been proposed, including one where composite fermions form Cooper pairs and condense, but the verdict on the nature of this sole even-denominator FQHE state is still out.

12.9 Aharonov-Bohm Effect

The discussion of the FQHE so far was based on quantum mechanics featuring many-body wave functions. In line with our general approach, we next consider the corresponding field theory. To see what ingredients are involved in such a description, we first in this section discuss the celebrated Aharonov-Bohm effect which describes how the presence of a fluxon affects the wave function of a charged particle.

Consider a spinless particle of electric charge $q = -e(< 0)$ that is adiabatically transported around an infinitely thin fluxon carrying a flux $\alpha\Phi_0$. The fluxon is described by the vector potential (12.49). In circling the fluxon, the electron's wave function acquires an Aharonov-Bohm phase, see Eq. (5.30),

$$\chi_{\rm AB}(\Gamma) = \exp\left(-\mathrm{i}\frac{e}{\hbar c}\oint_\Gamma \mathrm{d}\mathbf{x}\cdot\mathbf{A}\right). \tag{12.79}$$

By Stokes' law, the line integral can be written as a surface integral over the area $S(\Gamma)$ bounded by the loop Γ,

$$\oint_\Gamma \mathrm{d}\mathbf{x}\cdot\mathbf{A} = \int_{S(\Gamma)} \mathrm{d}^2x\,(\nabla\times\mathbf{A})^3 = -\alpha\Phi_0. \tag{12.80}$$

The minus sign on the right arises because the vector potential (12.49) leads to an induction field directed in the negative third direction. The phase factor thus becomes

$$\chi_{\rm AB}(\Gamma) = \mathrm{e}^{\mathrm{i}(e/\hbar c)\alpha\Phi_0} = \mathrm{e}^{2\pi\mathrm{i}\alpha}. \tag{12.81}$$

In the light of the presence of composite fermions in the FQHE, we next generalize the above example to charge-flux composites each consisting of a charge and a fluxon with α flux quanta. Let the composite particle at rest be denoted by 1, and the one traversing the path Γ by 2. The charge of particle 1 and the flux of particle 2 give rise to the phase factor (12.81). But since particle 1 carries also a charge and particle 2 also a flux, the same contribution arises twice, and whence the Aharonov-Bohm phase is

$$\chi_{\rm AB}(\Gamma) = \mathrm{e}^{2\mathrm{i}(e/\hbar c)\alpha\Phi_0} = \mathrm{e}^{4\pi\mathrm{i}\alpha}. \tag{12.82}$$

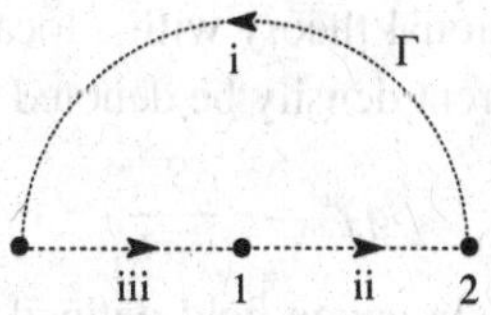

Fig. 12.6 The adiabatic interchange of two particles labeled by 1 and 2 proceeds in three steps labeled by i, ii, and iii.

In addition, the phase factor arising from the intrinsic statistics of the charges involved is to be included.

This result can be used to determine the statistics of a charge-flux composite. To this end, we adiabatically interchange the position of two such objects labeled by 1 and 2, see Fig. 12.6. First, particle 2 is brought to the left of particle 1 along a semicircle (i). Then, particle 1 is shifted to the old position of particle 2 (ii), and, finally, particle 2 is shifted to the old position of particle 1 (iii). Only in the first step, a phase factor emerges. Since the exchange involves a semicircle, the phase factor acquired by particle 2 when transported along that semicircle is half of what it is for a full circle. Hence, the Aharonov-Bohm phase picked up in the process of interchanging two charge-flux composites is

$$\chi_{\mathrm{AB}}(\Gamma) = \mathrm{e}^{\mathrm{i}(e/\hbar c)\alpha\Phi_0} = \mathrm{e}^{2\pi\mathrm{i}\alpha}. \tag{12.83}$$

Assuming that the charges are electrons, we have to include the intrinsic statistics factor $\exp(\mathrm{i}\pi)$ for fermions. For the statistics ς of a charge-flux composite we thus obtain

$$\varsigma = 1 + 2\alpha. \tag{12.84}$$

Comparison to the statistics $\varsigma = 1 + 2l$ found for a composite fermion as it appears in the FQHE, see Eq. (12.67), which corresponds to taking $\alpha = 2l$, reveals a mismatch of a factor of two in the last term in Eq. (12.84). In other words, a composite fermion is not simply a bound state of an electron and fluxons.

12.10 Chern-Simons Theory

In this section, a field theory of the FQHE is considered. Its most important ingredient is a Chern-Simons term which we already encountered in the context of a free electron gas placed in an external magnetic field, see Eq. (12.21).

Consider a (2+1)-dimensional theory with a local U(1) symmetry and let the corresponding conserved current density be denoted by j^μ, with

$$q j^\mu = -\frac{\delta S_\mathrm{m}}{\delta a^\mu}. \tag{12.85}$$

Here, a^μ ($\mu = 0, 1, 2$) is a U(1) gauge field defined in the plane which is not to be confused with the electromagnetic gauge field A^μ, and q is the charge with which the matter fields couple to this gauge field. The precise form of the action $S_\mathrm{m} = \int \mathrm{d}^3 x \mathcal{L}_\mathrm{m}$ governing the matter fields is not important for the moment. We assume that the gauge field a^μ, which for reasons that will become clear when we proceed is often called a *statistical gauge field*, is governed by a Chern-Simons term, i.e.,

$$\mathcal{L} = \mathcal{L}_\mathrm{m} + \frac{1}{2}\vartheta \epsilon^{\mu\nu\lambda} a_\mu \partial_\nu a_\lambda. \tag{12.86}$$

A Maxwell term, which is of higher order in derivatives, is not included. As remarked before, the Chern-Simons term, involving the antisymmetric Levi-Civita symbol $\epsilon^{\mu\nu\lambda}$, is special to three dimensions. The physical relevance of the Chern-Simons term follows from examining the field equation for the statistical gauge field

$$q j^\mu = \vartheta \epsilon^{\mu\nu\lambda} \partial_\nu a_\lambda. \tag{12.87}$$

The integrated zeroth component of this equation,

$$Q = \vartheta \Phi_a, \tag{12.88}$$

with $Q = q \int \mathrm{d}^2 x\, n$ the total charge and $\Phi_a = \int \mathrm{d}^2 x\, \nabla \times \mathbf{a}$ the "magnetic" flux associated with the statistical gauge field, reveals that the theory describes objects which carry both charge and flux. More specifically, an object of charge q also carries a flux q/ϑ. This is precisely what is needed to describe the composite fermions of the FQHE. Recall that in two space dimensions $\nabla \times \mathbf{a} \equiv \epsilon^{ij}\partial_i a^j$ ($i, j = 1, 2$) with $\epsilon^{12} = 1$.

To investigate the statistics of Chern-Simons charge-flux composites, note that the Noether current of the coupled theory,

$$q J^\mu \equiv -\frac{\delta S}{\delta a^\mu} = q j^\mu - \frac{1}{2}\vartheta \epsilon^{\mu\nu\lambda} \partial_\nu a_\lambda \tag{12.89}$$

consists of two parts. The first term is the contribution from the matter fields, while the second stems from the Chern-Simons term. Upon inserting the field equation (12.87), we see that the current of the coupled theory is half the current of the matter fields: $q J^\mu = \frac{1}{2} q j^\mu$. This is again a property needed to describe the composite fermions in the FQHE. Remember that an ordinary charge-flux composite has a statistics given by Eq. (12.84) which is not consistent with the statistics of composite fermions. On the other hand, because of the factor two appearing here, Chern-Simons charge-flux composites have a statistics that precisely matches that of composite fermions.

Given the observation in Sec. 12.6 that the composite fermions can in a first approximation be considered as void of mutual interactions, we now have all the elements to write down a field theory of the FQHE:

$$\mathcal{L} = \mathrm{i}\hbar\psi^* D_t\psi - \frac{\hbar^2}{2m}|\mathbf{D}\psi|^2 + \mu\psi^*\psi + \frac{1}{2}\vartheta\epsilon^{\mu\nu\lambda}a_\mu\partial_\nu a_\lambda. \tag{12.90}$$

Here, a^μ is the statistical gauge field, and $D_t \equiv \partial_t - \mathrm{i}(e/\hbar c)(A^0 - a^0)$ and $\mathbf{D} \equiv \nabla + \mathrm{i}(e/\hbar c)(\mathbf{A}+\mathbf{a})$ are the covariant derivatives which also include the statistical gauge field. The statistical gauge field has been rescaled such that it couples to the field ψ with the same charge as the electromagnetic vector potential, first introduced in Eq. (2.180). As usual, μ is the chemical potential and ψ a Grassmann field describing the electrons. In writing Eq. (12.90), we assumed that the system is spin-polarized, and that the spin-$\uparrow$ energy levels are too high to be occupied. Only spin-$\downarrow$ electrons need then be considered and the spin index on ψ can be omitted. To specify the coefficient ϑ of the Chern-Simons term we consider for definiteness a filling factor $\nu = i_{\mathrm{eff}}/(2li_{\mathrm{eff}} + 1)$. These FQHE states feature superfermions composed of the original electrons with $2l$ flux quanta attached to them. The field equation for the zeroth component of the statistical gauge field reads when integrated over the plane

$$N \equiv \int \mathrm{d}^2x\,\psi^*\psi = 2\pi\vartheta\frac{\hbar c^2}{e^2}\frac{\Phi_a}{\Phi_0}. \tag{12.91}$$

With the choice

$$\vartheta = -\frac{1}{4\pi}\frac{e^2}{\hbar c^2}\frac{1}{l} \tag{12.92}$$

each charge has associated with it the required $2l$ flux quanta.

We next calculate the conductivity from the Lagrangian density (12.90). From our study of the IQHE we know that ordinary electrons which fill the lowest i Landau levels generate a Chern-Simons term at the one-loop level with a coefficient

$$\vartheta_{\mathrm{IQHE}} = -i_{\mathrm{eff}}\frac{e^2}{2\pi\hbar c^2}, \tag{12.93}$$

see below Eq. (12.21). Transcribed to composite fermions, this implies the effective theory

$$\mathcal{L}_{\mathrm{eff}} = \frac{1}{2}\vartheta\epsilon^{\mu\nu\lambda}a_\mu\partial_\nu a_\lambda + \frac{1}{2}\vartheta_{\mathrm{IQHE}}\epsilon^{\mu\nu\lambda}(A_\mu + a_\mu)\partial_\nu(A_\lambda + a_\lambda), \tag{12.94}$$

where the first term is the classical contribution given by the last term in the Lagrangian (12.90), while the second term, with the coefficient given by Eq. (12.93) with i replaced with $i_{\rm eff}$, is induced by the one-fermion loop. The conductivity is found by calculating the induced current:

$$\frac{e}{c}\langle j_{\rm e}^{\mu}\rangle = \frac{\partial \mathcal{L}_{\rm eff}}{\partial A_\mu} = \vartheta_{\rm IQHE}\epsilon^{\mu\nu\lambda}\partial_\nu(A_\lambda + a_\lambda) \tag{12.95}$$

which by the field equation of the statistical gauge field,

$$\vartheta\epsilon^{\mu\nu\lambda}\partial_\nu a_\lambda + \vartheta_{\rm IQHE}\epsilon^{\mu\nu\lambda}\partial_\nu(A_\lambda + a_\lambda) = 0, \tag{12.96}$$

can be cast in the form

$$\frac{e}{c}\langle j_{\rm e}^{\mu}\rangle = -\frac{e^2}{2\pi\hbar c^2}\frac{i_{\rm eff}}{2li_{\rm eff}+1}\epsilon^{\mu\nu\lambda}\partial_\nu A_\lambda, \tag{12.97}$$

where on the right only the electromagnetic vector potential A_λ appears. The zeroth component of this equation,

$$n = \frac{e}{2\pi\hbar c}\frac{i_{\rm eff}}{2li_{\rm eff}+1}B \tag{12.98}$$

shows that the choice (12.92) leads to the required filling factor $\nu = i_{\rm eff}/(2li_{\rm eff}+1)$, while the ith component shows that the Hall conductivity has the observed form:

$$\langle j_{\rm e}^{i}\rangle = -\frac{e}{2\pi\hbar}\frac{i_{\rm eff}}{2li_{\rm eff}+1}\epsilon^{ij}E_j. \tag{12.99}$$

In other words, the Lagrangian (12.90) describing an electron gas in a strong external magnetic field coupled to a Chern-Simons term provides the field theory of the FQHE we were seeking. The connection with experiment, where the electron number rather than the chemical potential is kept fixed, can be established as for the IQHE by including random impurities.

12.11 Graphene

The latest member and rising star of the QHE family is graphene—a single atomic layer of carbon. The carbon atoms, each with one valence electron, form a two-dimensional honeycomb lattice, with has two sites per unit cell, see Fig. 12.7. The lattice can be considered as composed of two interpenetrating triangular lattices (A and B). Lattice A can be spanned by, for example, the two vectors

$$\mathbf{a}_1 = \frac{a}{2}\left(\sqrt{3},-3\right), \quad \mathbf{a}_2 = \frac{a}{2}\left(\sqrt{3},3\right), \tag{12.100}$$

where a is the lattice spacing. The vectors

$$\mathbf{v}_1 = a(0,-1), \quad \mathbf{v}_2 = \frac{a}{2}\left(\sqrt{3},1\right), \quad \mathbf{v}_3 = \frac{a}{2}\left(-\sqrt{3},1\right) \tag{12.101}$$

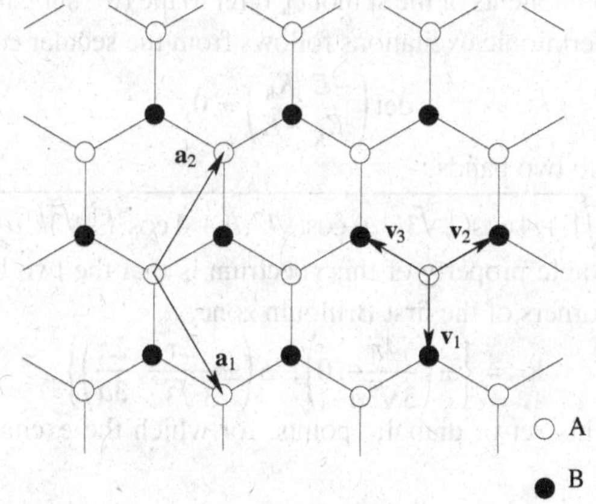

Fig. 12.7 Honeycomb lattice built from two interpenetrating triangular lattices (A and B). Open and solid circles mark sites on sublattice A and B, respectively.

connect a site on sublattice A to its nearest neighbor sites on sublattice B. The reciprocal lattice of sublattice A can be generated by the vectors

$$\mathbf{b}_1 = \frac{2\pi}{a}\left(\frac{1}{\sqrt{3}}, -\frac{1}{3}\right), \quad \mathbf{b}_2 = \frac{2\pi}{a}\left(\frac{1}{\sqrt{3}}, \frac{1}{3}\right), \tag{12.102}$$

satisfying $\mathbf{a}_i \cdot \mathbf{b}_j = 2\pi\delta_{ij}$ so that $e^{i\mathbf{b}_i \cdot \mathbf{a}_j} = 1$. The first Brillouin zone is a hexagon in momentum space with opposite sides identified.

With these conventions, the tight-binding Hamiltonian where only nearest-neighbor interactions are retained can be written as

$$H = -t \sum_{\mathbf{x}} \sum_{i=1}^{3} \left(a^\dagger_{\mathbf{x}} b_{\mathbf{x}+\mathbf{v}_i} + b^\dagger_{\mathbf{x}+\mathbf{v}_i} a_{\mathbf{x}}\right), \tag{12.103}$$

with t the hopping parameter and where the spin degree of freedom has been suppressed. The operator $a^\dagger_{\mathbf{x}}$ creates an electron at the site of sublattice A specified by the position vector $\mathbf{x}$, while $a_{\mathbf{x}}$ annihilates an electron at that site, and similar definitions for the creation ($b^\dagger$) and annihilation (b) operators on sublattice B. Repeating the steps leading to Eq. (7.8) in our treatment of the Hubbard model, we obtain for the Hamiltonian in momentum space

$$H = \sum_{\mathbf{k}} c^\dagger_{\mathbf{k}} \begin{pmatrix} 0 & K_{\mathbf{k}} \\ K^*_{\mathbf{k}} & 0 \end{pmatrix} c_{\mathbf{k}}, \qquad c_{\mathbf{k}} \equiv \begin{pmatrix} a_{\mathbf{k}} \\ b_{\mathbf{k}} \end{pmatrix}, \tag{12.104}$$

with $K_{\mathbf{k}} \equiv -t\,\sum_{i=1}^{3} e^{i\mathbf{k}\cdot\mathbf{v}_i}$ and where the sum over $\mathbf{k}$ is restricted to the first Brillouin zone. The two components of the spinor $c_{\mathbf{k}}$ refer to the two sublattices. The energy spectrum of the fermionic excitations follows from the secular equation

$$\det\begin{pmatrix} -E & K_{\mathbf{k}} \\ K^*_{\mathbf{k}} & -E \end{pmatrix} = 0, \tag{12.105}$$

which gives rise to two bands

$$E_{\mathbf{k}} = \pm t\sqrt{1 + 4\cos(\tfrac{1}{2}\sqrt{3}k^1 a)\cos(\tfrac{3}{2}k^2 a) + 4\cos^2(\tfrac{1}{2}\sqrt{3}k^1 a)}. \tag{12.106}$$

The most remarkable property of this spectrum is that the two bands touch each other at the six corners of the first Brillouin zone,

$$\mathbf{k}_{\mathrm{F}} = \left\{ \pm\left(\frac{4\pi}{3\sqrt{3}a}, 0\right),\ \pm\left(\pm\frac{2\pi}{3\sqrt{3}a}, \frac{2\pi}{3a}\right)\right\}, \tag{12.107}$$

see Fig. 12.8. This set of diabolic points, for which the excitation energy van-

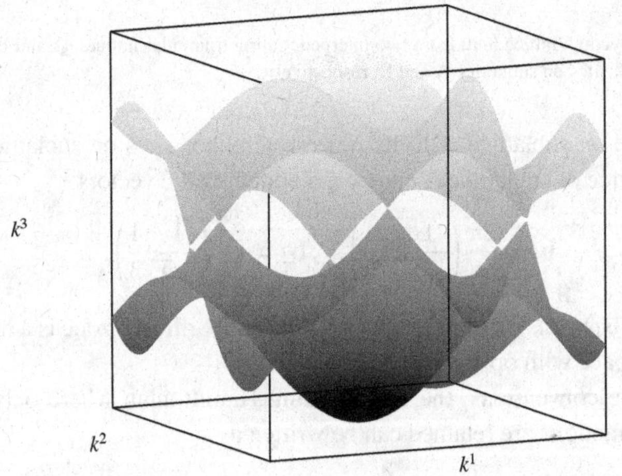

Fig. 12.8 Energy spectrum of the fermionic excitations in graphene.

ishes, forms the Fermi "surface" of the system. Only two degeneracy points are inequivalent since the other four are connected to these by multiples of the reciprocal lattice vectors $\mathbf{b}_1$ and $\mathbf{b}_2$. We may choose the first two listed in Eq. (12.107), which we denote by $\mathbf{k}_\pm$, as independent. When linearized around these Fermi points by setting $\mathbf{k} = \mathbf{k}_\pm + \tilde{\mathbf{k}}$, Eq. (12.104) takes the form of a massless Dirac theory on the lattice

$$H_\pm = v_{\mathrm{F}} \sum_{\mathbf{k}} c^\dagger_{\mathbf{k}} \begin{pmatrix} 0 & \mp\hbar\tilde{k}^1 + i\hbar\tilde{k}^2 \\ \mp\hbar\tilde{k}^1 - i\hbar\tilde{k}^2 & 0 \end{pmatrix} c_{\mathbf{k}}, \tag{12.108}$$

implying a linear rather than a quadratic fermionic excitation spectrum with the Fermi velocity $v_F = 3ta/2\hbar$ playing the role of the speed of light. The emerging one-particle Dirac Hamiltonian H_D can be put in the standard form, cf. Eq. (3.119)

$$H_D = v_F \boldsymbol{\alpha} \cdot \mathbf{p}, \tag{12.109}$$

with $\boldsymbol{\alpha} = (-\sigma^1, -\sigma^2)$ for the excitations around $\mathbf{k}_+$, and $\boldsymbol{\alpha} = (\sigma^1, -\sigma^2)$ for the excitations around $\mathbf{k}_-$, with σ^i $(i = 1, 2)$ the first two Pauli matrices. The third Pauli matrix σ^3 is, apart from a numerical factor $\pm\hbar/2$, the so-called *isospin* operator in the missing 3-direction. For relativistic planar electrons in vacuum, this operator would give the spin, but in the context of graphene this operator is unrelated to the spin of the charge carriers, which we have ignored, and refers instead to the presence of two sublattices. The two diabolic points lead to identical results and can therefore be accounted for by simply multiplying the relevant expressions by a factor of two. In the following, we focus on the diabolic point $\mathbf{k}_+$.

When graphene is placed in an external magnetic field, the spectrum of the Hamiltonian (12.109) with the minimal substitution $\mathbf{p} \to \mathbf{p} + (e/c)\mathbf{A}$ for electrons of charge $q = -e(< 0)$ splits up in discrete Landau levels. Because σ^3 anticommutes with the Hamiltonian, the energy levels are symmetric about $E = 0$. Specifically, if φ_n is an eigenstate with energy E_n,

$$v_F \boldsymbol{\alpha} \cdot \left(\mathbf{p} + \frac{e}{c}\mathbf{A}\right) \varphi_n = E_n \varphi_n, \tag{12.110}$$

also $\sigma^3 \varphi_n$ is an eigenstate of the Hamiltonian with eigenvalue $-E_n$, cf. Secs. 9.5 and 9.9. For zero energy, the two eigenvalue equations become degenerate, and the corresponding states can be chosen to be eigenstates of the isospin operator σ^3. For finite-energy states this is impossible. Explicitly, the Landau levels read

$$E_{\pm n} = \pm\sqrt{2n\hbar v_F^2 |eB|/c} \tag{12.111}$$

for $n = 1, 2, \cdots$ and $E_0 = 0$. Because the Zeeman splitting in graphene is much smaller than the spacing between the Landau levels, the Zeeman term can be ignored and the spin degrees of freedom can be accounted for by simply including a factor of 2 in the relevant expressions.

Following our approach to the IQHE, we consider for the moment a grand canonical ensemble specified by the chemical potential μ. The one-loop effective theory, obtained by integrating out the fermionic degrees of freedom, is given by Eq. (3.135) with the obvious changes,

$$\mathcal{L}_1 = \frac{|eB|}{4\pi\hbar c} \sum_{n=-\infty}^{\infty} |E_n - \mu|. \tag{12.112}$$

The chemical potential μ, which we take to be positive, is included by simply shifting all the energy levels, $E_n \to E_n - \mu$. Equation (12.112) is to be compared to its nonrelativistic counterpart (12.8) appropriate for electrons with a spectrum that depends quadratically on the momentum. The main difference is that for electrons with a linear spectrum, both positive- and negative-energy Landau levels appear, whereas for electrons with a quadratic spectrum, only positive-energy levels appear. Equation (12.112) leads to the average particle number density

$$\begin{aligned} n = \frac{\partial \mathcal{L}_1}{\partial \mu} &= \left[\sum_{n=1}^{\infty} \theta(\mu - E_n) + \frac{1}{2} \right] \frac{|eB|}{2\pi\hbar c} \\ &= \left(i + \tfrac{1}{2}\right) \frac{|eB|}{2\pi\hbar c}, \end{aligned} \tag{12.113}$$

where i denotes the number of filled Landau levels,

$$i \equiv \left[\frac{\mu^2 c}{2\hbar v_F^2 |eB|} \right] \tag{12.114}$$

and $[x]$ the integer-part function, returning the largest integer less than x. As before, we disregard the possibility that Landau levels coincide with the chemical potential, in which case the value of the step function in Eq. (12.113) becomes ambiguous. The expression (12.114) depends quadratically on μ rather than linearly as is the case for electrons with a quadratic spectrum, see Eq. (3.151). The main difference with the previous result (12.15) for electrons with a quadratic spectrum is that here, due to the presence of a zero-energy mode, the charge carrier density induced by an applied magnetic field is nonzero even when $i = 0$. The relative weight one-half carried by the zero-energy mode derives from the fact that this mode is unpaired, whereas the Landau levels with nonzero energy come in pairs, see Eq. (12.111), with $E_{+n}(> 0)$ and $E_{-n} = -E_{+n}$ contributing equally. By attaching a gate voltage to the substrate on which the graphene sits, experimentalists can inject additional electrons (or holes) into the sample and thereby tune the position of the Fermi energy so that additional Landau levels become occupied. Experiments have been carried out with charge carrier densities n of order 10^{12} cm^{-2}.

If, in addition to a magnetic field, also a uniform static electric field E is applied, a Hall current

$$\langle j_e^2 \rangle = -\left(i + \tfrac{1}{2}\right) eE/2\pi \tag{12.115}$$

arises. Without loss of generality, the electric field is chosen in the positive 1-direction. As Eq. (12.17), this result is obtained by simply multiplying n with the drift velocity (12.18). This results predicts a half-integer Hall conductivity σ_H,

$$\sigma_H = \left(i + \tfrac{1}{2}\right) e^2/2\pi\hbar, \tag{12.116}$$

with $i = 0, 1, 2, \cdots$. For graphene, this expression is to be multiplied by a factor of 4 to account for the spin degeneracy and the presence of two independent diabolic points. The connection with experiment, where the electron number rather than the chemical potential is kept fixed, can be established as for the IQHE by including random impurities.

Notes

(i) The IQHE was discovered by von Klitzing *et al.* (1980). For a first-hand account of the discovery, see [von Klitzing (2005)]. See von Klitzing's Nobel lecture [von Klitzing (1986)] for a clear discussion of the difference between resistivity and resistance.

(ii) The FQHE was discovered by Tsui *et al.* (1982). For lucid and authoritative accounts of the field, see the reviews by the discoverers [Tsui (1990); Stormer *et al.* (1999)], and [Stormer (1999)].

(iii) The celebrated wave function, describing the FQHE with filling factor $\nu = 1/m$, where m is an odd integer, was proposed by Laughlin (1983). For a review, see [Laughlin (1990)].

(iv) The fractional charges described by Laughlin's theory were shown to also carry fractional statistics by Arovas *et al.* (1984).

(v) Experimental evidence for fractional charges was provided in, for example, [Saminadayar *et al.* (1997)] and [de Picciotto *et al.* (1997)].

(vi) The reinterpretation of Laughlin's wave function in terms of composite fermions is due to Jain (1989). For a review, see [Jain (1992)].

(vii) The possibility that a charge-flux composite in two space dimensions may carry fractional statistics was pointed out by Wilczek (1982).

(viii) Our discussion of the Aharonov-Bohm phase [Aharonov and Bohm (1959)] that arises when exchanging two charge-flux composites is based on [Goldhaber *et al.* (1989)].

(ix) The even-denominator $\nu = \frac{5}{2}$ FQHE state was discovered by Willett *et al.* (1987).

(x) The physical relevance of composite fermions was established in [Kang *et al.* (1993)].

(xi) The Chern-Simons term for gauge theories in 2+1 dimensions was introduced by Jackiw and Templeton (1981) and Schonfeld (1981).

(xii) An effective *bosonic* field theory of the FQHE, featuring a Chern-Simons term, was first put froward by Girvin and MacDonald (1987) (see also [Girvin (1987)]) and improved upon by Zhang *et al.* (1989). For a review of this bosonic approach, see [Zhang (1992)]. The *fermionic* effective field theory, which we discuss, was put forward in [Lopez and Fradkin (1991)], see also the textbook [Fradkin (1991)].

(xiii) The band theory of graphite was developed in [Wallace (1947)], while the Landau levels in graphene were first calculated in [McClure (1956)].

(xiv) The emergence of (2+1)-dimensional QED in graphene was pointed out by Semenoff (1984) and further investigated by Haldane (1988).

(xv) The observation that an external field induces a vacuum current ($\mu = 0$) in (2+1)-dimensional QED is due to Niemi and Semenoff (1983).

(xvi) The expression (12.113) for the induced charge density in (2+1)-dimensional QED at finite density was first given by Lykken *et al.* (1990) in their study of anyon superconductivity. The expression (12.116) first appeared in [Gonzalez Felipe *et al.* (1990)].

(xvii) The half-integer QHE in graphene was predicted in [Schakel (1991)] and observed by Novoselov *et al.* (2005). Since then, the field is rapidly developing. For recent reviews, see [Geim and Novoselov (2007)] (experimental) and [Gusynin *et al.* (2007)] (theoretical).

(xviii) The Landau-level technique, which we use in our presentation, was applied earlier to (2+1)-dimensional QED by Hughes (1984) and Cea (1985).

Bibliography

Abers, E. S. and Lee, B. W. (1973). Gauge theories, *Phys. Rep.* **9**, pp. 1–141.

Abrahams, E. and Tsuneto, T. (1966). Time variation of the Ginzburg-Landau order parameter, *Phys. Rev.* **152**, pp. 416–432.

Abrikosov, A. A. (1957). On the magnetic properties of superconductors of the second group, *Zh. Eksp. Teor.* **32**, pp. 1442–1452, Engl. transl.: Sov. Phys. JETP **5**, 1174–1182 (1957).

Abrikosov, A. A. (1965). Electron scattering on magnetic impurities in metals and anomalous resistivity effect, *Physics* **2**, pp. 5–20.

Abrikosov, A. A. (1988). *Fundamentals of the Theory of Metals* (North-Holland, Amsterdam).

Abrikosov, A. A. and Gorkov, L. P. (1960). Theory of superconducting alloys with paramagnetic impurities, *Zh. Eksp. Teor.* **39**, pp. 1781–1796, Engl. transl.: Sov. Phys. JETP **12**, 1243–1253 (1961).

Abrikosov, A. A., Gorkov, L. P. and Dzyaloshinski, I. E. (1963). *Methods of Quantum Field Theory in Statistical Physics* (Dover, New York, N.Y.).

Adam, C. (1994). Instantons and vacuum expectation values in the Schwinger model, *Z. Phys.* **C63**, pp. 169–180.

Affleck, I. (2005). Non-Fermi liquid behavior in Kondo models, *J. Phys. Soc. Jpn.* **74**, pp. 59–66.

Aharonov, Y. and Bohm, D. (1959). Significance of electromagnetic potentials in the quantum theory, *Phys. Rev.* **115**, pp. 485–491.

Aitchison, I. J. R. (1987). Berry phases, magnetic monopoles and Wess-Zumino terms or how the Skyrmion got its spin, *Acta Phys. Polon.* **B18**, pp. 207–235.

Aitchison, I. J. R. and Fraser, C. M. (1984). Fermion loop contribution to skyrmion stability, *Phys. Lett.* **B146**, pp. 63–66.

Aitchison, I. J. R. and Fraser, C. M. (1985). Derivative expansions of fermion determinants: Anomaly induced vertices, Goldstone-Wilczek currents and Skyrme terms, *Phys. Rev.* **D31**, pp. 2605–2615.

Ambegaokar, V. and Kadanoff, L. P. (1961). Electromagnetic properties of superconductors, *Nuovo Cimento* **22**, pp. 914–935.

Anderson, P. W. (1958). Random-phase approximation in the theory of superconductivity, *Phys. Rev.* **112**, pp. 1900–1916.

Anderson, P. W. (1959). Theory of dirty superconductors, *J. Phys. Chem. Sol.* **11**, pp. 26–30.

Anderson, P. W. (1963). Plasmons, gauge invariance, and mass, *Phys. Rev.* **130**, pp. 439–442.

Anderson, P. W. (1966). Considerations on the flow of superfluid Helium, *Rev. Mod. Phys.* **38**, pp. 298–310.

Anderson, P. W. (1970). A poor man's derivation of scaling laws for the Kondo problem, *J. Phys.* **C3**, pp. 2436–2441.

Anderson, P. W. (1984). *Basic Notations of Condensed Matter Physics* (Benjamin-Cummings, Menlo Park, CA).

Anderson, P. W. and Brinkman, W. F. (1973). Anisotropic superfluidity in He^3: A possible interpretation of its stability as a spin-fluctuation effect, *Phys. Rev. Lett.* **30**, pp. 1108–1111.

Anderson, P. W. and Brinkman, W. F. (1975). Theory of anisotropic superfluidity in He^3, in J. G. M. Armitage and I. E. Farquhar (eds.), *The Helium Liquids* (Academic Press, New York, N.Y.), pp. 315–416.

Anderson, P. W. and Morel, P. (1961). Generalized Bardeen-Cooper-Schrieffer states and the proposed low-temperature phase of liquid He^3, *Phys. Rev.* **123**, pp. 1911–1934.

Arnold, P. and Tomášik, B. (2000). T_c for dilute Bose gases: Beyond leading order in $1/N$, *Phys. Rev.* **A62**, p. 063604.

Arovas, D., Schrieffer, J. R. and Wilczek, F. (1984). Fractional statistics and the quantum Hall effect, *Phys. Rev. Lett.* **53**, pp. 722–723.

Auer, J. and Ullmaier, H. (1973). Magnetic behavior of type-II superconductors with small Ginzburg-Landau parameters, *Phys. Rev.* **B7**, pp. 136–145.

Baaquie, B. E. (1982). New solution for the Schwinger model, *J. Phys.* **G8**, pp. 1621–1636.

Balatsky, A. V. (1987). Microscopically derived Wess-Zumino action for ^{3}He-A, *Phys. Lett.* **A123**, pp. 27–30.

Balatsky, A. V., Volovik, G. E. and Konyshev, V. A. (1986). On the chiral anomaly in superfluid ^{3}He-A, *Zh. Eksp. Teor. Fiz.* **90**, pp. 2038–2056, Engl. transl.: Sov. Phys. JETP **63**, 1194–1204 (1986).

Balian, R. and Werthamer, N. R. (1963). Superconductivity with pairs in a relative p wave, *Phys. Rev.* **131**, pp. 1553–1564.

Balibar, S. (2007). The discovery of superfluidity, *J. Low Temp. Phys.* **146**, pp. 441–470.

Bardeen, J., Cooper, L. N. and Schrieffer, J. R. (1957). Theory of superconductivity, *Phys. Rev.* **108**, pp. 1175–1204.

Baym, G., Blaizot, J.-P. and Zinn-Justin, J. (2000). The transition temperature of the dilute interacting Bose gas for N internal states, *Europhys. Lett.* **49**, pp. 150–155.

Beasley, M. R., Mooij, J. E. and Orlando, T. P. (1979). Possibility of vortex-antivortex pair dissociation in two-dimensional superconductors, *Phys. Rev. Lett.* **42**, pp. 1165–1168.

Beliaev, S. T. (1958). Energy spectrum of a non-ideal Bose gas, *Zh. Eksp. Teor. Fiz.* **34**, pp. 433–446, Engl. transl.: Sov. Phys. JETP **7**, 299–307 (1958).

Belitz, D. and Kirkpatrick, T. R. (1994). The Anderson-Mott transition, *Rev. Mod. Phys.* **66**, pp. 261–380.

Benson, K. M., Bernstein, J. and Dodelson, S. (1991). Phase structure and the effective potential at fixed charge, *Phys. Rev.* **D44**, pp. 2480–2497.

Berezinskii, V. L. (1970). Destruction of long-range order in one-dimensional and two-dimensional systems having a continuous symmetry group. I. Classical systems, *Zh.*

Eksp. Teor. Fiz. **59**, pp. 907–920, Engl. transl.: Sov. Phys. JETP **32**, 493–500 (1971).

Berezinskii, V. L. (1971). Destruction of long-range order in one dimensional and two-dimensional systems having a continuous symmetry group. II. Quantum systems, *Zh. Eksp. Teor. Fiz.* **61**, pp. 1144–1156, Engl. transl.: Sov. Phys. JETP **34**, 610–616 (1972).

Bernard, C. W. (1974). Feynman rules for gauge theories at finite temperature, *Phys. Rev.* **D9**, pp. 3312–3320.

Bernstein, J. and Dodelson, S. (1991). Relativistic Bose gas, *Phys. Rev. Lett.* **66**, pp. 683–686.

Berry, M. V. (1984). Quantal phase factors accompanying adiabatic changes, *Proc. R. Soc. London* **A392**, pp. 45–57.

Bettencourt, L. M. A. and Rivers, R. J. (1995). Interactions between U(1) cosmic strings: An analytical study, *Phys. Rev.* **D51**, pp. 1842–1853.

Bishop, D. J. and Reppy, J. D. (1980). Study of the superfluid transition in two-dimensional ^{4}He films, *Phys. Rev.* **B22**, pp. 5171–5185.

Blaha, S. (1976). Quantization rules for point singularities in superfluid ^{3}He and liquid crystals, *Phys. Rev. Lett.* **36**, pp. 874–876.

Bloch, F. (1930). Zur Theorie der Ferromagnetismus, *Z. Phys.* **61**, pp. 206–219.

Bloch, F. (1932). Zur Theorie des Austauschproblems und der Remanenzerscheinung der Ferromagnetika, *Z. Phys.* **74**, pp. 295–335.

Bodwin, G. T. and Kovacs, E. V. (1987). Lattice fermions in the Schwinger model, *Phys. Rev.* **D35**, pp. 3198–3218.

Bogoliubov, N. N. (1947). On the theory of superfluidity, *J. Phys. (Moscow)* **11**, pp. 23–32.

Bogoliubov, N. N. (1958a). A new method in the theory of superconductivity. I, *Zh. Eksp. Teor. Fiz.* **34**, pp. 58–65, Engl. transl.: Sov. Phys. JETP **7**, 41–46 (1958).

Bogoliubov, N. N. (1958b). On a new method in the theory of superconductivity, *Nuovo Cimento* **7**, pp. 794–805.

Bogoliubov, N. N., Tolmachev, V. V. and Shirkov, D. V. (1958). A new method in the theory of superconductivity, *Fortschr. Phys.* **6**, pp. 605–682.

Bohm, D. and Pines, D. (1953). A collective description of electron interactions. III. Coulomb interactions in a degenerate electron gas, *Phys. Rev.* **92**, pp. 609–625.

Brown, L. S. (1992). *Quantum Field Theory* (Cambridge University Press, Cambridge, U.K.).

Callan, C. G., Dashen, R. F. and Gross, D. J. (1977). Pseudoparticles and massless fermions in two dimensions, *Phys. Rev.* **D16**, pp. 2526–2534.

Callan, C. G. and Harvey, J. A. (1985). Anomalies and fermion zero modes on strings and domain walls, *Nucl. Phys.* **B250**, pp. 427–436.

Cao, T. Y. (1993). New philosophy of renormalization: From renormalization group equations to effective field theories, in L. M. Brown (ed.), *Renormalization: From Lorentz to Landau, and Beyond* (Springer-Verlag, Berlin), pp. 89–133.

Cao, T. Y. and Schweber, S. S. (1993). The conceptual foundations and philosophical aspects of renormalization theory, *Synthese* **97**, pp. 33–108.

Cea, P. (1985). Variational approach to (2+1)-dimensional QED, *Phys. Rev.* **D32**, pp. 2785–2793.

Chang, S.-J. (1990). *Introduction to Quantum Field Theory*, *Lecture Notes in Physics*, Vol. 29 (World Scientific, Singapore).

Coleman, S. R. (1975). Secret symmetry: An introduction to spontaneous symmetry breakdown and gauge fields, in A. Zichichi (ed.), *Laws of Hadronic Matter* (Academic Press, New York, N.Y.), pp. 138–215.

Coleman, S. R. (1977). Classical lumps and their quantum descendants, in A. Zichichi (ed.), *New Phenomena in Subnuclear Physics* (Plenum Press, New York, N.Y.), pp. 297–421.

Coleman, S. R. (1979). The uses of instantons, in A. Zichichi (ed.), *The Whys of Subnuclear Physics* (Plenum Press, New York, N.Y.), pp. 805–916.

Coleman, S. R. (1982). $1/N$, in A. Zichichi (ed.), *Pointlike Structures Inside and Outside Hadrons* (Plenum Press, New York, N.Y.), pp. 11–97.

Coleman, S. R. (1988). *Aspects of Symmetry: Selected Erice Lectures of Sidney Coleman* (Cambridge University Press, Cambridge, U.K.).

Coleman, S. R., Jackiw, R. and Susskind, L. (1975). Charge shielding and quark confinement in the massive Schwinger model, *Ann. Phys. (N.Y.)* **93**, pp. 267–275.

Combescot, R. and Dombre, T. (1986). Twisting in superfluid ^{3}He-A and consequences for hydrodynamics at $T = 0$, *Phys. Rev.* **B33**, pp. 79–90.

Continentino, M. A. (1994). Quantum scaling in many-body systems, *Phys. Rep.* **239**, pp. 179–213.

Cornell, E. A. and Wieman, C. E. (2002). Nobel Lecture: Bose-Einstein condensation in a dilute gas, the first 70 years and some recent experiments, *Rev. Mod. Phys.* **74**, pp. 875–893.

Crisan, M. (1989). *Theory of Superconductivity* (World Scientific, Singapore).

Cvitanović, P. (1983). *Field Theory* (Nordita, Copenhagen), URL `http://www.cns.gatech.edu/FieldTheory`.

Cyrot, M. (1973). Ginzburg-Landau theory for superconductors, *Rep. Prog. Phys.* **36**, pp. 103–158.

Dalfovo, F., Giorgini, S., Pitaevskii, L. P. and Stringari, S. (1999). Theory of Bose-Einstein condensation in trapped gases, *Rev. Mod. Phys.* **71**, pp. 463–512.

Das, A. (1999). *Finite Temperature Field Theory* (World Scientific, Singapore).

Das, A. and Karev, A. (1987). Derivative expansion and the solubility of two-dimensional models, *Phys. Rev.* **D36**, pp. 2591–2594.

Dashen, R. F., Hasslacher, B. and Neveu, A. (1975). Semiclassical bound states in an asymptotically free theory, *Phys. Rev.* **D12**, pp. 2443–2458.

de Gennes, P. G. (1989). *Superconductivity of Metals and Alloys*, Advanced Book Classics (Addison-Wesley Publishing Company, Reading, MA).

de Picciotto, R., Reznikov, M., Heiblum, M., Umansky, V., Bunin, G. and Mahalu, D. (1997). Direct observation of a fractional charge, *Nature* **389**, pp. 162–164.

DeMarco, B. and Jin, D. S. (1999). Onset of Fermi degeneracy in a trapped atomic gas, *Science* **285**, pp. 1703–1706.

Dirac, P. A. M. (1948). The theory of magnetic poles, *Phys. Rev.* **74**, pp. 817–830.

Dittrich, W. and Reuter, M. (1986). *Selected Topics in Gauge Theories*, *Lecture Notes in Physics*, Vol. 244 (Springer-Verlag, Berlin).

Dolan, L. and Jackiw, R. (1974). Symmetry behavior at finite temperature, *Phys. Rev.* **D9**, pp. 3320–3341.

Donoghue, J. F. (1994). General relativity as an effective field theory: The leading quantum corrections, *Phys. Rev.* **D50**, pp. 3874–3888.

Drechsler, M. and Zwerger, W. (1992). Crossover from BCS-superconductivity to Bose-condensation, *Ann. Phys. (Leipzig)* **1**, pp. 15–23.

Eagles, D. M. (1969). Possible pairing without superconductivity at low carrier concentrations in bulk and thin-film superconducting semiconductors, *Phys. Rev.* **186**, pp. 456–463.

Eckart, C. (1938). The electrodynamics of material media, *Phys. Rev.* **54**, pp. 920–923.

Edwards, M. and Burnett, K. (1995). Numerical solution of the nonlinear Schrödinger equation for small samples of trapped neutral atoms, *Phys. Rev.* **A51**, pp. 1382–1386.

Eltsov, V. B. and Krusius, M. (2000). Topological defects in ^{3}He superfluids, in Y. M. Bunkov and H. Godfrin (eds.), *Topological Defects and the Non-Equilibrium Dynamics of Symmetry Breaking Phase Transitions*, *NATO Science Series*, Vol. C549 (Kluwer Academic Publishers, Dordrecht, the Netherlands), pp. 325–344.

Euler, H. and Kockel, B. (1935). Über die Streuung von Licht an Licht nach der Diracschen Theorie, *Naturwissenschaften* **23**, pp. 246–247.

Fermi, E. (1928). Eine statistische Methode zur Bestimmung einiger Eigenschaften des Atoms und ihre Anwendung auf die Theorie des periodischen Systems der Elemente, *Z. Phys.* **A48**, pp. 73–79.

Fetter, A. L. and Walecka, J. D. (1971). *Quantum Theory of Many-Particle Systems* (McGraw-Hill, New York, N.Y.).

Feynman, R. P. (1948). Space-time approach to non-relativistic quantum mechanics, *Rev. Mod. Phys.* **20**, pp. 367–387.

Feynman, R. P. (1954). Atomic theory of the two-fluid model of liquid helium, *Phys. Rev.* **94**, pp. 262–277.

Fisher, D. S. and Fisher, M. P. A. (1988). Onset of superfluidity in random media, *Phys. Rev. Lett.* **61**, pp. 1847–1850.

Fisher, M. E. (1983). Scaling, universality and renormalization group theory, in F. J. W. Hahne (ed.), *Critical Phenomena*, *Lecture Notes in Physics*, Vol. 186 (Springer-Verlag, Berlin), pp. 1–139.

Fisher, M. E. (1998). Renormalization group theory: Its basis and formulation in statistical physics, *Rev. Mod. Phys.* **70**, pp. 653–681.

Fisher, M. P. A. (1990). Quantum phase transitions in disordered two-dimensional superconductors, *Phys. Rev. Lett.* **65**, pp. 923–926.

Fisher, M. P. A., Grinstein, G. and Girvin, S. M. (1990). Presence of quantum diffusion in two dimensions: Universal resistance at the superconductor-insulator transition, *Phys. Rev. Lett.* **64**, pp. 587–590.

Fisher, M. P. A., Weichman, P. B., Grinstein, G. and Fisher, D. S. (1989). Boson localization and the superfluid-insulator transition, *Phys. Rev.* **B40**, pp. 546–570.

Fradkin, E. (1991). *Field Theories of Condensed Matter Systems*, *Frontiers in Physics*, Vol. 82 (Addison-Wesley Publishing Company, Reading, MA).

Fradkin, E. S. (1967). Method of Green's functions in quantum field theory and in quantum statistics, in D. V. Skobel'tsyn (ed.), *Quantum Field Theory and Hydrodynamics* (Plenum Publishing Corporation, New York, N.Y.), pp. 1–132.

Fraser, C. M. (1985). Calculation of higher derivative terms in the one-loop effective Lagrangian, *Z. Phys.* **C28**, pp. 101–106.

Fujikawa, K. (1979). Path-integral measure for gauge-invariant fermion theories, *Phys. Rev.*

Lett. **42**, pp. 1195–1198.

Fujikawa, K. (1980). Path integral for gauge theories with fermions, *Phys. Rev.* **D21**, pp. 2848–2858.

Fujikawa, K. and Suzuki, H. (2004). *Path Integrals and Quantum Anomalies*, The International Series of Monographs on Physics (Oxford University Press, Oxford).

Gavoret, J. and Nozières, P. (1964). Structure of the perturbation expansion for the Bose liquid at zero temperature, *Ann. Phys. (N.Y.)* **28**, pp. 349–399.

Geim, A. K. and Novoselov, K. S. (2007). The rise of graphene, *Nature Materials* **6**, pp. 183–191.

Gell-Mann, M. and Brueckner, K. A. (1957). Correlation energy of an electron gas at high density, *Phys. Rev.* **106**, pp. 364–368.

Ginzburg, V. L. and Andryushin, E. A. (1994). *Superconductivity* (World Scientific, Singapore).

Ginzburg, V. L. and Landau, L. D. (1950). Theory of superconductivity, *Zh. Eksp. Teor. Fiz.* **20**, pp. 1064–1082.

Girvin, S. M. (1987). Summary, omissions an unanswered questions, in R. Prange and S. M. Girvin (eds.), *The Quantum Hall Effect* (Springer-Verlag, Berlin), pp. 381–399.

Girvin, S. M. and MacDonald, A. H. (1987). Off-diagonal long-range order, oblique confinement, and the fractional quantum Hall effect, *Phys. Rev. Lett.* **58**, pp. 1252–1255.

Girvin, S. M., Wallin, M., Cha, M.-C., Fisher, M. P. A. and Young, A. P. (1992). Universal conductivity at the superconductor-insulator transition in two-dimensions, *Prog. Theor. Phys. Suppl.* **107**, pp. 135–144.

Goff, W. E., Gaitan, F. and Stone, M. (1989). Berry's phase and the Wess-Zumino effective action in ^{3}He-A, *Phys. Lett.* **A136**, pp. 433–440.

Goldhaber, A. S., MacKenzie, R. and Wilczek, F. (1989). Field corrections to induced statistics, *Mod. Phys. Lett.* **A4**, pp. 21–31.

Goldman, A. M. and Marković, N. (1998). Superconductor-insulator transitions in the two-dimensional limit, *Phys. Today* **51**, pp. 39–44.

Goldstein, H. (1980). *Classical Mechanics*, 2nd edn. (Addison-Wesley Publishing Company, Reading, MA).

Goldstone, J. (1961). Field theories with "superconductor" solutions, *Nuovo Cimento* **19**, pp. 154–164.

Goldstone, J., Salam, A. and Weinberg, S. (1962). Broken symmetries, *Phys. Rev.* **127**, pp. 965–970.

Goldstone, J. and Wilczek, F. (1981). Fractional quantum numbers on solitons, *Phys. Rev. Lett.* **47**, pp. 986–989.

Gonzalez Felipe, R., Perez Martinez, A. and Perez Rojas, H. (1990). Relativistic quantum Hall conductivity for 3D and 2D electron plasma in an external magnetic field, *Mod. Phys. Lett.* **B4**, pp. 1103–1110.

Gorkov, L. P. (1959a). Microscopic derivation of the Ginzburg–Landau equations in the theory of superconductivity, *Zh. Eksp. Teor. Fiz.* **36**, pp. 1918–1923, Engl. transl.: Sov. Phys. JETP **9**, 1364–1367 (1959).

Gorkov, L. P. (1959b). Theory of superconducting alloys in a strong magnetic field near the critical temperature, *Zh. Eksp. Teor. Fiz.* **37**, pp. 1407–1416, Engl. transl.: Sov. Phys. JETP **10**, 998–1004 (1959).

Greiner, M., Mandel, O., Esslinger, T., Hänsch, T. W. and Bloch, I. (2002). Quantum phase transition from a superfluid to a Mott insulator in a gas of ultracold atoms, *Nature* **415**, pp. 39–44.

Greiner, M., Regal, C. A. and Jin, D. S. (2003). Emergence of a molecular Bose-Einstein condensate from a Fermi gas, *Nature* **426**, pp. 537–540.

Greiter, M., Wilczek, F. and Witten, E. (1989). Hydrodynamic relations in superconductivity, *Mod. Phys. Lett.* **B3**, pp. 903–918.

Griffin, A. (1999). A brief history of our understanding of BEC: From Bose to Beliaev, in M. Inguscio, S. Stringari and C. Wieman (eds.), *Bose-Einstein Condensation in Atomic Gases* (IOS Press, Amsterdam), pp. 1–12.

Gross, E. P. (1958). Classical theory of boson wave fields, *Ann. Phys. (N.Y.)* **4**, pp. 57–74.

Gross, E. P. (1961). Structure of a quantized vortex in boson systems, *Nuovo Cimento* **20**, pp. 454–477.

Gunton, J. D. and Buckingham, M. J. (1968). Condensation of the ideal Bose gas as a cooperative transition, *Phys. Rev.* **166**, pp. 152–158.

Gusynin, V. P., Sharapov, S. G. and Carbotte, J. P. (2007). AC conductivity of graphene: From tight-binding model to 2+1-dimensional quantum electrodynamics, *Int. J. Mod. Phys.* **B21**, pp. 4611–4658.

Haber, H. E. and Weldon, H. A. (1981). Thermodynamics of an ultrarelativistic ideal Bose gas, *Phys. Rev. Lett.* **46**, pp. 1497–1500.

Haber, H. E. and Weldon, H. A. (1982). Finite-temperature symmetry breaking as Bose-Einstein condensation, *Phys. Rev.* **D25**, pp. 502–525.

Hakonen, P. and Lounasmaa, O. V. (1987). Vortices in rotating superfluid He^3, *Phys. Today*, pp. 70–78.

Haldane, F. D. M. (1988). Model for a quantum Hall effect without Landau levels: Condensed-matter realization of the "parity anomaly", *Phys. Rev. Lett.* **61**, pp. 2015–2018.

Haldane, F. D. M. and Wu, Y.-S. (1985). Quantum dynamics and statistics of vortices in two-dimensional superfluids, *Phys. Rev. Lett.* **55**, pp. 2887–2890.

Hall, D. S. (2003). Resource letter; BEC-1: Bose-Einstein condensates in trapped dilute gases, *Am. J. Phys.* **71**, pp. 649–660.

Haussmann, R. (1993). Crossover from BCS superconductivity to Bose-Einstein condensation: A self-consistent theory, *Z. Phys.* **B91**, pp. 291–308.

Hearn, A. C. (2004). *Reduce*, URL `http://www.reduce-algebra.com`.

Hebard, A. F. and Paalanen, M. A. (1990). Magnetic-field-tuned superconductor-insulator transition in two-dimensional films, *Phys. Rev. Lett.* **65**, pp. 927–930.

Heisenberg, W. and Euler, H. (1936). Folgerungen aus der Diracschen Theorie des Positrons, *Z. Phys.* **98**, pp. 714–732.

Hertz, J. A. (1976). Quantum critical phenomena, *Phys. Rev.* **B14**, pp. 1165–1184.

Hoddeson, L., Schubert, H., Heims, S. J. and Baym, G. (1992). Collective phenomena, in L. Hoddeson, E. Braun, J. Teichmann and S. Weart (eds.), *Out of the Crystal Maze: Chapters from The History of Solid State Physics* (Oxford University Press, Oxford), pp. 489–616.

Huang, K. and Meng, H.-F. (1992). Hard-sphere Bose gas in random external potentials, *Phys. Rev. Lett.* **69**, pp. 644–647.

Huang, K. and Polonyi, J. (1991). Renormalization of the sine-Gordon model and nonconservation of the kink current, *Int. J. Mod. Phys.* **A6**, pp. 409–430.

Hubbard, J. (1963). Electron correlations in narrow energy bands, *Proc. R. Soc. London* **A276**, pp. 238–257.

Hubbard, J. (1964a). Electron correlations in narrow energy bands. II. The degenerate band case, *Proc. R. Soc. London* **A277**, pp. 237–259.

Hubbard, J. (1964b). Electron correlations in narrow energy bands. III. An improved solution, *Proc. R. Soc. London* **A281**, pp. 401–419.

Hugenholtz, N. M. (1965). Quantum theory of many-body systems, *Rep. Prog. Phys.* **28**, pp. 201–247.

Hugenholtz, N. M. and Pines, D. (1959). Ground-state energy and excitation spectrum of a system of interacting bosons, *Phys. Rev.* **116**, pp. 489–506.

Hughes, R. J. (1984). The effective action for photons in (2+1)-dimensions, *Phys. Lett.* **B148**, pp. 215–219.

Jackiw, R. (1985). Field theoretic investigations in current algebra, in S. B. Treiman, R. Jackiw, B. Zumino and E. Witten (eds.), *Current Algebra and Anomalies* (World Scientific, Singapore), pp. 81–210.

Jackiw, R. and Rebbi, C. (1976). Solitons with fermion number $\frac{1}{2}$, *Phys. Rev.* **D13**, pp. 3398–3409.

Jackiw, R. and Templeton, S. (1981). How super-renormalizable interactions cure their infrared divergences, *Phys. Rev.* **D23**, pp. 2291–2304.

Jackson, J. D. (1975). *Classical Electrodynamics*, 2nd edn. (Wiley, New York, N.Y.).

Jain, J. K. (1989). Composite-fermion approach for the fractional quantum Hall effect, *Phys. Rev. Lett.* **63**, pp. 199–202.

Jain, J. K. (1992). Microscopic theory of the fractional quantum Hall effect, *Adv. Phys.* **41**, pp. 105–146.

Jaksch, D., Bruder, C., Cirac, J. I., Gardiner, C. W. and Zoller, P. (1998). Cold bosonic atoms in optical lattices, *Phys. Rev. Lett.* **81**, pp. 3108–3111.

Jayewardena, C. (1988). Schwinger model on S^2, *Helv. Phys. Acta* **61**, pp. 636–711.

Jochim, S., Bartenstein, M., Altmeyer, A., Hendl, G., Riedl, S., Chin, C., Hecker Denschlag, J. and Grimm, R. (2003). Bose-Einstein condensation of molecules, *Science* **302**, pp. 2101–2103.

Jolicoeur, T. and Le Guillou, J. C. (1993). Abelian bosonization in a path integral framework, *Int. J. Mod. Phys.* **A8**, pp. 1923–1931.

Josephson, B. D. (1962). Possible new effects in superconductive tunneling, *Phys. Lett.* **1**, pp. 251–253.

Josephson, B. D. (1969). Weakly coupled superconductors, in R. D. Parks (ed.), *Superconductivity*, Vol. 1 (Marcel Decker, New York, N.Y.), pp. 423–448.

Kang, W., Stormer, H. L., Pfeiffer, L. N., Baldwin, K. W. and West, K. W. (1993). How real are composite fermions? *Phys. Rev. Lett.* **71**, pp. 3850–3853.

Kapusta, J. I. (1981). Bose-Einstein condensation, spontaneous symmetry breaking, and gauge theories, *Phys. Rev.* **D24**, pp. 426–439.

Kapusta, J. I. (1989). *Finite Temperature Field Theory*, Cambridge Monographs on Mathematical Physics (Cambridge University Press, Cambridge, U.K.).

Kashurnikov, V. A., Prokof'ev, N. V. and Svistunov, B. V. (2001). Critical temperature shift in weakly interacting Bose gas, *Phys. Rev. Lett.* **87**, p. 120402.

Kemoklidze, M. P. and Pitaevskii, L. P. (1966). On the dynamics of a superfluid Fermi gas, *Zh. Eksp. Teor. Fiz.* **50**, pp. 243–250, Engl. transl.: Sov. Phys. JETP, **23**, 160–164 (1966).

Ketterle, W. (2002). Nobel lecture: When atoms behave as waves: Bose-Einstein condensation and the atom laser, *Rev. Mod. Phys.* **74**, pp. 1131–1151.

Kleinert, H. (1989). *Gauge Fields in Condensed Matter, Vol. 1* (World Scientific, Singapore).

Kondo, J. (1964). Resistance minimum in dilute magnetic alloys, *Prog. Theor. Phys.* **32**, pp. 37–49.

Kopnin, N. B. (2002). Introduction to Ginzburg-Landau and Gross-Pitaevskii theories for superconductors and superfluids, *J. Low Temp. Phys.* **129**, pp. 219–262.

Kosterlitz, J. M. (1974). The critical properties of the two-dimensional *xy* model, *J. Phys.* **C7**, pp. 1046–1060.

Kosterlitz, J. M. and Thouless, D. J. (1973). Ordering, metastability and phase transitions in two-dimensional systems, *J. Phys.* **C6**, pp. 1181–1203.

Kragelöh, U. (1970). The intermediate state in type II superconductors, *Physica Status Solidi* **42**, pp. 559–576.

Krive, I. V. and Rozhavskii, A. S. (1987). Fractional charge in quantum field theory and solid state physics, *Usp. Fiz. Nauk.* **152**, pp. 33–74, Engl. transl.: Sov. Phys. Usp. **30**, 370–392 (1987).

Kubo, R. (1965). *Statistical Mechanics* (North-Holland, Amsterdam).

Lamb, H. (1932). *Hydrodynamics*, 6th edn. (Dover, New York, N.Y.).

Landau, L. D. (1930). Diamagnetismus der Metalle, *Z. Phys.* **A64**, pp. 629–637.

Landau, L. D. (1937). Zur Theorie der Phasenumwandlungen. I, *Phys. Z. Sowjetunion* **11**, pp. 26–47.

Landau, L. D. (1941). Theory of superfluidity of He II, *J. Phys. (Moscow)* **5**, pp. 71–90.

Landau, L. D. and Lifshitz, E. M. (1935). On the theory of the dispersion of magnetic permeability in ferromagnetic bodies, *Phys. Z. Sowjetunion* **8**, pp. 153–169.

Landau, L. D. and Lifshitz, E. M. (1975). *The Classical Theory of Fields, Course of Theoretical Physics*, Vol. II, 4th edn. (Pergamon Press, Oxford), translation from Russian, Nauka, Moscow, 1973.

Landsman, N. P. and van Weert, C. G. (1987). Real and imaginary time field theory at finite temperature and density, *Phys. Rep.* **145**, pp. 141–249.

Laughlin, R. B. (1983). Anomalous quantum Hall effect: An incompressible quantum fluid with fractionally charged excitations, *Phys. Rev. Lett.* **50**, pp. 1395–1398.

Laughlin, R. B. (1990). Fractional statistics in the quantum Hall effect, in F. Wilczek (ed.), *Fractional Statistics and Anyon Superconductivity* (World Scientific, Singapore), pp. 262–306.

Leggett, A. J. (1975). A theoretical description of the new phases of liquid ^{3}He, *Rev. Mod. Phys.* **47**, pp. 331–414.

Leggett, A. J. (1978). Theory of superfluidity in liquid ^{3}He, in J. Ruvalds and T. Regge (eds.), *Quantum Liquids* (North-Holland, Amsterdam), pp. 167–193.

Leggett, A. J. (1980). Diatomic molecules and Cooper pairs, in A. Pekalski and R. Przystawa (eds.), *Modern Trends in the Theory of Condensed Matter*, *Lecture Notes in Physics*, Vol. 115 (Springer-Verlag, Berlin), pp. 13–27.

Leggett, A. J. (1989). Low temperature physics: superconductivity and superfluidity, in P. Davies (ed.), *The New Physics* (Cambridge University Press, Cambridge, U.K.), pp. 268–288.

Leggett, A. J. (1999). Superfluidity, *Rev. Mod. Phys.* **71**, pp. S318–S323.

Leutwyler, H. (1994). Nonrelativistic effective Lagrangians, *Phys. Rev.* **D49**, pp. 3033–3043.

Lieb, E. H. (1963). Exact analysis of an interacting Bose gas. II. The excitation spectrum, *Phys. Rev.* **130**, pp. 1616–1624.

Lieb, E. H. and Liniger, W. (1963). Exact analysis of an interacting Bose gas. I. The general solution and the ground state, *Phys. Rev.* **130**, pp. 1605–1616.

Liu, Y. and Goldman, A. M. (1994). Superconductor-insulator transitions in two dimensions, *Mod. Phys. Lett.* **B8**, pp. 277–300.

London, F. (1938). The λ-phenomenon of liquid helium and the Bose-Einstein degeneracy, *Nature* **141**, pp. 643–644.

London, F. (1950). *Superfluids, Vol. 1: Macroscopic Theory of Superconductivity* (Wiley, New York, N.Y.).

London, F. (1954). *Superfluids, Vol. 2: Macroscopic Theory of Superfluid Helium* (Wiley, New York, N.Y.).

Lopez, A. and Fradkin, E. (1991). Fractional quantum Hall effect and Chern-Simons gauge theories, *Phys. Rev.* **B44**, pp. 5246–5262.

Lounasmaa, O. V. and Thuneberg, E. (1999). Vortices in rotating superfluid ^{3}He, *Proc. Natl. Acad. Sci. USA* **96**, pp. 7760–7767.

Lowenstein, J. H. and Swieca, J. A. (1971). Quantum electrodynamics in two dimensions, *Ann. Phys. (N.Y.)* **68**, pp. 172–195.

Lund, F. (1991). Defect dynamics for the nonlinear Schrödinger equation derived from a variational principle, *Phys. Lett.* **A159**, pp. 245–251.

Lykken, J. D., Sonnenschein, J. and Weiss, N. (1990). Anyonic superconductivity, *Phys. Rev.* **D42**, pp. 2161–2165.

Ma, S.-K. (1973). Critical exponents above T_c to $O(1/n)$, *Phys. Rev.* **A7**, pp. 2172–2187.

Ma, S.-K. (1976). *Modern Theory of Critical Phenomena, Frontiers in Physics*, Vol. 46 (Benjamin, Reading, MA).

Macke, W. (1950). Über die Wechselwirkungen im Fermi Gas: Polarisationsercheinungen, Korrelationsenergie, Elektronenkondensation, *Z. Naturforsch.* **A5**, pp. 192–208.

MacKenzie, R. (1984). *Vacuum Polarization by Solitons*, Ph.D. thesis, University of California, Santa Barbara.

Marini, M., Pistolesi, F. and Strinati, G. C. (1998). Evolution from BCS superconductivity to Bose condensation: Analytic results for the crossover in three dimensions, *Eur. Phys. J.* **B1**, pp. 151–159.

Matthews, P. T. and Salam, A. (1955). Propagators of quantized field, *Nuovo Cimento* **2**, pp. 120–134.

Mattuck, R. D. (1976). *A Guide to Feynman Diagrams in the Many-Body Problem*, 2nd edn. (McGraw-Hill, New York, N.Y.).

May, R. M. (1967). Exact equation of state for a 2-dimensional plasma, *Phys. Lett.* **A25**, p. 282.

McClure, J. W. (1956). Diamagnetism of graphite, *Phys. Rev.* **104**, pp. 666–671.

Meissner, W. and Ochsenfeld, R. (1933). Ein neuer Effekt bei Eintritt der Supraleitfähigkeit, *Naturwissenschaften* **21**, pp. 787–788.

Mermin, N. D. (1978). Superfluidity in helium-3, in J. Ruvalds and T. Regge (eds.), *Quantum Liquids* (North-Holland, Amsterdam), pp. 195–226.

Mermin, N. D. (1979). The topological theory of defects in ordered media, *Rev. Mod. Phys.* **51**, pp. 591–648.

Mermin, N. D. (1981). E pluribus boojum: the physicist as neologist, *Phys. Today* **34**, pp. 46–53.

Mermin, N. D. and Ho, T.-L. (1976). Circulation and angular momentum in the *A* phase of superfluid Helium-3, *Phys. Rev. Lett.* **36**, pp. 594–597.

Mineev, V. P. (1980). Topologically stable defects and solitons in ordered media, in I. M. Khalatnikov (ed.), *Physics Reviews, Soviet Scientific Reviews, Section A*, Vol. 2 (Harwood, Chur, Switzerland), pp. 173–246.

Mineev, V. P. (1983). Superfluid ^{3}He: Introduction to the subject, *Usp. Fiz. Nauk.* **139**, pp. 303–332, Engl. transl.: Sov. Phys. Usp. **26**, 160–175 (1983).

Nakahara, M. (2003). *Geometry, Topology and Physics*, 2nd edn., Graduate Student Series in Physics (IoP Publishing, Bristol, U.K.).

Nambu, Y. (1960). Quasi-particles and gauge invariance in the theory of superconductivity, *Phys. Rev.* **117**, pp. 648–663.

Nambu, Y. and Jona-Lasinio, G. (1961). Dynamical model of elementary particles based on an analogy with superconductivity. I, *Phys. Rev.* **122**, pp. 345–358.

Naón, C. M. (1985). Abelian and non-Abelian bosonization in the path-integral framework, *Phys. Rev.* **D31**, pp. 2035–2044.

Nelson, D. R. and Kosterlitz, J. M. (1977). Universal jump in the superfluid density of two-dimensional superfluids, *Phys. Rev. Lett.* **39**, pp. 1201–1205.

Nielsen, H. B. and Olesen, P. (1973). Vortex-line models for dual strings, *Nucl. Phys.* **B61**, pp. 45–61.

Nielsen, H. B. and Wirzba, A. (1987). The Cheshire cat principle applied to hybrid bag models, in J.-M. Richard, E. Aslanides and N. Boccara (eds.), *The Elementary Structure of Matter: Proceedings of the Workshop, Les Houches, France, March 24 – April 2, 1987* (Springer-Verlag, Berlin), pp. 72–100.

Nielsen, N. K. (1981). Asymptotic freedom as a spin effect, *Am. J. Phys.* **49**, pp. 1171–1178.

Nielsen, N. K. and Schroer, B. (1977a). Axial anomaly and Atiyah-Singer theorem, *Nucl. Phys.* **B127**, pp. 493–508.

Nielsen, N. K. and Schroer, B. (1977b). Topological fluctuations and breaking of chiral symmetry in gauge theories involving massless fermions, *Nucl. Phys.* **B120**, pp. 62–76.

Niemi, A. J. (1987). Quantum holonomy, in M. Jezabek and M. Praszatowicz (eds.), *Proceedings of the Workshop on Skyrmions and Anomalies, Krakow, Poland* (World Scientific, Singapore), pp. 408–434.

Niemi, A. J. and Semenoff, G. W. (1983). Axial-anomaly-induced fermion fractionization and effective gauge-theory actions in odd-dimensional space-times, *Phys. Rev. Lett.* **51**, pp. 2077–2080.

Niemi, A. J. and Semenoff, G. W. (1985). Quantum holonomy and the chiral gauge anomaly, *Phys. Rev. Lett.* **55**, pp. 927–930.

Novoselov, K. S., Geim, A. K., Morozov, S. V., Jiang, D., Katsnelson, M. I., Grigorieva, I. V., Dubonos, S. V. and Firsov, A. A. (2005). Two-dimensional gas of massless Dirac fermions in graphene, *Nature* **438**, pp. 197–200.

Nozières, P. (1974). The Kondo problem: Fancy mathematical techniques versus simple physical ideas, in M. Krusius and M. Vuorio (eds.), *Proceedings of the 14th International Conference on Low Temperature Physics*, Vol. 5 (North-Holland, Amsterdam), pp. 339–374.

Nozières, P. and Pines, D. (1990). *The Theory of Quantum Liquids, Vol. II*, Advanced Book Classics (Addison-Wesley Publishing Company, Reading, MA).

Nozières, P. and Schmitt-Rink, S. (1985). Bose condensation in an attractive fermion gas: From weak to strong coupling superconductivity, *J. Low Temp. Phys.* **59**, pp. 195–211.

Pathria, R. K. (1996). *Statistical Mechanics*, 2nd edn. (Butterworth-Heinemann, Oxford).

Pearl, J. (1964). Current distribution in superconducting films carrying quantized fluxoids, *Appl. Phys. Lett.* **5**, pp. 65–66.

Pearl, J. (1965). Distinctive properties of quantized vortices in superconducting films, in J. G. Daunt, D. O. Edwards, F. J. Milford and M. Yaqub (eds.), *Proceedings of the 9th International Conference on Low Temperature Physics* (Plenum, New York, N.Y.), pp. 566–570.

Peierls, R. E. (1955). *Quantum Theory of Solids*, The International Series of Monographs on Physics (Oxford University Press, Oxford).

Pethick, C. J. and Smith, H. (2002). *Bose-Einstein Condensation In Dilute Gases* (Cambridge University Press, Cambridge, U.K.).

Piette, B. M. A. G., Schroers, B. J. and Zakrzewski, W. J. (1995). Multi-solitons in a two-dimensional Skyrme model, *Z. Phys.* **C65**, pp. 165–174.

Pines, D. (1953). A collective description of electron interactions. IV. Electron interaction in metals, *Phys. Rev.* **92**, pp. 626–636.

Pines, D. and Nozières, P. (1966). *The Theory of Quantum Liquids* (Benjamin, New York, N.Y.).

Pitaevskii, L. P. (1958). Phenomenological theory of superfluidity near the λ point, *Zh. Eksp. Teor. Fiz.* **35**, pp. 408–415, Engl. transl.: Sov. Phys. JETP, **8**, 282–287 (1959).

Pitaevskii, L. P. (1959). Properties of the spectrum of elementary excitations near the disintegration threshold of the excitations, *Zh. Eksp. Teor. Fiz.* **36**, pp. 1168–1178, Engl. transl.: Sov. Phys. JETP **9**, 830–837 (1959).

Pitaevskii, L. P. (1961). Vortex lines in an imperfect Bose gas, *Zh. Eksp. Teor. Fiz.* **40**, pp. 646–651, Engl. Transl.: Sov. Phys. JETP, **13**, 451–454 (1961).

Polyakov, A. M. (1977). Quark confinement and topology of gauge theories, *Nucl. Phys.* **B120**, pp. 429–458.

Polyakov, A. M. (1987). *Gauge fields and strings*, *Contemporary Concepts in Physics*, Vol. 3 (Harwood, Chur, Switzerland).

Popov, V. N. (1983). *Functional Integral in Quantum Field Theory and Statistical Mechanics*, *Mathematical Physics and Applied Mathematics*, Vol. 8 (Reidel, Dordrecht, the Netherlands), translation from Russian, Atomizdat, Moscow, 1976.

Popov, V. N. (1987). *Functional Integrals and Collective Excitations*, Cambridge Monographs on Mathematical Physics (Cambridge University Press, Cambridge, U.K.), partly based on lectures given at the Institut für Theoretische Physik, Freie Universität Berlin, 1978.

Rajaraman, R. (1982). *Solitons and Instantons: An Introduction to Solitons and Instantons in Quantum Field Theory* (North-Holland, Amsterdam).

Ramakrishnan, T. V. (1989). Superconductivity in disordered thin films, *Phys. Scr.* **T27**, pp. 24–30.

Ramond, P. (1981). *Field Theory: A Modern Primer*, *Frontiers in Physics*, Vol. 51 (Addison-Wesley Publishing Company, Reading, MA).

Randeria, M., Duan, J.-M. and Shieh, L.-Y. (1990). Superconductivity in a two-dimensional Fermi gas: Evolution from Cooper pairing to Bose condensation, *Phys. Rev.* **B41**, pp. 327–343.

Regal, C. A., Greiner, M. and Jin, D. S. (2004). Observation of resonance condensation of fermionic atom pairs, *Phys. Rev. Lett.* **92**, p. 040403.

Roskies, R. and Schaposnik, F. A. (1981). Comment on Fujikawa's analysis applied to the Schwinger model, *Phys. Rev.* **D23**, pp. 558–560.

Sá de Melo, C. A. R., Randeria, M. and Engelbrecht, J. R. (1993). Crossover from BCS to Bose superconductivity: Transition temperature and time-dependent Ginzburg-Landau theory, *Phys. Rev. Lett.* **71**, pp. 3202–3205.

Saint-James, D., Sarma, G. and Thomas, E. J. (1969). *Type II Superconductivity*, *International Series of Monographs in Natural Philosophy*, Vol. 17 (Pergamon Press, Oxford).

Sakhel, A. R. and Glyde, H. R. (2004). Excitations and Bose-Einstein condensation in liquid He^4, *Phys. Rev.* **B70**, p. 144511.

Salzberg, A. M. and Prager, S. (1963). Equation of state for a two-dimensional electrolyte, *J. Chem. Phys.* **38**, pp. 2587–2587.

Saminadayar, L., Glattli, D. C., Jin, Y. and Etienne, B. (1997). Observation of the $e/3$ fractionally charged Laughlin quasiparticle, *Phys. Rev. Lett.* **79**, pp. 2526–2529.

Savit, R. (1980). Duality in field theory and statistical systems, *Rev. Mod. Phys.* **52**, pp. 453–487.

Schakel, A. M. J. (1989). *On Broken Symmetries in Fermi Systems*, Ph.D. thesis, Universiy of Amsterdam, the Netherlands.

Schakel, A. M. J. (1990). On the effective theory of a BCS system at zero temperature, *Mod. Phys. Lett.* **B4**, pp. 927–934.

Schakel, A. M. J. (1991). Relativistic quantum Hall effect, *Phys. Rev.* **D43**, pp. 1428–1431.

Schakel, A. M. J. (1994). Effective theory of bosonic superfluids, *Int. J. Mod. Phys.* **B8**, pp. 2021–2039.

Schakel, A. M. J. (1996). Effective field theory of ideal-fluid hydrodynamics, *Mod. Phys. Lett.* **B10**, pp. 999–1010.

Schakel, A. M. J. (2003). Zeta function regularization of infrared divergences in Bose-Einstein condensation, *J. Phys. Stud.* **7**, pp. 140–155.

Schaposnik, F. A. (1978). Pseudoparticles and confinement in the two-dimensional Abelian Higgs model, *Phys. Rev.* **D18**, pp. 1183–1191.

Schaposnik, F. A. (1986). Chiral symmetry in the path-integral approach, Lectures given at Centro Brasileiro de Pesquisas Físicas, Rio de Janeiro, Brazil, December 1986.

Schiff, L. I. (1968). *Quantum Mechanics*, 3rd edn. (McGraw-Hill, New York, N.Y.).

Schmid, A. (1966). A time-dependent Ginzburg-Landau equation and its application to the problem of resistivity in the mixed state, *Physik Kondensierten Materie* **5**, pp. 302–317.

Schonfeld, J. F. (1981). A mass term for three-dimensional gauge fields, *Nucl. Phys.* **B185**, pp. 157–171.

Schrieffer, J. R. (2004). Novel quantum numbers in condensed matter physics, *Curr. Appl.*

Phys. **4**, pp. 465–472.

Schrieffer, J. R. and Tinkham, M. (1999). Superconductivity, *Rev. Mod. Phys.* **71**, pp. S313–S317.

Schweber, S. S. (1993). Changing conceptualization of renormalization theory, in L. M. Brown (ed.), *Renormalization: From Lorentz to Landau, and Beyond* (Springer-Verlag, Berlin), pp. 137–166.

Schwinger, J. (1951). On gauge invariance and vacuum polarization, *Phys. Rev.* **82**, pp. 664–679.

Schwinger, J. (1962). Gauge invariance and mass. II, *Phys. Rev.* **128**, pp. 2425–2429.

Semenoff, G. W. (1984). Condensed-matter simulation of a three-dimensional anomaly, *Phys. Rev. Lett.* **53**, pp. 2449–2452.

Shangina, E. L. and Dolgopolov, V. T. (2003). Quantum phase transitions in two-dimensional systems, *Phys.-Usp.* **46**, pp. 777–787.

Shapere, A. D. and Wilczek, F. (eds.) (1989). *Geometric Phases in Physics*, *Advanced Series in Mathematical Physics*, Vol. 5 (World Scientific, Singapore).

Shivamoggi, B. K. (1998). *Theoretical Fluid Dynamics*, 2nd edn. (Wiley, New York, N.Y.).

Shopova, D. V. and Uzunov, D. I. (2003). Some basic aspects of quantum phase transitions, *Phys. Rep.* **379**, pp. 1–67.

Skyrme, T. H. R. (1961). A non-linear field theory, *Proc. R. Soc. London* **A260**, pp. 127–138.

Smith, H. (1983). The Lindhard function and the teaching of solid state physics, *Phys. Scr.* **28**, pp. 287–293.

Sondhi, S. L., Girvin, S. M., Carini, J. P. and Shahar, D. (1997). Continuous quantum phase transitions, *Rev. Mod. Phys.* **69**, pp. 315–333.

Spałek, J. (ed.) (2000). *New Quantum States, Phases and Quantum Phase Transitions, and Computational Methods Approach to Quantum Macro States and Particle Production*, *Acta Phys. Polon.* **B31**, pp. 2803–3135.

Speight, J. M. (1997). Static intervortex forces, *Phys. Rev.* **D55**, pp. 3830–3835.

Stanley, H. E. (1999). Scaling, universality, and renormalization: Three pillars of modern critical phenomena, *Rev. Mod. Phys.* **71**, pp. S358–S366.

Stone, M. (1986). The Born-Oppenheimer approximation and the origin of Wess-Zumino terms: Some quantum mechanical examples, *Phys. Rev.* **D33**, pp. 1191–1194.

Stone, M. and Gaitan, F. (1987). Topological charge and chiral anomalies in Fermi superfluids, *Ann. Phys. (N.Y.)* **178**, pp. 89–109.

Stormer, H. L. (1999). Nobel lecture: The fractional quantum Hall effect, *Rev. Mod. Phys.* **71**, pp. 875–889.

Stormer, H. L., Tsui, D. C. and Gossard, A. C. (1999). The fractional quantum Hall effect, *Rev. Mod. Phys.* **71**, pp. S298–S305.

Stückelberg, E. C. G. (1938). Die Wechselwirkungskräfte in der Elektrodynamik und in der Feldtheorie der Kernkräfte (I), *Helv. Phys. Acta* **11**, pp. 225–299.

Su, W. P., Schrieffer, J. R. and Heeger, A. J. (1979). Solitons in polyacetylene, *Phys. Rev. Lett.* **42**, pp. 1698–1701.

Su, W. P., Schrieffer, J. R. and Heeger, A. J. (1980). Soliton excitations in polyacetylene, *Phys. Rev.* **B22**, pp. 2099–2111.

't Hooft, G. and Veltman, M. J. G. (1972). Regularization and renormalization of gauge fields, *Nucl. Phys.* **B44**, pp. 189–213.

't Hooft, G. and Veltman, M. J. G. (1974). Diagrammar, *NATO Adv. Study Inst. Ser. B Phys.* **4**, pp. 177–322.

Takahashi, Y. (1957). On the generalized Ward identity, *Nuovo Cimento* **6**, pp. 371–375.

Takahashi, Y. (1986). Canonical quantization and generalized Ward relations, in F. Mancini (ed.), *Quantum Field Theory* (North-Holland, Amsterdam), pp. 19–37.

Takahashi, Y. (1988). Towards the many-body theory with the Galilei invariance as a guide. I, *Fortschr. Phys.* **36**, pp. 63–81.

Thomas, L. H. (1927). The calculations of atomic fields, *Proc. Cambridge Philos. Soc.* **23**, pp. 542–598.

Tisza, L. (1938). Transport phenomena in helium II, *Nature* **141**, p. 913.

Trebin, H.-R. (1982). The topology of non-uniform media in condensed matter physics, *Adv. Phys.* **31**, pp. 195–254.

Tsui, D. C. (1990). Correlation and localization of two-dimensional electrons in a strong magnetic field, *Mod. Phys. Lett.* **B4**, pp. 301–310.

Tsui, D. C., Stormer, H. L. and Gossard, A. C. (1982). Two-dimensional magnetotransport in the extreme quantum limit, *Phys. Rev. Lett.* **48**, pp. 1559–1562.

Uzunov, D. I. (1981). On the zero temperature critical behavior of the nonideal Bose gas, *Phys. Lett.* **A87**, pp. 11–15.

Uzunov, D. I. and Shopova, D. V. (eds.) (1999). *Correlations, Coherence and Order* (Plenum Press, New York, N.Y.).

Veltman, M. J. G. and Williams, D. N. (1993). *Schoonschip '91*, URL `hep-ph/9306228`.

Vermaseren, J. A. M. (2007). *Form*, URL `http://www.nikhef.nl/~form`.

Villain, J. (1975). Theory of one- and two-dimensional magnets with an easy magnetization plane. II. The planar, classical, two-dimensional magnet. *J. Phys. (Paris)* **36**, pp. 581–590.

Vollhardt, D. and Wölfle, P. (1990). *The Superfluid Phases of Helium-3* (Taylor and Francis, London, U.K.).

Volovik, G. E. (1984). Superfluid properties of ^{3}He-A, *Usp. Fiz. Nauk.* **143**, pp. 73–109, Engl. transl.: Sov. Phys. Usp. **27**, 363–384 (1984).

Volovik, G. E. (1986a). Chiral anomaly and the law of conservation of momentum in ^{3}He-A, *Pis'ma Zh. Eksp. Teor. Fiz.* **43**, pp. 428–431, Engl. transl.: JETP Lett. **43**, 551–554 (1986).

Volovik, G. E. (1986b). Wess-Zumino action for the orbital dynamics ^{3}He-A, *Pis'ma Zh. Eksp. Teor. Fiz.* **44**, pp. 144–148, Engl. transl.: JETP Lett. **44**, 185–189 (1986).

Volovik, G. E. (1991). *Exotic properties of superfluid ^{3}He, Series in Modern Condensed Matter Physics*, Vol. 1 (World Scientific, Singapore).

Volovik, G. E. (2001). Superfluid analogies of cosmological phenomena, *Phys. Rep.* **351**, pp. 195–348.

Volovik, G. E. and Mineev, V. P. (1977). Investigation of singularities in superfluid He^3 and in liquid crystals by the homotopic topology methods, *Zh. Eksp. Teor. Fiz.* **72**, pp. 2256–2274, Engl. transl.: Sov. Phys. JETP, **45**, 1186–1196 (1977).

Volovik, G. E. and Mineev, V. P. (1981). Orbital angular momentum and orbital dynamics: ^{3}He-A and the Bose liquid, *Zh. Eksp. Teor. Fiz.* **81**, pp. 989–1000, Engl. transl.: Sov. Phys. JETP **54**, 524–530 (1981).

von Klitzing, K. (1986). The quantized Hall effect, *Rev. Mod. Phys.* **58**, pp. 519–531.

von Klitzing, K. (2005). 25 years of Quantum Hall Effect (QHE): A personal view on the discovery, physics and applications of this quantum effect, in B. Doucot, V. Pasquier, B. Duplantier and V. Rivasseau (eds.), *The Quantum Hall Effect*, *Prog. Math. Phys.*, Vol. 45 (Birkhäuser, Basel), pp. 1–21.

von Klitzing, K., Dorda, G. and Pepper, M. (1980). New method for high-accuracy determination of the fine-structure constant based on quantized Hall resistance, *Phys. Rev. Lett.* **45**, pp. 494–497.

Wallace, P. R. (1947). The band theory of graphite, *Phys. Rev.* **71**, pp. 622–634.

Ward, J. C. (1950). An identity in quantum electrodynamics, *Phys. Rev.* **78**, p. 182.

Wegner, F. J. (1976). Electrons in disordered systems. Scaling near the mobility edge, *Z. Phys.* **B25**, pp. 327–337.

Weinberg, S. (1986). Superconductivity for particular theorists, *Prog. Theor. Phys. Suppl.* **86**, pp. 43–53.

Weinberg, S. (1996). *The Quantum Theory of Fields, Vol. II: Modern Applications* (Cambridge University Press, Cambridge, U.K.).

Wen, X. G. and Zee, A. (1988). Spin waves and topological terms in the mean-field theory of two-dimensional ferromagnets and antiferromagnets, *Phys. Rev. Lett.* **61**, pp. 1025–1028.

Wigner, E. (1938). Effects of electron interaction on the energy levels of electrons in metals, *Trans. Faraday Soc.* **34**, pp. 678–685.

Wilczek, F. (1982). Magnetic flux, angular momentum, and statistics, *Phys. Rev. Lett.* **48**, pp. 1144–1146.

Wilczek, F. (2002). Some basic aspects of fractional quantum numbers, in N. E. Bonesteel and L. P. Gorkov (eds.), *Selected Papers of J. Robert Schrieffer: In Celebration of His 70th Birthday* (World Scientific, Singapore), pp. 135–151.

Willett, R., Eisenstein, J. P., Stormer, H. L., Tsui, D. C., Gossard, A. C. and English, J. H. (1987). Observation of an even-denominator quantum number in the fractional quantum Hall effect, *Phys. Rev. Lett.* **59**, pp. 1776–1779.

Wilson, K. G. (1975). The renormalization group: Critical phenomena and the Kondo problem, *Rev. Mod. Phys.* **47**, pp. 773–840.

Witten, E. (1983). Global aspects of current algebra, *Nucl. Phys.* **B223**, pp. 422–432.

Witten, E. (1985). Superconducting strings, *Nucl. Phys.* **B249**, pp. 557–592.

Yourgrau, W. and Mandelstam, S. (1979). *Variational Principles in Dynamics and Quantum Theory* (Dover, New York, N.Y.).

Zhang, S. C. (1992). The Chern-Simons-Landau-Ginzburg theory of the fractional quantum Hall effect, *Int. J. Mod. Phys.* **B6**, pp. 803–807.

Zhang, S. C., Hansson, T. H. and Kivelson, S. (1989). Effective-field-theory model for the fractional quantum Hall effect, *Phys. Rev. Lett.* **62**, pp. 82–85.

Zuk, J. A. (1987). Induced fermionic charges from topological actions, in M. Jezabek and M. Praszatowicz (eds.), *Proceedings of the Workshop on Skyrmions and Anomalies, Krakow, Poland* (World Scientific, Singapore), pp. 449–474.

Zwierlein, M. W., Stan, C. A., Schunck, C. H., Raupach, S. M. F., Gupta, S., Hadzibabic, Z. and Ketterle, W. (2003). Observation of Bose-Einstein condensation of molecules, *Phys. Rev. Lett.* **91**, p. 250401.

Zwierlein, M. W., Stan, C. A., Schunck, C. H., Raupach, S. M. F., Kerman, A. J. and Ketterle, W. (2004). Condensation of pairs of fermionic atoms near a Feshbach resonance, *Phys. Rev. Lett.* **92**, p. 120403.

Author Index

Index